High Performance

'69 Cortina-Lotus

Workshop Manual

Ford of Britain

© **Ford Motor Company Ltd. 1969**

This book is published by Brooklands Books Limited under licence and is based upon text and illustrations protected by Copyright and first published in 1969 by Ford Motor Company Ltd and may not be reporduced, transmitted or copied by any means without the prior written permission of Ford of Europe Incorporated.

CORTINA - LOTUS

CONTENTS

FOREWORD

    Replacement Parts

    Vehicle Identification

SECTION	SUBJECT
1	WHEELS, HUBS AND TYRES
2	BRAKING SYSTEM
3	STEERING GEAR AND LINKAGE
4	REAR AXLE AND DRIVESHAFT
5/1	FRONT SUSPENSION
5/2	REAR SUSPENSION
6/1	ENGINE
6/2	EXHAUST SYSTEM
7	CLUTCH AND GEARBOX (MANUAL)
8	COOLING SYSTEM
9	FUEL SYSTEM
10	ELECTRICAL SYSTEM
	Comprising: CHARGING SYSTEM
	STARTING SYSTEM
	IGNITION SYSTEM
	INSTRUMENTS, CONTROLS AND ANCILLARIES
	WIRING
11/1	ACCESSORIES — FITTING INSTRUCTIONS
11/2	ACCESSORIES — RADIO (PUSH-BUTTON)
11/3	ACCESSORIES — RADIO (MANUAL)
12	BODYWORK
13	LUBRICATION AND MAINTENANCE
14	SPECIFICATIONS, SERVICING AND REPAIR DATA
15	SPECIAL TOOLS

Ford policy is one of continuous improvement, and the right to change prices, specifications and equipment at any time without notice is reserved.

CORTINA - LOTUS

FOREWORD

This Manual gives repair and adjustment procedures applicable to the 1969 Cortina-Lotus model. It includes component illustrations wherever possible, technical specifications and details of any special tools or equipment that may be required. *Supplements are included, where necessary, to enable the manual to be used with earlier models.*

When the special tools or equipment mentioned are not available, changes may be necessary to some of the dismantling and reassembly procedures. **In certain instances use of special equipment is essential if a satisfactory repair is to be ensured.** For this reason we strongly advise owners and operators to allow Ford Dealerships to carry out the more complex repairs, e.g. the overhaul of gearbox or differential assemblies.

Repair operations tabulated in this manual are of three types:—

(i) *Basic Operation* – This can be complete by itself or as a prelude to additional work. Generally a basic operation describes the replacement of a component or assembly.

(ii) *Subsidiary Operation* – This is additional work related to a basic operation. For example, overhauling an assembly that has already been removed.

(iii) *Group Operation* – This is a more extensive operation which combines one basic and one or more subsidiary operations. Because the basic and subsidiary operations are themselves already given in detail, the group operation merely lists which of the basic and subsidiary operations are required to successfully perform the group operation.

Examples:—

Basic Operation

    OPERATION 1202-A   FRONT WHEEL BEARINGS – EACH WHEEL – ADJUST

Subsidiary Operation

    OPERATION 1202-A1   Extra: lubricate front wheel bearings

Group Operation

    OPERATION 1202-B   **FRONT WHEEL BEARINGS – LUBRICATE AND ADJUST**
                               (Includes OPS 1202-A and A1)

In most cases, the basic operation number corresponds to the part number of the component being serviced, i.e. 1107 is the part number of a wheel stud and OP 1107-A refers to wheel studs.

It is not, however, practical to use part numbers for body operations. Therefore, bodywork operations commence at 1000 and are prefixed by the letter M.

Replacement Parts

Behind all Ford products are the vast resources of Ford of Britain Parts Division from which replacement parts and reconditioned units are supplied to markets throughout the world. Remember, these parts are made to the same exacting standards as the original factory fitted components. For this reason always insist that only genuine FoMoCo parts are used as service replacements.

CORTINA - LOTUS

Vehicle Identification

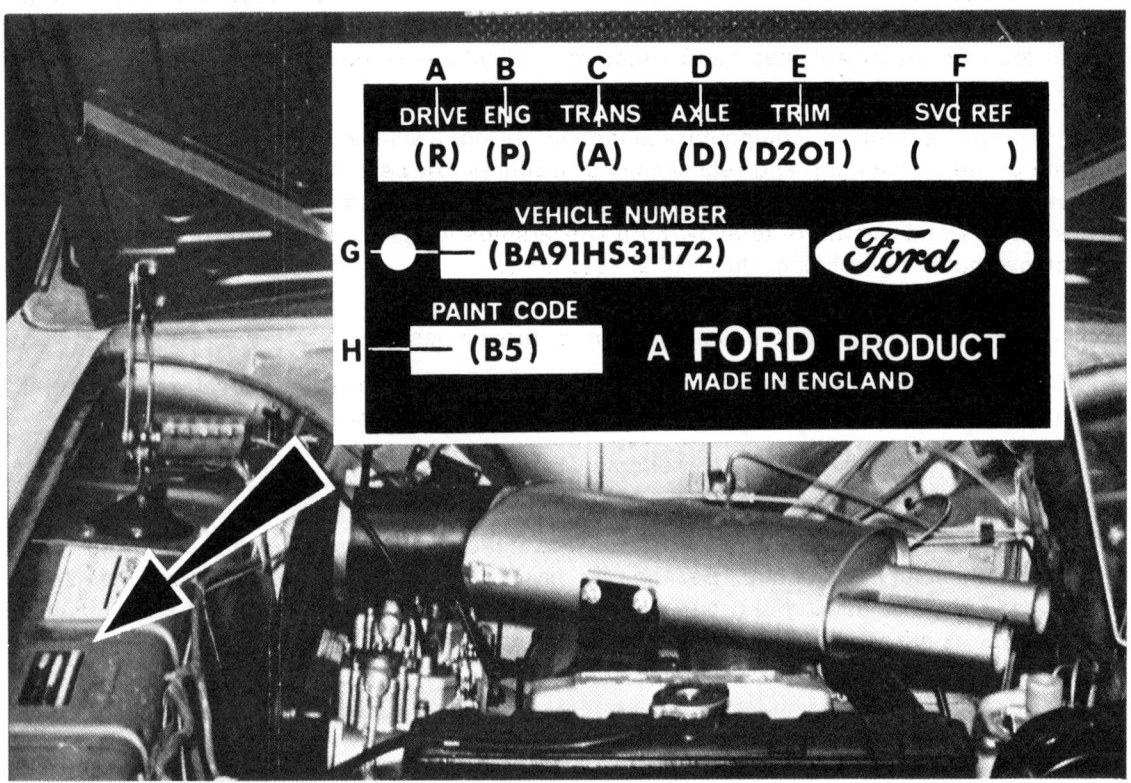

A vehicle identification plate is riveted to the top of the engine side apron panel and gives details of the type of engine, transmission, axle and body. The useful codes can be broken down as follows:

A: *Drive Code:*

    R — Right-Hand Drive
    L — Left-Hand Drive

B: *Engine Code:*

    P — 1560 c.c. Twin Overhead Camshaft

C: *Transmission Code:*

    A — Floor change manual

D: *Rear Axle Code:*

    D — 3.777 : 1

E: *Trim Combination Number*

This code, which consists of an alphabetical prefix and three digits (e.g. D.201) indicates the colour and type of material used within the vehicle. Refer to the Soft Trim Parts List for details.

F: *S.V.C. Reference*

This signifies the date of manufacture when a vehicle is shipped unassembled from the source company for assembly in another factory.

CORTINA - LOTUS

G: *Vehicle Number*

The vehicle number consists of eleven symbols, both numerical and alphabetical, arranged in five sections.

1. Product Source	2. Assembly Plant	3. Body Type	4. Assembly Code	5. Sequence No.
B	A	91	HS	31172

1. Product Source

    "B" signifies that the vehicle was assembled in Britain. When a vehicle is shipped unassembled the letter "C" or "D" is shown to denote the territory in which it is assembled.

2. Assembly Plant

    "A" denotes Dagenham.

3. Body Type

    "91" signifies that body type is 2 door saloon.

4. Assembly Code

    This is a two-figure alphabetical date code for factory use only.

5. Sequence Number

    This will be a five-figure number within the range 00001 to 99999 and upon reaching the end of a particular range, the sequence number will revert to the original number for that respective model.

H: *Paint Code*

This code consists of either one or two groups of letters, for single and two tone body colours respectively. The colour, or combination of colours, may be determined from the following table:

A2	— Silver Fox	B5	— Fern Green
A3	— Blue Mink	AB	— Ermine White
A5	— Aquatic Jade	BJ	— Anchor Blue
A6	— Saluki Bronze	BN	— Red II
A8	— Light Orchid	BV	— Beige
AQ	— Aubergine	CG	— Light Blue
B4	— Amber Gold	CR	— Light Green

CORTINA - LOTUS

1
WHEELS, HUBS AND TYRES

CORTINA - LOTUS

SECTION INDEX

GENERAL DESCRIPTION

QUICK REFERENCE DATA

SERVICE AND REPAIR OPERATIONS

OPERATION	1015-A	WHEEL ASSEMBLIES – BOTH FRONT OR BOTH REAR – BALANCE
"	1015-A1	Extra: remaining road wheels – balance
"	1015-B	**WHEEL ASSEMBLIES – FOUR ROAD WHEELS – BALANCE** (Includes OPS 1015-A and A1)
"	1015-C	WHEEL ASSEMBLIES – ALL FIVE – BALANCE
"	1107-A	FRONT WHEEL STUD – ONE – RENEW
"	1107-A1	Extra: remaining studs – renew
"	1107-B	**FRONT WHEEL STUDS – ALL – ONE WHEEL – RENEW** (Includes OPS 1107-A and A1)
"	1110-A	FRONT WHEEL HUB ASSEMBLY – ONE – REMOVE AND INSTALL
"	1110-A1	Extra: remaining front wheel hub assembly – remove and install (front of vehicle on stands)
"	1110-B	**FRONT WHEEL HUB ASSEMBLIES – BOTH – REMOVE AND INSTALL** (Includes OPS 1110-A and A1)
"	1117-A	WHEEL STUD – REAR – ONE – RENEW
"	1117-A1	Extra: remaining studs – renew (brake drum removed)
"	1117-B	**REAR WHEEL STUDS – ALL – ONE WHEEL – RENEW** (Includes OPS 1117-A and A1)
"	1125-A	BRAKE DISC – ONE – REMOVE AND INSTALL
"	1125-A1	Extra: remaining front brake disc – remove and install (front of vehicle on stands)
"	1125-B	**BRAKE DISCS – BOTH – REMOVE AND INSTALL** (Includes OPS 1125-A and A1)
"	1126-A	BRAKE DRUM – REAR – ONE – REMOVE AND INSTALL
"	1126-A1	Extra: remaining rear brake drum – remove and install (rear of vehicle on stands)
"	1126-B	**BRAKE DRUMS – REAR – BOTH – REMOVE AND INSTALL** (Includes OPS 1126-A and A1)
"	1202-A	WHEEL BEARING – ONE FRONT WHEEL – ADJUST
"	1202-A1	Extra: remaining front wheel bearing – adjust (front of vehicle on stands)
OPERATION	1202-A2	Extra: bearings – one front wheel – lubricate (front of vehicle on stands)
"	1202-A3	Extra: wheel bearing cups – each front hub – renew (front hub removed)
"	1202-B	**WHEEL BEARINGS – BOTH FRONT WHEELS – ADJUST** (Includes OPS 1202-A and A1)
"	1202-C	**WHEEL BEARINGS – ONE FRONT WHEEL – LUBRICATE AND ADJUST** (Includes OPS 1202-A and A2)
"	1202-D	**WHEEL BEARINGS – BOTH FRONT WHEELS – LUBRICATE AND ADJUST** (Includes OPS 1202-A, A1 and A2×2)
"	1202-E	**WHEEL BEARING CUPS – ONE FRONT WHEEL – RENEW** (Includes OPS 1202-A, A2 and A3)
"	1202-F	**WHEEL BEARING CUPS – BOTH FRONT WHEELS – RENEW** (Includes OPS 1202-A, A1, A2×2 and A3×2)
"	1225-A	WHEEL BEARING – ONE – REAR – REMOVE AND INSTALL
"	1225-A1	Extra: remaining rear wheel bearing – remove and install (rear of vehicle on stands)
"	1225-B	**WHEEL BEARINGS – BOTH REAR – REMOVE AND INSTALL** (Includes OPS 1225-A and A1)

CORTINA - LOTUS

GENERAL DESCRIPTION

Pressed steel disc type wheels, size 5½J×13, with radial ply tubeless tyres, size 165×13, are fitted to this model. Tyre pressures should be maintained at 24 p.s.i. (1.69 kg/sq. cm.) front and rear under normal conditions and the pressures increased to 30 p.s.i. (2.09 kg/sq. cm.) when motoring at high speed. The spare wheel is mounted on the left-hand side of the luggage compartment and is retained by a clamp.

The front wheel hubs are each mounted on two taper roller bearings which thus permit free rotation under side loads (such as occur when cornering) as well as normal vertical loads.

The rear hub bearings consist of ball races pressed onto the outer ends of the axle shafts. These bearings incorporate built-in oil seals.

It is suggested that when fitting a new tubeless tyre, a new snap-in valve is also fitted. The valve is made to last the life of the tyre, but beyond that time fatigue of the valve rubber body is likely to impair the air seal at the rim hole.

TYRES AND TYRE CARE

Original Equipment Tyres

The original equipment tyres have undergone extensive testing by the Ford Motor Co. and the tyre manufacturers who produce them.

It is essential for safety that the recommended inflation pressures are always maintained. When tyre tread depth is less than 1 mm., the tyre must (by legal requirement) be renewed. (U.K. only, other territories may have different regulations.)

Radial ply tyres will visually appear under-inflated at the correct recommended pressures shown over-page. This is normal and they should never be inflated beyond the recommended pressures. Under no circumstances fit radial-ply tyres on the front with conventional cross-ply on the rear, nor should both types be fitted to one axle.

Tyre Care

Tyre tread life varies from car to car because of driving conditions. Aside from good driving practices, the most important factor in obtaining maximum tread life is maintaining proper tyre pressure. Pressures lower than recommended may affect vehicle handling. Over-inflated tyres reduce ride comfort by magnifying rather than absorbing road shocks and are more vulnerable to damage from road surface impacts.

QUICK REFERENCE DATA

PERIODIC SERVICE ATTENTION

Daily and weekly or before high speed motoring
    Check tyre pressures and inspect tyres

At first and every subsequent 3,000 miles (5000 km.) or three months (whichever occurs first)
    Correct tyre pressures and inspect tyres

DATA

Wheel size	5½J×13
Tyre size	165×13
Tyre pressures—normal speed conditions	24 p.s.i. (1.69 kg/sq. cm.)
—high speed conditions	30 p.s.i. (2.09 kg/sq. cm.)

Tightening Torques

Wheel nuts	50 to 55 lb. ft. (6.91 to 7.60 kg.m.)
Brake calliper to front suspension unit	45 to 50 lb. ft. (6.22 to 6.94 kg.m.)
Front brake disc to hub	30 to 34 lb. ft. (4.15 to 4.70 kg.m.)
Front wheel bearing adjusting nut	See text
Axle shaft bearing retainer bolts	15 to 18 lb. ft. (2.08 to 2.48 kg.m.)

CORTINA - LOTUS

SERVICE AND REPAIR OPERATIONS

OP 1015-A WHEEL ASSEMBLIES – BOTH FRONT OR BOTH REAR – BALANCE (INCLUDES STATIC AND DYNAMIC BALANCE)

The following instructions indicate the general principles to be followed for wheel balancing. The method of balancing, however, will vary for machines of different manufacture. For specific details refer to the equipment manufacturer's instructions.

A – ON THE CAR

1. Jack up the front or rear of the car as required and fit chassis stands.
2. Check that wheel bearings, suspension ball joints, etc., are in correct order, and that no excess play is present, and that the tyres are correctly inflated.
3. Position the spin-up motor in correct relation to the wheel to be balanced.
4. Remove any wheel weights present.
5. Fit the wheel adaptor, or fit the pick-up onto the suspension, as close to the wheel centre as possible.
6. Spin up the wheel, and balance according to the manufacturer's instructions.
7. If applicable, move the pick-up to the brake back plate or calliper.
8. Again spin up and balance the wheel.
9. Repeat for other wheel.

B – OFF THE CAR

1. Jack up the front or rear of the car as required, fit chassis stands and remove the wheels.
2. Remove any wheel weights present, and check that the tyres are correctly inflated.
3. Mount the wheel on the balancer.
4. Statically balance the wheel by allowing it to spin so that the heaviest sector rests at the bottom, and applying suitable weights.
5. Spin up, and dynamically balance according to the manufacturer's instructions.
6. If necessary, repeat static balance.
7. Remove the wheel, mount the other wheel and balance.

OP 1015-A1 EXTRA: REMAINING ROAD WHEELS – BALANCE (INCLUDES STATIC AND DYNAMIC BALANCE) – (Includes OP 1015-A)

OP 1015-B WHEEL ASSEMBLIES – FOUR ROAD WHEELS – BALANCE (INCLUDES STATIC AND DYNAMIC BALANCE) – (Includes OPS 1015-A and A1)

OP 1015-C WHEEL ASSEMBLIES – ALL FIVE – BALANCE (INCLUDES STATIC AND DYNAMIC BALANCE)

1. Jack up the car and fit chassis stands all round.
2. Remove the road wheels and spare wheel.
3. Remove any wheel weights present, and check that the tyres are correctly inflated.
4. Mount one wheel on the balancer.
5. Statically balance the wheel by allowing it to spin so that the heaviest sector rests at the bottom, and applying suitable weights.
6. Spin up and dynamically balance according to the manufacturer's instructions.
7. If necessary, repeat static balance.
8. Remove the wheel and repeat for the other four wheels.

OP 1107-A FRONT WHEEL STUD – ONE – RENEW

To Remove

1. Remove the hub cap, slacken the wheel nuts, jack up the front of the car and fit stands.
2. Detach the road wheel.
3. Drive out the wheel stud.

To Install

4. Locate the new stud in its splined hole and draw it into position using a spacer and the wheel nut reversed, i.e. tapered face outwards.
5. Replace the wheel.
6. Jack up, remove the stands and lower the car to the ground.
7. Tighten the wheel nuts and fit the hub cap.

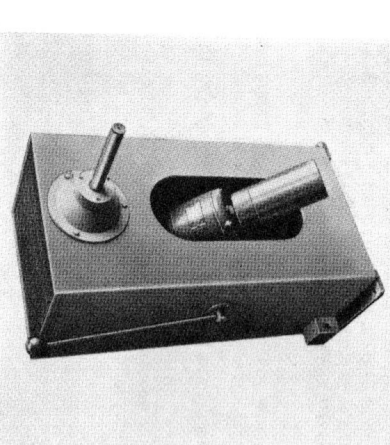

On the Car

Off the Car

Wheel Balancing Equipment

OP 1107-A1 EXTRA: REMAINING STUDS – RENEW

Repeat sub-paragraphs 3 and 4 above.

OP 1107-B FRONT WHEEL STUDS – ALL – ONE WHEEL – RENEW
(Includes OPS 1107-A and A1)

OP 1110-A FRONT WHEEL HUB ASSEMBLY – ONE – REMOVE AND INSTALL

Tools Required

550	Universal Handle
P.1029	Front hub grease seal replacer
P.2012	Brake line plugs
	Dial gauge

To Remove

1. Remove the hub cap and slacken the wheel nuts.
2. Jack up the front of the car and fit chassis stands.
3. Remove the road wheel.
4. Break the fluid supply to the front brake calliper at the pipe bracket on the suspension unit. Fit a brake line plug, P.2012, to each of the opened pipe ends.
5. Detach the brake calliper from its mounting lugs on the base of the suspension unit after bending back the lock tabs and unscrewing the securing bolts.
6. Remove the dust cap from the centre of the hub. Withdraw the split pin, remove the nut retainer, adjusting nut, thrust washer and outer bearing cone.
7. Remove the hub and disc assembly from the wheel spindle.
8. Bend back the lock tabs, unscrew the bolts and separate the disc from the hub assembly.
9. Discard the hub assembly.

To Install

NOTE – A new hub assembly is supplied complete with cups and cones. The cups are in position in the hub and the cones are separate. It should be noted that the cups and cone must be of the same manufacture. This should be checked, the manufacturer's name can be read off the cone and the initial letter of the name is stamped on the hub (i.e. T=Timken, S=Skefco).

10. Clean the mating faces of the hub and disc, these must be scrupulously clean if disc run-out is to be avoided.
11. Align the run-out marks and assemble the disc to the hub using new locking plates under the bolts. Tighten the bolts to a torque of 30 to 34 lb. ft. (4.15 to 4.70 kg.m.) and bend up the tabs on the locking plates.
12. Pack the hub with lithium base grease and fit the inner bearing cone.
13. Using Tool No. P.1029 and the 550 handle, fit the grease seal with the lip towards the bearing.
14. Fit the hub and disc assembly to the wheel spindle and assemble the outer bearing, thrust washer and bearing adjusting nut. Tighten the adjusting nut to a torque of 27 lb. ft. (3.73 kg.m) whilst rotating the hub and disc assembly. Slacken the nut back 90° and then fit the nut retainer so as to align a slot in the retainer with the hole in the spindle. Fit a new split pin but do not bend up at this stage.
15. Check the disc run-out using a dial gauge. The maximum permissible run-out, measured near the periphery of the disc is 0.004 in. (0.10 mm.) total indicated run-out. If this is excessive, examine the mating faces of the hub and disc for dirt or distortion and check the bearings and cups for damage.
16. Bend up the split pin and fit the dust cap to the wheel spindle.
17. Fit the calliper to its mounting lugs using new lock tabs and tighten the securing bolts to a torque of 45 to 50 lb. ft. (6.22 to 6.94 kg.m.).
18. Reconnect the hydraulic fluid pipe to the calliper assembly and tighten the union securely.
19. Replace the road wheel.
20. Jack up the car, remove the stands and lower the car to the ground.
21. Tighten the wheel nuts and replace the hub cap.
22. Bleed the brakes as described in OP 2000-A.

OP 1110-A1 EXTRA: REMAINING FRONT WHEEL HUB ASSEMBLY – REMOVE AND INSTALL (FRONT OF VEHICLE ON STANDS)

Repeat sub-operations 3 to 19 above.

OP 1110-B FRONT WHEEL HUB ASSEMBLY – BOTH – REMOVE AND INSTALL
(Includes OPS 1110-A and A1)

Front Hub Assembly - Exploded

CORTINA - LOTUS

OP 1117-A WHEEL STUD – REAR – ONE – RENEW

To Remove

1. Remove the hub cap, slacken the wheel nuts, chock the front wheels, jack up the rear of the car and fit stands.
2. Detach the road wheel.
3. Remove the brake drum after unscrewing the countersunk head screw and releasing the handbrake.
4. Drive out the wheel stud.

To Install

5. Locate the new stud in its splined hole and draw it into position using a spacer and the wheel nut reversed, i.e. tapered face outwards.
6. Replace the wheel.
7. Jack up, remove the stands and lower the car to the ground.
8. Tighten the wheel nuts and fit the hub cap.
9. Apply the handbrake and remove the chocks from the front wheels.

OP 1117-A1 EXTRA: REMAINING STUDS – RENEW (BRAKE DRUM REMOVED)

Repeat sub-operations 4 and 5 above.

OP 1117-B REAR WHEEL STUDS – ALL – ONE WHEEL – RENEW
(Includes OPS 1117-A and A1)

Brake Drum Securing Screw

OP 1125-A BRAKE DISC – ONE – REMOVE AND INSTALL

Tools Required

550	Universal Handle
P.1029	Front hub grease seal replacer
P.2012	Brake line plugs
	Dial gauge

To Remove

1. Remove the hub cap and slacken the wheel nuts.
2. Jack up the front of the car and fit chassis stands.
3. Remove the road wheel.
4. Break the fluid supply to the front brake calliper at the pipe bracket on the suspension unit. Fit a brake line plug, P.2012, to each of the opened pipe ends.
5. Detach the brake calliper from the mounting lugs on the base of the suspension unit after bending back the lock tabs and unscrewing the securing bolts.
6. Remove the dust cap from the centre of the hub. Withdraw the split pin, remove the nut retainer, adjusting nut, thrust washer and outer bearing cone.
7. Remove the hub and disc assembly from the wheel spindle.
8. Bend back the lock tabs, unscrew the bolts and separate the disc from the hub assembly.
9. Discard the disc.

To Install

10. Clean the mating faces of the hub and disc, these must be scrupulously clean if disc run-out is to be avoided.
11. Align the run-out marks and assemble the disc to the hub using new locking plates under the bolts. Tighten the bolts to a torque of 30 to 34 lb. ft. (4.15 to 4.70 kg.m.) and bend up the tabs on the locking plates.

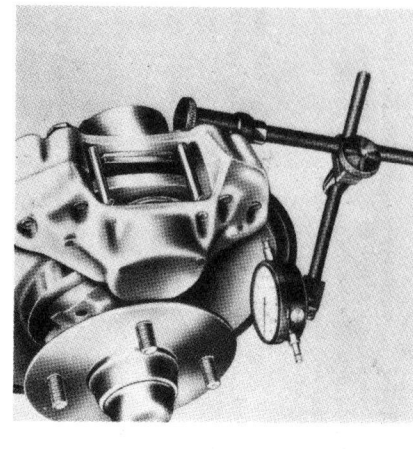

Checking Brake Disc Run-Out

CORTINA - LOTUS

OP 1125-B BRAKE DISCS – BOTH – REMOVE AND INSTALL
(Includes OPS 1125-A and A1)

OP 1126-A REAR BRAKE DRUM – REMOVE AND INSTALL

To Remove

1. Remove the hub cap and slacken wheel nuts, chock the front wheels, jack up the rear of the car and fit chassis stands.
2. Remove the road wheel.
3. Remove the brake drum after unscrewing the countersunk head screw and releasing the handbrake.
4. Back off the brake adjustment mechanism by turning the serrated ratchet wheel against the pawl. To do this it is necessary to lift the pawl out of engagement with the ratchet.

To Install

5. Install the new brake drum and secure with the countersunk head screw.
6. Replace the road wheel.
7. Jack up, remove the stands and lower the car to the ground.
8. Tighten wheel nuts and replace the hub cap.
9. Operate the handbrake several times to bring the brake shoes into their correct adjustment relative to the brake drum.
10. Apply the handbrake and remove chocks from the front wheels.

OP 1126-A1 EXTRA: REMAINING REAR BRAKE DRUM – REMOVE AND INSTALL
(rear of vehicle on stands)

Repeat sub-operations 2 to 6 above.

OP 1126-B BRAKE DRUMS – REAR – BOTH – REMOVE AND INSTALL
(Includes OPS 1126-A and A1)

OP 1202-A FRONT WHEEL BEARING – ONE FRONT WHEEL – ADJUST

To Adjust

1. With the handbrake applied, jack up the front of the car and fit chassis stands.
2. Remove the hub cap and prise out the dust cap from the end of the wheel spindle.
3. Remove the split pin and detach the adjusting nut retainer.
4. Adjust the bearing by tightening the adjusting nut to a torque of 27 lb. ft. (3.73 kg.m.) whilst rotating the road wheel.
5. Slacken back the nut 90° and then fit the nut retainer so that a slot in the retainer lines up with the hole in the spindle.
6. Fit a new split pin and bend back the legs. Tap the dust cap into place and replace the hub cap.
7. Jack up, remove the stands and lower the car to the ground.

---

12. Pack the hub with lithium base grease and, if removed previously, refit the inner bearing cone.
13. Using Tool No. P. 1029 and the 550 handle, fit the grease seal with the lip towards the bearing.
14. Fit the hub and disc assembly to the wheel spindle and assemble the outer bearing, thrust washer and bearing adjusting nut. Tighten the adjusting nut to a torque of 27 lb. ft. (3.73 kg.m.) whilst rotating the hub and disc assembly. Slacken the nut back 90° and then fit the nut retainer so as to align a slot in the retainer with the hole in the spindle. Fit a new split pin but do not bend up at this stage.
15. Check the disc run-out using a dial gauge. The maximum permissible run-out, measured near the periphery of the disc, is 0.004 in. (0.10 mm.) total indicated run-out. If this is excessive, examine the mating faces of the hub and disc for dirt or distortion and check the bearings and cups for damage.
16. Bend up the split pin and fit the dust cap to the wheel spindle.
17. Fit the calliper to its mounting lugs using new lock tabs and tighten the securing bolts to a torque of 45 to 50 lb. ft. (6.22 to 6.94 kg.m.).
18. Reconnect the hydraulic fluid pipe to the calliper asembly and tighten the union securely.
19. Replace the road wheel.
20. Jack up the car, remove the stands and lower the car to the ground.
21. Tighten the wheel nuts and replace the hub cap.
22. Bleed the brakes as described in OP. 2000-A.

OP 1125-A1 EXTRA: REMAINING FRONT BRAKE DISC – REMOVE AND INSTALL
(front of vehicle on stands)

Repeat sub-operations 3 to 19 above.

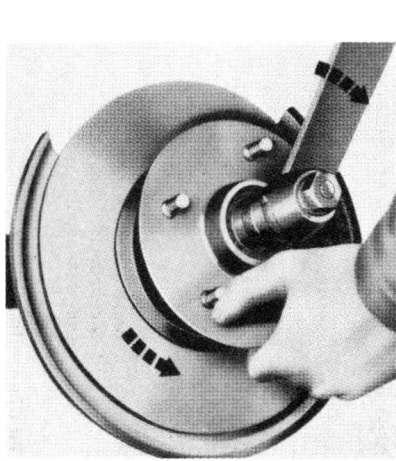

Adjusting the Front Wheel Bearings

CORTINA - LOTUS

OP 1202-A1 EXTRA: REMAINING FRONT WHEEL BEARING – ADJUST
(front of vehicle on stands)

Repeat sub-operations 2 to 6 above.

OP 1202-A2 EXTRA: BEARINGS – ONE FRONT WHEEL – LUBRICATE
(front of vehicle on stands)

Tools Required

P.1029   Front grease seal replacer
P.2012   Brake line plugs
550   { Dial gauge
          Universal handle

To Dismantle

1. Break the fluid supply to the front brake calliper at the pipe bracket on the suspension unit. Fit a brake line plug, Tool No. 2012, to each of the opened pipe ends.
2. Detach the brake calliper from the mounting lugs on the base of the suspension unit after bending back the lock tabs and unscrewing the securing bolts.
3. Remove the dust cap from the centre of the hub. Withdraw the split pin, remove the nut retainer, adjusting nut, thrust washer and outer bearing cone.
4. Remove the hub and disc assembly from the wheel spindle.
5. Prise out the grease seal and remove the inner bearing cone.
6. Wash out the hub and the bearing cones. During this operation take care not to get any grease, etc., on the surface of the disc.

To Reassemble

7. Pack the bearings and the hub with lithium base grease. Leave sufficient space for the wheel spindle to pass through the hub. Fit inner bearing cone.
8. Using Tool No. P.1029 and the 550 handle, fit the grease seal with the lip towards the bearing
9. Fit the hub and disc assembly to the wheel spindle and assemble the outer bearing, thrust washer and bearing adjusting nut. Tighten the adjusting nut to a torque of 27 lb. ft. (3.37 kg. m.) whilst rotating the hub and disc assembly. Slacken the nut back 90° and then fit the nut retainer so as to align a slot in the retainer with the hole in the spindle. Fit a new split pin but do not bend up at this stage.
10. Check the disc run-out using a dial gauge. The maximum permissible run-out, measured near the periphery of the disc is 0.004 in. (0.10 mm.) total indicated run-out. If this is excessive examine the mating faces of the hub and disc for dirt or distortion and check the bearings and cups for damage.
11. Bend up the split pin and fit the dust cap to the wheel spindle.
12. Fit the calliper to its mounting lugs using new lock tabs and tighten the securing bolts to a torque of 45 to 50 lb. ft. (6.22 to 694 kg. m.).
13. Reconnect the hydraulic fluid pipe to the calliper assembly and tighten the union securely.
14. Replace the road wheel.
15. Bleed the brakes as described in OP 2000-A.

OP 1202-A3 EXTRA: WHEEL BEARING CUPS – EACH FRONT HUB – RENEW
(front hub removed)

Tools Required

PT. 1024   Front bearing cups remover and replacer (Main Tool)
P. 1024-3 Adaptors for PT. 1024.

1. Remove the inner and outer bearing cup, using PT. 1024.
2. Replace new cups using the same tools.
   NOTE – When replacing cones **always** fit new bearing cups ensuring that the cups and bearing cones are of the same manufacture.

OP 1202-B WHEEL BEARINGS – BOTH FRONT WHEELS – ADJUST
(Includes OPS 1202-A and A1)

OP 1202-C WHEEL BEARINGS – ONE FRONT WHEEL – LUBRICATE AND ADJUST
(Includes OPS 1202-A and A2)

OP 1202-D WHEEL BEARINGS – BOTH FRONT WHEELS – LUBRICATE AND ADJUST
(Includes OPS 1202-A, A1 and A2×2)

OP 1202-E WHEEL BEARING CUPS – ONE FRONT WHEEL – RENEW
(Includes OPS 1202-A, A2 and A3)

OP 1202-F WHEEL BEARING CUPS – BOTH FRONT WHEELS – RENEW
(Includes OPS 1202-A, A1, A2×2 and A3×2)

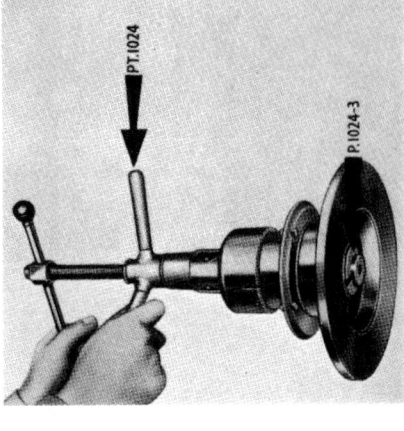

Refitting the Bearing Cups

CORTINA - LOTUS

OP 1225-A WHEEL BEARING – ONE – REAR – REMOVE AND INSTALL

Tools Required

CPT.3072   Slide hammer
P.4090-2 & 6   Axle shaft bearing remover
370   Universal taper base for press
P.4084   Spring indicator

To Remove

1. Remove the hub cap, slacken the wheel nuts, chock the front wheels, jack up the rear of the car, fit stands and remove the road wheel.
2. Release the handbrake inside the car.
3. Remove the brake drum securing screw and pull off the drum. (Ensure handbrake is released.)
4. Remove the four bolts and spring washers securing the bearing retainer plate to the axle casing. These bolts are accessible through holes in the axle shaft flange.
5. Secure the base of Tool No. P.3072 to the axle shaft flange with the wheel nuts. Use the slide hammer of the tool to draw out the axle shaft.
6. Locate the adaptors (Tool No. P.4090-6) and a slave ring between the bearing and axle shaft flange. Support the assembly in the base plate (Tool No. 370) of a hydraulic press and push the axle shaft out of the bearing.

To Install

7. Locate the bearing retainer plate and the bearing on the axle shaft with the oil seal side towards the splined end.
8. Support the assembly in the bed of a hydraulic press on a spacer ring, adaptors (Tool No. P.4090-2) and slave ring.

Fitting a Front Hub Grease Seal

9. Fit a spring indicator (Tool No. P. 4084) to the ram and press the bearing onto the axle shaft shoulder. A minimum pressure of 1,200 lb. (544 kg.) should be required. A lower pressure indicates an incorrect fit.
10. Use the same tools as in operations 6 and 7 and fit the bearing collar to abut the bearing. A minimum pressure of 1,500 lb. (680 kg.) should be required.
11. Insert the axle shaft into the casing and engage the splines in the differential side gear. Tap the shaft fully home.
12. Fit the four bolts and spring washers to secure the bearing retainer plate; torque to 15 to 18 lb. ft. (2.08 to 2.48 kg.m.).
13. Replace the brake drum and refit the securing screw.
14. Replace the road wheel and apply the handbrake.
15. Jack up, remove the stands and lower the car to the ground. Remove the chocks from the front wheels.
16. Tighten wheel nuts and fit the hub cap.

OP 1225-A1 EXTRA: REMAINING REAR WHEEL BEARING – REMOVE AND INSTALL
(rear of vehicle on stands)

1. Remove hub cap, slacken wheel nuts and remove the road wheel.
2. Repeat sub-operations 3 to 14 above.

OP 1225-B WHEEL BEARINGS – BOTH REAR – REMOVE AND INSTALL
(Includes OPS 1225-A and A1)

Axle Shaft and Bearing Assembly

2
BRAKING SYSTEM

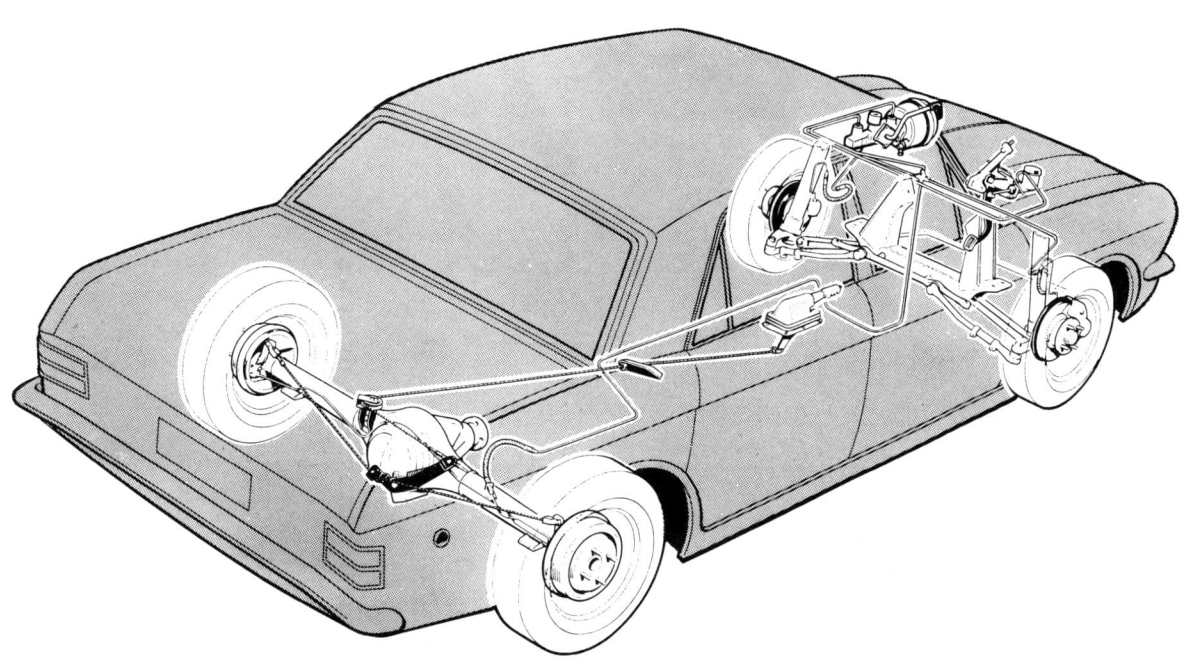

CORTINA - LOTUS

SECTION INDEX

GENERAL DESCRIPTION

QUICK REFERENCE DATA

BRAKE PIPES

SERVICE AND REPAIR OPERATIONS

OPERATION	2000-A	BRAKING SYSTEM – BLEED (SINGLE LINE)
,,	2000-A	BRAKING SYSTEM – BLEED (DUAL LINE)
,,	2000-D	BRAKE SHOES AND PADS – ALL – RENEW
,,	2000-E	BRAKE HYDRAULIC FLUID – RENEW
,,	2000-F	BRAKE HYDRAULIC SEALS AND FLEXIBLE HOSES – RENEW
,,	2004-A	BRAKE SERVO UNIT – REMOVE AND INSTALL
,,	2004-A2	Extra: brake servo unit – overhaul
,,	2004-C	BRAKE SERVO UNIT – OVERHAUL (Includes OPS 2004-A and A2)
,,	2005-A	BRAKE SERVO FILTER – RENEW
,,	2005-B	BRAKE SERVO NON-RETURN VALVE – RENEW
,,	2010-A	FRONT BRAKE CALLIPER – ONE SIDE – REMOVE AND INSTALL
,,	2010-A1	Extra: remaining front calliper – remove and install
,,	2010-A2	Extra: calliper pistons and/or seals – renew
,,	2010-A3	Extra: splash shield – remove and install
,,	2010-A4	Extra: brake disc – remove and install
,,	2010-B	FRONT BRAKE CALLIPERS – BOTH – REMOVE AND INSTALL (Includes OPS 2010-A and A1)
,,	2010-C	FRONT BRAKE CALLIPER PISTONS AND/OR SEALS – ONE – RENEW (Includes OPS 2010-A and A2)
,,	2010-D	FRONT BRAKE CALLIPER PISTONS AND/OR SEALS – BOTH CALLIPERS – RENEW (Includes OPS 2010-A, A1 and A2×2)
,,	2010-E	FRONT BRAKE CALLIPER SPLASH SHIELD – ONE – REMOVE AND INSTALL (Includes OPS 2010-A and A3)
,,	2010-F	FRONT BRAKE CALLIPER SPLASH SHIELDS – BOTH – REMOVE AND INSTALL (Includes OPS 2010-A and A3×2)
,,	2010-G	FRONT BRAKE DISC – ONE – REMOVE AND INSTALL (Includes OPS 2010-A and A4)
OPERATION	2010-H	FRONT BRAKE DISCS – BOTH – REMOVE AND INSTALL (Includes OPS 2010-A and A4×2)
,,	2018-A	FRONT BRAKE PADS – ALL – INSPECT AND/OR RENEW
,,	2075-A	FOUR-WAY UNION – RENEW
,,	2078-A	HYDRAULIC FLEXIBLE HOSE – ANY ONE – RENEW
,,	2078-A1	Extra: hydraulic flexible hose – each additional – renew
,,	2140-A	BRAKE MASTER CYLINDER – REPLACE (SINGLE LINE)
,,	2140-A	BRAKE MASTER CYLINDER – REPLACE (DUAL LINE) – L.H.D.
,,	2140-A	BRAKE MASTER CYLINDER – REPLACE (DUAL LINE) – R.H.D.
,,	2140-A1	Extra: brake master cylinder – overhaul
,,	2140-B	BRAKE MASTER CYLINDER – OVERHAUL (Includes OPS 2140-A and A1)
,,	2145-A	BRAKE SERVO FLEXIBLE VACUUM PIPE – RENEW
,,	2220-B	REAR BRAKE SHOES AND/OR RETRACTING SPRINGS – ONE SIDE – RENEW
,,	2220-B1	Extra: brake shoes and/or retracting springs – second side – renew
,,	2220-B3	Extra: rear wheel cylinder – one – remove and install
,,	2220-B4	Extra: rear wheel cylinder – overhaul
,,	2220-B5	Extra: rear brake carrier plate – remove and install
,,	2220-C	REAR BRAKE SHOES AND/OR RETRACTING SPRINGS – ALL – RENEW (Includes OPS 2220-B and B1)
,,	2220-F	REAR WHEEL CYLINDER – ONE SIDE – REMOVE AND INSTALL (Includes 2220-B and B3)
,,	2220-G	REAR WHEEL CYLINDERS – BOTH – REMOVE AND INSTALL (Includes OPS 2220-B, B1 and B3×2)
,,	2220-H	REAR WHEEL CYLINDER – ONE SIDE – OVERHAUL (Includes OPS 2220-B, and B3 and B4)
,,	2220-J	REAR WHEEL CYLINDERS – BOTH – OVERHAUL (Includes OPS 2220-B, B1, B3×2 and B4×2)
,,	2220-K	REAR BRAKE CARRIER PLATE – ONE SIDE – REMOVE AND INSTALL (Includes OPS 2220-B, B3 and B5)

OPERATION	2220-L	**REAR BRAKE CARRIER PLATES – BOTH – REMOVE AND INSTALL** (Includes OPS 2220-B, B1, B3×2 and B5×2)
,,	2493-A	VALVE AND SWITCH ASSEMBLY – REMOVE AND INSTALL
,,	2493-A1	Extra: valve and switch assembly – overhaul
,,	2493-B	**VALVE AND SWITCH ASSEMBLY – OVERHAUL** (Includes OPS 2493-A and A1)
,,	2800-A	HANDBRAKE LINKAGE – ADJUST
,,	2811-A	HANDBRAKE LEVER ASSEMBLY – REMOVE AND INSTALL
,,	2841-A	HANDBRAKE TRANSVERSE CABLE – RENEW
,,	2853-A	HANDBRAKE PRIMARY CABLE – RENEW

GENERAL DESCRIPTION

Hydraulically operated servo assisted brakes, discs on the front, drums on the rear, are fitted to all four wheels.

The front disc brakes are similar to those used on the New Cortina G.T. model utilising forward facing callipers.

The rear drum brakes are of leading and trailing shoe design. A "self-adjust" mechanism is incorporated and this automatically compensates for wear of the shoe lining material.

The hydraulic system may be either dual line or conventional single line. The dual line system provides separate hydraulic circuits for the front and rear brakes. Should one circuit fail the other circuit is unaffected and the car can still be stopped.

The suspended vacuum type servo is mounted in the front left-hand corner of the engine compartment with single line brake systems.

This is duplicated in the dual line system, the second servo unit being mounted at the rear of the engine compartment.

The handbrake, which operates on the rear wheels only, is applied by means of a floor mounted lever. The handbrake lever operates through an open cable system incorporating a relay lever on the rear axle and its operation causes the "self-adjust" mechanism in the rear brakes to function as and when necessary.

QUICK REFERENCE DATA

PERIODIC SERVICE ATTENTION

Weekly
    Check brake fluid level in reservoir

At first 600 miles (1,000 km.)
    Top-up brake fluid reservoir
    Inspect brake hoses and lines for signs of leaking or chafing

At first 3,000 and every subsequent 3,000 miles (5,000 km.) or three months
(whichever occurs first)
    Check the brake pads and shoes for wear and blow clean

Every 6,000 miles (10,000 km.) or six months
(whichever occurs first)
    Lubricate handbrake linkage
    Remove road wheels, check brake pads and shoes for wear
    Check self-adjusting mechanism and blow clean

Every 24,000 miles (40,000 km.) or two years
(whichever occurs first)
    Renew brake fluid
    Renew brake servo filter

Every 36,000 miles (60,000 km.) or four years
    It is advisable to renew all brake seals, flexible hoses and brake fluid

CORTINA - LOTUS

DATA

Front Brakes

Disc diameter	9.625 in. (24.45 cm.)
Disc run-out (maximum)	0.0035 in. (0.089 mm.) TIR
Pad swept area (total)	20.64 sq. in. (133.2 sq. cm.)
Pad colour coding	One Green Spot

Rear Brakes

Drum diameter and width	9 × 1.75 in. (22.9 × 4.45 cm.)
Shoe swept area (total)	99.0 sq. in. (639 sq. cm.)
Wheel cylinder diameter	0.70 in. (1.78 cm.)

General

Master cylinder diameter (single line system)	0.75 in. (1.91 cm.)
Master cylinder diameter (dual line system)	0.875 in. (2.22 cm.)
Braking ratio	67.9%F 32.1%R

Servo

Type	Suspended vacuum
Boost ratio	2.04 : 1

Tightening Torques

	lb. ft.	kg.m.
Brake calliper to front suspension unit	45 to 50	6.22 to 6.91
Brake disc to hub	30 to 34	4.15 to 4.70
Rear brake plate to axle housing	15 to 18	2.07 to 2.49
Hydraulic unions	7 to 8	1.0 to 1.1
Bleed valves	5 to 7	0.7 to 1.0

Brake Pipe Chart

Part Number	Length Inches	Length cms.	End Fittings A	End Fittings B	End Fittings C
3050E-2263-A	38.25	97.5	1	1	—
3050E-2264-A	29.75	75.5	1	1	—
3050E-2265-A	89.50	227.0	—	2	—
69BB-2265-A-A	89.50	227.0	—	2	—
3016E-2286-A	53.00	134.6	2	—	—
3014E-2A009-A	13.25	33.7	1	1	—
3016E-2K018-A	10.00	25.4	1	1	—
3016E-2K019-A	10.00	25.4	1	1	—
3020E-2K467-C	55.00	139.7	2	—	—
3050E-2K468-A	36.00	91.5	—	2	—
3051E-2K470-A	39.50	100.5	—	1	1
3051E-2K471-A	29.75	75.5	1	1	—
3051E-2K475-A	15.00	38.1	1	1	—
69BB-2K475-A-A	15.25	38.7	1	1	—
3050E-2K485-A	65.50	166.4	1	1	—
3051E-2K485-A	36.25	92.1	1	1	—
3051E-2K486-A	35.50	90.2	2	—	1
3050E-2K486-A	40.50	102.9	2	—	1
3051E-2K487-A	36.00	91.5	1	—	1
3051E-2K488-A	22.25	56.6	2	—	1
3050E-2K488-A	14.75	37.5	2	—	1
69BB-2L162-B-A	9.00	22.9	—	—	2
3050E-2L170-A	28.50	72.4	1	1	—
3051E-2L170-A	23.50	59.7	1	1	—
69BB-2L188-B-A	85.25	216.6	—	—	1

Brake Pipe End Fittings

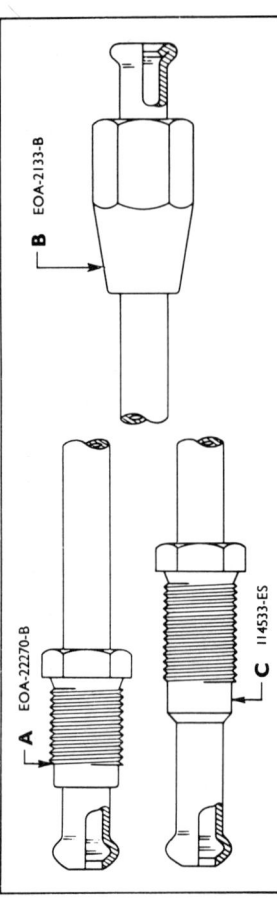

A EOA-22270-B
B EOA-2133-B
C II4533-ES

BRAKE PIPES

To eliminate impractical and expensive stock holding, brake pipes are not serviced individually. They should be fabricated in line with the information in the chart overleaf.

All brake pipes must be manufactured from bundy tubing having a wall thickness of 0.028 in., (0.7 mm.) (EM-900D-1 is a suitable material). **On no account must fuel pipe be used as it is designed to operate at a much lower pressure and will fail if used as brake pipe.**

FORMING PIPES

Two types of flare are used on piping:

(1) Single Flare
(2) Double Flare

Either of these flares can be formed, using the same tool, and this process is described below.

Flaring a Pipe (Split-Die Type)

Single Flare	Ops. 1 to 5, 7 and 8
Double Flare	Ops. 1 to 8

1. Cut off and straighten the required length of pipe. (A pipe cutting tool will simplify obtaining a clean and square cut.)

2. Square off the ends of the pipe with a file and chamfer the end of the pipe to be flared.

3. Select the split die for the pipe being used and insert the die into the tapered hole in the body.

4. Push the bundy tube through the die until the pipe is flush with the face of the die. Lock the pipe in this position by tightening the wing nut securely.

5. The punches are marked Op. 1 and Op. 2. Slide the first operation punch into the hole in the centre of the body and tighten the screw well home to form the single flare.

6. Release the screw and replace the first operation punch with the second operation punch and tighten the screw to form the double flare.

7. Release the screw, wing nut, punch and dies.

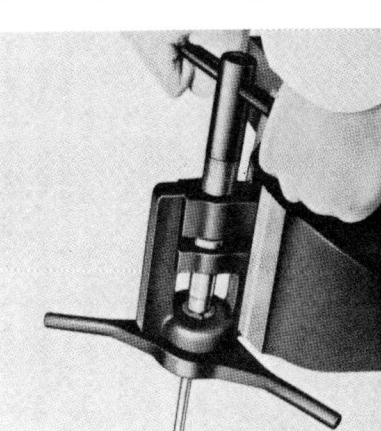

1
Before Flaring
Chamfer Edges

2
First Operation
Single Flare

3
Second Operation
Double Flare

Flaring Sequence

8. Remove the pipe and inspect the flare for cracks or poor flare form. If any doubt exists about the flare it should be cut off and the process repeated.

The finished flare must be square with the pipe, free from any cracks and have a smooth mating surface to ensure a leakproof connection.

Flaring a Pipe (Flaring Bar-Type)

Single Flare	Ops. 1 to 6 and 8
Double Flare	Ops. 1 to 8

1. Cut off and straighten the required length of pipe. (A pipe cutting tool will simplify obtaining a clean and square cut.)

2. Square off the ends of pipe with a file and chamfer the end of the pipe to be flared.

3. Insert the pipe through its appropriate ribbed hole in the bar assembly until the end of the pipe protrudes approximately the thickness of the respective adaptor above the bar, or flush with the bar, depending on the tool used.

4. Fit the adaptor on to the pipe and slide the bar into the yoke. Lock the bar in position with the pipe beneath the yoke screw.

5. Form the single flare by screwing the yoke screw well home.

6. Release the screw and remove the adaptor.

7. Form the double flare by screwing down the yoke screw again, with second adaptor fitted, depending on the tool used.

8. Release the screw, bar and flared pipe. Inspect the flare for cracks or poor flare form and if any doubt exists about the flare, it should be cut off and the process repeated.

Bending the Pipe

Any suitable bundy tube bending tool should be used in the manner prescribed by the tool manufacturer. It is also possible to hand form bends around a suitable former, which can be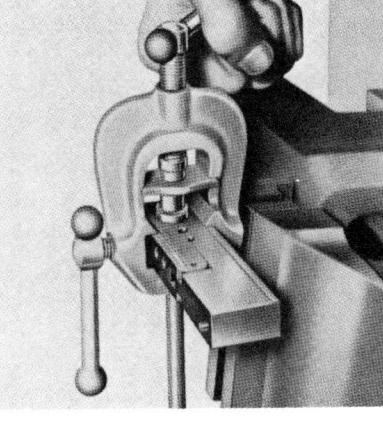

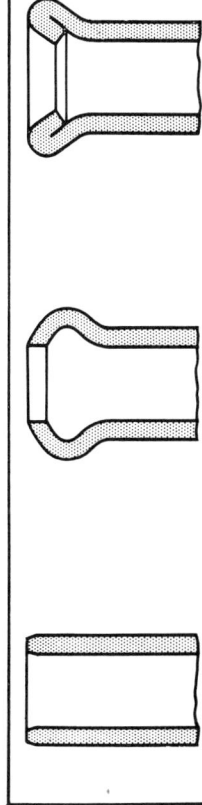

Split Die Type

Flaring-Bar Type

Pipe Flaring

manufactured to suit the different size pipes. When forming a pipe by hand it is essential that the pipe is not kinked as it will be then weakened, and may fracture when fitted to the car.

When special formers are made for specific applications, always keep the surfaces smooth and the shapes free from edges which could damage the tube.

If it is necessary to form a sharp bend very close to the end of the pipe, the difficulty arises of bending the tube with the use of a male and female connection, alternatively, screw several Brake Pipe Plugs, Tool No. P.2012 together and screw into the pipe connection. It is then possible to grip the extension and form the bend.

MINIMUM PERMISSIBLE BENDING RADII

UP TO 45°				46° TO 90°				91° TO 180°				181° TO 360°			
HAND BEND		DIE BEND		HAND BEND		DIE BEND		HAND BEND		DIE BEND		HAND BEND		DIE BEND	
in.	mm.	in.	mm.	in.	mm.	in.	mm.	in.	mm.	in.	mm.	in.	mm.	in.	mm.
$\frac{3}{8}$	9.5	$\frac{3}{16}$	4.8	$\frac{3}{4}$	19.1	$\frac{3}{8}$	9.5	$\frac{7}{8}$	22.2	$\frac{7}{16}$	11.1	$\frac{7}{8}$	22.2	$\frac{7}{16}$	11.1

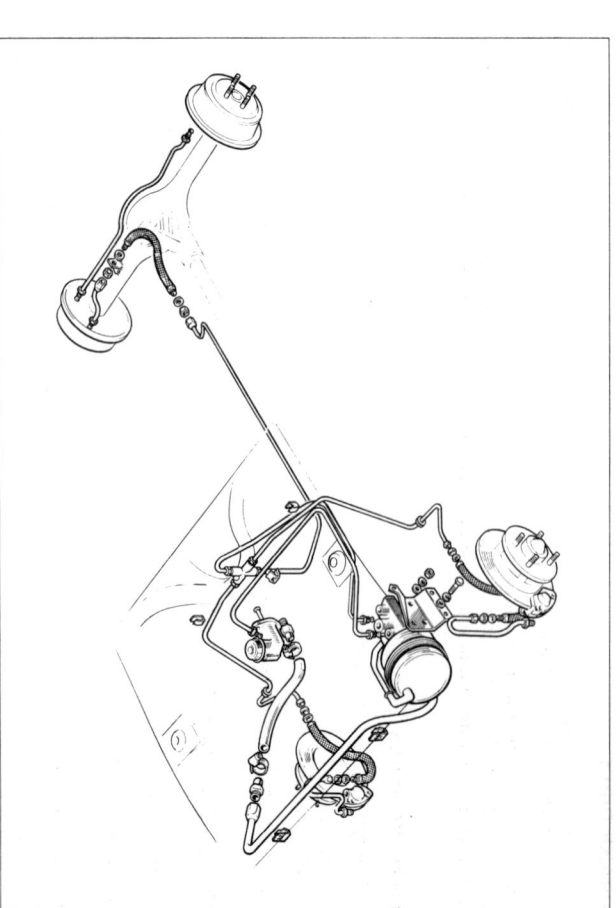

Braking System — Single Line

SERVICE AND REPAIR OPERATIONS

OP 2000-A   BRAKING SYSTEM – BLEED (SINGLE LINE)

Tools Required

P.2006   Brake bleeding tubes

1. Remove the reservoir filler cap and top-up the reservoir.
2. Remove the rubber dust cap from the right-hand front bleed nipple and fit a bleed tube Tool No. P.2006.
3. Place the other end of the tube in a glass jar containing a small quantity of approved brake fluid. During the bleeding operation the end of the pipe must be kept immersed in the fluid.
4. Unscrew the bleed valve about half-a-turn, depress the brake pedal fully and then allow it to return to its "off" position. Brake fluid and/or air should have been pumped into the jar; if not, unscrew the bleed valve further.
5. Pause for an instant (about three seconds) to allow full recuperation of the master cylinder.
6. Continue depressing the brake pedal, pausing after each return stroke, until the fluid entering the jar is clean and free from air bubbles.
7. Press the pedal down to the floor and hold it there whilst the bleed valve is tightened. The correct tightening torque is 5 to 7 lb. ft. (0.7 to 1.0 kg.m.), DO NOT OVERTIGHTEN.
8. During the bleeding operation keep the master cylinder topped-up with approved fluid.
9. Remove the tube and replace the rubber dust cap.
10. Bleed the left-hand front brake in the same way.
11. Bleed the left-hand rear brake in the same way.
12. Re-bleed the front right-hand brake to check that air has not been drawn in through the master cylinder.
13. Top-up the master cylinder and replace the filler cap after checking that the vent hole is clear.

OP 2000-A   BRAKING SYSTEM – BLEED (DUAL LINE)

Tools Required

P.2006   Brake bleeding tubes
        Valve and Switch Assembly piston centralising tool (see text)

NOTE – The procedure for bleeding the brakes is similar to that used with standard brakes. The only difference is that the piston in the Valve and Switch Assembly (where fitted) has to be held in the central position. If this is not done, it will be very difficult to get the warning light to go out and stay out.

1. Centralise the piston in the valve and switch assembly (where fitted) using a screwdriver with the blade modified in line with dimensions given. The tool should be inserted through the aperture in the base of the assembly after the rubber has been removed.

CORTINA - LOTUS

2. Remove the reservoir cap and top-up the reservoir ensuring that both halves are filled.

3. Remove the rubber dust cap from the right-hand front bleed nipple and fit a bleed tube P.2006.

4. Place the other end of the tube in a glass jar containing a small quantity of approved brake fluid. During the bleeding operation the end of the pipe must be kept immersed in brake fluid.

5. Unscrew the bleed valve about half-a-turn, depress the brake pedal fully and then allow it to return to its "off" position. Brake fluid and/or air should have been pumped into the jar, if not, unscrew the bleed valve further.

6. Pause for a moment (about three seconds) to allow full recuperation of the master cylinder.

7. Continue depressing the brake pedal, pausing after each return stroke, until the fluid entering the jar is clean and free from air bubbles.

8. Press the pedal down to the floor and hold it there whilst the bleed valve is tightened. The correct tightening torque is 5 to 7 lb. ft. (0.7 to 1.0 kg.m.). DO NOT OVERTIGHTEN.

9. During the bleeding operation keep the master cylinder topped-up with approved fluid.

10. Remove the tube and replace the rubber dust cap.

11. Bleed the left-hand front brake in the same way.

12. Bleed the left-hand rear brake in the same way.

13. Top-up the master cylinder reservoir and replace the cap after checking that the vent hole is free.

14. Remove the tool from the valve and switch assembly (where fitted) and, with the ignition switched on, depress the brake pedal several times. The brake warning light should not illuminate.

15. With the ignition still switched on, operate the test switch, the warning light should illuminate.

OP 2000-D BRAKE SHOES AND PADS — ALL — RENEW

This is equivalent to Operations 2018-A and 2220-C.

OP 2000-E BRAKE HYDRAULIC FLUID — RENEW

This is the same as Operation 2000-A, Braking System Bleed, except that the system must be bled at each point until clean, new fluid is expelled through each bleed valve.

OP 2000-F BRAKE HYDRAULIC SEALS AND FLEXIBLE HOSES — RENEW

Tools Required

P.2012   Brake line plugs (small)
P.2031   Brake line plugs (large)

1. Open the bonnet and fit wing covers.

2. Disconnect the brake master cylinder push rod from the pedal by unscrewing the locknut and withdrawing the shouldered bolt.

3. Detach the fluid pipes from the master cylinder and plug the pipes.

4. Remove the master cylinder by unscrewing the securing nuts.

5. Dismantle the master cylinder (see Op 2140-A1). Clean the cylinder and inspect the bore, reassemble using new seals.

Bleed Valve

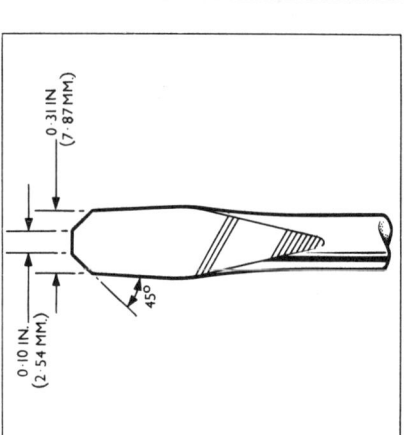

Valve and Switch Assembly
Piston Centralising Tool

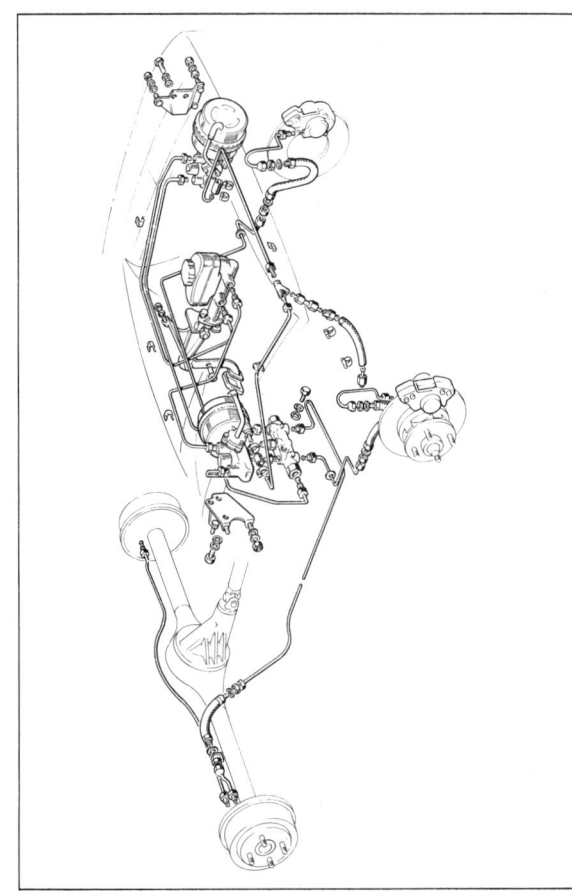

Braking System — Dual Line — L.H.D.

CORTINA - LOTUS

6. Replace the master cylinder and secure with the two nuts and spring washers.
7. Reconnect the fluid pipes to the master cylinder.
8. Fit the push rod to the brake pedal and secure.
9. Jack up the car front and rear and fit chassis stands. Remove the road wheels.
10. Remove the two callipers and the two rear drums.
11. Remove the brake shoes and retracting springs. Detach the wheel cylinders and dismantle. Discard the seals, inspect all other components. Reassemble using new seals. Dismantle the callipers and overhaul, see OP 2010-A2, fitting new piston seals.
12. Replace the wheel cylinders on the brake back plates. Refit the brake shoes and retracting springs.
13. Replace the callipers and rear brake drums.
14. Detach, in turn, the two front flexible hoses and the rear flexible hose. Discard the hoses.
15. Fit new hoses at each of the three locations.
16. On dual line braking systems, remove the valve and switch assembly, where fitted, and overhaul (refer to OP 2493-A1) fitting new hydraulic seals.
17. Bleed the braking system until clean, new fluid is expelled through each bleed valve, see OP 2000-A.
18. Replace the road wheels.
19. Jack up, remove the chassis stands and lower the car to the ground.
20. Remove the wing covers and close the bonnet.

OP 2004-A BRAKE SERVO UNIT – REMOVE AND INSTALL

Tools Required

P.2012 Brake line plugs

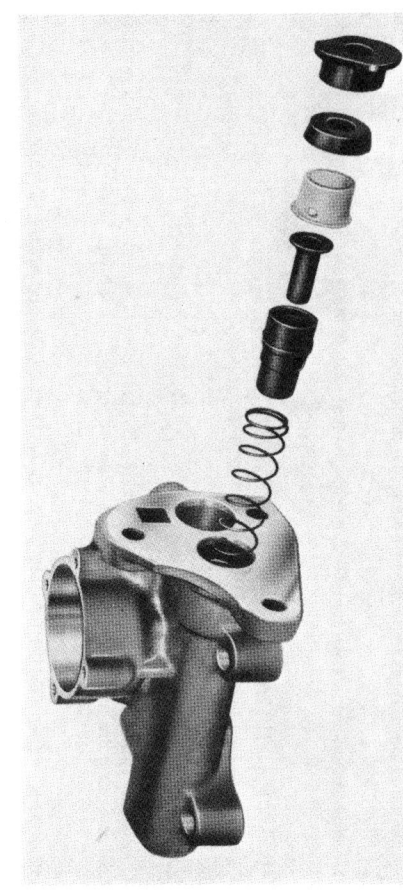

Location of the Front Servo Unit

To Remove

1. Open the bonnet and fit wing covers.
2. Disconnect the vacuum line to the servo unit.
3. Disconnect the hydraulic pipe-lines from the servo unit and plug the pipe ends.
4. Remove the bolts securing the servo unit to the bracket and remove the unit.

To Install

5. Locate the servo unit to the bracket and retain it with bolts.
6. Remove the brake line plugs and reconnect the hydraulic pipe lines to the servo.
7. Bleed the braking system as detailed in OP 2000-A.

OP 2004-A2 EXTRA: BRAKE SERVO UNIT – OVERHAUL

To Dismantle

1. Remove the filter cover retaining screw and remove the cover, filter element and base washer.
2. Break the spot weld securing the crimped band to the front shell.
3. Lightly clamp the shells of the unit in the protected jaws of a vice then cut the plate joining the ends of the crimped band, ensuring no encroachments is made into the shells.
4. With a "second pair of hands" holding the shells together, release the unit from the vice and slowly allow the shells to separate.

NOTE – The internal spring which is compressed at this stage and is now separating the shells is easily controllable by hand.

5. Remove the front shell, diaphragm, piston and spring. Separate the diaphragm and piston.
6. Remove the three screws and copper washers securing the reinforcing plate to the lower shell.

Output Valve – Exploded

CORTINA - LOTUS

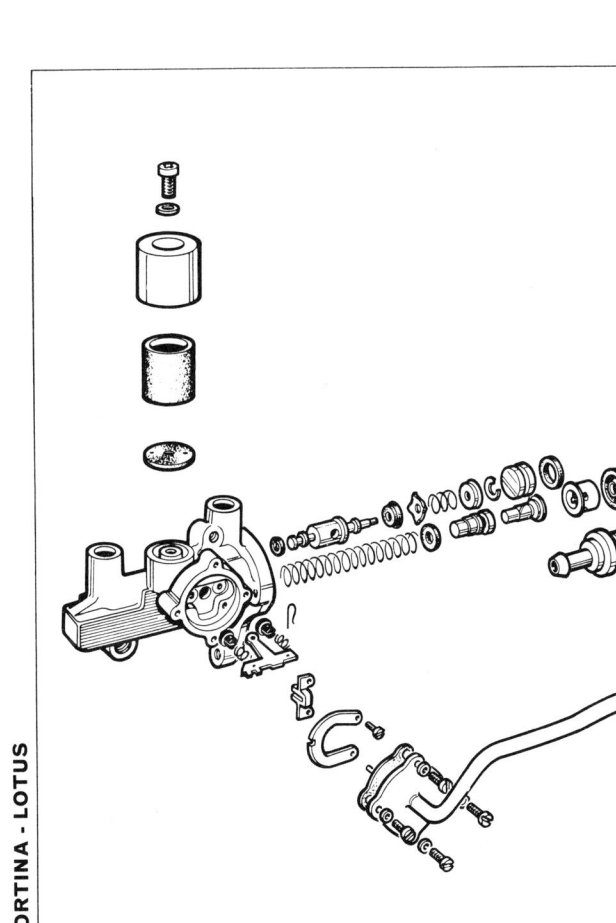

Brake Servo Unit — Exploded

CORTINA - LOTUS

7. Remove the reinforcing plate ensuring that the black collar, located beneath the plate and having three flats, is retained in position by finger pressure.

   NOTE – Failure to retain the collar in position will result in the collar and output valve being ejected from the unit with consequent loss of/or damage to components.

8. Carefully remove the black collar, seal, spacer, output valve and spring.

9. Separate the valve and spring then withdraw the anti-knock valve from the bore of the output valve.

10. Remove the rear shell and gasket from the slave and control cylinder.

11. Remove the four screws and washers securing the air/vacuum tube to the valve chest. Remove the gasket.

12. Remove the two screws to release the bridge and "U" spring. Withdraw the twin valve assembly.

13. Withdraw the plug and control valve assembly from its bore.

14. Depress the spring on the control valve, remove the retaining clip, dished collar, spring and four-sided spring seat.

To Reassemble

NOTE – It is recommended that a repair kit containing seals and a replacement crimped band is used. Ensure that all bores and component parts are scrupulously clean. Lubricate all seals prior to fitting with approved brake fluid.

15. View the control valve with the small diameter to the right, then fit the smaller diameter seal to the left hand spigot, and the seal with the larger diameter to the right-hand spigot and abutting the shoulder.

    NOTE – The seals are correctly fitted when the seal lips are facing away from each other.

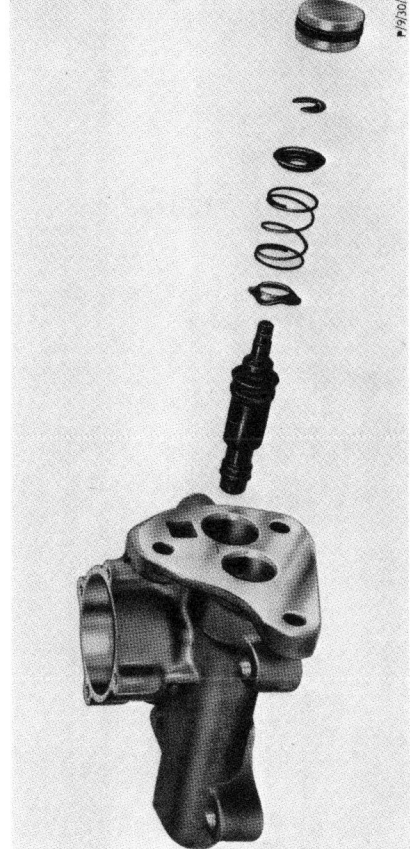

Control Valve — Exploded

CORTINA - LOTUS

16. Locate the four-sided spring seat, concave face first, on the right-hand spigot of the control piston followed by the spring and collar, convex face first, and retain with clip.

17. Locate the control piston in the upper bore of the slave and control cylinder.

18. Renew the seal on the control piston bore plug then locate the plug in the bore.

19. Locate the twin air valve assembly in the valve chest with the slotted valve – arm towards the sealed end of the casting.

20. Position the "U" spring in the valve chest.

21. Locate the bridge piece over the twin valve arm and on top of the "U" spring. Secure with two screws.

22. Position a new gasket on the valve chest followed by the air/vacuum tube and secure with four screws and washers ensuring that the tube is towards the open end of the slave and control cylinder.

23. Locate the slave and control cylinder in the protected jaws of a vice, open end uppermost, then fit a gasket with its intermediate diameter hole aligning with the square bore in the cylinder casting.

24. Locate the rear shell on the gasket with the elongated hole over the square bore in the casting. Temporarily and lightly secure with one bolt and copper washer.

25. Remove the old seal from the output valve and fit a new seal noting that the seal lip is towards the end of the valve having the smaller diameter.

26. Locate the anti-knock valve within the bore of the output valve.

27. Position the spring on the output valve then lower the assembly, spring first, into the bore of the casting.

28. Fit the spacer, flange first, followed by a new seal, lip towards the bore. Locate the collar on the valve and press the assembly into the bore and retain with the reinforcing plate held by finger pressure.

29. Remove the screw fitted in sub-operation 24, slide the reinforcing plate into position with the hole having the larger diameter over the square bore in the casting. Secure with three screws and copper washers, tightening the screws to a torque of 10 to 12 lb. ft. (1.383 to 1.659 kg.m.).

30. Position a new diaphragm over the piston then the assembly within the front shell.

31. Locate the spring, large diameter first, within the piston.

32. Lower the spring, piston, diaphragm and front shell onto the rear shell, bring the two half shells together, aligning the rubber elbow with the air/vacuum tube, and secure with new crimped band.

33. Spot weld the crimped band to the front shell in one place only.

34. Fit new filter base washer, filter element and the filter cover and secure with screw and spring washer.

OP 2004-C   BRAKE SERVO UNIT – OVERHAUL
(Includes OPS 2004-A and A2)

OP 2005-A   BRAKE SERVO FILTER – RENEW

To Remove

1. Raise the bonnet and fit wing covers.

2. Remove the outer body retaining screw and remove the outer body and filter.

To Install

3. Locate a new element and the filter body on the servo unit and secure with retaining screw.

4. Remove wing covers and close the bonnet.

OP 2005-B   BRAKE SERVO NON-RETURN VALVE – RENEW

To Remove

1. Lever the non-return valve from the rear shell by inserting a large screwdriver between the rubber grommet and valve.

2. Remove the grommet ensuring that it does not drop into the vacuum chamber.

To Install

3. Fit a grommet into the rear shell.

4. Lubricate the ribs of the non-return valve and push the valve fully home in the grommet.

Servo Filter Location

Vacuum Non-Return Valve

CORTINA - LOTUS

OP 2010-A FRONT BRAKE CALLIPER – ONE SIDE – REMOVE AND INSTALL

Tools Required

P 2012 Brake line plugs

To Remove

1. Remove the hub cap and slacken the wheel nuts.
2. With the handbrake applied, jack up the front of the car and fit stands.
3. Remove the road wheel.
4. Remove the brake pads after withdrawing the shield, retaining pins and clips. If it is intended to overhaul the calliper unit, depress the brake pedal to bring the pistons into contact with the disc and thus facilitate removal of the pistons.
5. Detach the hydraulic pipe from the union on the rear of the calliper and fit a brake line plug, Tool No. P 2012, to each opened end.
6. Bend up the lock tabs and remove the two calliper retaining bolts and detach the calliper assembly.

To Install

7. Replace the calliper assembly, using a new locking plate and tighten the bolts to a torque of 45 to 50 lb. ft. (6.22 to 6.91 kg.m.). Bend up the lock tabs.
8. Recouple the hydraulic pipe to the union on the rear of the calliper.
9. Push the pistons sufficiently into their bores to accommodate the pads and fit the pads and shims.
10. Secure the pads in position with the retaining pins and clips. Refit the shield.
11. Replace the road wheel.
12. Remove the stands and lower the car to the ground.
13. Tighten the wheel nuts and replace the hub cap.
14. Bleed the brakes, see OP 2000-A.

OP 2010-A1 EXTRA: REMAINING FRONT CALLIPER – REMOVE AND INSTALL

Repeat sub-operations 3 to 11 of OP 2010-A.

OP 2010-A2 EXTRA: CALLIPER PISTONS AND/OR SEALS – RENEW

NOTE – The calliper is made in two paired halves, which are bolted together. Under no circumstances should the two halves be separated.

To Dismantle

1. Partially remove the piston from one cylinder bore and remove the sealing bellows from its location in the lower part of the piston skirt. Piston removal can be facilitated by using air pressure or low hydraulic pressure. Withdraw the piston.
2. Pull the sealing bellows from its location in the annular ring machined in the cylinder bore.
3. Withdraw the piston sealing ring from the cylinder bore.

Front Brake Assembly

4. Repeat these operations for the other cylinder.
5. Wash the pistons and piston bores in commercial alcohol, methylated spirit or approved brake fluid. Do NOT use a mineral base fluid such as petrol, paraffin, carbon tetrachloride, etc.
6. Ensure that the pistons and their bores are free from score marks and are fit for further service.

To Reassemble

7. Assemble a piston seal to the annular groove provided in the cylinder.
8. Refit the rubber bellows to the cylinder with the lip that is turned outwards fitting in the groove provided in the cylinder.
9. Place the piston, crown first, through the rubber sealing bellows and into the cylinder. Care should be exercised when carrying out this operation as the piston may tend to damage the rubber bellows.
10. When the piston is located in the cylinder fit the inner edge of the bellows in the annular groove provided in the piston skirt.
11. Push the piston as far down the cylinder bore as possible.
12. Repeat the operations for the other cylinder.

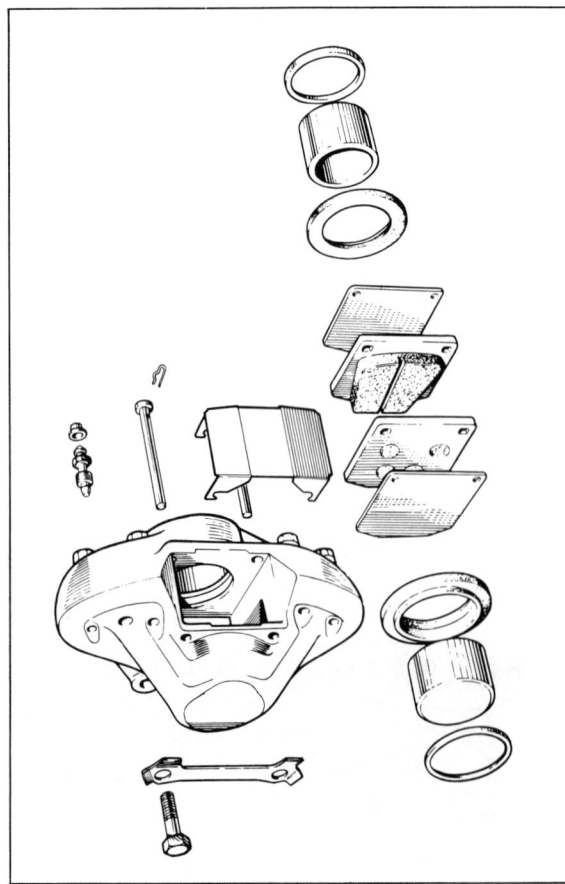

Calliper Assembly — Exploded

OP 2010-A3 EXTRA: SPLASH SHIELD – REMOVE AND INSTALL

To Remove

1. Lever the dust cap from its aperture in the end of the hub.
2. Remove the split pin and adjusting nut retainer, then unscrew the adjusting nut and remove the thrust washer and the outer bearing cone.
3. Detach the hub and disc assembly from the wheel spindle.
4. Remove the splash shield after bending back the integral lock tabs and removing the securing screws.

To Install

5. Bolt the splash shield to the base of the front suspension unit and bend up the locking tabs.
6. Replace the hub and disc assembly on the wheel spindle and fit the outer bearing cone, the thrust washer and the adjusting nut.
7. Tighten the nut to a torque of 27 lb. ft. (3.73 kg.m.) whilst rotating the disc to ensure proper seating of the bearings.
8. Slacken back the nut 90° and fit the nut retainer so that one of the retainer castellations lines up with the split pin hole in the wheel spindle. Fit a new split pin.
9. Tap the dust cap back into place in the end of the hub.

OP 2010-A4 EXTRA: BRAKE DISC – REMOVE AND INSTALL

Tools Required
P 4008   Dial gauge

To Remove

1. Lever the dust cap from its aperture in the end of the hub.
2. Remove the split pin and adjusting nut retainer, then unscrew the adjusting nut and remove the thrust washer and the outer bearing cone.
3. Detach the hub and disc assembly from the wheel spindle.
4. Separate the brake disc from the hub after bending back the bolt lock tabs and unscrewing the bolts. Discard the locking plates and the bolts.

To Replace

5. Thoroughly clean the mating faces of the hub and disc – this is most important.
6. Align the mating marks, place the disc on the hub and fit two new locking plates and four new bolts.
7. Tighten the bolts to a torque of 30 to 34 lb. ft. (4.15 to 4.70 kg.m.) and bend up the locking tabs to secure.
8. Replace the hub and disc assembly on the wheel spindle and fit the outer bearing cone, the thrust washer and the adjusting nut.
9. Tighten the nut to a torque of 27 lb. ft. (3.73 kg.m.) whilst rotating the disc to ensure proper seating of the bearings.

CORTINA - LOTUS

10. Slacken back the nut 90° and fit the nut retainer so that one of the retainer castellations lines up with the split pin hole in the wheel spindle. Fit a new split pin.

11. Check the disc run-out (relative to the axis of the spindle body) as follows:—

   (a) Disconnect the track rod from the steering arm at its outer end, after removing the split pin and castellated nut and separating the ball joint.

   (b) Using a dial gauge (Tool No. P 4008) attached to the steering arm, check the run-out of the disc.

   This figure must be within 0.0035 in. (0.089 mm.), total indicator reading. Should a reading in excess of this be recorded, the cause of this excessive run-out, i.e. a worn or distorted disc, dirt between the disc and hub faces, or mal-alignment of the hub bearings, etc., must be eliminated.

   (c) Remove the dial gauge and refit the track rod end, tightening the castellated nut to 18 to 22 lb. ft. (2.48 to 3.04 kg.m.) securing with a new split pin.

12. Tap the dust cap back into place in the end of the hub.

OP 2010-B FRONT BRAKE CALLIPERS – BOTH – REMOVE AND INSTALL
(Includes OPS 2010-A and A1)

OP 2010-C FRONT BRAKE CALLIPER PISTONS AND/OR SEALS – ONE CALLIPER – RENEW
(Includes OPS 2010-A and A2)

OP 2010-D FRONT BRAKE CALLIPER PISTONS AND/OR SEALS – BOTH – RENEW
(Includes OPS 2010-A, A1 and A2×2)

OP 2010-E FRONT BRAKE CALLIPER SPLASH SHIELD – ONE – REMOVE AND INSTALL
(Includes OPS 2010-A and A2)

OP 2010-F FRONT BRAKE CALLIPER SPLASH SHIELD – BOTH – REMOVE AND INSTALL
(Includes OPS 2010-A, A1 and A3×2)

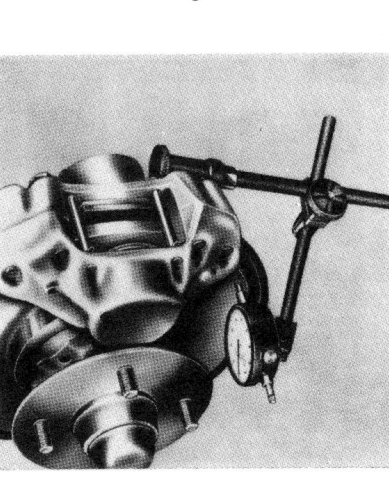

Checking Brake Disc Run-out

OP 2010-G FRONT BRAKE DISC – ONE – REMOVE AND INSTALL
(Includes OPS 2010-A and A4)

OP 2010-H FRONT BRAKE DISCS – BOTH – REMOVE AND INSTALL
(Includes OPS 2010-A, A1 and A4×2)

OP 2018-A FRONT BRAKE PADS – ALL – INSPECT AND/OR RENEW

To Remove

1. With the handbrake applied, jack up the front of the car and fit stands. Remove the front road wheels.

2. Remove the shield from the front of the pads then pull out the retaining pin clips, withdraw the retaining pins and remove the brake pads and shims using, if necessary, a pair of thin-nosed pliers.

   NOTE – Prior to fitting new pads, check that they are the correct type for the Cortina-Lotus. Also, ensure that the pads and the disc are free from grease, oil or dirt.

To Replace

3. To enable the new pads to be fitted, push the pistons into their bores. This action will cause fluid to be returned to the master cylinder which, if it has recently been topped-up, may overflow. To avoid this, examine the fluid level and, if necessary, remove a quantity of fluid.

4. Fit new brake pads and shims ensuring that both are correctly fitted.

5. Refit the retaining pins and secure with the retaining pin clips.

6. Operate the brake pedal several times to bring the pads into the correct adjustment. Check that the pads are free to move slightly, this indicates that the retaining pins are not fouling the pad.

7. Replace the road wheels and lower the car to the ground.

8. Tighten the wheel nuts and replace the hub cap.

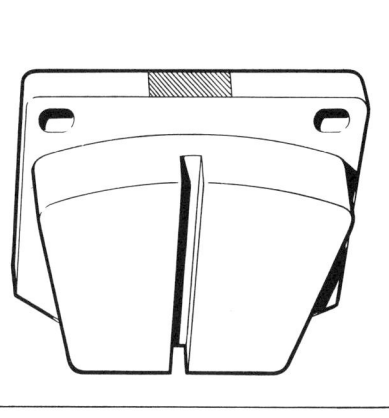

Brake Pad Colour Coding - Green

Brake Pedal and Master Cylinder — Single Line

OP 2075-A   FOUR-WAY UNION – RENEW

Tools Required

P 2012   Brake line plugs

1. Disconnect the brake pipes from the union and plug the end of the pipe from the master cylinder.
2. Remove the bolt securing the union to the engine compartment.
3. Reconnect the hydraulic pipes. Be careful not to overtighten the unions.
4. Replace the union and secure with the retaining bolt.
5. Bleed the braking system as described in OP 2000-A.

OP 2078-A   HYDRAULIC FLEXIBLE HOSE – ANY ONE – RENEW

Tools Required

P 2012   Brake line plugs (if required)

Front

1. With the handbrake applied jack up the front of the car and fit stands.
2. Disconnect one end of the flexible tube from the bracket on the suspension unit and the other end from underneath the wheel arch.
3. Connect the new flexible tubing, ensuring that the locknuts and both securing brackets are tight.
4. Jack up, remove the stand and lower the car to the ground.
5. Bleed the braking system as described in OP 2000-A.

Rear

1. Jack up the rear of the car and fit stands.
2. Detach the flexible tube from the bracket on the underbody and the bracket on the right-hand side of the axle casing.
3. Connect the new flexible tubing, ensuring that the locknuts and both securing brackets are tight.
4. Jack up, remove the stands and lower the car to the ground.
5. Bleed the braking system as described in OP 2000-A.

OP 2140-A   BRAKE MASTER CYLINDER – REPLACE (SINGLE LINE)

Tools Required

P 2012   Brake line plugs

To Remove

1. Disconnect the brake master cylinder push rod from the pedal by unscrewing the locknut and withdrawing the shouldered bolt.

OP 2140-A BRAKE MASTER CYLINDER – REPLACE – (DUAL LINE) – R.H.D.

Tools Required

P.2012   Brake line plugs

1. Disconnect the brake master cylinder push rod from the operating lever assembly by unscrewing the locknut and withdrawing the shouldered bolt.
2. Detach the two fluid lines, not from the reservoir, by unscrewing the union nuts; use blanking plugs to prevent dirt entering the lines.
3. Locate a suitable container beneath the reservoir then detach the lines from the reservoir.
4. Unscrew the two nuts and spring washers securing the master cylinder to the bulkhead then withdraw the cylinder.
5. If a new master cylinder is being fitted transfer reservoir lines from the old master cylinder.
6. Refit the master cylinder to its mounting bracket and secure with two spring washers and nuts.
7. Reconnect the pipe lines to the master cylinder and the lines to the reservoir.
8. Reconnect the master cylinder push rod to the actuating lever, passing the shouldered bolt through the actuating lever and then through the pedal. Fit the locknut.
9. Fill the reservoir with clean approved fluid, Part No. ME-3833-F and bleed the system as described in OP 2000-A.
   Check the action of the brakes on road test.

OP 2140-A1   EXTRA: BRAKE MASTER CYLINDER – OVERHAUL

NOTE – Unless otherwise stated the following sub-operations apply to L.H.D. and R.H.D., single and dual line, master cylinders.

To Dismantle

1. Pull off the rubber boot, remove the circlip and detach the push rod.
2. L.H.D. dual line system only.
   (a) Remove the two crosshead screws securing the reservoir to the body. Turn the reservoir.
   (b) Remove and discard the sealing washer adjacent to the primary recuperating valve.
   (c) Using a suitable hexagon key unscrew the plug which retains the primary recuperating valve.
   (d) Lift out and discard the primary recuperating valve assembly.
3. Fit plastic (or other suitable) plugs to the two outlet ports and block off the primary recuperating valve aperture.

---

2. Detach the fluid line by unscrewing the union nut, using a blanking plug to prevent dirt entering the line.
3. Disconnect the air cleaner hose, remove the air box cover and the air box.
4. Remove the carburettor air horn retaining plates, the horns and the breather pipe.
5. Withdraw the master cylinder after unscrewing the two nuts and spring washers securing the master cylinder to the bulkhead.
6. Empty the contents of the fluid reservoir into a waste container.

To Replace

7. Refit the master cylinder to the engine bulkhead, replace the two spring washers and nuts and tighten securely.
8. Reconnect the fluid pipe, tighten the union nut securely, but do not overtighten.
9. Refit the breather pipe, air horns and retaining plates.
10. Refit the air box, air box cover and air cleaner hose.
11. Reconnect the brake master cylinder push rod to the pedal by passing the shouldered bolt through the push rod and then the pedal. Fit the locknut.
12. Fill the master cylinder reservoir with clean approved fluid, Part No. ME-3833-F, and then bleed the system as described in OP 2000-A.
    Check the action of the brakes on road test.

OP 2140-A   BRAKE MASTER CYLINDER – REPLACE (DUAL LINE) – L.H.D.

Tools Required

P.2012   Brake line plugs

To Remove

1. Disconnect the brake master cylinder push rod from the pedal by unscrewing the locknut and withdrawing the shouldered bolt.
2. Detach the two fluid lines by unscrewing the union nuts, using blanking plugs to prevent dirt entering the lines.
3. Withdraw the master cylinder after unscrewing the two nuts and spring washers securing the master cylinder to the bulkhead.
4. Empty the contents of the fluid reservoir into a waste container.

To Replace

5. Refit the master cylinder to the engine bulkhead, replace the two spring washers and nuts and tighten securely.
6. Reconnect the two fluid pipes, tighten the union nuts securely, but do not overtighten.
7. Reconnect the brake master cylinder push rod to the pedal by passing the shouldered bolt through the push rod and then the pedal. Fit the locknut.
8. Fill the master cylinder reservoir with clean approved fluid, Part No. ME-3833-F, and then bleed the system as described in OP 2000-A.
   Check the action of the brakes on road test.

CORTINA - LOTUS

4. Using an air line, blow air into the secondary recuperating valve aperture and expel the primary piston, the primary piston spring and the secondary piston and recuperating valve assembly from the cylinder bore.

5. Remove the primary piston seal from the primary piston.

6. The secondary piston is held in the spring retainer by a tab which engages under a shoulder on the front of the piston. Carefully lift this tab and remove the piston.

7. Compress the spring and move the retainer to one side to release the end of the secondary recuperating valve stem from the retainer.

8. Slide the valve spacer and shim off the valve stem.

9. Remove the rubber valve seal and the secondary piston seal and discard.

10. Wash all parts in methylated spirit, commercial alcohol or approved brake fluid. Do not use mineral base oils such as petrol or carbon tetrachloride.

11. Inspect the pistons and cylinder bore for score marks and other parts for damage. Renew any parts that appear unsuitable for further service. If either of the pistons or the cylinder bore are defective in any way, a new master cylinder assembly must be fitted.

To Reassemble

12. Fit a new seal to the secondary piston.

13. Replace the shim washer on the valve stem together with the seal spacer so that the legs of the spacer are towards the valve seal. Ensure that the shim is fitted concentrically on the rear shoulder of the valve stem so that its convex face abuts the shoulder flange.

14. Fit the return spring over the valve stem and insert the spring retainer into the end of the return spring. Compress the spring and engage the boss on the valve stem in its recess in the spring retainer.

15. Insert the spigot end of the secondary piston into the spring retainer and secure by pressing down the tab so that it locates against the shoulder of the piston.

16. Dip the secondary piston and recuperating valve assembly in approved brake fluid and enter it into the cylinder with the valve leading.

17. Fit a new seal to the primary piston.

18. Enter the primary piston spring into the cylinder.

19. Dip the primary piston in approved brake fluid and enter it into the cylinder with the drilled end leading.

20. Fit the push rod so that it fits into the recess in the end of the primary piston, press the guide washer into position and secure the complete assembly with the circlip. Check that the push rod moves freely.

21. L.H.D. Dual line system only.
    (a) Fit the primary recuperating valve into its aperture, moving the push rod as necessary to gain access.

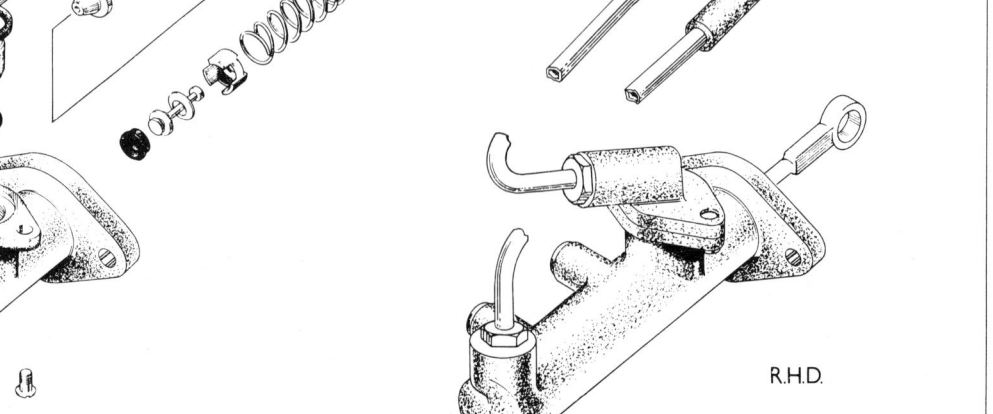

Master Cylinder and Reservoir — Dual Line

(b) Screw the retaining plug into position using a suitable hexagon key. Tighten to a torque of 35 to 45 lb. ft. (4.9 to 6.2 kg.m.).

(c) Cycle the push rod and ensure that the primary recuperating valve "tips" open when the push rod is fully withdrawn and that the valve closes when the push rod is pushed in.

(d) Fit a new rubber sealing washer in the recuperating valve entry port.

(e) Fit the reservoir and secure with the two crosshead screws.

22. After installation in the car and bleeding the brakes, depress the brake pedal for about ten seconds and then examine the master cylinder to ensure that there are no signs of fluid leakage.

OP 2140-B   BRAKE MASTER CYLINDER – OVERHAUL
(Includes OPS 2140-A and A1)

OP 2154-A   BRAKE SERVO FLEXIBLE VACUUM PIPES – RENEW

To Remove

1. Open the bonnet and fit wing covers.
2. Slacken the clips securing the ends of the pipe to the adaptor on the servo unit, the inlet manifold and the connecting metal pipe.
3. Remove the vacuum pipes and pull off the clips.

To Install

4. Fit the clips to the new pipes and push onto the adaptors on the servo, inlet manifold and the connecting metal pipe.
5. Tighten the clips.
6. Remove the wing covers and close the bonnet.

OP 2220-B   REAR BRAKE SHOES AND/OR RETRACTING SPRINGS – ONE SIDE – RENEW

NOTE – The rear brake shoes should be inspected for wear at 3,000 miles (5,000 km.) intervals. There should always be at least $\frac{1}{32}$ in. (0.793 mm.) of lining material above the rivet heads. If the linings are worn so that there is less than $\frac{1}{32}$ in. they should be renewed. Also, if the linings are contaminated by oil or grease, it is preferable to renew them rather than attempt to clean them.

To Remove

1. Remove the hub cap, slacken off the wheel nuts, jack up the rear end, fit stands and remove the wheel.

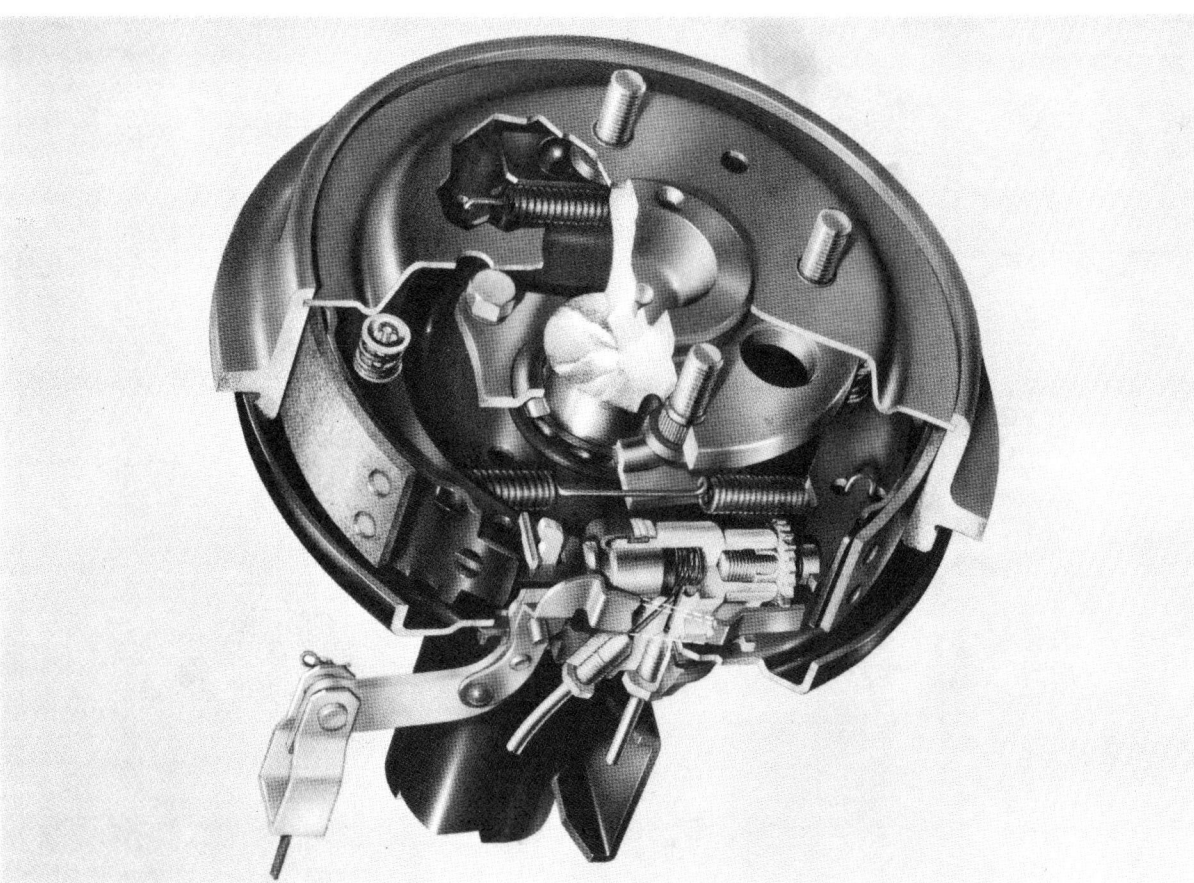

Rear Brake Assembly

CORTINA - LOTUS

2. Ensure that the handbrake is fully released then remove the pan head screw securing the brake drum to the half-shaft and remove the drum.

3. Remove the shoe holding down springs, one on each shoe, by turning the top washer through 90° and pulling off the washer and spring.

4. Disengage each shoe from its location slot in the fixed pivot and the wheel cylinder and remove the shoes. To prevent the piston falling out of the wheel cylinder it should be held in with a clip or a rubber band round the cylinder.

5. Remove the retracting springs from the brake shoes.

6. Withdraw the ratchet wheel assembly from the wheel cylinder and rotate the wheel until it abuts the slot head bolt shoulder. This moves the brake shoes (when refitting) to the fully off adjustment. If this is not done difficulty may be encountered when refitting the brake drum.

To Install

7. Assemble the retracting springs between the two shoes. They must be fitted from the drum side of the shoe.

8. Smear grease (EM 1C 18) on the brake shoe support pads, brake shoe pivots, the ratchet wheel threads and the spigot and face of ratchet wheel which abuts the wheel cylinder.

9. Fit the shoes assembly to the backplate by positioning the rear shoe in its location on the fixed pivot and over the handbrake link.

10. Secure the shoes to the backplate with the holding down spindles and retain the spindles with a washer, spring and another washer. Turn the top washer through 90°.

11. Check to ensure that the shoes are firmly seated and that the springs are not binding on the backplate or the slave cylinder and that the cable adjustment allows ratchet indexing finger to return to the fully off position.

12. Replace the brake drum and secure with the pan head screw.

13. Refit the wheel, remove the stands and lower the car to the ground. Tighten the wheel nuts and replace the hub cap.

14. Operate the handbrake sufficiently to bring the brakes into correct adjustment.

15. Check the operation of the brakes on road test.

OP 2220-B1 EXTRA BRAKE SHOES AND/OR RETRACTING SPRINGS – SECOND SIDE RENEW

Repeat sub-operations 1 and 3 to 13 of OP 2220-B

OP 2220-B2 EXTRA REAR WHEEL CYLINDER – ONE – REMOVE AND INSTALL

Tools Required

P 2012   Brake line plugs

Rear Brake - Exploded

CORTINA - LOTUS

To Remove

1. Disconnect the brake fluid pipe (two on right-hand brake plate) and fit brake line plugs. Care must be taken to avoid ingress of dirt to the cylinder.
2. Remove the spring clip and clevis pin from the handbrake link on the inside of the brake plate.
3. Prise the rubber boot on the rear of the wheel cylinder away from the brake plate and remove. Pull off the two 'U' shaped retainers securing the cylinder to the brake plate.
4. Remove the wheel cylinder and handbrake link.

To Install

5. Rotate the ratchet wheel on its shaft so that when the shoes are assembled they will be on the "fully off" limit of their adjustment. Replace the ratchet wheel assembly in its recess in the end of the wheel cylinder body, after applying a smear of grease to the threads and the spigot and face of ratchet wheel which abuts the wheel cylinder.
6. Smear the brake plate with grease (EM 1C 18) in the area where the wheel cylinder slides.
7. Replace the handbrake link and wheel cylinder in the aperture in the brake plate. Ensure that the pivot on the handbrake link is correctly located in the slot in the wheel cylinder body.
8. Secure the wheel cylinder to the brake plate, using the 'U' shaped spring retainer and the 'U' shaped flat retainer. Note that the spring retainer is fitted from the handbrake link end of the wheel cylinder and the flat retainer from the other.
9. Fit the rubber boot over the wheel cylinder and the handbrake link. Ensure that the wheel cylinder can slide in the carrier plate. Check that the handbrake link operates the self-adjusting mechanism.

NOTE – Make sure that the correct ratchet wheel and screw assembly is fitted to the wheel cylinder. There is a right-handed threaded screw on the right-hand wheel cylinder and a left-handed threaded screw on the left-hand wheel cylinder.

Fitting the Wheel Cylinder Retainers

10. Grease the brake shoe support pads, using zinc oxide grease.
11. Reconnect the handbrake linkage to the handbrake link, using a clevis pin and retain in position with a spring clip.
12. Remove the brake line blanking plug from the brake line and fit the pipe to the wheel cylinder.
13. After refitting the brake shoes, the brake drum and the road wheel (part of OP 2220-A), bleed the braking system as described in OP 2000-A.

OP 2220-B3   EXTRA: REAR WHEEL CYLINDER – OVERHAUL

To Dismantle

1. Remove the boot retainer, prise off the boot and withdraw the piston, complete with seal, from the wheel cylinder bore.
2. Detach the seal from the piston.
3. Remove the return spring from the cylinder bore.
4. Remove the ratchet wheel and screw assembly from the other end of the wheel cylinder.

To Reassemble

All parts should be washed in clean brake fluid and inspected for wear or damage, any components that are not considered fit for further service should be discarded. Avoid contamination by dirt during assembly.

5. Dip the piston and seal in approved brake fluid and reassemble them. Fit the seal to the piston with the flat face of the seal adjacent to the piston rear shoulder.
6. Fit the return spring in the wheel cylinder bore.

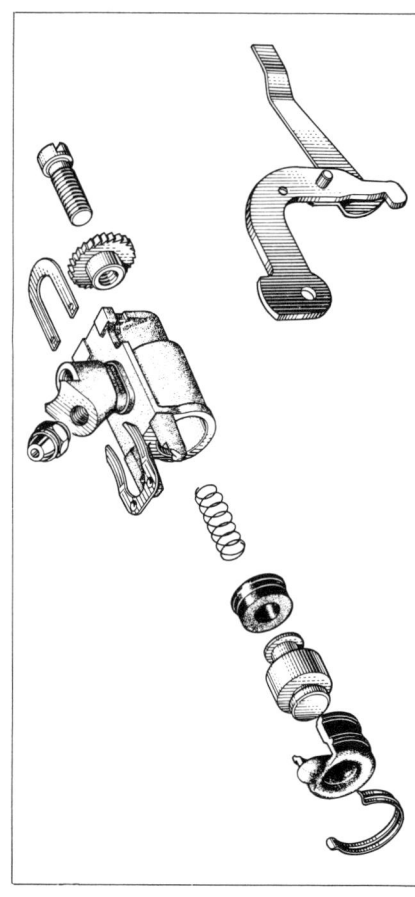

Rear Wheel Cylinder - Exploded

7. Dip the piston and seal assembly in approved brake fluid and insert into the cylinder bore seal end first.
8. Locate the rubber boot on the wheel cylinder and fit the retainer.
9. Replace the ratchet wheel and screw assembly in the wheel cylinder. Rotate the ratchet wheel until it abuts the shoulder of the slot head bolt. This ensures that the brakes are in the fully "off" position.

OP 2220-B4 EXTRA: REAR BRAKE CARRIER PLATE – REMOVE AND INSTALL

Tools Required

P 3072 Slide hammer (main tool)
P 3072-4 Rear axle shaft assembly remover and replacer (adaptor)

To Remove

1. Rotate the axle shaft flange to gain access to two of the retaining bolts securing the axle shaft retaining plate and back plate to the axle casing. Rotate the axle shaft through 90° and remove the other two bolts and spring washers.
2. Withdraw the axle shaft, using the slide hammer and adaptor, Tool Nos. P 3072 and P 3072-4. Remove the brake plate.

To Install

3. Position the brake plate on the axle casing and replace the axle shaft and retainer.
4. Secure the axle shaft retainer and the back plate to the rear axle housing, using four bolts and spring washers.

OP 2220-C REAR BRAKE SHOES AND/OR RETRACTING SPRINGS – ALL – RENEW
(Includes OPS 2220-B and B1)

OP 2220-F REAR WHEEL CYLINDER ASSEMBLY – ONE SIDE – REMOVE AND INSTALL
(Includes OPS 2220-B and B3)

OP 2220-G REAR WHEEL CYLINDER ASSEMBLY – BOTH – REMOVE AND INSTALL
(Includes OPS 2220-B, B1 and B3×2)

OP 2220-H REAR WHEEL CYLINDER ASSEMBLY – ONE SIDE – OVERHAUL
(Includes OPS 2220-B, B3 and B4)

OP 2220-J REAR WHEEL CYLINDER ASSEMBLY – BOTH – OVERHAUL
(Includes OPS 2220-B, B1, B3×2 and B4×2)

OP 2220-K REAR BRAKE CARRIER PLATE – ONE SIDE – REMOVE AND INSTALL
(Includes OPS 2220-B, B3 and B5)

OP 2220-L REAR BRAKE CARRIER PLATE – BOTH – REMOVE AND INSTALL
(Includes OPS 2220-B, B1, B3×2 and B5×2)

OP 2493-A VALVE AND SWITCH ASSEMBLY (WHERE FITTED) – REMOVE AND INSTALL

Tools Required

P.2012 Brake line plugs

To Remove

1. Disconnect the five brake pipes from the ports on the valve and switch assembly. Plug the ends of the two pipes from the master cylinder.
2. Disconnect the wiring from the switch.
3. Unscrew the bolt securing the assembly to the rear of the engine compartment. Remove the assembly.

To Replace

4. Position the valve and switch assembly in place on the engine compartment rear bulkhead and loosely fit the securing bolt.
5. Reconnect the hydraulic pipes. Be careful not to overtighten the unions.
6. Tighten the bolt securing the assembly to the bulkhead.
7. Reconnect the electrical wiring to the switch.

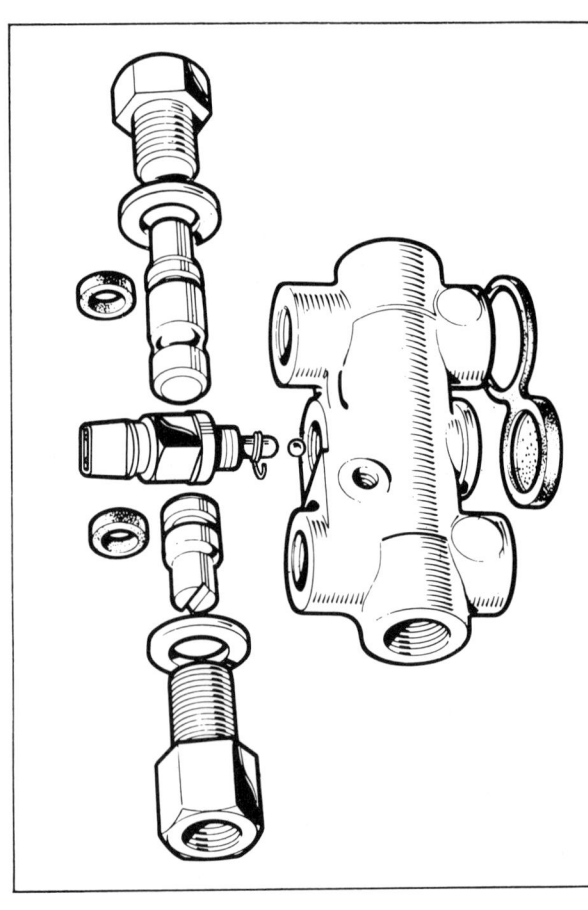

Valve and Switch Assembly — Exploded

CORTINA - LOTUS

OP 2493-A1   EXTRA: VALVE AND SWITCH ASSEMBLY (WHERE FITTED) – OVERHAUL

To Dismantle

1. Unscrew the end plug, discard the copper gasket.
2. Unscrew the adaptor, discard the copper gasket.
3. Unscrew the switch assembly.
4. Push the pistons out of the bore, taking extreme care to avoid damaging the bore surface finish.
5. Remove the seals from the pistons and discard.
6. Pull off the dust cover and discard.
7. Wash all parts in methylated spirit, commercial alcohol or approved brake fluid. Do not use mineral base oils such as petrol, paraffin or carbon tetrachloride.

To Reassemble

8. Fit new seals to the piston so that the larger diameter of each seal is adjacent to the slotted end of the piston, see illustration.
9. Dip the pistons in clean brake fluid.

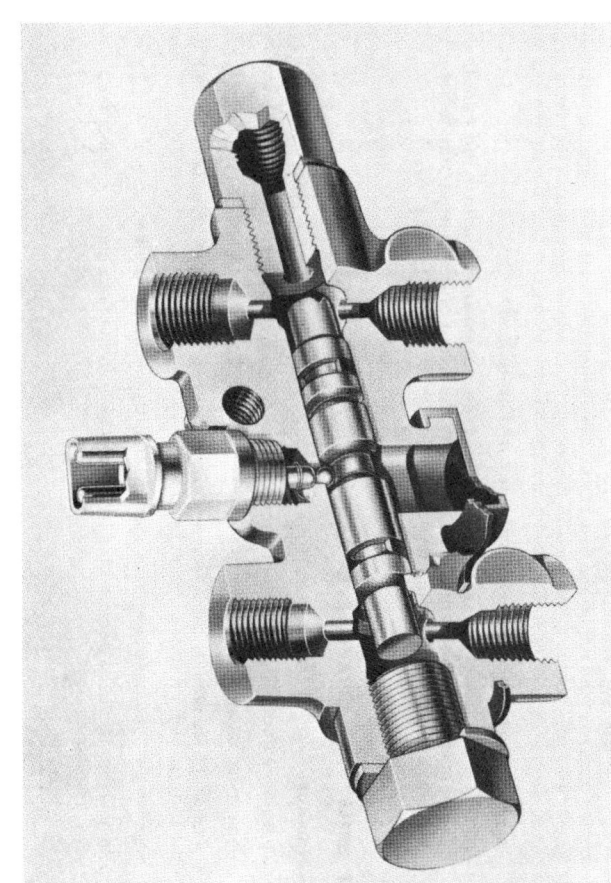

Valve and Switch Assembly — Sectioned

10. Insert the longer piston into the bore (slotted end outermost) until the groove is opposite the switch plunger aperture.
11. Screw the switch plunger into position and tighten to a torque of 2 to 2.5 lb. ft. (0.28 to 0.34 kg.m.).
12. Insert the shorter piston into the bore (slotted end outermost).
13. Fit a new copper gasket to the adaptor and screw the adaptor into the end of the assembly adjacent to the mounting bolt hole.
14. Fit a new copper gasket to the end plug and screw it into the other end of the assembly.
15. Tighten plug and adaptor to a torque of 16 to 20 lb. ft. (2.22 to 2.80 kg.m.).
16. Fit a new dust cover.

OP 2493-B   VALVE AND SWITCH ASSEMBLY (WHERE FITTED) – OVERHAUL
(Includes OPS 2493-A and A1)

OP 2800-A   HANDBRAKE LINKAGE – ADJUST

To Adjust

1. Jack up the rear of the car, fit stands under the body jacking points and release the handbrake.
   NOTE – Prior to commencing the adjustment, check that the primary cable follows its correct run and is properly located in the guides. Also ensure that all cable guides are well greased.
2. Adjust the effective length of the primary cable. Slacken the locknut on the end of the cable adjacent to the relay lever on the rear axle. Adjust the nut until the primary cable has no slack in it and the relay lever is just clear of the stop on the banjo casing.

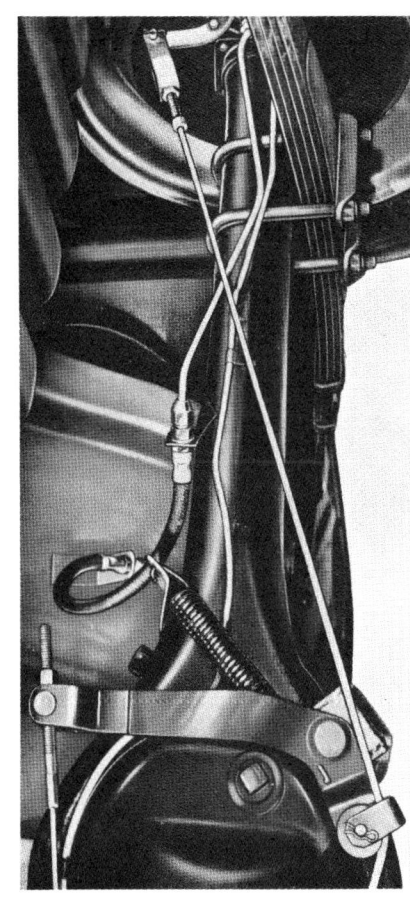

Handbrake Linkage Adjustment Points

CORTINA - LOTUS

3. Adjust the effective length of the transverse cable. Slacken the locknut on the end of the cable adjacent to the right-hand rear brake. Check that the handbrake operating levers are in the fully "off" position, (i.e. back on their stops) and adjust the cable so that there is no slack. Check that the operating levers are still on their stops and then tighten the locknut.

4. Jack up, remove the stands and lower the car to the ground.

NOTE – Once correctly set it should only be necessary to adjust the handbrake system in the event of replacing parts, or in the event of lost motion due to wear on handbrake linkage components.

OP 2811-A   HANDBRAKE LEVER ASSEMBLY – REMOVE AND INSTALL

To Remove

1. Jack up the front of the car and fit stands. Chock the rear wheels and release the handbrake.
2. Remove carpet around handbrake area.
3. Disconnect the primary cable from the end of the handbrake lever protruding beneath the car by removing the spring clip and clevis pin.
4. Remove the handbrake lever boot after unscrewing the six self-tapping screws.
5. Remove the two bolts securing the handbrake lever to the floor and lift out the handbrake lever assembly.

To Replace

6. Fit the handbrake lever in position on the floor and secure with the two bolts.
7. Pass the boot over the handbrake lever and secure with the six self-tapping screws. Ensure that the boot and retainer are correctly located.
8. Replace carpet.
9. Attach the primary cable to the end of the handbrake lever by fitting the clevis pin (which should be smeared with grease) and securing with the spring clip.
10. Jack up, remove the stands and lower the car to the ground. Apply the handbrake and remove the rear wheel chocks.
11. Check primary cable adjustment, correct if necessary by using adjustment procedure OP 2800-A.

OP 2841-A   HANDBRAKE TRANSVERSE CABLE – RENEW

To Remove

1. Chock the front wheels, release the handbrake, jack up the rear of the car and fit stands.
2. Remove the spring clip and clevis pin securing the cable to the left-hand rear brake.
3. Detach the cable from the right-hand rear brake by releasing the locknut and unscrewing the cable from the clevis.
4. Remove the relay lever pulley by withdrawing the spring clip and pivot pin.

To Replace

5. Grease the sides of the pulley and the pulley pivot pin.
6. Lay the cable in the relay lever flange so that the threaded end of the cable is adjacent to the right-hand rear brake. Fit the relay lever pulley and secure with the pivot pin and spring clip.
7. Screw the threaded portion of the cable into the clevis on the right-hand rear brake but do not tighten.
8. Secure the other end of the cable to the left-hand rear brake and fit the clevis pin (which should be smeared with grease) and the spring clip.
9. Adjust the effective length of the cable by screwing it into the clevis and then tightening the locknut. When correctly adjusted, the cable should be taut but not so tight that the handbrake operating levers on the brake carrier plates are pulled off their stops.
10. Jack up, remove the stands and lower the car to the ground. Apply the handbrake and remove the front wheel chocks.

OP 2853-A   HANDBRAKE PRIMARY CABLE – REPLACE

To Remove

1. Chock the front wheels, jack up the rear of the car, fit stands and release the handbrake.

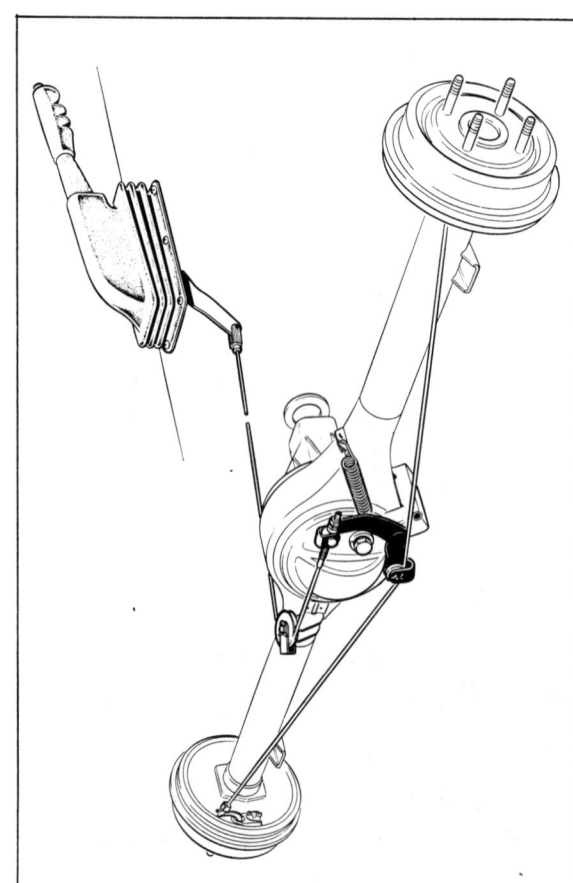

Handbrake Layout

BRAKING SYSTEM

(1967-1968)

2. From underneath the car, unscrew the nuts securing the end of the primary cable to the relay lever on the rear of the axle casing.

3. Detach the primary cable from the end of the handbrake lever protruding beneath the car by removing the spring clip and clevis pin.

4. Free the cable from its guides on the underbody and remove it from beneath the car.

To Replace

5. Attach the primary cable to the end of the handbrake lever by fitting the clevis pin (which should be smeared with grease) and securing with the spring clip.

6. Apply grease to the cable guides mounted on the underbody and thread the cable through the guides.

7. Reconnect the primary cable to the relay lever after passing it around the pulley and retainer by threading it through the pivot pin (which should be smeared with grease), fitting the spacer and securing with the two nuts. Adjust the primary cable, see OP 2800-A (adjustment procedure).

8. Jack up, remove the stands and lower the car to the ground. Apply the handbrake and remove the front wheel chocks.

Important

After carrying out any repair work on the braking system (or any other work that involves disturbing part of the braking system) perform a road or roller test to ensure that the brakes function satisfactorily.

In addition to the normal checks made to ensure satisfactory performance after any service attention to the brakes, the dual brake warning light system, where fitted, must be checked for correct operation in line with the following procedure:—

1. Switch on the ignition.

2. Depress the test switch momentarily, the warning light should illuminate. If it does not, there is an electrical fault (blown bulb, broken connection, etc.).

3. Depress the brake pedal several times, the warning light should not illuminate. If it does, there is a fault in the hydraulic system of the brakes.

4. Switch off the ignition.

CORTINA - LOTUS

SYSTEM VARIATIONS

The braking systems fitted to the '67 – '68 vehicles varies in the following respects when compared to a '69 vehicle in that:—

1. The dual line system was fitted to L.H.D. vehicles only.
2. A brake servo unit of different design was incorporated.
3. The brake pipe-line layout differs.
4. The handbrake was dash mounted.

SERVICE AND REPAIR OPERATIONS

With the exception of the procedures described in this supplement, operations on '67 – '68 vehicles may be carried out as detailed in Section 2.

Brake Pipe Chart

Part Number	LENGTH Inches	cms.	END FITTINGS A	B	C
3014E-2263-B	38.25	97.5	1	1	–
3014E-2264-B	29.75	75.5	1	1	–
3014E-2265-B	89.50	227.0	–	2	–
3020E-2265-A	89.50	227.0	–	2	–
69BB-2265-A-A	89.50	227.0	–	1	–
3014E-2286-A	53.00	134.6	1	–	–
3014E-2A009-A	13.25	33.7	1	1	–
3016E-2K018-B	10.00	25.4	1	1	–
3016E-2K019-A	10.00	25.4	1	1	–
3020E-2K467-C	55.00	139.7	2	–	1
3050E-2K468-A	36.00	91.5	–	1	1
3015E-2K470-A	39.50	100.5	–	1	1
3040E-2K470-A	39.50	100.5	–	1	1
3015E-2K471-A	29.75	75.5	–	1	1
3034E-2K471-A	29.75	75.5	–	1	1
3051E-2K475-A	15.25	38.7	–	1	1
3051E-2K475-A	15.00	38.1	–	1	1
3050E-2K485-A	65.50	166.4	1	1	–
3051E-2K485-A	36.25	92.1	1	1	–
3050E-2K486-A	40.50	102.9	1	1	–
3051E-2K486-A	35.50	90.2	1	1	–
3051E-2K487-A	36.00	91.5	1	1	–
3051E-2K488-A	22.25	56.6	1	1	–
3050E-2K488-A	14.75	37.5	1	1	–
69BB-2L162-A-A	10.00	25.4	–	1	–
3050E-2L170-A	28.50	72.4	1	1	–
3051E-2L170-A	23.50	59.7	1	1	–
69BB-2L188-A-A	6.25	15.9	–	1	–

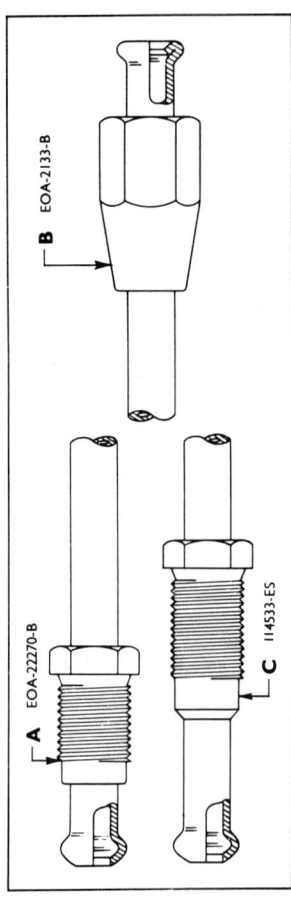

Brake Pipe End Fittings

CORTINA - LOTUS

SERVICE AND REPAIR OPERATIONS

OP 2004-C BRAKE SERVO UNIT – OVERHAUL

To Dismantle

1. Remove the filter outer body retaining screw and remove the outer body and filter.

2. Unscrew the eight set screws and square nuts from the power cylinder cover.

   NOTE – Care must be taken when carrying out this operation as this cover, through the medium of the power piston, takes the load off the piston return spring. Also, two of these screws connect the air/vacuum transfer pipe to the vacuum power cylinder casing.

3. Remove the cover, large gasket, piston assembly and return spring.

4. Place the casting portion of the servo unit in the protected jaws of a vice with the power cylinder uppermost. Remove the bolts and copper washers securing the power cylinder to the servo body.

5. Remove the reinforcement plate. Note the holes in the plate are to align with the holes in the chamber of the valve body on reassembly.

6. Unscrew the four set screws retaining the air vacuum transfer tube to the valve chest. Pull the other end of this pipe from the grommet located at the vacuum power cylinder flange. Leave the rubber grommet and retaining plate in position.

   NOTE – There is a gasket located between the valve chest and the air/vacuum transfer tube.

7. Detach the vacuum power cylinder, the metal push rod guide bush from the slave cylinder, and the socket.

   NOTE – It is necessary to carry out all the above operations before any replacement of operating valves, hydraulic seals and pistons is contemplated.

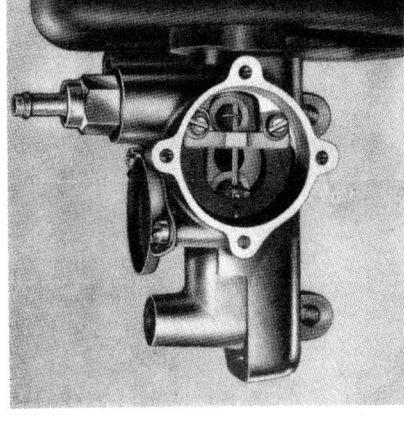

The Valve Chest

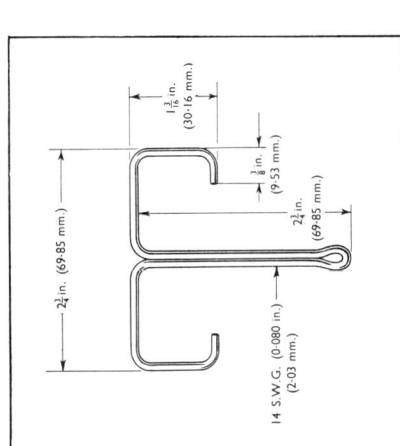

Slave Spring Compressor

Servo Unit - Exploded

8. Remove the two screws securing the bridge piece and horseshoe-shaped valve spring to the valve chest.

9. Apply slight pressure with the thumb against the plug at the vacuum power cylinder end of the control piston, plug. This will release the valve mechanism "T" piece, allowing the two valves to be withdrawn complete with the "T" piece.

10. Remove the hydraulic stationary seal from the slave cylinder bore noting that the seal lips are facing inwards.

11. Remove the nylon seal spacer noting that its rearward end engages between the lips of the hydraulic stationary seal previously removed.

12. To facilitate removal of the circlip in the slave cylinder bore, press down on the slave cylinder to remove the spring pressure on the circlip. A suitable compressor tool to retain the piston in this position is illustrated.

    Insert the spigot end of the tool in the slave cylinder bore to compress the return spring, then position the two arms of the tool so that they are hooked under the servo body flange. This will retain the piston and the tool leaving the operator's hands free to remove the circlip.

13. Using a pair of long-nosed circlip pliers, remove the circlip from the bore. Care should be exercised when carrying out this operation otherwise the bore of this slave cylinder may easily become scored.

14. Remove the retaining washer, slave piston and piston return spring from the bore. Should any difficulty occur when trying to remove this slave piston, suitably blank off the end of this cylinder with a piece of clean rag and "blow" these parts out of the cylinder bore with the aid of an air-line.

15. Remove the assembly of the plug and seal from the power cylinder end of the control cylinder bore.

16. Remove the control piston complete with the piston return spring.

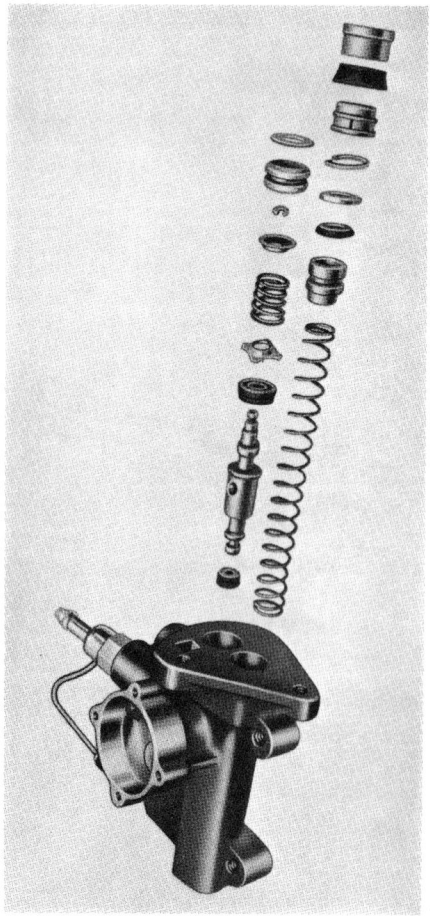

Slave and Control Cylinder

17. If required, dismantle the control piston. Remove the circlip, spring seat, spring and additional spring seat. Remove both the large and small diameter hydraulic seals, after ensuring that replacements are available if required.

18. Place a piece of clean rag over the end of the control cylinder bore and utilise an air-line to remove the secondary piston complete with seal.

To Reassemble

19. Refit the smaller diameter seal to the shorter spigot end of the control piston.

20. Fit the larger diameter seal at the other end of the piston adjacent to the shoulder and with the lips facing the circlip groove in the piston.

    NOTE – The seals have been correctly assembled when the sealing lips face away from one another.

21. Fit the large diameter spring seat, concave face first, then the spring followed by the second spring seat, convex face first, then secure with a circlip.

22. Refit the control piston to the cylinder bore with the smaller diameter seal first, the transverse hole aligning with the hole in the valve chest and the large hydraulic seal with its lips facing the operator.

23. Press the assembly of the plug and seal into the end of the control cylinder. By pushing on the plug ensure that the control piston will move and will be effectively returned by its return spring, also check that the transverse hole in the primary piston aligns with the drilling in the valve chest.

24. Place the slave piston return spring in the cylinder bore.

25. Check that the slave piston hydraulic seal is fitted so that the lip faces the spigot end of this piston, and insert this assembly into the cylinder, spigot end first, to locate in the return spring.

26. Refit the retaining washer and circlip.

    To check the function of the slave piston and return spring it is advisable to insert the push rod in the slave piston and press down allowing the return spring to return the piston. If satisfactory, remove the push rod.

27. Insert the nylon seal spacer with the spigot end outwards, refit the hydraulic stationary seal so that the lips are towards the nylon seal spacer just previously fitted, and refit the push rod guide bush, spigot end first.

28. Locate the air and vacuum valves on the "T" piece securing them in place with the spring clips.

29. Locate the ball end of the "T" piece in the transverse drilling provided in the control piston. NOTE-The small recess in the arm of this "T" piece must be located at the opposite side of the valve chest to the two securing screw holes for the valve mechanism.

30. To ensure complete engagement of the ball end with the control piston it may be necessary to press the "T" piece downwards whilst pushing the control piston forward through the medium of the plug located in the end of the control cylinder.

31. Locate the horseshoe-shaped spring retainer in the groove provided in the "T" piece, fit the bridge piece with the two "Legs" towards the centre of the valve chest and secure the complete assembly with two set screws and lockwashers.

Check that the nylon valves are seating and unseating by pressing the control cylinder plug.

32. Place the gasket over the operating cylinder casting so that all the bolt holes align.

33. Place the casting in the protected jaws of the vice and locate the vacuum power cylinder and reinforcement plate, aligning them with the guide bush, then fit the three bolts and copper washers. Tighten the bolts to a torque of 10 to 12 lb. ft. (1.38 to 1.66 kg.m.).

34. Locate the power piston return spring in the cylinder, and locate the assembly of the push rod and piston in the cylinder. Locate the gasket and cover plate on the power piston and press down ensuring that the push rod passes through the push rod guide bush in the end of the valve cylinder. Check that there is no sign of binding when the power piston is moved in the cylinder.

35. Align the holes of the cover plate and gasket with the holes in the vacuum power cylinder and secure with eight set screws and square nuts.

36. Fit a gasket to the valve chest, and push the transfer pipe through the rubber grommet and retaining plate in the flange of the vacuum power cylinder.

37. Secure the flanged end of the transfer pipe to the valve chest using four set screws and lockwashers.

38. Fit the filter element and outer body and retain with screw.

OP 2800-A   HANDBRAKE LINKAGE – ADJUST

To Adjust

1. Jack up the rear of the car, fit stands under the body jacking points and release the handbrake.

   NOTE – Prior to commencing the adjustment, check that the primary cable follows its correct run and is properly located in the guides, including the metal guide on the engine rear bulkhead. Also ensure that all cable guides are well greased.

2. Adjust the effective length of the primary cable. Slacken the locknut on the end of the cable adjacent to the relay lever on the rear axle. Adjust the nut until the primary cable is taut and check that the relay lever is correctly positioned by measuring from the inner face of the pulley mounting bracket to the centre of the trunnion mounting. When the primary cable is correctly adjusted, the dimension should be 13.68 in. (34.8 cm.).

   N.B. Maximum allowed variation is ± 0.25 in. (0.63 cm.).

3. Adjust the effective length of the transverse cable. Slacken the locknut on the end of the cable adjacent to the right-hand rear brake. Check that the handbrake operating levers are in the fully "off" position, (i.e. back on their stops) and adjust the cable so that it is taut. Check that the operating levers are still on their stops and then tighten the locknut.

4. Adjust the effective length of the primary cable. Slacken the locknut on the end of the cable adjacent to the relay lever on the rear axle. Adjust the nut until the primary cable is taut and check that the brake operating levers have not been pulled off their stops.

5. Jack up, remove the stands and lower the car to the ground.

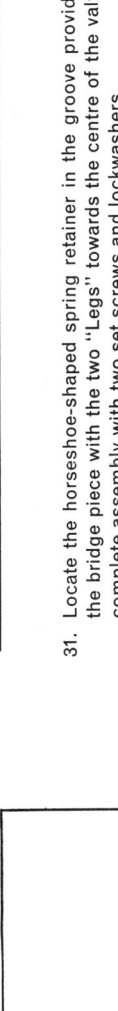

Braking System — Single Line

Braking System — Dual Line — L.H.D.

OP 2811-B HANDBRAKE LEVER ASSEMBLY – REMOVE AND INSTALL

To Remove

1. Chock the front wheels, jack up the rear of the car, fit stands, and release the handbrake.
2. From inside the car, remove the two bolts, spring and flat washers securing the rearmost end of the handbrake lever to the underside of the facia panel. Also, remove the nut, spring and flat washers securing the forward end of the lever assembly to the engine compartment rear bulkhead.
3. Remove the handbrake lever assembly and detach the primary cable from the relay lever.

To Install

4. Refit the primary cable to the handbrake lever and place the lever assembly in position beneath the facia panel with the primary cable passing through the aperture in the engine compartment rear bulkhead.
5. Secure the lever assembly to the engine compartment rear bulkhead and to the underside of the facia panel. Ensure that the rubber boot locates correctly in its aperture.

   NOTE – Ensure that the cable passes through the metal guide on the engine compartment rear bulkhead between the two front pulleys.
6. Adjust the handbrake linkage, see OP 2800-A.
7. Jack up, remove the stands and lower the car to the ground. Apply the handbrake and remove the front wheel chocks.

Primary Cable Adjustment Dimension

13·68 IN (34·8 CM)

OP 2841-A   HANDBRAKE TRANSVERSE CABLE – RENEW

To Remove

1. Chock the front wheels, release the handbrake, jack up the rear of the car and fit stands.
2. Remove the spring clip and clevis pin securing the cable to the left-hand rear brake.
3. Detach the cable from the right-hand rear brake by releasing the locknut and unscrewing the cable from the clevis.
4. Remove the relay lever pulley by withdrawing the spring clip and pivot pin.

To Install

5. Grease the sides of the pulley and the pulley pivot pin.
6. Lay the cable in the relay lever flange so that the threaded end of the cable is adjacent to the right-hand rear brake. Fit the relay lever pulley and secure with the pivot pin and spring clip.
7. Screw the threaded portion of the cable into the clevis on the right-hand rear brake but do not tighten.
8. Secure the other end of the cable to the left-hand rear brake and fit the clevis pin (which should be smeared with grease) and the spring clip.
9. Adjust the effective length of the cable by screwing it into the clevis and then tightening the locknut. When correctly adjusted, the cable should be taut but not so tight that the handbrake operating levers on the brake carrier plates are pulled off their stops.
10. Jack up, remove the stands and lower the car to the ground. Apply the handbrake and remove the front wheel chocks.

OP 2853-A   HANDBRAKE PRIMARY CABLE – RENEW

To Remove

1. Chock the front wheels, jack up the rear of the car, fit stands and release the handbrake.
2. From underneath the car, unscrew the nuts securing the end of the primary cable to the relay lever on the rear of the axle casing. Detach the cable and free it from the cable guides on the underbody.
3. Detach the cable from the handbrake lever assembly and withdraw the cable together with the grommet.

To Install

4. Fit the primary cable to the handbrake lever. Ensure that the grommet is correctly located.
5. Thread the primary cable over its pulley wheels and through its guides to the relay lever on the rear axle casing.
   NOTE – Ensure that the cable passes through the metal guide on the engine compartment rear bulkhead between the two front pulleys.
6. Connect the cable to the relay lever by threading it through the pivot pin (which should be smeared with grease), fitting the spacer and securing with the two nuts.
7. Adjust the handbrake linkage, see Op. 2800-A.
8. Jack up, remove the stands and lower the car to the ground. Apply the handbrake and remove the front wheel chocks.

Important

After carrying out any repair work on the dual line braking system (or any other work that involves disturbing part of the braking system) perform a road or roller test to ensure that the brakes function satisfactorily.

In addition to the normal checks made to ensure satisfactory performance after any service attention to the brakes, the dual brake warning light system, where fitted, must be checked for correct operation in line with the following procedure:—

1. Switch on the ignition.
2. Depress the test switch momentarily, the warning light should illuminate. If it does not, there is an electrical fault (blown bulb, broken connection, etc.).
3. Depress the brake pedal several times, the warning light should not illuminate. If it does, there is a fault in the hydraulic system of the brakes.
4. Switch off the ignition.

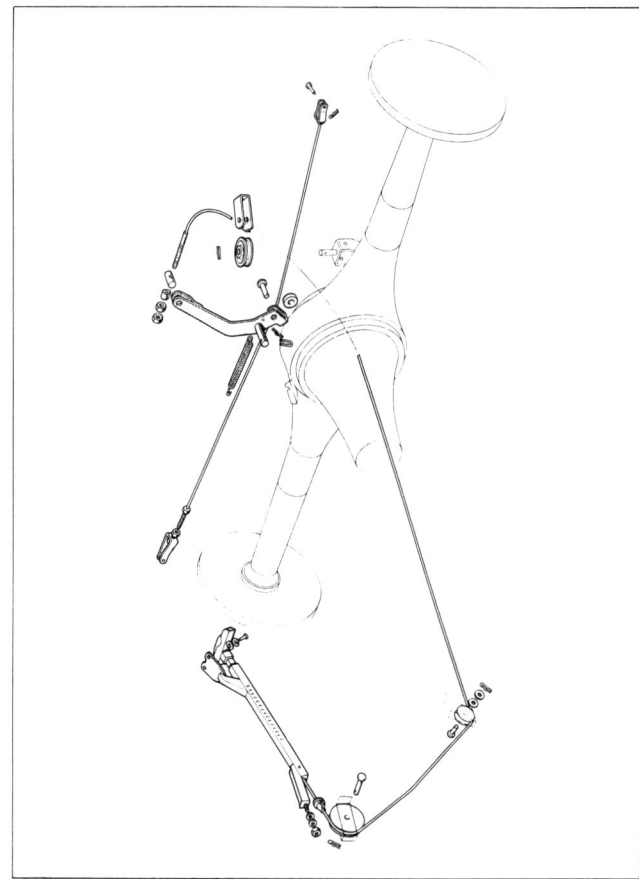

Handbrake Layout

CORTINA - LOTUS

3
STEERING GEAR AND LINKAGE

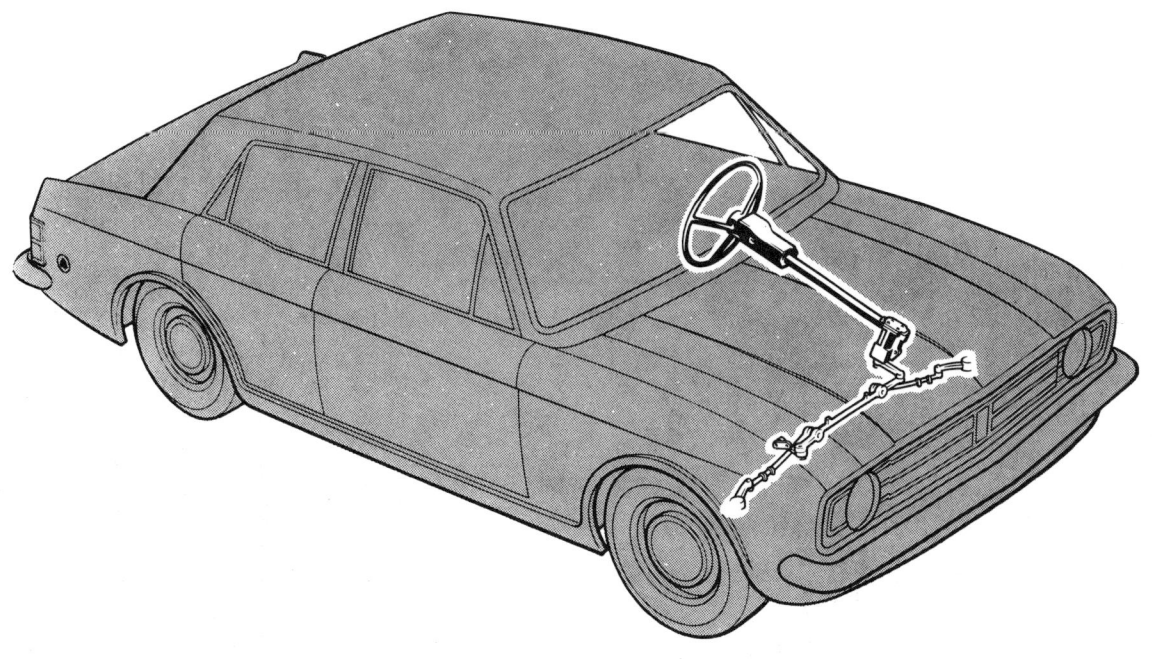

CORTINA - LOTUS

SECTION INDEX

STEERING GEAR

GENERAL DESCRIPTION

QUICK REFERENCE DATA

SERVICE AND REPAIR PROCEDURES

OPERATION	3498-A	STEERING GEAR – REMOVE AND INSTALL
,,	3498-A1	Extra: steering gear – adjust
,,	3498-A5	Extra: steering gear – overhaul
,,	3498-B	**STEERING GEAR – ADJUST** (Includes OPS 3498-A and A1)
,,	3498-E	**STEERING GEAR ASSEMBLY – OVERHAUL** (Includes OPS 3498-A, A1 and A5)
,,	3575-A	ROCKER SHAFT – PRE-LOAD – ADJUST
,,	3590-A	DROP ARM – REMOVE AND INSTALL
,,	3591-A	ROCKER SHAFT OIL SEAL – REMOVE AND INSTALL (Steering Gear in situ)
,,	3600-A	STEERING WHEEL AND/OR INDICATOR CAM – REMOVE AND INSTALL

STEERING LINKAGE

GENERAL DESCRIPTION

QUICK REFERENCE DATA

SERVICE AND REPAIR PROCEDURES

OPERATION	3000-A	TOE-IN AND WHEEL LOCK ANGLES – CHECK
,,	3000-A1	Extra: toe-in and wheel lock angles – adjust
,,	3000-A2	Extra: wheel alignment – check
,,	3000-B	**FRONT WHEEL TOE-IN AND WHEEL LOCK ANGLES – ADJUST** (Includes OPS 3000-A and A1)
,,	3000-C	**FRONT WHEEL ALIGNMENT – CHECK** (Includes OPS 3000-A and A2)
,,	3000-D	**FRONT WHEEL ALIGNMENT – CHECK AND ADJUST** (Includes OPS 3000-A, A1 and A2)
,,	3130-A	STEERING ARM – ONE SIDE – REMOVE AND INSTALL
,,	3130-A1	Extra: remaining steering arm – remove and install
,,	3130-B	**STEERING ARMS – BOTH – REMOVE AND INSTALL** (Includes OPS 3130-A and A1)
,,	3289-A	TRACK ROD END – ONE – RENEW
,,	3289-A1	Extra: second track rod end (same side) – renew
,,	3289-B	TRACK ROD ENDS – ALL – RENEW
,,	3301-A	DROP ARM TO IDLER ARM – RENEW
,,	3351-A	IDLER ARM – RENEW
,,	3535-A	STEERING LOCK STOPS – ADJUST

CORTINA - LOTUS

STEERING GEAR

GENERAL DESCRIPTION

Movement of the steering wheel is transmitted by a solid shaft to a worm and nut steering gear, which is mounted in the engine compartment attached to the sidemember. The nut is of the recirculatory ball type having thirteen balls.

The steering ratio in the straight-ahead position is 16.4:1; however, the ratio varies between the straight-ahead and extreme lock positions.

The construction of the steering gear provides for two adjustments:—

(a) Rocker shaft adjustment: this is by means of a stud and locknut on the steering box top cover plate.

(b) Steering shaft adjustment: this is by means of shims between the rear faces of the steering box and the steering column flange.

The rocker shaft may be adjusted with the steering gear installed in the car, but it must be removed to adjust the steering shaft.

QUICK REFERENCE DATA

PERIODIC SERVICE ATTENTION

The oil level in the steering box should be checked at the first 600 miles (1,000 km.) and hence at every 3,000 miles (5,000 km.) and if necessary, topped-up with an approved S.A.E. 90 E.P. oil to the bottom of the combined filler and level plug hole.

DATA

Steering shaft adjustment is by means of shims. Shim details are as follows:—

105E-3592-B	Paper	0.010 in. (0.254 mm.)
105E-3595-A	Steel	0.004 in. (0.102 mm.)
105E-3595-B	Steel	0.010 in. (0.254 mm.)
3014E-3595-A	Steel	0.002 in. (0.051 mm.)

Tightening Torques

Drop arm nut	60 to 80 lb. ft. (8.3 to 11.1 kg.m.)
Steering wheel nut	25 to 30 lb. ft. (3.4 to 4.1 kg.m.)
Steering gear to sidemember	20 to 25 lb. ft. (2.7 to 3.4 kg.m.)
Steering gear top cover	18 to 20 lb. ft. (2.5 to 2.7 kg.m.)
Steering linkage to drop arm	25 to 30 lb. ft. (3.4 to 4.1 kg.m.)

IMPORTANT

Following the completion of any service work normal test procedures should be used to ensure that a satisfactory repair has been performed. In addition, as a precautionary measure:—

Where Fitted

1. Check the Dual Brake warning light by depressing the test switch with the ignition "ON".

SERVICE AND REPAIR OPERATIONS

OP 3498-A STEERING GEAR ASSEMBLY – REMOVE AND INSTALL

To Remove

1. Disconnect the battery.
2. Prise off the steering wheel centre emblem.
3. Remove the steering wheel retaining nut and lift off the steering wheel.
4. Remove the screws securing the column shrouds together. Separate the shrouds and remove.
5. Disconnect the wires supplying the indicator switch, etc. at the multi-point connector.
6. Remove the indicator switch by undoing the two cross-head screws securing it to the bracket on the column.
7. Remove the parcel tray which is secured by two screws each side of the car.
8. Roll back the carpet and unscrew the two screws securing the floor plate.
9. Detach the brake pedal return spring from the bracket on the column.
10. Remove the bolt securing the steering column to the pedal bracket and the two bolts securing the top of the column to the underside of the facia and remove the 'U' brackets.
11. Jack up the front of the car and fit chassis stands to the front jacking point.
12. Release the lower end of the drop arm from the drop arm idler arm rod.
13. Undo the three self-locking nuts securing the steering gear assembly to the sidemember. Push out the bolts. Note that there is a flat washer under each nut and under each bolt head.
14. Remove the floor plate and rubber draught excluder by lifting it over the top of the steering column.

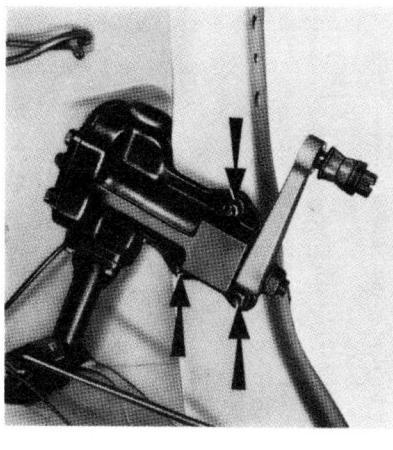

Steering Gear Mounting Bolts

15. On right-hand drive cars remove the steering gear assembly by withdrawing it from beneath the car. On left-hand drive cars it is necessary to remove the clutch slave cylinder prior to removing the steering gear, which has to be removed through the top of the engine compartment. During this operation avoid having the steering box above the steering wheel end of the column for long, as this allows oil to run down the column and out through the end.

16. Rotate the drop arm so that the steering gear is in the straight-ahead position, note the position of the indicator cancelling cam and then remove it from the column.

To Install

17. Check that the drop arm is in the straight-ahead position and then tap the indicator cam into place on its spline on the top of the steering shaft. It should be fitted in the same position as that from which it was removed, i.e. with the cam pointing towards the indicator switch.

18. Place the steering gear in position and slide the draught excluder and floor plate over the top of the column.

19. Fit the bolts and locknuts securing the steering gear assembly to the sidemember. Ensure that a flat washer is fitted under each nut and each bolt head. Tighten the bolts to a torque of 20 to 25 lb. ft. (2.8 to 3.5 kg.m.).

20. Assemble the drop arm to the drop arm to idler arm rod. The conical rubber bushes should be lubricated (bore only) prior to assembly. Fit the washer and castellated nut, tighten the nut to a torque of 25 to 30 lb. ft. (3.4 to 4.1 kg.m.). Fit a new split pin.

21. Jack up, remove the stands and lower the car to the ground.

22. Fit the bolt securing the column to the pedal bracket.

23. Locate the 'U' brackets around the column and secure them to the underside of the facia panel.

24. Fit the brake pedal return spring to the bracket on the steering column.

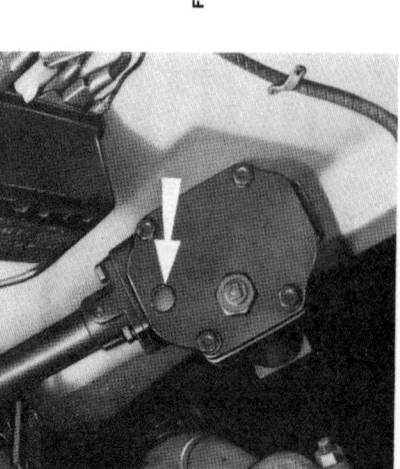

Filler Plug

25. Locate the draught excluder and floor plate in position at the base of the steering column and secure with two screws.

26. Fit the parcel tray in position and secure with two screws each side of the car.

27. Assemble the indicator switch to the bracket on the steering column upper end. Check that the indicator cam is correctly positioned, if necessary temporarily fit the steering wheel to check that the cam operates the switch properly.

28. Plug the wires supplying the indicator switch into the multi-point connector.

29. Fit the column shrouds, check that they are correctly located and do not foul the steering wheel.

30. Locate the steering wheel on its spline, check that the drop arm is in the straight-ahead position and fit a new nut. Tighten the nut to a torque of 25 to 30 lb. ft. (3.4 to 4.1 kg.m.).

31. Tap the steering wheel motif into place in the steering wheel centre.

32. Reconnect the battery and check the operation of the indicators, horn, etc.

33. Check the steering box oil level and top-up if necessary.

OP 3498-A1 EXTRA: STEERING GEAR – ADJUST
(Includes removing the Steering Gear from the car)

Tools Required

Dial gauge
Pull scale

1. Remove the steering gear from the car, as in OP 3498-A.

2. Remove the four bolts securing the steering column flange to the steering box, drain the lubricant and detach the column. Remove the old shim pack and discard the gaskets.

   NOTE – Be careful not to disturb the steering shaft or bearings.

3. Position a new gasket and two 0.010 in. (0.254 mm.) shims on the rear face of the steering box and fit the steering column, securing it with the four bolts.

4. Using a dial gauge mounted on the top of the steering column, measure the steering shaft end-float.

5. Remove the steering column and withdraw the shim pack.

6. Assemble a new shim pack so that the overall end-float is 0.001 in. (0.025 mm.) and fit it to the steering box and check that the end-float is 0.001 in. (0.025 mm.).

7. Remove the column again and withdraw a 0.004 in. (0.102 mm.) thick shim to give a preload of between 0.002 in. (0.051 mm.) and 0.004 in. (0.102 mm.).

8. Coat the four bolts with a suitable sealer and finally assemble the column to the box.

9. Remove the oil filler plug and pour the correct quantity (420 c.c.) of oil into the box.

10. Fit the steering wheel to the top of the steering shaft and fit the pull scale to the periphery of the steering wheel at its junction with a spoke.

11. Remove the adjusting stud and locknut from the steering box top cover and clean all traces of old sealer from the stud threads.

CORTINA - LOTUS

12. Coat the stud with sealer and assemble it to the steering box. Fit the locknut loosely.
13. Adjust the stud so that an effort of $1\frac{3}{4}$ to 2 lb. (0.79 to 0.91 kg.) is required on the pull scale to turn the steering wheel from the straight-ahead position.
14. Tighten the locknut and check the adjustment.
15. Replace the steering gear in the car, as in OP 3498-A.

OP 3498-A5   EXTRA: STEERING GEAR – OVERHAUL

Tools Required

P 3041-C   Drop arm remover
P 3087   Steering rocker shaft bush and oil seal replacer
(details 'a' & 'b')
P 3089   Steering rocker shaft bush broaching kit
(details 'a' & 'b')
Pull scale
Dial gauge

To Dismantle

1. If the lubricant has not been drained, remove the rubber filler plug and empty the oil into a suitable waste container.
2. Mark the drop arm and rocker shaft for reference on reassembly, unscrew the retaining nut and remove the washer. Using Tool No. P 3041-C, pull the drop arm off the rocker shaft. If necessary use a hammer on the domed nut to jar the drop arm off. NEVER hammer the drop arm.
3. Remove the four bolts and spring washers and lift off the top cover plate.
4. Detach the roller from the peg on the steering nut and pull out the rocker shaft.

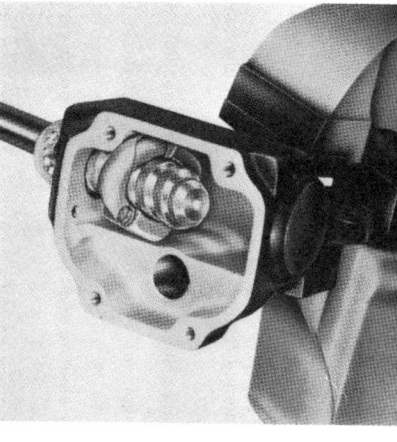

Removing the Top Cover

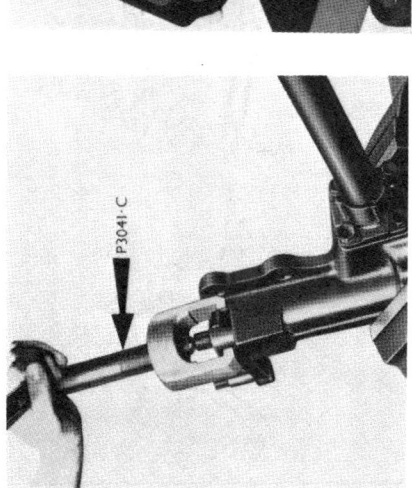

Removing the Drop Arm

Section 3 — 8

5. Unscrew the four bolts and spring washers securing the steering column flange to the rear face of the steering box. Pull the steering column and flange off the steering shaft and remove the gaskets and shim pack.
6. Withdraw the spacer from the rear of the steering box.
7. Carefully pull the steering shaft rearwards, this will dislodge the upper bearing track and the twelve ball bearings, take care not to lose any. Remove the bearing track and retrieve the balls from both upper and lower bearings.
8. Unscrew the steering nut from the steering shaft and collect the thirteen balls contained inside the nut.
9. Remove the steering shaft.
10. Remove the lower bearing track by tapping the rear face of the steering box on a block of wood.
11. Remove and discard the rocker shaft oil seal.
12. Inspect the rocker shaft bush and if it is unserviceable remove it by using a $\frac{3}{8}$ in. B.S.P. tap screwed into the bush and then forcing the tap and bush out of the bore.
13. If necessary, remove the bearing assembly from the top of the steering column.
14. Clean all the components and examine all parts for wear, replacing those which appear unserviceable.

NOTE – The ball bearings fitted in this gear assembly are not all the same size. Those fitted in the upper and lower steering shaft bearings (24 balls, 12 in each bearing) are $\frac{7}{32}$ in. (5.56 mm.) diameter, while those fitted in the steering nut (13 balls) are $\frac{5}{16}$ in. (7.94 mm.) diameter.

To Reassemble

15. If the rocker shaft bush has been removed, install a new one with the open end of the oil groove facing towards the steering shaft. Tool No. P 3087 should be used to fit the bush using detail 'd' to support the box and detail 'a' to drive the bush into position.

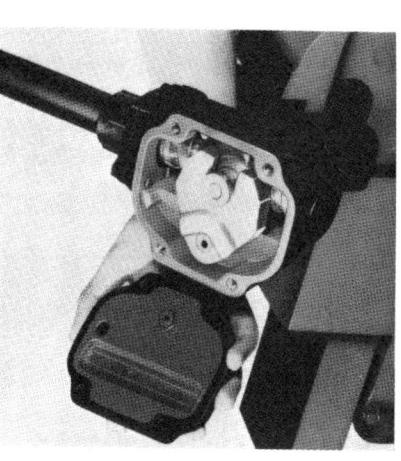

The Steering Nut

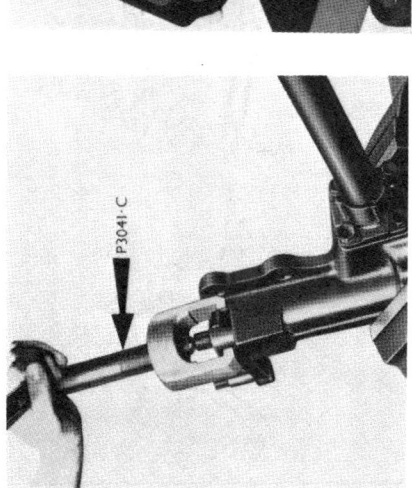

Assembling the Upper Bearing

Section 3 — 9

CORTINA - LOTUS

16. Broach the rocker shaft bush to size using Tool No. P 3089 and details 'a' and 'b'.

17. When the bush is fitted and reamed to size clean the box thoroughly to remove any traces of swarf.

18. Install a new oil seal in the housing using Tool No. P 3087. The sharp inner edge of the seal should be towards the interior of the steering box.

19. Replace the lower bearing track in the housing and retain twelve $\frac{7}{32}$ in. (5.56 mm.) diameter balls in the track using grease.

20. Assemble the thirteen balls into the steering nut using grease to retain them in position.

21. Fit the upper bearing track over the steering shaft and assemble the bearings to it, retaining them with grease.

22. With the housing clamped in a vice, pass the steering shaft into the housing and carefully screw the nut onto the worm.

23. Locate the steering shaft in the lower bearing and carefully push the upper bearing into position behind the worm.

24. If the bearing assembly has been removed from the top of the steering column, replace it in its position after lightly smearing the inside diameter with grease.

25. Fit the spacer behind the upper bearing track and then fit the steering column using new gaskets either side of the original shim pack.

26. Secure the steering column with the four bolts and lockwashers, tighten them gradually whilst rotating the shaft. Any binding indicates that the shim thickness is insufficient.

27. Fit a dial gauge to the top of the column and check the end-float. Obtain an end-float reading of 0.001 in. (0.025 mm.) by adding or removing shims as required.

28. Remove the column again and withdraw a 0.004 in. (0.102 mm.) thick shim to give a preload of between 0.002 in. (0.051 mm.) and 0.004 in. (0.102 mm.).

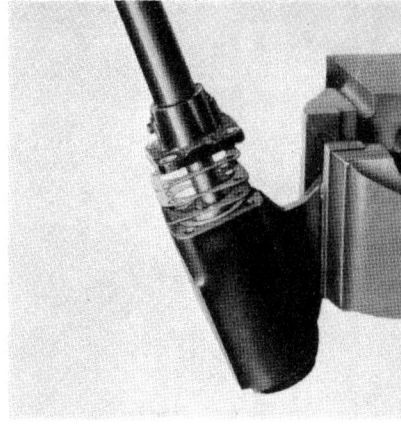

Steering Shaft Shims

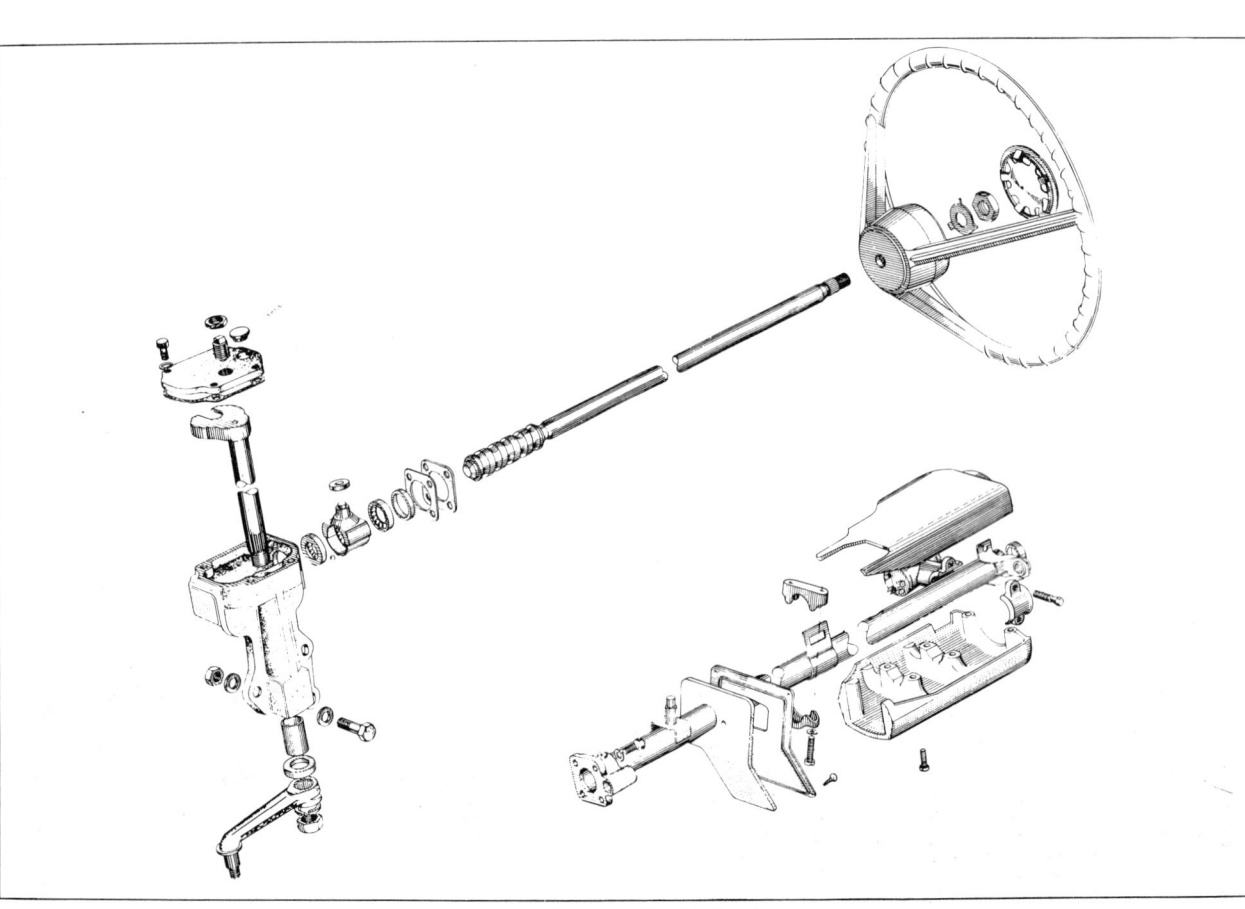

Steering Gear Assembly—Exploded

29. Coat the four bolts with a suitable sealer and finally assemble the column to the box, tighten the four bolts to a torque of 12 to 15 lb. ft. (1.66 to 2.07 kg.m.).

30. Slide the rocker shaft into its bore in the housing and fit the roller so that it engages with the peg on the nut.

31. Fit a gasket to the top face of the steering box and position the cover plate on top. Coat the four bolts with a suitable sealer and secure the cover plate to the housing tightening the bolts to a torque of 18 to 20 lb. ft. (2.5 to 2.7 kg.m.).

32. Fill the steering box with the correct quantity (420 c.c.) of S.A.E. 90 E.P. lubricant.

33. Temporarily fit the steering wheel to the top of the steering shaft and fit a pull scale to the periphery of wheel at its junction with a spoke.

34. Remove the adjusting stud and locknut from the steering box top cover and clean all traces of old sealer from the stud threads. Coat the stud with sealer, assemble it to the cover plate and loosely fit the locknut.

35. Adjust the stud so that an effort of $1\tfrac{3}{4}$ to 2 lbs. (0.79 to 0.91 kg.) is required to turn the steering wheel from the straight-ahead position.

36. Tighten the locknut, recheck the adjustment and then remove the steering wheel.

37. Refit the drop arm to the rocker shaft, checking that the mating marks align correctly.

38. Fit the washer and retaining nut. Tighten the nut to a torque of 60 to 80 lb. ft. (8.3 to 11.1 kg.m.).

OP 3498-B STEERING GEAR – ADJUST
(Includes OPS 3498-A and A1)

OP 3498-E STEERING GEAR – OVERHAUL
(Includes OPS 3498-A, A1 and A5)

Measuring the Steering Shaft End-Float

OP 3575-A ROCKER SHAFT PRE-LOAD – ADJUST
(Steering gear in situ)

Tools Required
Pull scale

1. With handbrake applied, jack up the car and fit stands.
2. Disconnect the steering linkage from the drop arm by removing the split pin and nut.
3. Centralise the steering gear in the straight-ahead position. The drop arm should point straight ahead, i.e. parallel to the steering shaft.
4. Attach a pull scale to the periphery of the steering wheel at its junction with a spoke.
5. Remove the adjusting stud and locknut from the steering box top cover and clean all traces of old sealer from the stud threads.
6. Coat the stud with sealer and refit it to the steering box. Loosely fit the locknut.
7. Adjust the stud so that an effort of $1\tfrac{3}{4}$ to 2 lb. (0.79 to 0.91 kg.) is required on the pull scale to turn the steering wheel from the straight-ahead position.
8. Tighten the locknut and check the adjustment.
9. Reconnect the steering linkage to the drop arm, securing it with a castellated nut. Tighten the nut to a torque of 25 to 30 lb. ft. (3.4 to 4.1 kg.m.) and fit a new split pin.
10. Jack up the car, remove the stands and lower the car to the ground.

OP 3590-A DROP ARM – REMOVE AND INSTALL

Tools Required
P 3041-C Drop arm remover

To Remove
1. With the handbrake applied, jack up the car and fit stands.
2. Remove the split pin and castellated nut and detach the drop arm from the drop arm to idler arm rod.
3. Mark the drop arm and rocker shaft for reference on reassembly and remove the retaining nut and washer. Using Tool No. P 3041-C, pull the drop arm off the rocker shaft. If necessary a hammer can be used on the domed part of the tool to jar the drop arm off. NEVER hammer the drop arm.

To Install
4. Refit the drop arm to the rocker shaft spline checking that the mating marks align correctly.
5. Fit the washer and retaining nut. Tighten the nut to a torque of 60 to 80 lb. ft. (8.3 to 11.1 kg.m.).
6. Reconnect the steering linkage to the drop arm securing it with a castellated nut. Tighten the nut to a torque of 25 to 30 lb. ft. (3.4 to 4.7 kg.m.) and fit a new split pin.
7. Jack up the car, remove the stands and lower the car to the ground.

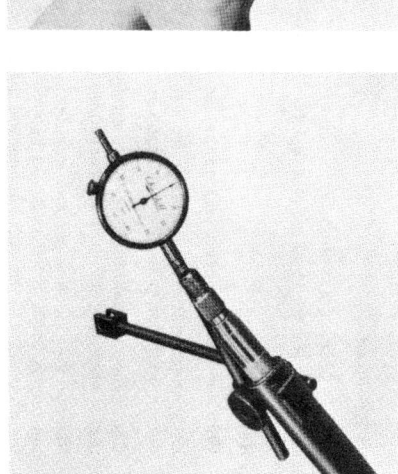

Fitting the Rocker Shaft and Roller

CORTINA - LOTUS

OP 3591-A   ROCKER SHAFT OIL SEAL – REMOVE AND INSTALL
(Steering gear in situ)

Tools Required

P 3041-C   Drop arm remover
P 3085   Rocker shaft oil seal replacer (in situ)

To Remove

1. With handbrake applied, jack up the car and fit stands.
2. Disconnect the steering linkage from the drop arm by removing the split pin and nut and detaching the drop arm from the idler arm rod.
3. Centralise the steering gear in the straight-ahead position and mark the drop arm and the rocker shaft for reference on reassembly.
4. Remove the retaining nut and washer and, using Tool No. P 3041-C, pull the drop arm off the rocker shaft. If necessary, a hammer can be used on the domed part of the tool to jar the drop arm off. NEVER hammer the drop arm.
5. Using a suitable pointed instrument, lever the existing oil seal out of the recess in the steering gear housing.

To Install

6. Install a new oil seal in the housing, using Tool No. P 3085. The sharp inner edge of the seal should be towards the interior of the steering box.
7. Refit the drop arm to the rocker shaft, checking that the mating marks align correctly.
8. Fit the washer and retaining nut. Tighten the nut to a torque of 60 to 80 lb. ft. (8.3 to 11.1 kg.m.).
9. Reconnect the steering linkage to the drop arm securing it with a castellated nut. Tighten the nut to a torque of 25 to 30 lb. ft. (3.4 to 4.1 kg.m.) and fit a new split pin.
10. Jack up the car, remove stands and lower car to the ground.
11. Check the steering box oil level and top-up if necessary.

OP 3600-A   STEERING WHEEL AND/OR INDICATOR CAM – REMOVE AND INSTALL

To Remove

1. Set the road wheels in the straight-ahead position.
2. Prise off the steering wheel centre emblem.
3. Remove the steering wheel retaining nut.
4. Remove the steering wheel.
5. Note the position of the lobe on the indicator cancelling cam and pull the cam off its spline.

To Install

6. Replace the cancelling cam ensuring that the cam lobe is midway between the cancelling "fingers" with the road wheels in the straight-ahead position.
7. Replace the steering wheel in the straight-ahead position.
8. Fit a new retaining nut onto the steering shaft.
9. Tighten the steering wheel nut to a torque of 25 to 30 lb. ft. (3.4 to 4.1 kg.m.).
10. Tap the steering wheel centre emblem into position.

CORTINA - LOTUS

STEERING LINKAGE

GENERAL DESCRIPTION

The steering linkage transmits movement of the steering rocker shaft to each front wheel.

The drop arm, which is attached to the rocker shaft, connects to one end of the drop arm to idler arm rod (drag link). The idler arm forms an idling link at the other end of this rod and is parallel to the drop arm at all times. Two adjustable track rods connect the ends of the drop arm to idler arm rod to the steering arms located at the base of each suspension unit.

This linkage incorporates seven joints, five of the joints are of the ball stud type, pre-packed with lubricant and sealed, the other two joints are of the "loose" polyurethane bush type.

All seven joints are greased for life and therefore no routine attention is required.

QUICK REFERENCE DATA

PERIODIC SERVICE ATTENTION

Every 6,000 miles (10,000 km.) or six months (whichever occurs first)

Check steering linkages for wear

Tightening Torques

Steering arm to suspension unit	30 to 35 lb. ft. (4.2 to 4.8 kg.m.)
Drop arm and idler arm to drop arm to idler arm rod	25 to 30 lb. ft. (3.5 to 4.1 kg.m.)
Idler arm joint	25 to 30 lb. ft. (3.5 to 4.1 kg.m.)
Track rod clamps	15 to 18 lb. ft. (2.1 to 2.4 kg.m.)
Track rod ball joints	18 to 22 lb. ft. (2.5 to 3.0 kg.m.)
Idler arm bracket to body sidemember	20 to 25 lb. ft. (2.8 to 3.5 kg.m.)

DATA

Wheel Alignment

Castor	$-0°$ 45' to $+0°$ 45'
Camber	0° 15' to 1° 45'
King pin inclination	7° 11' to 8° 41'
Front lock wheel angle with back lock set at 20°	19° 00' to 20° 30'
Toe-in	0.06 to 0.25 (1.52 to 6.35 mm.)
Turning circle	31 ft. (9.44 m.)

SERVICE AND REPAIR OPERATIONS

OP 3000-A TOE-IN AND WHEEL LOCK ANGLES – CHECK

Equipment Required

Churchill 96 Track Gauge
(The following procedure is based on this equipment, if other equipment is used follow the manufacturer's instructions)

Turntables and rear wheel blocks
Pedal Depressor

1. Position the toe-in gauge and check the toe-in by rolling the car so that the road wheels revolve through 180°. The toe-in should be 0.06 to 0.25 in. (1.52 to 6.35 mm.).

2. Position the car so that the front wheels are on turntables and the rear wheels are on blocks. Set the turntable calibration scales to zero. Apply the brakes using the pedal depressor.

3. Turn the front wheels so that a back lock angle of 20° is shown on one wheel. (Due to the straight-ahead bias of the suspension unit mounts, it is necessary to hold the steering wheel in position.) Read the front lock angle of the other wheel, it should be between 19° 00' and 20° 30'.

   N.B.—If the steering is turned onto left lock, the left front wheel indicates the back lock angle and the right front wheel the front lock angle. The converse applies on right lock.

4. Repeat this for the other steering lock.

5. Remove the car from the turntables and the wooden blocks.

Incorrect Wheel Lock Angles

If the wheel lock angles are incorrect or uneven when compared with the figures obtained on the other lock, first check the toe-in. If this is correct, examine the track rods which should be approximately the same length. If they are appreciably different (i.e. ¼ in., 6.35 mm.) the wheel lock angles will be adversely affected. It should be noted that minor differences in track rod length are acceptable and, in fact, sometimes necessary to compensate for production tolerances in the build up of the suspension and steering assembly.

If the toe-in and the track rod lengths are satisfactory, examine the steering arms and track rods for distortion. The steering linkage ball joints should be also checked for wear or looseness.

OP 3000-A1 EXTRA: TOE-IN AND WHEEL LOCK ANGLES – ADJUST

Equipment Required

Churchill 96 Track Gauge
(The following procedure is based on this equipment, if other equipment is used follow the manufacturer's instructions)

Turntables and rear wheel blocks
Pedal Depressor

1. Slacken the clamps on the track rod outer ends adjacent to the ball joints.

CORTINA - LOTUS

2. Using the toe-in gauge, check, and if necessary, set the toe-in to within 0.06 to 0.25 in. (1.52 to 6.35 mm.).

3. Check that the track rods are approximately equal in length.

4. Position the car on turntables and set the calibration scales. Apply the brakes using the pedal depressor.

5. Turn the front wheels so that a back lock angle of 20° is shown on one wheel. (Due to the straight-ahead bias of the top mounts, it is necessary to hold the road wheel in position.) Read the front lock angle of the other wheel, it should be between 19° 00' and 20° 30'.

   N.B. – If the steering is turned onto left lock, the left front wheel indicates the back lock angle, and the right front wheel the front lock angle. The converse applies on right lock.

6. Repeat this for the other steering lock.

7. If these are incorrect or unequal, alter the track rod lengths. To maintain the toe-in figure, the track rods should be turned equal amounts, i.e. if one track rod is turned so that its top moves a certain amount forwards, then the top of the other track rod must also be turned the same amount forwards. Recheck the wheel lock angles.

8. Drive the car off the turntables and recheck the toe-in.

9. Tighten the clamps on the track rod outer ends.

NOTE – If either the wheel lock angles or the toe-in is adjusted, then both the toe-in and the wheel lock angles MUST be rechecked when the adjustment is completed.

OP 3000-A2   EXTRA: WHEEL ALIGNMENT – CHECK
(Includes measuring Castor, Camber and King Pin Inclination)

Equipment Required

Churchill 121 Wheel Alignment Gauge
(The following procedure is based on this equipment, if other equipment is used follow the manufacturer's instructions)

Turntables and rear wheel blocks

Pedal Depressor

Before any alignment checks are made the following points should be checked and, if necessary, corrected:—

(a) Correct inflation of the tyres

(b) Wheels for true running

(c) Front wheel bearing adjustment

(d) Stabiliser bar brackets to body crossmember nuts for tightness

(e) All ball joints for excess play

(f) Front suspension springs for correct seating

---

1. Position the car with the front wheels on locked turntables and the rear wheels on wooden blocks so that the car is level.

2. Unlock the turntables.

3. Attach the alignment gauge to the left-hand front wheel. Level the gauge until the bubble aligns with the zero on the camber gauge.

4. Turn the gauge through 90° and read off the camber (0° 15' to 1° 45').

5. Turn the road wheel onto 20° back lock, turn the gauge through 90° (i.e. normal to the wheel) and align the zero on the castor scale with the bubble. Turn the road wheel onto 20° front lock and read the castor off the scale (−0° 45' to + 0° 45').

6. Fit the pedal depressor to lock the brakes. Turn the road wheel onto 20° back lock and, with the gauge parallel to the wheel, align the zero on the k.p.i. scale with the bubble. Turn the road wheel onto 20° front lock and read off the king pin inclination (7° 11' to 8° 41').

7. Fit the gauge to the right front wheel and repeat operations 3, 4, 5 and 6. Remove the pedal depressor.

The castor, camber and king pin inclination angles are not adjustable, but the following points should be checked in cases where the actual reading differs from that specified.

Incorrect Castor Angle

Check that the stabiliser bar "U" clamps are secure and that the "U" clamp retaining bolts are tight.

Incorrect Camber or King Pin Inclination Angles

If the king pin inclination angle is correct, but the camber angle is wrong the wheel spindle should be checked for distortion.

If the king pin inclination and the camber angles are both wrong, check the track control arm for distortion and the track control arm ball joint for looseness and excess wear. The track control arm to crossmember mounting should also be checked for wear and distortion.

Zeroing the Gauges   Checking the Camber Angle

CORTINA - LOTUS

Group Operations

OP 3000-B FRONT WHEEL TOE-IN AND WHEEL LOCK ANGLES – ADJUST
(Includes OPS 3000-A and A1)

OP 3000-C FRONT WHEEL ALIGNMENT – CHECK
(Includes OPS 3000-A and A2)

OP 3000-D FRONT WHEEL ALIGNMENT – CHECK AND ADJUST
(Includes OPS 3000-A, A1 and A2)

OP 3130-A STEERING ARM – ONE SIDE – REMOVE AND INSTALL

Tools Required

P 3073-9 Steering ball joint separator
P 2012 Brake line plugs

To Remove

1. Remove the hub cap and slacken the wheel nuts.
2. Jack up the front of the car and fit chassis stands.
3. Remove the road wheel.
4. Break the fluid supply to the front brake calliper at the pipe bracket on the suspension unit. Fit a brake line plug, P 2012, to each of the opened pipe ends.
5. Detach the brake calliper from its mounting lugs on the base of the suspension unit after bending back the lock tabs and unscrewing the securing bolts.
6. Remove the dust cap from the centre of the hub. Withdraw the split pin, remove the nut retainer, adjusting nut, thrust washer and outer bearing cone.
7. Remove the hub and disc assembly from the wheel spindle.
8. Remove the split pin and nut securing the track rod end to the steering arm.
9. Detach the track rod end from the steering arm using tool No. P 3073-9.
10. Remove the split pins and nuts securing the steering arm to the suspension unit. Remove the steering arm.

To Install

11. Secure the steering arm to the front suspension unit with the nuts and bolts and tighten to a torque of 30 to 35 lb. ft. (4.2 to 4.8 kg.m.). Fit new split pins.
12. Assemble the track rod end to the steering arm. Tighten the castellated nut to a torque of 18 to 22 lb. ft. (2.5 to 3.0 kg.m.) and fit a new split pin.
13. Fit the hub and disc assembly to the wheel spindle and assemble the outer bearing, thrust washer and bearing adjustment nut. Tighten the adjusting nut to a torque of 27 lb. ft. (3.7 kg.m.) whilst rotating the hub and disc assembly. Slacken the nut back 90° and then fit the nut retainer so as to align a slot in the retainer with the hole in the spindle. Fit a new split pin but do not bend up at this stage.
14. Bend up the split pin and fit the dust cap to the wheel spindle.
15. Fit the calliper to its mounting lugs using new lock tabs and tighten the securing bolts to a torque of 45 to 50 lb. ft. (6.3 to 6.9 kg.m.).
16. Reconnect the hydraulic fluid pipe to the calliper assembly and tighten the union securely.
17. Replace the road wheel.
18. Bleed the brakes as described in OP 2000-A.
19. Jack up the car, remove the stands and lower the car to the ground.
20. Tighten the wheel nuts and fit the hub caps.

OP 3130-A1 EXTRA: REMAINING STEERING ARM – REMOVE AND INSTALL
(Repeat Operations 1 and 3 to 20, OF 3130-A)

OP 3130-B STEERING ARMS – BOTH – REMOVE AND INSTALL
(Includes OPS 3130-A and A1)

OP 3289-A TRACK ROD END – ONE – RENEW
(Includes setting toe-in)

Tools Required

P 3073-9 Steering joint taper separator

To Remove

1. With the handbrake applied, jack up the front of the car and fit stands under the body jacking points.
2. Remove the split pin and nut securing the track rod end to either the steering arm or the drop arm to idler arm rod.

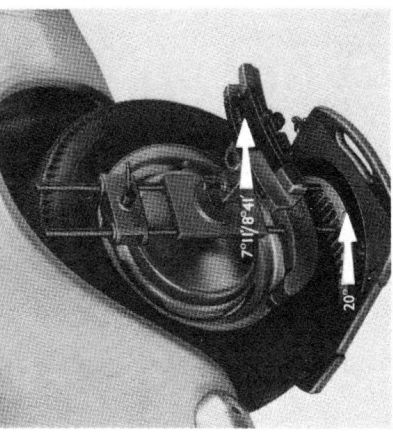

Checking the King Pin Inclination

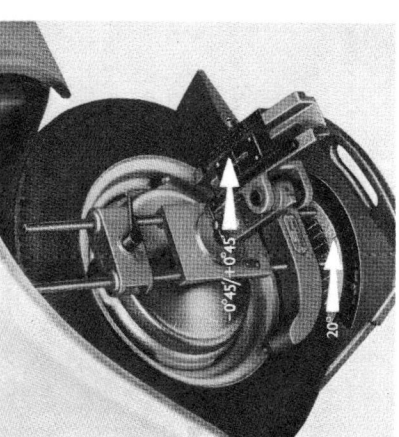

Checking the Castor Angle

3. Break the taper joint using the separator Tool No. P 3073-9.

4. Slacken the track rod clamp and unscrew the end, noting the number of turns required to free it from the sleeve.

To Install

5. Screw the new end into the sleeve, using the same number of turns as was required to remove the old track rod end.

6. Position the clamp against the indentation on the connecting sleeve, align the clamp slots with the slots in the sleeve and nip the clamp bolts, do not fully tighten at this stage.

7. Assemble the track rod to either the steering arm or the drop arm to idler arm rod. Tighten the castellated nut to a torque of 18 to 22 lb. ft. (2.5 to 3.0 kg.m.) and fit a new split pin.

8. Remove the stands and lower the car to the ground.

9. Check the toe-in and adjust if necessary.

10. Tighten the track rod clamp bolt to a torque of 15 to 18 lb. ft. (2.1 to 2.4 kg.m.).

OP 3289-A1 EXTRA: SECOND TRACK ROD END (Same side) – RENEW

To Remove

1. Remove the split pin and nut and detach the other track rod end.

To Install

2. Reconnect the track rod end and fit the retaining nut tightening it to a torque of 18 to 22 lb. ft. (2.5 to 3.0 kg.m.). Fit a new split pin.

Steering Linkage (one side)

The Steering Linkage—Exploded

CORTINA - LOTUS

OP 3289-B TRACK ROD ENDS - ALL - RENEW
(Includes adjusting toe-in and back lock angles)

Tools Required
P 3073-9 Steering joint taper separator
Tracking gauge
Turntables

To Remove
1. With the handbrake applied, jack up the front of the car and fit chassis stands.
2. Remove the split pins and nuts and detach the track rod assemblies from the steering arms and the drop arm to idler arm rod.
3. Slacken the connecting sleeve clamps and unscrew the track rod ends, noting the number of turns required to release them.

To Install
4. Screw new track rod ends into the connecting sleeve using the same number of turns that were required to remove the old ends. Position the clamps against the indentations and nip the bolts.
5. Reconnect the track rods to the steering arms and to the drop arm to idler arm rod. Secure with new castellated nuts, tighten the nuts to a torque of 18 to 22 lb. ft. (2.5 to 3.0 kg.m.) and fit new split pins.
6. Jack up the car, remove the chassis stands and lower the car to the ground.
7. Using the tracking gauge check and if necessary adjust the toe-in to within 0.06 to 0.25 in. (1.52 to 6.35 mm.).
8. Position the car on turntables and set the calibration scales.

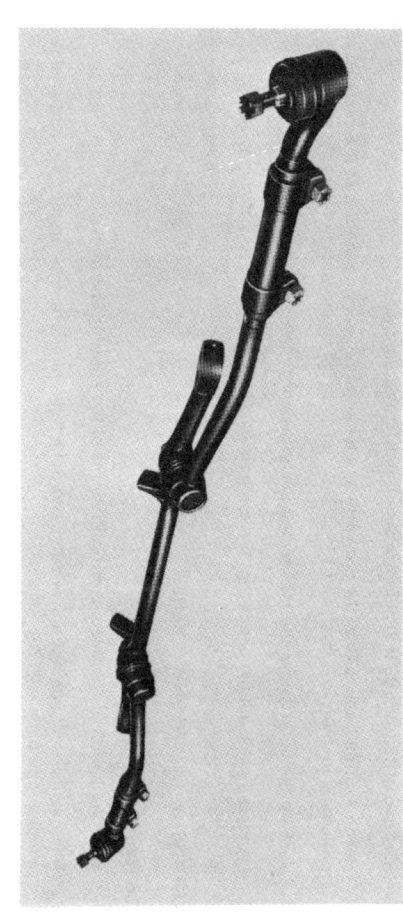

Drop Arm to Idler Arm Rod and Track Rods

9. Turn the front wheels so that a 20° back lock angle is obtained on one wheel and then read off the front lock angle on the other wheel. This should be within 19° 00' to 20° 30'.
10. Repeat this operation for the other steering lock.
11. If these are incorrect or unequal, alter the track rod lengths by releasing the clamps and turning the connecting sleeves. Each track rod sleeve must be turned the same amount.
12. Drive the car off the turntables and recheck the toe-in adjustment.
13. Reposition the car on turntables and recheck the wheel lock angles.

NOTE – If either the wheel lock angles or the toe-in is adjusted, then both the toe-in and the wheel lock angles must be rechecked when the adjustment is completed.

14. Tighten the track rod clamp bolts to a torque of 15 to 18 lb. ft. (2.1 to 2.4 kg.m.).

OP 3301-A DROP ARM TO IDLER ARM ROD - RENEW

Tools Required
P 3073-9 Steering joint taper separator
Tracking gauge

To Remove
1. With the handbrake applied, jack up the front of the car and fit chassis stands.
2. Disconnect both track rods from the drop arm to idler arm rod, using Tool No. P 3073-9 after removing the split pins and nuts.
3. Remove the split pins, nuts and washers and detach the drop arm and idler arm, remove the drop arm to idler arm rod complete with conical bushes and washers.

To Install
4. Fit new polyurethene bushes and nylon washers, if damaged, to the drop arm to idler arm rod, lubricating their internal diameters and top surfaces with 0.14 oz. of ESA-MIC-75 A grease or equivalent.

The Idler Arm

5. Replace the drop arm to idler arm rod securing them with castellated nuts, tighten them to a torque of 18 to 22 lb. ft. (2.5 to 3.0 kg.m.) and fit new split pins.

6. Reconnect the track rods to the drop arm to idler arm rod securing them with the castellated nuts, tighten them to a torque of 18 to 22 lb. ft. (2.5 to 3.0 kg.m.) and fit new split pins.

7. Jack up the car, remove the stands and lower the car to the ground.

8. Using the track gauge, check that the toe-in is 0.06 to 0.25 in. (1.52 to 6.35 mm.).

OP 3351-A IDLER ARM – RENEW

To Remove

1. With the handbrake applied, jack up the front of the car and fit stands.

2. Remove the split pin, nut and washer and detach the idler arm from its bracket on the sidemember complete with rubber seal.

3. Remove the split pin, nut and washer and detach the idler arm from the drop arm to idler arm rod. Retrieve the polyurethane bushes.

To Install

4. Lubricate the bores of the polyurethane bushes with ESA-MIC-75 A grease and assemble the idler arm to the drop arm to idler arm rod. Fit the washer and nut, tighten the nut to a torque of 25 to 30 lb. ft. (3.5 to 4.1 kg.m.) and fit a new split pin.

5. Reconnect the idler arm to its bracket on the sidemember using a new rubber seal. Retain with a castellated nut, tighten the nut to a torque of 25 to 30 lb. ft. (3.5 to 4.1 kg.m.) and fit a new split pin.

6. Jack up the car, remove the stands and lower the car to the ground.

Steering Lock Stop Adjustment

OP 3535-A STEERING LOCK STOPS – ADJUST

Tools Required
Turntables

1. Drive the car onto turntables, slacken the locknuts on both steering stop bolts and screw the bolts inwards towards the centre-line of the car.

2. Turn one road wheel until the minimum clearance between the tyre sidewall and the stabiliser bar is 1.34 in. (3.41 cm.).

3. Adjust the appropriate steering stop bolt so that it contacts the lug on the drop arm to idler arm rod and tighten the locknut.

4. & 5. Repeat operations 2 and 3 for the other road wheel.

6. Recheck the adjustments.

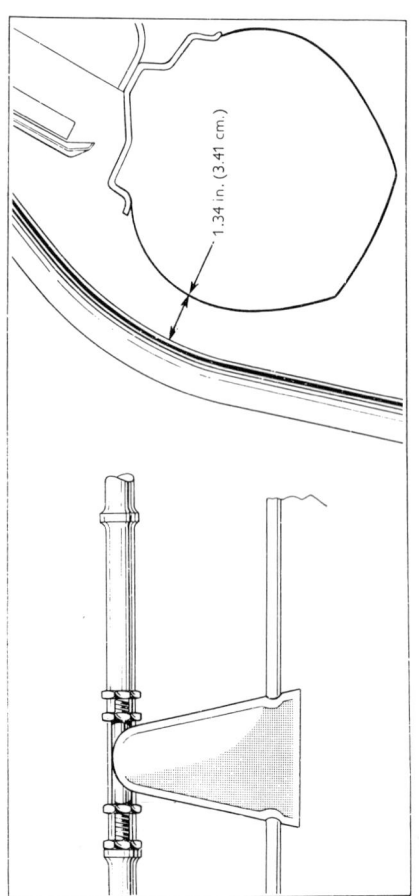

CORTINA - LOTUS

4
REAR AXLE

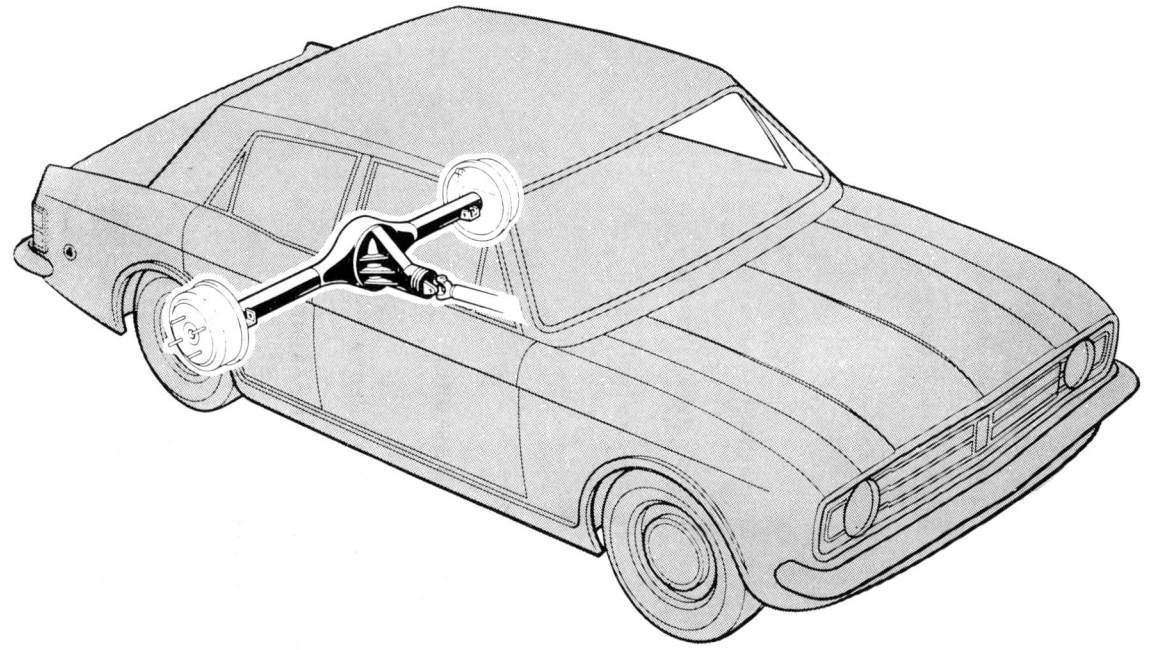

CORTINA - LOTUS

SECTION INDEX

GENERAL DESCRIPTION

QUICK REFERENCE DATA

SERVICE AND REPAIR OPERATIONS

OPERATION	4003-A	REAR AXLE ASSEMBLY – REMOVE AND INSTALL
"	4003-A1	Extra: rear axle housing – remove and install (rear axle assembly removed)
"	4003-A2	Extra: rear spring assembly – one – remove and install (rear axle assembly removed)
"	4003-A3	Extra: rear shock absorber – one – remove and install (rear axle assembly removed)
"	4010-A	**REAR AXLE HOUSING – RENEW** (Includes OPS 4003-A and A1)
"	4235-A	**REAR AXLE SHAFT AND BEARING ASSEMBLY – ONE – REMOVE AND INSTALL**
"	4235-A1	Extra: remaining rear axle shaft and bearing assembly – remove and install (rear of vehicle on stands)
"	4235-A2	Extra: rear axle shaft or bearing – one – renew (axle shaft and bearing assembly removed)
"	4235-A3	Extra: differential assembly – remove and install (axle shafts removed)
"	4235-A4	Extra: differential assembly – overhaul (differential assembly removed)
"	4235-B	**REAR AXLE SHAFT AND BEARING ASSEMBLY – BOTH – REMOVE AND INSTALL** (Includes OPS 4235-A and A1)
"	4235-C	**REAR AXLE SHAFT OR BEARING – ONE – RENEW** (Includes OPS 4235-A and A2)
"	4235-D	**REAR AXLE SHAFTS OR BEARINGS – BOTH – RENEW** (Includes OPS 4235-A, A1 and A2×2)
"	4235-E	**DIFFERENTIAL ASSEMBLY – REMOVE AND INSTALL** (Includes OPS 4235-A, A1 and A3)
"	4235-F	**DIFFERENTIAL ASSEMBLY – OVERHAUL** (Includes OPS 4235-A, A1, A3 and A4)
"	4602-A	DRIVE SHAFT ASSEMBLY – REMOVE AND INSTALL
"	4602-A1	Extra: universal joint – one – overhaul (driveshaft removed)
"	4602-A2	Extra: drive pinion oil seal – renew (driveshaft removed)
"	4602-A3	Extra: centre bearing – remove and install (driveshaft assembly removed)
OPERATION	4602-B	**DRIVESHAFT UNIVERSAL JOINT – ONE – OVERHAUL** (Includes OPS 4602-A and A1)
"	4602-C	**DRIVESHAFT UNIVERSAL JOINTS – TWO – OVERHAUL** (Includes OPS 4602-A and A1×2)
"	4602-D	**DRIVESHAFT UNIVERSAL JOINTS – THREE – OVERHAUL** (Includes OPS 4602-A and A1×3)
"	4676-A	**DRIVE PINION OIL SEAL – RENEW** (Includes remove and adjust differential assembly
"	4676-B	**DRIVE PINION OIL SEAL – RENEW (DIFFERENTIAL ASSEMBLY IN SITU)** (Includes OPS 4602-A and A2)
"	4817-D	**CENTRE BEARING – REMOVE AND INSTALL** (Includes OPS 4602-A and A3)

CORTINA - LOTUS

GENERAL DESCRIPTION

The rear axle is of the semi-floating type with a hypoid crown wheel and pinion and a two pinion differential. The standard ratio is 3.7 : 1.

The crown wheel and pinion are mounted in the differential carrier which is bolted to the front of the axle housing. The pinion is mounted on two taper roller bearings which are pre-loaded after collapsing a tubular spacer set between them. The crown wheel is bolted to the differential case which also runs on two taper roller bearings. These bearings are pre-loaded by spreading the differential carrier. In addition to the above pre-load settings the only other adjustment is the pinion depth of mesh in the crown wheel, controlled by a selective spacer between the pinion head and the rear taper roller bearing.

The axle shafts are splined to the differential side gears and run in ball races in the axle casing at their outer ends. The ball races have a built-in oil seal and no separate seal is fitted in the axle casing.

The tubular drive shaft is in two pieces having a centre bearing supported by a carrier bolted to the floor pan. The front section incorporates a universal joint and the centre bearing and the rear section has a universal joint. The drive shaft is splined to the gearbox output shaft and bolted to a flange fitted to the pinion shaft. The universal joints are pre-lubricated and sealed.

QUICK REFERENCE DATA

PERIODIC SERVICE ATTENTION

The combined oil filler and level plug is situated on the rear of the banjo housing which is welded to the axle casing.

Every 6,000 miles (10,000 km.), the oil level should be checked and topped-up as required with the car on level ground. Only hypoid gear oil of S.A.E. 90 grade may be used.

If a new crown wheel and pinion are fitted, then the special running-in oil EM-2C-29 must be used to fill the axle.

DATA

Axle ratio – standard	3.7 : 1
– optional (available through Ford Competitions Dept.) ...	3.9, 4.1, 4.4 and 4.7 : 1
Oil capacity	2 Imp. pints (2.5 U.S. pints, 1.1 litres)
Grade of lubricant	S.A.E. 90 hypoid gear oil
Initial fill lubricant	EM-2C-29

Tightening Torques

Crown wheel to differential case bolts	50 to 55 lb. ft. (6.913 to 7.604 kg.m.)
Differential carrier to axle housing nuts	25 to 30 lb. ft. (3.5 to 4.0 kg.m.)
Differential bearing locking plate bolts	12 to 15 lb. ft. (1.659 to 2.074 kg.m.)
Differential bearing cap bolts	45 to 50 lb. ft. (6.221 to 6.931 kg.m.)
Axle shaft bearing retainer bolts	15 to 18 lb. ft. (2.074 to 2.489 kg.m.)
Universal joint flange to pinion flange	15 to 18 lb. ft. (2.074 to 2.489 kg.m.)

SERVICE AND REPAIR OPERATIONS

OP 4003-A   REAR AXLE ASSEMBLY – COMPLETE WITH BRAKES – REPLACE

Tools Required

P.2012   Brake line plugs

To Remove

1. Remove the rear hub caps and slacken the wheel nuts.
2. Jack up the rear of the car, fit stands and remove the wheels.
3. Release the handbrake inside the car and disconnect the cable at the lever on the rear of the axle casing.
4. Disconnect the drive shaft at the pinion flange after marking the pinion and drive shaft flanges to ensure correct alignment on refitting. Move the drive shaft to one side.
5. Disconnect the brake pipe at the junction on the axle and fit brake line plugs (Tool No. P.2012).
6. Jack up under the centre of the axle and disconnect the shock absorber lower mounting bolts and the rear end of the radius arms. Lower the jack.
7. Remove the rear spring "U" bolt nuts and detach the clips and plates.
8. Lift the axle and remove it from the car through the wheel arch.

To Replace

9. Replace the axle and slide it through the wheel arch into position on the springs. Ensure that the spigot on the springs fits in the hole in the axle mounting pads.
10. Fit the "U" bolts over the axle, slide on the lower plates and fit the locknuts; torque to 18 to 26 lb. ft. (2.5 to 3.6 kg.m.).
11. Jack up under the centre of the axle and fit the shock absorbers into the lower mounting brackets and the radius arms to their brackets. Fit the securing bolts and locknuts. Lower the jack.
12. Reconnect the drive shaft to the pinion flange aligning the marks.
13. Remove the brake line plugs and reconnect the pipes THROUGH the bracket on the axle casing.
14. Reconnect the handbrake cable to the lever on the axle and adjust the cable as necessary by moving the handbrake levers on the backplates in and out.
15. Bleed the braking system.
16. Refit the road wheels, jack up the car, remove the stands and lower the car to the ground.
17. Tighten the wheel nuts and replace the hub caps. Recheck torque of 'U' bolt locknuts.

OP 4003-A1   EXTRA: REAR AXLE HOUSING – RENEW (REAR AXLE ASSEMBLY REMOVED)

Tools Required

P.2012   Brake line plugs
P.3072   Axle shaft remover

CORTINA - LOTUS

To Dismantle

1. Disconnect the hydraulic pipes at the brake expanders and fit brake line plugs (Tool No. P.2012).
2. Remove the transverse hydraulic pipe from the clips on the axle housing, and remove the pipe.
3. Mount the axle assembly in a suitable vice.
4. Remove the slave cylinder protection plate from the axle casing.
5. Remove the brake drum securing screw and pull off the brake drum.
6. Remove the four bolts and spring washers securing the bearing retainer plate to the axle casing. These bolts are accessible through holes in the axle shaft flange.
7. Secure the base of Tool No. P.3072 to the axle shaft flange with the wheel nuts. Use the slide hammer of the tool to knock out the axle shaft.
8. Repeat operations 4 to 7 for the other side.
9. Unscrew the eight nuts holding the differential carrier assembly to the axle casing, and allow the oil to drain.
10. Withdraw the carrier assembly from the casing, and remove the bolts.
11. Remove the eight brake plate bolts, four each side, and remove the brake plates.
12. Remove the rear axle filler and level plug, and the breather.

To Reassemble

13. Refit the rear axle breather, brake plates and eight brake plate bolts.
14. Replace the eight differential carrier bolts and fit a new gasket.
15. Replace the carrier assembly, and tighten the nuts to a torque of 50 to 55 lb. ft. (6.91 to 7.60 kg.m.).

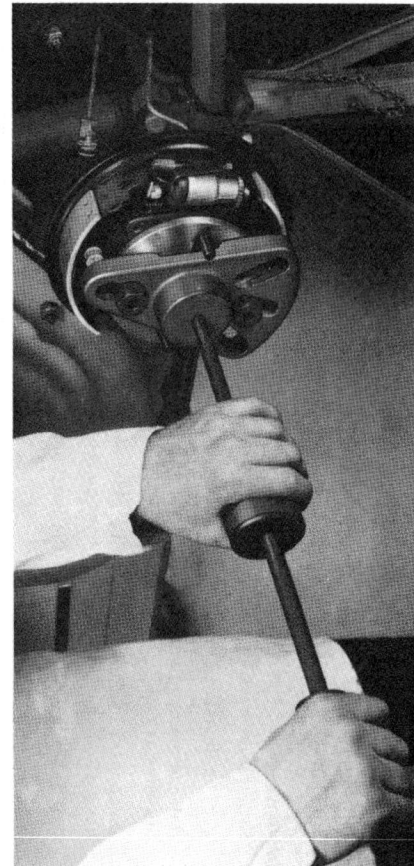

Withdrawing an Axle Shaft

Rear Axle Assembly – Exploded View

16. Insert the axle shaft into the housing and engage the splines in the differential side gear. Tap the shaft fully home.
17. Fit the four bolts and spring washers to secure the bearing retainer plate; torque to 15 to 18 lb. ft. (2.07 to 2.49 kg.m.).
18. Replace the brake drum and fit the securing screw.
19. Refit the slave cylinder protection plate on the axle casing.
20. Repeat operations 16 to 19 for the other side.
21. Remove the brake line plugs and reconnect the hydraulic pipes at the brake expanders.
22. Replace the transverse brake pipe behind the clips on the axle casing.

OP 4003-A2   EXTRA: REAR SPRING ASSEMBLY – ONE – REMOVE AND INSTALL (REAR AXLE ASSEMBLY REMOVED)

Tools Required

P.5029-A   Spring shackle bush remover and replacer

To Remove

1. Remove the rear shackle nuts and detach the combined shackle bolt and plate assemblies. Remove the four rubber bushes.
2. Unscrew the nut from the front mounting, and withdraw the through bolt.
3. Remove the spring assembly.
4. Using Tool No. P.5029 pull the bush out of the front eye.

To Install

5. Replace the front mounting bush using Tool No. P.5029 to ensure that the bush is fitted squarely in its aperture.
6. Locate the front of the spring in its body mounting bracket. Fit the through bolt and lightly assemble the nut. Do not tighten at this stage.
7. Fit new rubber bushes to the rear spring "eye" and the apertures in the body for the rear shackle.
8. Locate the spring in position and assemble the rear shackle bolt and plate assemblies. Fit the nuts, but again do not tighten.
9. Lower the car so that the weight is resting on its rear wheels.
10. Tighten the front hanger bolt to 25 to 30 lb. ft. (3.46 to 4.15 kg.m.) and the rear shackle nuts to 8 to 10 lb. ft. (1.10 to 1.40 kg.m.).

OP 4003-A3   EXTRA: REAR SHOCK ABSORBER – ONE – REMOVE AND INSTALL

To Remove

1. From inside the luggage compartment remove the two nuts from the top of the shock absorber, using a second spanner on the flats on the piston rod to prevent it turning.
2. Lift off the top steel washer, the rubber bush and the lower steel washer.
3. Remove the shock absorber from the car.
4. Remove the rubber bush and steel washer from the top of the shock absorber.

To Install

5. Assemble the large steel washer (lipped edge upwards) and the rubber bush to the top of the shock absorber.
6. Fit the lower end of the shock absorber into the bracket on the axle and line up the holes in the bracket with the hole in the shock absorber. Fit the bolt, lockwasher and nut, tighten the nut to a torque of 40 to 45 lb. ft. (5.54 to 6.22 kg.m.).
7. Extend the shock absorber upwards and pass it through the mounting aperture in the body. Position the cupped washer (large opening upwards) followed by the rubber bush and dished washer over the piston rod.
8. Fit the first nut and tighten to a torque of 15 to 20 lb. ft. (2.0 to 2.75 kg.m.), remembering to prevent the piston rod rotating. If the torque wrench permits, this can be done by holding the top of the piston rod, using a ¼ in. A/F open-ended spanner across the flats or, alternatively, if the torque wrench prevents access, grip the upper part of the shock absorber body (this requires the assistance of a second mechanic).
9. Fit the locknut and tighten securely.
10. Jack up the rear of the car, remove the chassis stands and lower the car to the ground. Remove the chocks from the front wheels.

OP 4010-A   REAR AXLE HOUSING – RENEW
(Includes OPS 4003-A and A1)

OP 4235-A   REAR AXLE SHAFT AND BEARING ASSEMBLY – ONE – REMOVE AND INSTALL

Tools Required

PT.3072   Axle shaft remover

These operations may be carried out with the axle either in or out of the car.

To Remove

1. Remove the brake drum securing screw and pull off the drum. (Ensure handbrake is released.)
2. Remove the four bolts and spring washers securing the bearing retainer plate to the axle casing. These bolts are accessible through holes in the axle shaft flange.
3. Secure the base of Tool No. PT.3072 to the axle shaft flange with the wheel nuts. Use the slide hammer of the tool to knock out the axle shaft.

To Replace

4. Insert the axle shaft into the casing and engage the splines in the differential side gear. Tap the shaft fully home.
5. Fit the four bolts and spring washers to secure the bearing retainer plate; torque to 15 to 18 lb. ft. (2.08 to 2.48 kg.m.).
6. Replace the brake drum and refit the securing screw.

CORTINA - LOTUS

OP 4235-A1 EXTRA: REMAINING REAR AXLE SHAFT AND BEARING ASSEMBLY - REMOVE AND INSTALL

Tools Required

P.3072   Axle shaft remover

These operations may be carried out with the axle either in or out of the car.

To Remove

1. Remove the brake drum securing screw and pull off the drum. (Ensure handbrake is released.)
2. Remove the four bolts and spring washers securing the bearing retainer plate to the axle casing. These bolts are accessible through holes in the axle shaft flange.
3. Secure the base of Tool No. P.3072 to the axle shaft flange with the wheel nuts. Use the slide hammer of the tool to knock out the axle shaft.

To Replace

4. Insert the axle shaft into the casing and engage the splines in the differential side gear. Tap the shaft fully home.
5. Fit the four bolts and spring washers to secure the bearing retainer plate; torque to 15 to 18 lb. ft. (2.08 to 2.48 kg.m.).
6. Replace the brake drum and refit the securing screw.

OP 4235-A2 EXTRA: REAR AXLE SHAFT OR BEARING - ONE - RENEW

Tools Required

P.4090-2 & 6   Axle shaft bearing remover
370   Universal taper base
P.4084   Spring indicator

To Remove

1. Locate the adaptors (Tool No. P.4090-6) and a slave ring between the bearing and axle shaft flange. Support the assembly in the base plate (Tool No. 370) of a hydraulic press and push the axle shaft out.

To Replace

2. Locate the bearing retainer plate and the bearing on the axle shaft with the oil seal side towards the splined end.
3. Support the assembly in the bed of a hydraulic press on a spacer ring, adaptors (Tool No. P.4090-2) and slave ring.
4. Fit a spring indicator (Tool No. P.4084) to the ram and press the bearing onto the axle shaft shoulder. A minimum pressure of 1,200 lb. (544 kg.) should be required. A lower pressure indicates an incorrect fit.
5. Use the same tools as in operations 3 and 4 and fit the bearing collar to abut the bearing. A minimum pressure of 1,500 lb. (680 kg.) should be required.

OP 4235-A3 EXTRA: DIFFERENTIAL ASSEMBLY - REMOVE AND INSTALL

To Remove

1. If the axle is in the car, disconnect the drive shaft from the pinion flange having first marked the pinion and drive shaft flanges to ensure correct alignment on refitting. Move the drive shaft to one side.
2. Unscrew the eight nuts holding the differential carrier assembly to the axle casing. Allow the oil to drain.
3. Withdraw the carrier assembly from the casing.

To Replace

4. Clean the mating faces of the carrier assembly and axle casing. Fit a new gasket over the studs.
5. Refit the differential carrier assembly to the casing and replace the nuts; torque to 20 lb. ft. (2.8 kg.m.).
6. Fill the axle with S.A.E. 90 hypoid gear oil or if a new crown wheel and pinion have been fitted fill with the special running-in oil EM-2C-29.
7. Reconnect drive shaft to the pinion flange aligning the mating marks.

OP 4235-A4 EXTRA: DIFFERENTIAL ASSEMBLY - OVERHAUL

NOTE – If a new pinion or differential taper roller bearing is fitted, then a new cone **and** a new cup should be used. Mating cones and cups should be of the same manufacture (i.e. Timken, Skefco, etc.).

Tools Required

P.4077-A   Mounting adaptor
P.4079   Differential bearing adjuster
P.4028   Pinion flange holding wrench
P.4015-A   Pinion bearing cup remover
CP.4000   Hand press
P.4000-28   Split ring
P.4000-27A   Split ring
P.4080   Thrust button
P.4013-A   Pinion bearing cup replacer
P.4013-3   Pinion bearing cup replacer (adaptor)
P.4075-4   Dummy pinion
CP.4030   Drive pinion bearing preload gauge
P.4030-1   Adaptor
P.4075   Depth of mesh gauge
P.4009   Cap spread gauge
P.4008   Backlash gauge

To Dismantle

1. Mount the assembly on a dismantling stand with an adaptor (Tool No. P.4077-A).
2. Check for mating marks on the differential adjusting nut caps and if necessary mark them

CORTINA - LOTUS

for correct reassembly. Slacken the small centre bolts and pull back the lock tabs. Back-off the adjusting nuts with the special spanner (Tool No. P.4079) and then remove the cap bolts and detach the caps.

3. Lift out the crown wheel and differential case assembly, keeping the bearing cones and cups together as a pair.

4. Relieve the staking securing the pinion nut, then hold the pinion flange (use Tool No. P.4028) and unscrew the nut. Pull the flange from the pinion. Remove the pinion with the tubular spacer and rear bearing cone from the carrier.

5. Prise the pinion oil seal from the carrier or, if required, the pinion front bearing cup and oil seal can be driven out by locating the legs of the driver (Tool No. P.4015-A) through the carrier and up against the bearing in the notches provided.

The rear bearing cup can be removed by inserting the driver from the other end of the carrier.

6. Locate the pinion in the split ring (Tool No. P.4000-28) with the lips behind the bearing cone. Mount the assembly in a hand press (Tool No. CP.4000) and after checking that the bearing cage is not trapped, press out the pinion. Remove the spacer from behind the pinion head.

7. Remove the six bolts securing the crown wheel to the differential case. Support the crown wheel on wood blocks in the bed of a press and press out the differential case (use thrust button P.4080).

8. Press out the differential pinion gear shaft lock-pin from the crown wheel side of the differential case. Push out the shaft.

9. Rotate the pinion gears around the side gears and extract them from the case. Remove the side gears, the two spherical and two flat thrust washers from the case.

10. Locate the differential bearing cones in the adaptors (Tool No. P.4000-27A) and in a support ring in the bed of a press. Press off the cones, one at a time.

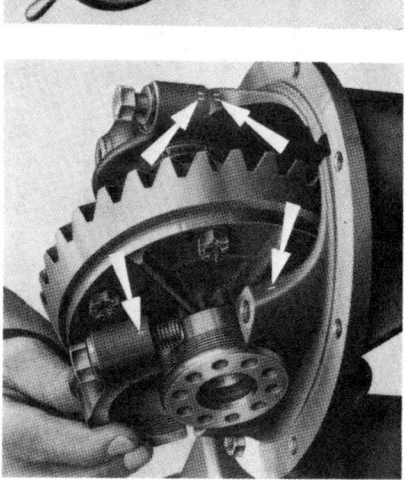

Differential Bearing Cap Mating Marks

Removing Pinion Bearing Cups

To Reassemble

11. Lubricate the flat thrust washers and position them on the flanges of the side gears. Fit the side gears in the differential case.

Lubricate the spherical thrust washers and position them on the rear of the pinion gears. Position the pinions in the cut-outs and rotate the side gears to draw them into place.

Check that the thrust washers are not misplaced.

12. Push the pinion gear shaft through the case and gears so that the lock-pin holes line up. Fit the tapered end of the lock-pin from the differential side of the case and tap it right home. Lightly peen the case to retain the pin.

13. Examine the mating faces of the crown wheel and the differential case, removing any burrs by lightly stoning. Locate the crown wheel on the case and enter two suitable long bolts through the case into the crown wheel to ensure correct alignment.

14. Place the crown wheel, teeth downwards, on wood blocks in the bed of a press. Place a thrust button (Tool No. P.4080) on the case and bring down the ram to press on the crown wheel.

15. Whilst the assembly is in the press, replace the bearing cones using the thrust button (Tool No. P.4080).

16. Remove the two pilot bolts and fit the six self-locking bolts; torque to 35 lb. ft. (4.8 kg.m.).

17. Place the pinion rear bearing cup on Tool No. P.4013-3 and pass it through the differential carrier from the rear. Assemble the front bearing cup to the tool and fit the front adaptor and wing nut. Tighten the wing nut to pull the bearings fully home.

Refitting the Crown Wheel

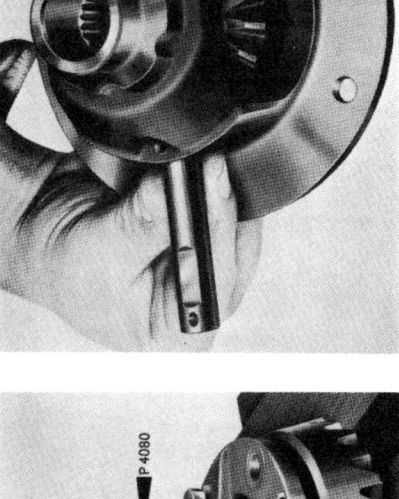

Replacing or Removing the Pinion Gear Shaft

18. Select the pinion bearing shim.

(a) Slide the rear bearing cone onto the dummy pinion (Tool No. P.4075-4) and fit it in the carrier.

(b) Fit the front bearing cone to the dummy pinion, slide on the pinion flange and screw on the nut.

(c) Hold the pinion flange (use Tool No. P.4028) and tighten the nut whilst rocking the pinion to ensure correct seating of the bearings. When there is slight bearing drag, fit the pre-load gauge (Tool No's. CP.4030 and P.4030-1) to the flange and measure the running torque by allowing the gauge weight to drop through the horizontal position.

Tighten the nut until a pre-load of 9 to 11 lb. in. (0.104 to 0.127 kg.m.) is obtained. If the figures specified are exceeded, slacken the nut to remove all pre-load and recommence tightening.

(d) Set the dial gauge of Tool No. P.4075 to zero by sliding the setting button across the under face.

(e) Clean the differential bearing location in the carrier and fit the gauge (with adaptors P.4075-4) so that the plunger rests on the dummy pinion. Rock the gauge backwards and forwards slightly to obtain a minimum reading.

(f) To the reading obtained add the amount 0.100 in. (2.54 mm.) which is the gauge compensation figure and to allow for bearing expansion when it is pressed onto the pinion, subtract 0.001 in. (0.025 mm.). Examine the shaft of the pinion to be fitted, between the bearing locations, any variance from standard size will be indicated by a painted figure. If it is a "+" figure it should be subtracted and if "—" it should be added.

For example—

Gauge Reading	...	0.055 in. (1.397 mm.)
plus Gauge Compensation	...	0.100 in. (2.540 mm.)
		0.155 in. (3.937 mm.)
minus allowance for bearing expansion	...	0.001 in. (0.025 mm.)
		0.154 in. (3.912 mm.)
plus/minus pinion marking (—2)	...	0.002 in. (0.051 mm.)
		0.156 in. (3.963 mm.)

The shim required may now be selected from the following list:—

Pinion Bearing Shims

105E-4672-A	0.1304 to 0.1308 in. (3.312 to 3.322 mm.)
105E-4672-B	0.1314 to 0.1318 in. (3.337 to 3.347 mm.)
105E-4672-C	0.1324 to 0.1328 in. (3.363 to 3.373 mm.)
105E-4672-D	0.1334 to 0.1338 in. (3.388 to 3.398 mm.)
105E-4672-E	0.1344 to 0.1348 in. (3.414 to 3.424 mm.)
105E-4672-F	0.1354 to 0.1358 in. (3.439 to 3.449 mm.)
105E-4672-G	0.1364 to 0.1368 in. (3.465 to 3.474 mm.)
105E-4672-H	0.1374 to 0.1378 in. (3.490 to 3.500 mm.)
105E-4672-J	0.1384 to 0.1388 in. (3.515 to 3.525 mm.)
105E-4672-K	0.1394 to 0.1398 in. (3.541 to 3.551 mm.)
105E-4672-L	0.1404 to 0.1408 in. (3.566 to 3.576 mm.)
105E-4672-M	0.1414 to 0.1418 in. (3.595 to 3.600 mm.)
105E-4672-N	0.1424 to 0.1428 in. (3.605 to 3.627 mm.)

(g) Remove the dummy pinion and gauge.

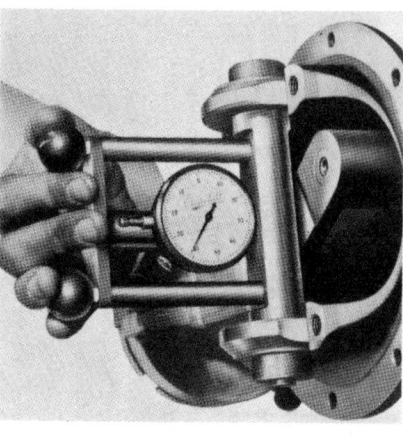

Pinion Depth Gauge

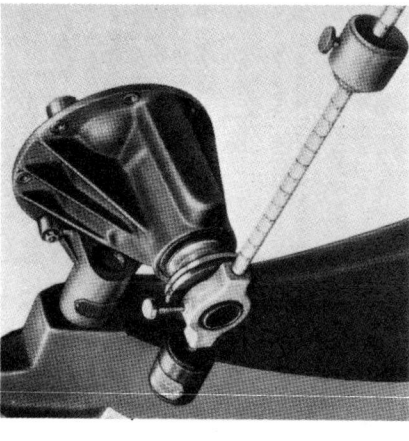

Pinion Bearing Pre-Load Gauge

Fitting the Pinion

19. Fit the selected shim to the pinion with the internal chamfer towards the gear teeth. Fit the rear bearing cone and support the assembly in the split rings (Tool No. P.4000-28). Press the bearing fully home, taking care not to damage the bearing cage.

20. Fit the front bearing cone to its cup in the carrier and locate a new oil seal, previously immersed in oil, in the carrier throat with the lip towards the bearing. Pass Tool No. P.4013 through the carrier and fit adaptor No. P.4013-3 with its flat face against the seal. Tighten the wing nut to pull the seal fully home. Lightly oil the seal.

21. Fit the pinion in the carrier with a new collapsible spacer. Fit the drive flange and a new retaining nut. Tighten the nut until just a slight end-float can be felt on the pinion.

22. Fit the pre-load gauge (Tools CP.4030 and P.4030-1) to the drive flange and measure the oil seal drag (usually around 5 lb. in. (0.06 kg.m.)). To the figure obtained add the specified bearing pre-load of 9 to 11 lb. in. (0.104 to 0.127 kg.m.) which will give the correct figure required.

23. Gradually and carefully tighten the drive flange nut, rocking the pinion to seat the bearings, until the required pre-load is obtained. Frequent checks must be taken with the pre-load gauge and if the maximum pre-load is exceeded the collapsible spacer must be renewed and the operation recommenced.

24. Stake the nut to the pinion when the correct pre-load is obtained.

25. Locate the differential bearing cups on their cones and position the differential assembly in the carrier housing.

26. Refit the bearing caps with the mating marks aligned and replace the bolts so that they just nip the caps in position. Screw in the adjusting nuts (use Tool No. P.4079), whilst rotating the crown wheel, until there is just a slight backlash.

27. Bolt a spread gauge (Tool No. P.4009) to the centre bolt hole of a bearing cap and fit an inverted bearing cap lock-tab to the other cap. Ensure that the dial plunger rests against the lock-tab and set the gauge to zero.

Correct Tooth Marking

Crown Wheel Backlash and Cap Spread Gauges

Incorrect Tooth Marking and Method of Correction

Heavy Flank Contact (*left*)
In this case the area of contact is below the centre line of the tooth, and the condition should be rectified by moving the pinion away from the crown wheel, using a thinner shim behind the pinion. Reset the backlash and differential bearing pre-load.

Heavy Face Contact (*right*)
In this case the area of contact is above the centre line of the tooth, due to the pinion being too far away from the crown wheel. Use a thicker pinion bearing shim to lower the contact area and reset the backlash and differential bearing pre-load.

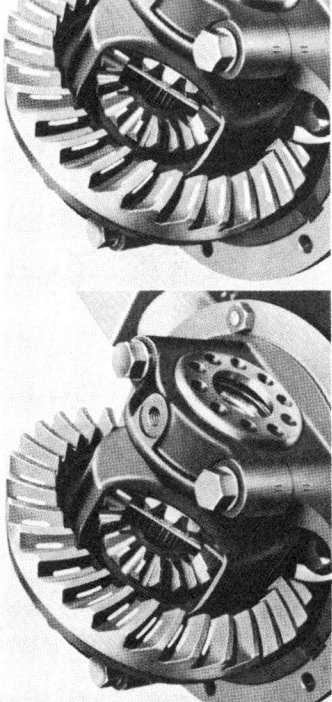

Contact on Toe (*left*)
When the area of contact is running off the toe of the pinion, move the crown wheel away from the pinion. Slacken the crown wheel adjusting nut and screw in the differential side nut an equal amount. It may also be necessary to use a thicker shim behind the pinion in order to keep the backlash within the correct limits.

Contact on Heel (*right*)
In this case the crown wheel is too far out from the pinion. Slacken the differential side-adjusting nut and tighten the crown wheel side nut, re-check the backlash and differential bearing pre-load readings. If the backlash is reduced below the minimum specified, use a thinner shim behind the pinion and, using a new collapsible spacer, readjust pinion bearing pre-load.

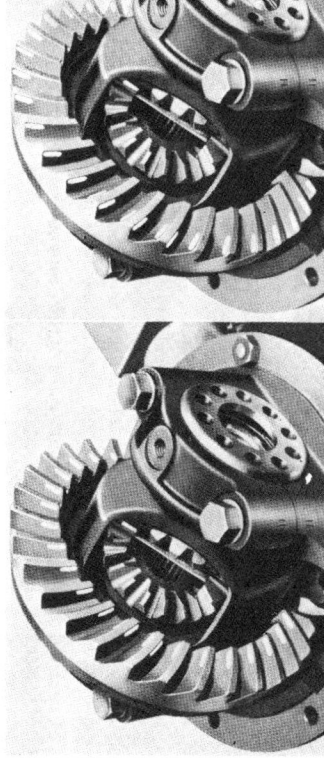

28. Mount a backlash gauge (Tool No. P.4008) to a suitable hole in the carrier flange so that the dial plunger rests against a tooth and at right angles to it. Zero the gauge.

29. Screw in the adjusting nuts (use Tool No. P.4079) until a backlash of 0.001 to 0.002 in. (0.025 to 0.050 mm.) is indicated when rocking the crown wheel. Swing the backlash gauge out of position.

30. Screw in the adjusting nut on the differential side, whilst rotating the crown wheel, until a constant cap spread of 0.005 to 0.007 in. (0.127 to 0.178 mm.) is indicated.

31. Swing the backlash gauge back in position and zero the dial. Hold the pinion and rock the crown wheel. The backlash should now be 0.005 to 0.007 in. (0.127 to 0.178 mm.). If it is outside these limits, adjust the position of the crown wheel relative to the pinion by slackening the adjusting nut on one side and tightening the nut on the other side so that the cap spread remains unaltered. The final tightening must always be made to the nut on the crown wheel side.

32. Refit the lock-tabs (left and right-hand offsets are available) and torque the lock-tab bolts to 15 lb. ft. (2.1 kg.m.). Torque the cap bolts to 45 to 50 lb. ft. (6.2 to 6.9 kg.m.).

33. Apply a thin coating of red or yellow ochre to the crown wheel teeth. Fit the axle shafts and hold them to apply a load, and rotate the pinion in both directions.

Check the tooth patterns on both sides of the gears and rectify if incorrect. (Four methods of correction are illustrated.)

34. Place a new gasket on the axle housing and refit the differential assembly. Replace the locknuts.

OP 4235-B REAR AXLE SHAFT AND BEARING ASSEMBLY – BOTH – REMOVE AND INSTALL
(Includes OPS 4235-A and A1)

OP 4235-C REAR AXLE SHAFT OR BEARING – ONE – RENEW
(Includes OPS 4235-A and A2)

OP 4235-D REAR AXLE SHAFTS OR BEARINGS – BOTH – RENEW
(Includes OPS 4235-A, A1 and A2×2)

OP 4235-E DIFFERENTIAL ASSEMBLY – REMOVE AND INSTALL
(Includes OPS 4235-A, A1 and A3)

OP 4235-F DIFFERENTIAL ASSEMBLY – OVERHAUL
(Includes OPS 4235-A, A1, A3 and A4)

OP 4602-A DRIVE SHAFT ASSEMBLY – REMOVE AND INSTALL

To Remove

1. Jack up the rear of the car and fit chassis stands.

2. Mark the drive shaft and pinion flanges for correct alignment on reassembly and then remove the four nuts and bolts.

3. Remove the bolts and washers securing the bearing assembly to the floor pan.

4. Lower the drive shaft assembly and withdraw it from the gearbox.

To Install

5. Slide the front yoke into the gearbox, engaging the mainshaft splines, taking care not to damage the oil seal or bearing in the extension housing.

6. Secure the centre bearing carrier to the floor pan.

7. Lift the rear of the driveshaft assembly, align the mating marks on the driveshaft and pinion flanges, fit the four bolts and spring washers and tighten to a torque of 15 to 18 lb. ft. (2.074 to 2.489 kg.m.).

8. Check the gearbox oil level.

OP 4602-A1 EXTRA: UNIVERSAL JOINT – ONE – OVERHAUL

The universal joint spider, bearings, oil seals and retainers are serviced as a kit.

1. Extract each spider bearing snap ring and remove the bearing cups and rollers by gently tapping the yoke at each bearing.

2. Remove the spider and detach the oil seal and seal retainer from each spider journal.

3. To reassemble, fit new oil seals to the retainers and locate them on the shoulders of the spider journals with the oil seals outwards. Position the spider in the drive shaft yoke and assemble the needle rollers in each bearing cup. Pack the bearings with a multi-purpose lithium base grease, leaving an air space to allow for expansion of the grease when warm. Refit the bearings, tapping them squarely into place. Take care not to dislodge the needle rollers.

4. Similarly, refit the other half of the joint.

5. Refit the snap rings to each bearing.

OP 4602-A2 EXTRA: DRIVE PINION OIL SEAL – RENEW

Tools Required

P.4028 Flange holding wrench
CP.4030 Drive pinion bearing pre-load gauge
P.4030-1 Adaptor

To Remove

1. Release the handbrake and ensure that the rear brakes are not binding.

2. Relieve the staking securing the pinion nut and, holding the pinion flange using Tool No. P.4028, unscrew the nut.

3. Remove the pinion flange and the pinion flange oil seal.

CORTINA - LOTUS

To Install

4. Fit a new pinion flange oil seal previously immersed in oil.
5. Fit the drive flange and a new retaining nut. Tighten the nut until a small amount of end-float remains at the pinion shaft (approx. 0.002 to 0.005 in.).
6. Fit the pre-load gauge, Tool No. CP.4030 and P.4030-1, to the drive flange and measure the torque required to turn the pinion drive flange. (This will check torque required to turn the differential case and gear assembly and the pinion oil seal drag.)

   NOTE – The car must be horizontal to the kerb loaded condition for correct operation of Tool No. CP.4030.
7. Carefully tighten the drive flange nut to give 4½ to 8 lb. in. (0.052 to 0.092 kg.m.) torque plus the torque measured in sub-operation 6. Frequent checks must be made with the pre-load gauge to prevent the above torque figure being exceeded. If the figure is exceeded the pinion bearing collapsible spacer must be renewed.
8. Stake the nut to the pinion when the correct torque has been attained.
9. Lower the car to the ground and check the rear axle oil level.

OP 4602-A3 EXTRA: CENTRE BEARING – REMOVE AND INSTALL
(Driveshaft Assembly Removed)

To Remove

1. Mark the centre universal yoke and front universal yoke for correct re-alignment on re-assembly. Bend back the lock tab in the centre of the universal joint yoke and slacken the restraining bolt. Remove the "U" shaped plate and separate the two halves of the driveshaft.
2. Remove the driveshaft and bearing assembly from the rubber insulator.
3. Bend back the six tabs securing the rubber insulator into its carrier, and remove.
4. With a suitable two-legged puller remove the bearing and protective caps from the driveshaft.

To Install

5. Drive the ball bearing and protective caps onto the driveshaft, using a tube of suitable diameter.
6. Insert the rubber insulator into its carrier with the boss upwards. Bend the six tabs on the carrier back over the beaded edge of the rubber insulator.
7. Slide the carrier and the rubber insulator over the bearing assembly.
8. Screw the securing bolt with a new lock tab onto the end of the front driveshaft, leaving enough space to allow for the "U"-shaped plate.
9. Line up the mating marks on the two universal joint yokes and assemble the driveshaft. Insert the "U"-shaped plate, with the smooth surface facing the fork, under the securing bolt head and tighten the bolt to a torque of 25 to 30 lb. ft. (3.5 to 4.0 kg.m.). Bend up the lock tab.

OP 4602-B DRIVESHAFT UNIVERSAL JOINT – ONE – OVERHAUL
(Includes OPS 4602-A and A1)

Drive Shaft – Exploded

4
REAR AXLE
(1967-1968)

CORTINA - LOTUS
Supplement

OP 4602-C	DRIVESHAFT UNIVERSAL JOINTS – TWO – OVERHAUL (Includes OPS 4602-A and A1×2)
OP 4602-D	DRIVESHAFT UNIVERSAL JOINTS – THREE – OVERHAUL (Includes OPS 4602-A and A1×3)
OP 4676-A	DRIVE PINION OIL SEAL – RENEW (Includes remove and adjust differential assembly)
OP 4676-B	DRIVE PINION OIL SEAL – RENEW (DIFFERENTIAL ASSEMBLY IN SITU) (Includes OPS 4602-A and A2)
OP 4817-D	CENTRE BEARING – REMOVE AND INSTALL (Includes OPS 4602-A and A3)

CORTINA - LOTUS

REAR AXLE AND DRIVE SHAFT VARIATIONS

While the rear axle internals remain basically the same the drive shaft differs considerably in that:—

1. The rear half of the drive shaft incorporates a sliding joint. A grease point is provided to enable the joint to be lubricated every 3,000 miles.
2. The centre bearing is of different design with the bearing located by two circlips.
3. Insulators are fitted to the bearing carrier.

SERVICE AND REPAIR OPERATIONS

With the exception of the procedures described in the supplement, operations on '67 - '68 vehicles may be carried out as detailed in Section 4.

CORTINA - LOTUS

SERVICE AND REPAIR OPERATIONS

OP 4602-A   DRIVE SHAFT ASSEMBLY – REMOVE AND INSTALL

To Remove

1. Jack up the rear of the car and fit chassis stands.
2. Mark the drive shaft and pinion flanges for correct alignment on reassembly and then remove the four nuts and bolts.
3. Remove the four nuts, bolts and washers securing the drive shaft flange to the coupling shaft flange and remove the drive shaft.

To Install

4. Align the marks on the drive shaft and pinion flanges and fit the four bolts and locknuts.
5. Lift up the front flange of the shaft and locate it on the coupling shaft flange. Fit the four bolts, nuts and washers.
6. Jack up the car, remove the stands and lower the car to the ground.

OP 4817-A   COUPLING SHAFT AND CENTRE BEARING ASSEMBLY – REMOVE AND INSTALL

To Remove

1. Jack up the car and fit chassis stands all round.
2. Remove the four nuts, bolts and washers securing the front drive shaft flange to the coupling shaft flange. Lay the drive shaft to one side.
3. Remove the nuts and bolts securing the bearing carrier to the floor pan and the nuts and washers securing the insulators to the carrier. Lower the coupling shaft and bearing assembly and withdraw the yoke from the gearbox.
4. Remove the two nuts and washers securing the bearing carrier to the insulators and remove the carrier.

To Install

5. Fit the bearing carrier on the insulator blocks and replace the nuts and washers.
6. Locate the yoke of the shaft in the gearbox and raise the shaft into position. Secure the bearing carrier to the floor pan with the nuts and bolts and replace the nuts and washers on the insulator blocks.
7. Fit the four bolts, washers and nuts securing the drive shaft flange to the coupling shaft flange.
8. Jack up the car, remove the stands and lower the car to the ground. Check the gearbox oil level.

CORTINA - LOTUS

OP 4817-A2 EXTRA: COUPLING SHAFT AND CENTRE BEARING ASSEMBLY – OVERHAUL

To Remove

1. Remove the split pin and unscrew the castellated nut securing the coupling shaft to the coupling shaft flange. Pull the flange off the shaft using a suitable puller. Remove the woodruff key from its location in the shaft.
2. Remove the circlip from its location above the bearing.
3. Bend back the bearing shield sufficiently to remove the circlip from behind the bearing.
4. Using a suitable press and adaptor, push the bearing and shaft out of the bearing housing.
5. Using a suitable puller pull the bearing off the shaft. Remove the bearing shield from the shaft.

To Replace

6. Push the bearing into the bearing housing using a suitable adaptor and press.
7. Fit a circlip above and below the bearing.
8. Place a new bearing shield on the shaft. Push the bearing and bearing housing on to the shaft using a suitable press and adaptor.
9. Position the woodruff key in its locating key-way on the shaft and locate the flange on the shaft aligning the key with its mating slot. Fit the castellated nut and tighten it to a torque of 75 to 100 lb. ft. (10.37 to 13.83 kg.m.). Fit a new split pin.

OP 4817-A3 EXTRA: BEARING CARRIER INSULATORS – RENEW

1. Remove the nuts and washers securing the insulators to the bearing housing. Remove the insulators.
2. Fit the new insulators and replace the nuts and washers.

OP 4817-C COUPLING SHAFT AND/OR CENTRE BEARING – OVERHAUL
(Includes OPS 4817-A and A2)

OP 4843-A BEARING CARRIER INSULATORS – REPLACE

To Remove

1. Jack up the car and fit chassis stands all round.
2. Remove the nuts and bolts securing the bearing carrier to the floor pan. Slacken the nuts and washers securing the carrier to the rubber insulators and remove the carrier.
3. Remove the nuts and washers securing the insulators to the bearing housing and remove the insulators.

To Replace

4. Fit the insulators on the bearing housing and replace the nuts and washers.
5. Locate the bearing carrier in position on the insulators and replace the nuts and washers. Fit the bolts and nuts securing the carrier to the floor pan.
6. Jack up the car, remove the stands and lower the car to the ground.

Drive Shaft – Exploded

CORTINA - LOTUS

5/1
FRONT SUSPENSION

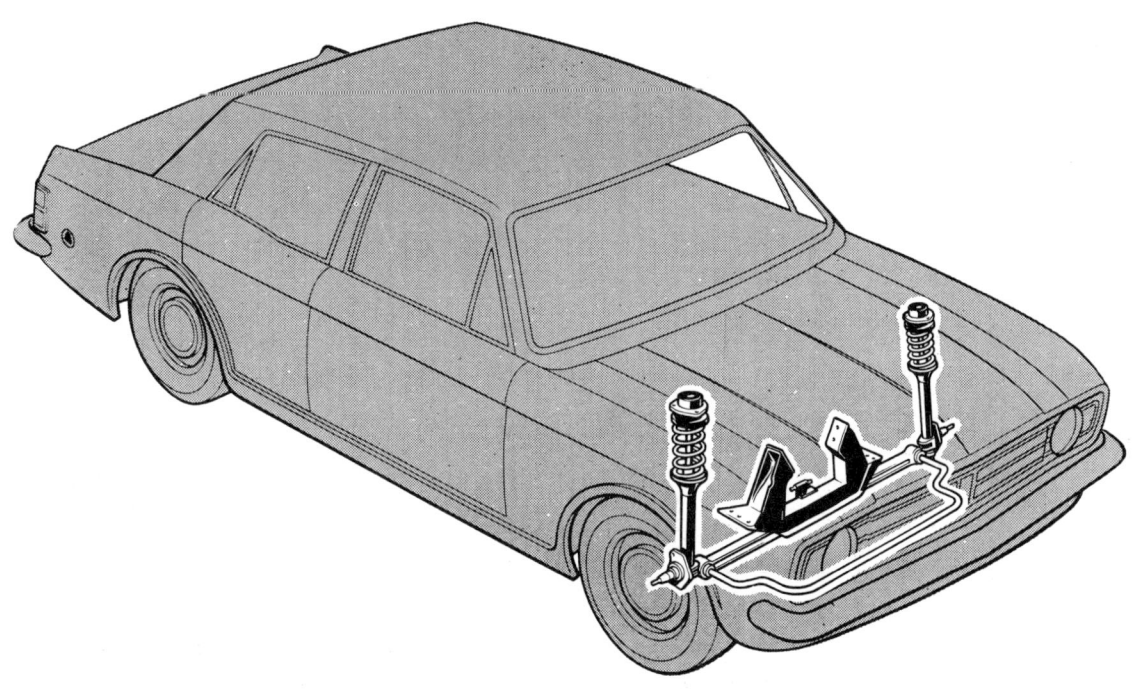

CORTINA - LOTUS

SECTION INDEX

GENERAL DESCRIPTION

QUICK REFERENCE DATA

SERVICE AND REPAIR OPERATIONS

OPERATION	3063-A	TRACK CONTROL ARM INNER BUSH – ONE SIDE – RENEW
,,	3063-A1	Extra: remaining track control arm inner bush – renew
,,	3063-B	**TRACK CONTROL ARM INNER BUSHES – ALL – RENEW** (Includes OPS 3063-A and A1)
,,	3078-A	TRACK CONTROL ARM ASSEMBLY – ONE – REMOVE AND INSTALL
,,	3078-A1	Extra: remaining track control arm assembly – remove and install
,,	3078-B	**TRACK CONTROL ARM ASSEMBLIES – BOTH – REMOVE AND INSTALL** (Includes OPS 3078-A and A1)
,,	3482-A	STABILISER BAR CLAMPS – REMOVE AND INSTALL
,,	3482-A1	Extra: stabilizer bar mounting bushes – renew
,,	3530-A	STABILISER BAR – REMOVE AND INSTALL
,,	3530-A1	Extra: stabiliser bar to track control arm bushes – renew
,,	3530-B	STABILISER BAR TO TRACK CONTROL ARM BUSHES – RENEW (Includes OPS 3530-A and A1)
,,	5310-A	SUSPENSION COIL SPRING – REMOVE AND INSTALL
,,	5310-A1	Extra: remaining suspension coil spring – remove and install
,,	5310-B	**SUSPENSION COIL SPRING – BOTH – REMOVE AND INSTALL** (Includes OPS 5310-A and A1)
,,	5325-A	BUMP RUBBER – ONE SIDE – RENEW
,,	5325-A1	Extra: remaining bump rubber – renew
,,	5325-B	**BUMP RUBBER – BOTH – RENEW** (Includes OPS 5325-A and A1)
,,	5425-A	FRONT SUSPENSION TOP MOUNT ASSEMBLY – RENEW
,,	5425-A1	Extra: remaining front suspension top mount assembly – renew
,,	5425-B	**FRONT SUSPENSION TOP MOUNT ASSEMBLY – BOTH – RENEW** (Includes OPS 5425-A and A1)
,,	5431-A	FRONT SUSPENSION ASSEMBLY – ONE SIDE – REMOVE AND INSTALL
,,	5431-A1	Extra: remaining front suspension assembly – remove and install
,,	5431-A2	Extra: suspension unit – one – renew
,,	5431-A3	Extra: front suspension unit – overhaul
,,	5431-B	**FRONT SUSPENSION UNIT ASSEMBLIES – BOTH – REMOVE AND INSTALL** (Includes OPS 5431-A and A1)
,,	5431-C	**FRONT SUSPENSION UNIT – ONE – RENEW** (Includes OPS 5431-A and A2)
,,	5431-D	**FRONT SUSPENSION UNITS – BOTH – RENEW** (Includes OPS 5431-A, A1 and A2×2)
,,	5431-E	**FRONT SUSPENSION UNIT – ONE – OVERHAUL** (Includes OPS 5431-A and A3)
,,	5431-F	**FRONT SUSPENSION UNITS – BOTH – OVERHAUL** (Includes OPS 5431-A, A1 and A3×2)

CORTINA - LOTUS

GENERAL DESCRIPTION

The front suspension utilises shock absorbers surrounded by large coil springs. Lateral movement of each front wheel is controlled by the track control arm and fore and aft movement is controlled by the stabiliser bar. Vertical movement of the wheel is limited by a rebound stop inside the suspension unit and by the suspension spring reaching the limit of its compression. A rubber bump stop is fitted around the suspension unit piston rod and this comes into operation before the spring is fully compressed.

The suspension mounting points are all rubber insulated to minimise the transmission of road noise and vibration to the body and interior. A "compliance device" is incorporated in the stabiliser bar to track control arm mounting. Its function is to permit the wheel a small amount of fore and aft movement and thus reduce the shock loading on the steering linkage when the wheel hits a sudden irregularity in the road surface.

The toe-in and toe-out on turns (wheel-lock angles) are adjustable. But the camber, castor and king pin inclination are set in manufacture and are not adjustable.

It should be noted that the front suspension units/shock absorbers are sealed and therefore do not require periodic topping-up.

QUICK REFERENCE DATA

DATA

Coil spring identification colour	Blue
Coil spring diameter (mean)	4.808 in. (11.19 cm.)
Wire diameter	0.492 in. (12.49 mm.)
Number of coils	$6\frac{3}{4}$
Free spring length	10.33 in. (26.24 cm.)
Spring length fitted	5.93 in. (15.06 cm.) initial, 5.68 in. (14.33 cm.) minimum length settled
Spring rate	139 lb./in. (24.99 kg./cm.)

Tightening Torques

	lb./ft.	kg./m.
Suspension unit upper mounting bolts	15 — 18	2.07 — 2.49
Spindle to top mount assembly	28 — 32	3.9 — 4.4
Track control arm ball stud nut	30 — 35	4.15 — 4.84
*Stabiliser bar attachment clamps	15 — 18	2.07 — 2.49
*Stabiliser bar to track control arm nut	15 — 45	2.07 — 6.22
*Track control arm inner bushing	22 — 27	3.04 — 3.73
Front suspension crossmember to body sidemember	25 — 30	3.46 — 4.15

*These to be tightened with the weight of the car on its wheels.

SERVICE AND REPAIR OPERATIONS

OP 3063-A TRACK CONTROL ARM INNER BUSHES — ONE SIDE — RENEW

To Remove

1. Jack up the front of the car and fit chassis stands.
2. Remove the self-locking nut and flat washer from the rear of the track control arm pivot bolt and push out the bolt.
3. Pull the inner end of the track control arm downwards and remove the rubber bushes.

To Install

4. Assemble new bushes to the recess in the track control arm.
5. Position the track control arm so that the pivot bolt can be fitted. Slide the pivot bolt into position from the front and fit the flat washer and self-locking nut from the rear. Do not tighten the nut at this stage.
6. Jack up, remove the chassis stands and lower the car to the ground, and tighten the nut to a torque of 22 to 27 lb. ft. (3.04 to 3.73 kg.m.).

OP 3063-A1 EXTRA: REMOVING TRACK CONTROL ARM INNER BUSHES — RENEW

Repeat Operations 2 to 5 OP 3063-A

OP 3063-B TRACK CONTROL ARM INNER BUSHES — ALL — RENEW
(Includes OPS 3063-A and A1)

OP 3078-A TRACK CONTROL ARM ASSEMBLY — ONE — REMOVE AND INSTALL

To Remove

1. Jack up the front of the car and fit chassis stands.
2. Remove the split pin and unscrew the castellated nut securing the track control arm to the stabiliser bar. Pull off the large dished washer.
3. Remove the self-locking nut and flat washer from the rear of the track control arm pivot bolt and release the inner end of the track control arm.
4. Remove the split pin and unscrew the nut securing the track control arm ball joint to the base of the suspension unit and separate the joint.

To Install

5. Assemble the track control arm ball stud to the base of the suspension unit, tighten to a torque of 30 to 35 lb. ft. (4.15 to 4.84 kg.m.) and fit a new split pin.
6. Position the track control arm so that it locates correctly over the stabiliser bar and then secure the inner end. Slide the pivot bolt into position from the front and fit the flat washer and self-locking nut from the rear. Tighten the nut to a torque of 22 to 27 lb. ft. (3.04 to 3.73 kg.m.) when the car is resting on the ground.

CORTINA - LOTUS

7. Assemble the rear dished washer to the end of the stabiliser bar. Ensure that the washer is fitted the correct way round. Fit the castellated nut and, when the car is resting on the ground, tighten to a torque of 15 to 45 lb. ft. (2.07 to 6.22 kg.m.). Fit a new split pin.

8. Jack up the car and remove the chassis stands.

OP 3078-A1 EXTRA: REMAINING TRACK CONTROL ARM ASSEMBLY – REMOVE AND INSTALL

Repeat Operations 2 to 7, OP 3078-A.

OP 3078-B TRACK CONTROL ARM ASSEMBLY – BOTH – REMOVE AND INSTALL
(Includes OPS 3078-A and A1)

OP 3482-A STABILISER BAR CLAMPS – REMOVE AND INSTALL

1. With the handbrake applied, jack up the front of the car and fit stands.
2. Bend back the lock tab and remove two bolts securing one clamp to the body member.
3. Replace the clamp and refit the bolts, using a new lock tab. Do not fully tighten the bolts at this stage.
4. Repeat Operations 2 and 3 for the second clamp.
5. Jack up the car, remove the stands and lower the car to the ground. Tighten the clamp bolts to a torque of 15 to 18 lb. ft. (2.07 to 2.49 kg.m.) and turn up the lock tabs.

OP 3482-A1 EXTRA: STABILISER BAR MOUNTING BUSHES – RENEW

To Remove

1. Withdraw the split pin and unscrew the stabiliser bar nut which secures the end of the stabiliser bar to the track control arm. Remove the large washer.
2. Pull the end of the stabiliser bar forward and remove from the track control arm.
3. Withdraw the sleeve and large washer from the end of the stabiliser bar.
4. Remove the two mounting bushes by sliding them along the stabiliser bar and off the end.

To Install

5. Replace the bushes on the stabiliser bar, and slide them along until they are beneath the clamp bolt holes in the body members.
6. Replace the large washer and sleeve onto the end of the stabiliser bar, ensuring that the washer is the right way round.
7. Position the end of the stabiliser bar in the track control arm.
8. Assemble the large washer to the end of the stabiliser bar and secure with the castellated nut. When the car has been lowered onto its wheels, tighten the nut to a torque of 15 to 45 lb. ft. (2.07 to 6.22 kg.m.) and secure with a new split pin.

Stabiliser Bar and Track Control Arm

CORTINA - LOTUS

OP 3530-A STABILISER BAR – REMOVE AND INSTALL

To Remove

1. Jack up the front of the car and fit chassis stands under the front jacking points.
2. Remove the two attachment clamps from the front of the stabiliser bar after bending back the lock tabs and removing the four bolts.
3. Withdraw the split pins and unscrew the stabiliser bar nuts which secure the ends of the stabiliser bar to the track control arms. Remove the nuts and pull off the large washers.
4. Pull the stabiliser bar forward and remove.
5. Withdraw the sleeve and large washer from each end of the stabiliser bar.
6. Remove the stabiliser bar mounting bushes.

To Install

7. Replace the stabiliser bar mounting bushes by sliding them along the bar until they are beneath the clamp body bolt holes in the body bracket.
8. Assemble a large washer and sleeve to each end of the stabiliser bar and insert the stabiliser bar through the apertures in the track control arms. Ensure that washer is located correctly.
9. Assemble the large washers to the ends of the stabiliser bar and secure with the castellated nuts. When the car has been lowered onto its wheels tighten the nuts to a torque of 15 to 45 lb. ft. (2.07 to 6.22 kg.m.) and secure with new split pins.
10. Locate the stabiliser bar attachment clamps.
11. Secure the bar to the mounting points, using two new locking washers and two bolts on each clamp. When the car is resting on the ground, tighten the bolts to a torque of 15 to 18 lb. ft. (2.07 to 2.49 kg.m.). Turn up the tabs of the locking washers.
12. Jack up the car, remove the stands and lower the car to the ground.

OP 3530-A1 EXTRA: STABILISER BAR TO TRACK CONTROL ARM BUSHES – RENEW

To Remove

1. Push the stabiliser bar sleeve from the centre of the bush.
2. Cut the bush from the track control arm with a sharp knife.

To Install

NOTE – New stabiliser bar bushes are supplied in one piece. To facilitate ease of fitting the bush should be cut in half, using a sharp knife (not a hacksaw). Care should be taken to cut the bush at right angles to its axis in the centre central portion.

3. Fit the two halves of the bush into the track control arm and push the metal sleeve through both halves.

OP 3530-B STABILISER BAR TO TRACK CONTROL ARM BUSHES – RENEW
(Includes OPS 3530-A and A1)

OP 5310-A SUSPENSION COIL SPRING – REMOVE AND INSTALL

Tools Required

P.5045   Coil spring adjustable restrainers

To Remove

1. Open the bonnet and fit wing covers.
2. Measure the distance the threaded end of the piston rod protrudes above the piston rod nut. Slacken the piston rod nut.
3. Jack up the front of the car, fit chassis stands and remove the road wheel.
4. Fit the adjustable spring restrainers, Tool No. P.5045, to the front suspension spring.
5. Using a "bottle" jack under the outer end of the track control arm, support the suspension unit.
6. Remove the piston rod nut and the cranked retainer.
7. Remove the three bolts securing the top mount assembly to the side apron panel.
8. Push the piston rod downwards as far as it will go.
9. Lift off the top mount assembly, the dished washer and the spring upper seat assembly from the top of the spring and manoeuvre the spring out of its location.
10. Mount the spring in a vice and note the position of the spring restrainer jaws relative to the ends of the spring. Remove the adjustable spring restrainers.

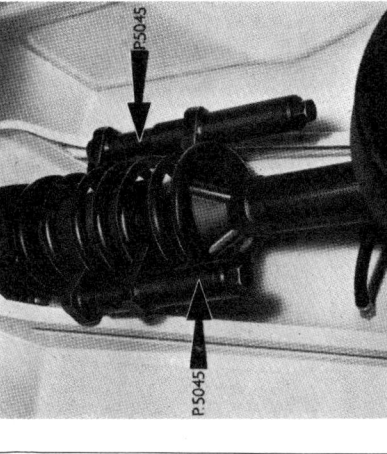

Adjustable Spring Retainers

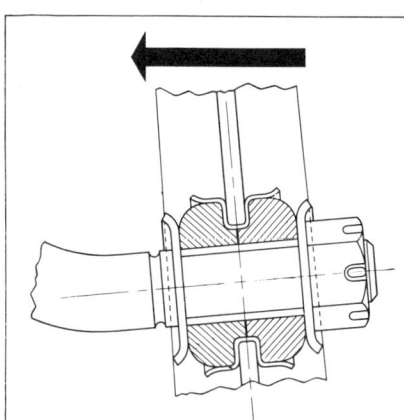

Stabiliser Bar Bush

CORTINA - LOTUS

To Install

11. Fit the new spring into a vice and compress with the adjustable spring restrainers. Ensure that the jaws of the spring restrainers are correctly located relative to the ends of the spring.

    NOTE — Check the colour coding of the replacement spring against the chart in the Quick Reference Data at the beginning of this section. Also check that the other spring is the same. Never mix springs, this will give unequal ride heights each side of the car and upset the handling characteristics.

12. Manoeuvre the spring into position and locate it on the lower spring seat. Push the piston rod upwards and fit the upper spring seat so that it locates on the flats on the piston rod.

13. Fit the dished washer, convex side upwards, on to the piston rod.

14. Push the piston rod further upwards if necessary and fit the top mount assembly.

15. With the steering in the straight-ahead position fit the cranked retainer so that the upper tang faces inwards, i.e. towards the engine, and assemble the securing nut. Do not tighten the nut at this stage.

16. Locate the top mount assembly and secure it to the side apron panel with the three bolts. Tighten the bolts to a torque of 15 to 18 lb. ft. (2.07 to 2.49 kg.m.).

17. Replace the road wheel after removing the spring restrainers.

18. Jack up, remove the chassis stands and the "bottle" jack from under the track control arm. Lower the car to the ground.

19. Tighten the piston rod nut to a torque of 28 to 32 lb. ft. (3.9 to 4.4 kg.m.) whilst holding the spring upper seat to prevent it rotating. Check that the cranked retainer still faces inwards and that the piston rod thread protrusion is the same as that measured in sub-operation 2.

    NOTE — It is essential that the road wheels are kept in the straight-ahead position when the top nut is tightened. If the wheels are turned onto lock, this will cause the top mount rubber to be twisted when the wheels are subsequently turned into the straight-ahead position. This puts a bias into the steering so that the car continually pulls to one side.

20. Remove the wing covers and close the bonnet.

OP 5310-A1 EXTRA: REMAINING SUSPENSION COIL SPRING – REMOVE AND INSTALL

Repeat sub-operations 2 to 17 above.

OP 5310-B SUSPENSION COIL SPRING – BOTH – REMOVE AND INSTALL
(Includes OPS 5310-A and A1)

OP 5325-A BUMP RUBBER – RENEW

Tools Required

P.5045   Coil spring adjustable restrainers

To Remove

1. Open the bonnet and fit wing covers.

2. Measure the distance the threaded end of the piston rod protrudes above the piston rod nut. Slacken the piston rod nut.

3. Jack up the front of the car, remove the road wheels and fit chassis stands.

4. Fit the adjustable spring restrainers, Tool No. P.5045, to the front suspension spring.

5. Position a "bottle" jack under the outer end of the track control arm to support the suspension unit.

6. Remove the piston rod nut and the cranked retainer.

7. Remove the three bolts securing the top mount assembly to the side apron panel.

8. Push the piston rod downwards as far as it will go.

9. Lift off the top mount assembly, the dished washer and the spring upper seat assembly from the top of the spring and pull the bump rubber off the piston rod.

To Install

10. Fit the bump rubber to the piston rod and push it down as far as it will go.

11. Fit the spring upper seat on the spring and pass the piston rod upwards so that it locates in the upper seat. Fit the dished washer, convex side uppermost, onto the piston rod.

12. Pull the piston rod upwards and fit the top mount assembly.

13. With the road wheels in the straight-ahead position, fit the cranked retainer so that the upper tang faces inwards, i.e. towards the engine, and assemble the securing nut. Do not tighten the nut at this stage.

14. Locate the top mount assembly and secure it to the side apron panel with the three bolts. Tighten the bolts to a torque of 15 to 18 lb. ft. (2.07 to 2.49 kg.m.).

15. Remove the spring restrainers and refit the road wheel.

16. Jack up, remove the chassis stands, and the "bottle" jack from under the track control arm. Lower the car to the ground.

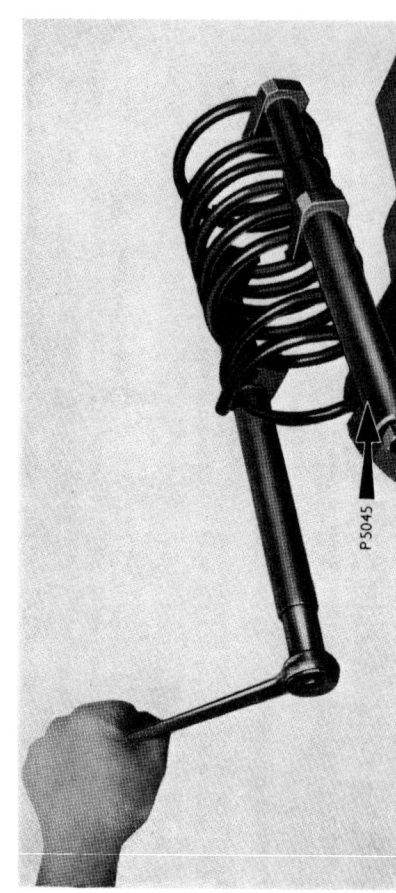

Compressing a Spring with the Adjustable Spring Restrainers

CORTINA - LOTUS

17. Tighten the piston rod nut to a torque of 28 to 32 lb. ft. (3.9 to 4.4 kg.m.) whilst holding the spring upper seat to prevent it rotating. Check that the cranked retainer still faces inwards and that the piston rod protrusion is the same as that measured in sub-operation 2.

    NOTE — It is essential that the front wheels are kept in the straight-ahead position while the top nut is tightened.

18. Remove the wing covers and close the bonnet.

OP 5325-A1 EXTRA: REMAINING BUMP RUBBER – RENEW

Repeat sub-operations 2 to 15 above.

OP 5325-B BUMP RUBBER – BOTH – RENEW
(Includes OPS 5325-A and A1)

OP 5425-A FRONT SUSPENSION TOP MOUNT ASSEMBLY – RENEW

Tools Required

P.5045   Coil spring adjustable restrainers

To Remove

1. Open the bonnet and fit wing covers.
2. Measure the distance the threaded end of the piston rod protrudes above the piston rod nut. Slacken the piston rod nut.
3. Jack up the front of the car, remove the road wheel and fit chassis stands.
4. Fit the adjustable spring restrainers, Tool No. P.5045, to the front suspension spring.
5. Position a "bottle" jack under the outer end of the track control arm to support the suspension unit.
6. Remove the piston rod nut and the cranked retainer.
7. Push the piston rod downwards as far as it will go.
8. Remove the three bolts securing the top mount assembly to the reinforced side apron panel. Detach the upper mount assembly.

To Install

9. Check that the spring upper seat and the dished washer are correctly located on the piston rod.
10. Push the piston rod upwards and fit the top mount assembly to the piston rod.
11. With the road wheels in the straight-ahead position, fit the cranked retainer so that the upper tang faces inwards, i.e. towards the engine, and assemble the securing nut. Do not tighten the nut at this stage.
12. Locate the mount assembly and secure it to the side apron with the three bolts. Tighten the bolts to a torque of 15 to 18 lb. ft. (2.07 to 2.49 kg.m.).
13. Remove the spring restrainers and refit the road wheel.

Section 5/1 — 12   1 . 1969

---

14. Jack up, remove the chassis stands and the "bottle" jack from under the track control arm. Lower the car to the ground.
15. Tighten the piston rod nut to a torque of 28 to 32 lb. ft. (3.9 to 4.4 kg.m.) whilst holding the spring upper seat to prevent it rotating. Check that the cranked retainer still faces inwards and that the piston rod protrusion is the same as that measured in sub-operation 2.

    NOTE — It is essential that the front wheels are kept in the straight-ahead position while the top nut is tightened.

16. Remove the wing covers and close the bonnet.

OP 5425-A1 EXTRA: REMAINING FRONT SUSPENSION TOP MOUNT ASSEMBLY – RENEW

Repeat sub-operations 2 to 13 above.

OP 5425-B FRONT SUSPENSION TOP MOUNT ASSEMBLY – BOTH – RENEW
(Includes OPS 5425-A and A1)

OP 5431-A FRONT SUSPENSION ASSEMBLY – ONE SIDE - REMOVE AND INSTALL

Tools Required

P.5045   Coil spring adjustable restrainers
P.2012   Brake line plugs

To Remove

1. Open the bonnet and fit wing covers.
2. Jack up the front of the car and fit chassis stands.
3. Remove the road wheel.
4. Detach the hydraulic flexible brake pipe from the bracket on the suspension unit. Fit brake line plugs (Tool No. P.2012) to each of the opened unions.
5. Fit the adjustable spring restrainers (Tool No. P.5045) to the front suspension spring.
6. Disconnect the track rod from the steering arm after removing the split pin and nut.
7. Detach the track control arm from the base of the suspension unit after removing the split pin and nut.
8. Remove the three bolts securing the top mount assembly to the side apron panel and remove the suspension assembly complete with the brake calliper.

To Install

9. Reposition the suspension assembly and secure with the three bolts through the side apron panel, to the top mount assembly. Tighten the bolts to a torque of 15 to 18 lb. ft. (2.07 to 2.49 kg.m.).
10. Assemble the track control arm ball stud to the base of the suspension unit, tighten the securing nut to a torque of 30 to 35 lb. ft. (4.15 to 4.85 kg.m.) and fit a new split pin.
11. Fit the track rod end into the locating hole in the steering arm. Fit the castellated nut, tighten to a torque of 18 to 22 lb. ft. (2.5 to 3.0 kg.m.) and lock with a new split pin.

1 . 1969   Section 5/1 — 13

12. Remove the brake line plugs and reconnect the flexible brake pipe to the bracket on the suspension unit.
13. Remove the spring restrainers and check visually to ensure correct location of the spring.
14. Bleed the brakes, see Operation 2000-A.
15. Replace the road wheel.
16. Jack up and remove the chassis stands.
17. Remove the wing covers and close the bonnet.

OP 5431-A1   EXTRA: REMAINING FRONT SUSPENSION ASSEMBLY – REMOVE AND INSTALL

Repeat sub-operations 3 to 13 of OP 5431-A.

OP 5431-A2   EXTRA: SUSPENSION UNIT ASSEMBLY – RENEW

To Renew

1. Unscrew the piston rod nut and remove the cranked retainer.
2. Detach the top mount and lift off the dished washer, spring upper seat, suspension spring and the bump rubber, if fitted.
3. Remove the brake pads after removing the clips, retaining pins and shims.
4. Bend back the lock tabs on the calliper securing bolts, remove the bolts and detach the calliper.
5. Remove the hub and disc assembly after detaching the dust cap, adjusting nut, etc.
6. Remove the disc splash shield.
7. Detach the steering arm.
8. Replace the steering arm, tighten the securing nuts to a torque of 30 to 34 lb. ft. (4.2 to 4.7 kg.m.).
9. Fit the hub and disc assembly onto the wheel spindle. Assemble the outer bearing, the thrust washer and fit the adjusting nut. Tighten the nut to a torque of 27 lb. ft. (3.73 kg.m.) whilst rotating the hub to seat the bearings. Slacken the nut back 90° then fit the retainer and secure with a new split pin. Tap the dust cap into place.
10. Fit the calliper assembly to its mounting lugs on the suspension unit using a new locking plate under the bolts. Tighten the bolts to a torque of 45 to 50 lb. ft. (6.22 to 6.91 kg.m.) and bend up the lock tabs.
11. Replace the brake pads and shims, secure with the retaining pins and clips.
12. Reassemble the spring, etc., to the suspension unit; pull the piston rod fully upwards and fit the bump rubber, suspension spring, spring upper seat and dished washer, convex side upwards, to the piston rod.
13. Assemble the top mount and the cranked retainer. Fit the securing nut loosely.
14. Tighten the securing nut to a torque of 28 to 32 lb. ft. (3.9 to 4.4 kg.m.) when the suspension unit is assembled into the car and the car is on the ground. When tightening the piston rod nut, the road wheels must be in the straight-ahead position and the cranked retainer must face inwards, i.e. towards the engine.

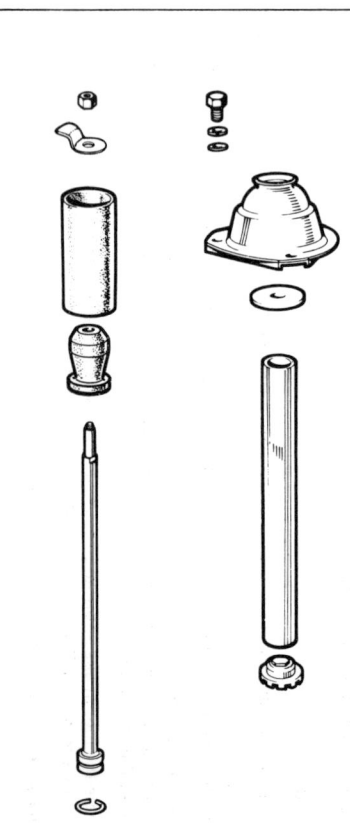
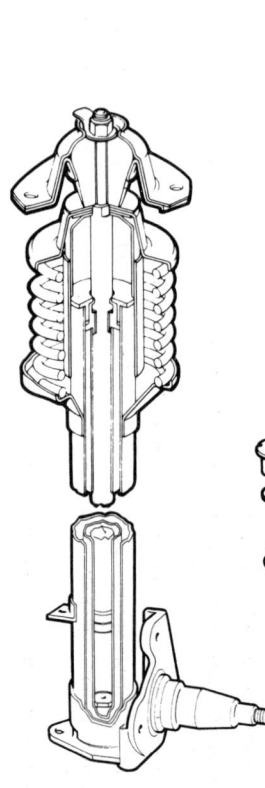
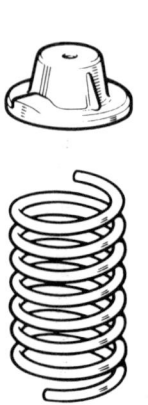

Front Suspension Unit—Exploded

CORTINA - LOTUS

OP 5431-A3 EXTRA: FRONT SUSPENSION UNIT – OVERHAUL

Tools Required

P.5041-A   Front suspension unit bump stop platform wrench
P.5042   Gland and bush guide

To Dismantle

1. Unscrew the piston rod nut and remove the cranked retainer.

2. Detach the top mount and lift off the dished washer, spring upper seat, suspension spring and the bump rubber, if fitted.

3. Relieve the staking, then using Tool No. P.5041-A unscrew the bump stop platform and lift it off the suspension unit. It should be noted that the bump stop platform incorporates a scraper seal which cleans the piston rod during operation. This seal is not serviced separately.

4. Remove the "O" ring from above the upper guide and gland assembly.

5. Ensure that the top edge of the machined area of the piston rod is free from burrs, etc., by using a suitable stone. This is most essential. If this is not done the action of removing or replacing the gland and bush assembly will inevitably damage the coated bearing surface of the bush.

6. Lift the piston rod upwards until the gland and bush assembly is clear of the outer casing. Slide the gland assembly off the rod.

7. Empty the fluid into a suitable waste container.

8. Pull the piston rod complete with piston, cylinder and compression valve out of the outer casing.

9. Remove the piston rod from the cylinder by pushing the compression valve out of the base and then pushing the rod downwards and withdrawing it from the cylinder.

10. If necessary remove the piston ring from the piston.

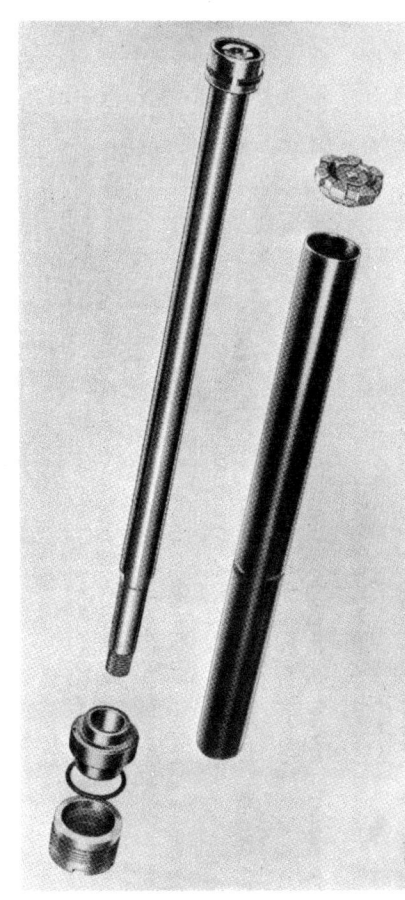

Front Suspension Unit – Internal Components

11. Wash all components in a suitable bath and then examine them for wear or damage.
N.B. Do not remove the piston from the piston rod, these components are serviced as an assembly, they are not available separately. This is because the piston incorporates a valve and the assembly procedure is critical if the valve is to function correctly.

To Reassemble

12. If the piston ring has been removed, fit a new one to the piston.

13. Insert the piston rod into the cylinder and push the compression valve into the base of the cylinder.

14. Carefully pass the cylinder and piston rod assembly into the outer casing. Fill the unit with the correct quantity of fluid, 350 c.c. Ensure that the correct fluid is used, M-100502-E.

15. Fit the gland and bush guide P.5043 onto the end of the piston rod and slide the gland and bush over the guide. Push it down until it contacts the end of the cylinder and the complete internal assembly is below the top of the outer casing.

16. Place the rubber "O" ring on the top of the gland and bush assembly and locate it correctly round the bore of the outer casing.

17. Screw the bump stop platform into the top of the outer casing. Using Tool No. P.5041-A, tighten to a torque of 55-60 lb. ft.

18. Reassemble the spring, etc., to the suspension unit; pull the piston rod fully upwards and fit the bump rubber, suspension spring, spring upper seat and dished washer, convex side upwards to the piston rod.

19. Assemble the top mount and the cranked retainer. Fit the securing nut loosely.

20. Tighten the securing nut to a torque of 28 to 32 lb. ft. (3.9 to 4.4 kg.m.) when the suspension unit is assembled into the car and the car is on the ground. When tightening the piston rod nut the road wheels must be in the straight-ahead position and the cranked retainer must face inwards, i.e. towards the engine.

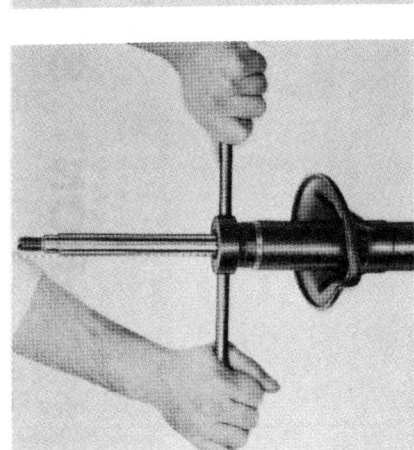

Bump Stop Platform Removal

Replacing the Gland and Bush Assembly

CORTINA - LOTUS
Supplement

FRONT SUSPENSION

5/1

(1967 - 1968)

CORTINA - LOTUS

OP 5431-B FRONT SUSPENSION UNIT ASSEMBLIES – BOTH – REMOVE AND INSTALL
(Includes OPS 5431-A and A1)

OP 5431-C FRONT SUSPENSION UNIT – ONE – RENEW
(Includes OPS 5431-A and A2)

OP 5431-D FRONT SUSPENSION UNITS – BOTH – RENEW
(Includes OPS 5431-A, A1 and A2×2)

OP 5431-E FRONT SUSPENSION UNIT – ONE – OVERHAUL
(Includes OPS 5431-A and A3)

OP 5431-F FRONT SUSPENSION UNITS – BOTH – OVERHAUL
(Includes OPS 5431-A, A1 and A3×2)

FRONT SUSPENSION

The front suspension units fitted to the 1967-'68 vehicles incorporated a bearing in the top mount assembly in lieu of the bonded rubber bush currently in use.

SERVICE AND REPAIR OPERATIONS

With the exception of the following procedure, operations on '67 - '68 may be carried out as detailed in Section 5/1.

SERVICE AND REPAIR OPERATIONS

OP 5310-A SUSPENSION COIL SPRING – REMOVE AND INSTALL

Tools Required

P.5045   Coil spring adjustable restrainers
P.5025   Front suspension upper bearing locknut wrench
P.5026   Front suspension upper bearing locknut torque wrench adaptor

To Remove

1. Fit spring restrainers, Tool No. P.5045, to the suspension spring.

2. With the handbrake applied, jack up the front of the car and fit chassis stands.

3. Remove the road wheel.

4. Open the bonnet, fit a wing cover and remove the plastic cover over the top bearing.

5. Using the front suspension upper bearing locknut wrench, Tool No. P.5025, unscrew the upper bearing retaining nut and push the piston rod downwards.

6. From underneath the car push the piston rod into the suspension unit as far as it will go. Lift off the upper spring seat and manoeuvre the spring out of its location.

7. Mount the spring in a vice and note the position of the spring restrainer jaws relative to the ends of the spring. Remove the spring restrainers.

To Install

8. Fit the new spring into a vice and compress with the adjustable spring restrainers. Ensure that the jaws of the restrainers are correctly located relative to the ends of the spring.

   NOTE – Check the colour coding of the replacement spring. Also check that the other spring is the same. Never mix springs, this will give unequal ride heights each side of the car and upset the handling characteristics.

9. Manoeuvre the spring into position and locate it on the lower spring seat. Force the piston rod upwards and fit the upper spring seat so that the "D" shaped hole locates on the flat on the piston rod.

10. Push the piston rod further upwards and pass it through the top mount and bearing assembly.

11. Fit the retaining nut using Tool No. P.5025 and tighten to a torque of 45 to 55 lb. ft. (6.22 to 7.6 kg.m.) using Tool No. P.5026.

12. Refit the plastic dust cover, remove the wing cover and close the bonnet.

13. Replace the road wheel and lower the car to the ground. Tighten the wheel nuts.

14. Remove the spring restrainers.

CORTINA - LOTUS

Front Suspension Unit — Exploded

OP 5325-A BUMP RUBBER – REMOVE AND INSTALL

Tools Required

P.5045   Coil spring adjustable restrainers

To Remove

1. Fit spring restrainers, Tool No. P.5045, to the suspension spring.
2. With the handbrake applied, jack up the front of the car and fit chassis stands.
3. Remove the road wheel.
4. Open the bonnet, fit a wing cover and remove the plastic cover over the top bearing.
5. Using the front suspension upper bearing locknut wrench (P.5025), unscrew the upper bearing retaining nut and push the piston rod downwards.
6. From underneath the car push the piston rod into the suspension unit as far as it will go. Lift off the upper spring seat and pull the bump rubber off the piston rod.

To Install

7. Slide the bump rubber onto the piston rod and pull the piston upwards so that the upper spring seat can be fitted.
8. Pull the piston rod further upwards and carefully push it through the upper mounting.
9. Fit the retaining nut using Tool No. P.5025 and tighten to a torque of 45 to 55 lb. ft. (6.22 to 7.6 kg.m.) using Tool No. 5026.
10. Refit the plastic dust cover, remove the wing cover and close the bonnet.
11. Replace the road wheel.
12. Jack up, remove the chassis stands and lower the car to the ground. Tighten the wheel nuts.
13. Remove the spring restrainers.

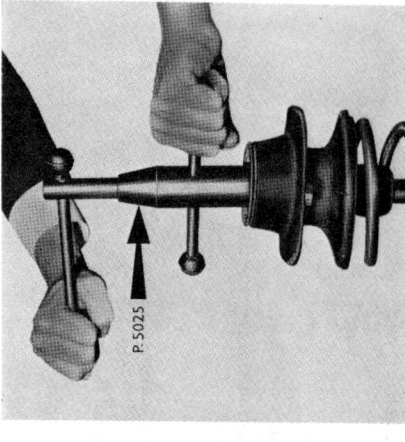

Removing/Replacing the Upper Bearing Locknut

Upper Mount and Bearing Assembly

CORTINA - LOTUS

OP 5425-A FRONT SUSPENSION UPPER MOUNT AND BEARING ASSEMBLY – REMOVE AND INSTALL

Tools Required

P.5045　Coil spring adjustable restrainers
P.5025　Front suspension upper bearing locknut wrench
P.5026　Front suspension upper bearing locknut torque wrench adaptor

To Remove

1. Fit spring restrainers, Tool No. P.5045, to the suspension springs.
2. With the handbrake applied, jack up the front of the car and fit chassis stands.
3. Remove the road wheel.
4. Open the bonnet, fit a wing cover and remove the plastic cover over the top bearing.
5. Using Tool No. P.5025, unscrew the upper bearing retaining nut and push the piston rod downwards.
6. Remove the three bolts securing the upper mounting and bearing assembly to the reinforced side apron panel. Detach the mounting and bearing assembly.

To Install

7. Fit the new mounting and bearing assembly over the piston rod. Push the piston rod upwards so that it passes through the bearing.
8. Loosely fit the retaining nut in position.
9. Rotate the upper bearing assembly as necessary so that the three mounting holes align with the holes in the body reinforcement. Fit the three bolts and tighten them to a torque of 15 to 18 lb. ft. (2.1 to 2.5 kg.m.).
10. Tighten the retaining nut, using Tool No. P.5026, to a torque of 45 to 55 lb. ft. (6.22 to 7.6 kg.m.) and fit the plastic dust cover.
11. Replace the road wheel.

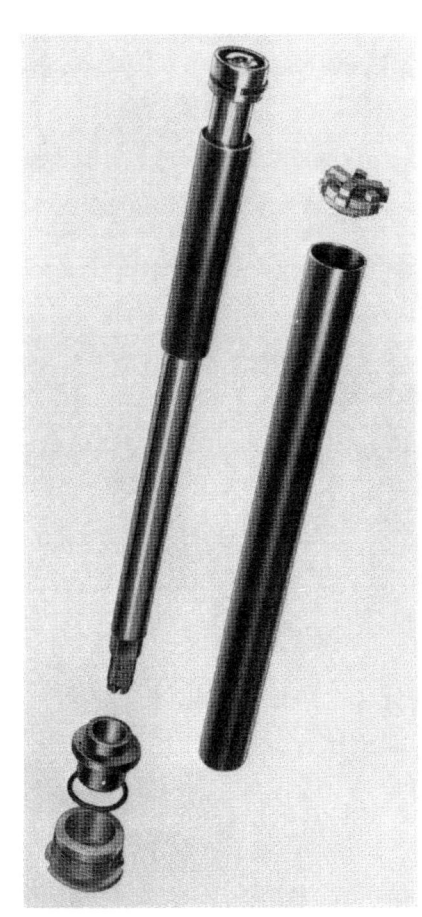

Suspension Unit Internal Components

CORTINA - LOTUS

5/2
REAR SUSPENSION

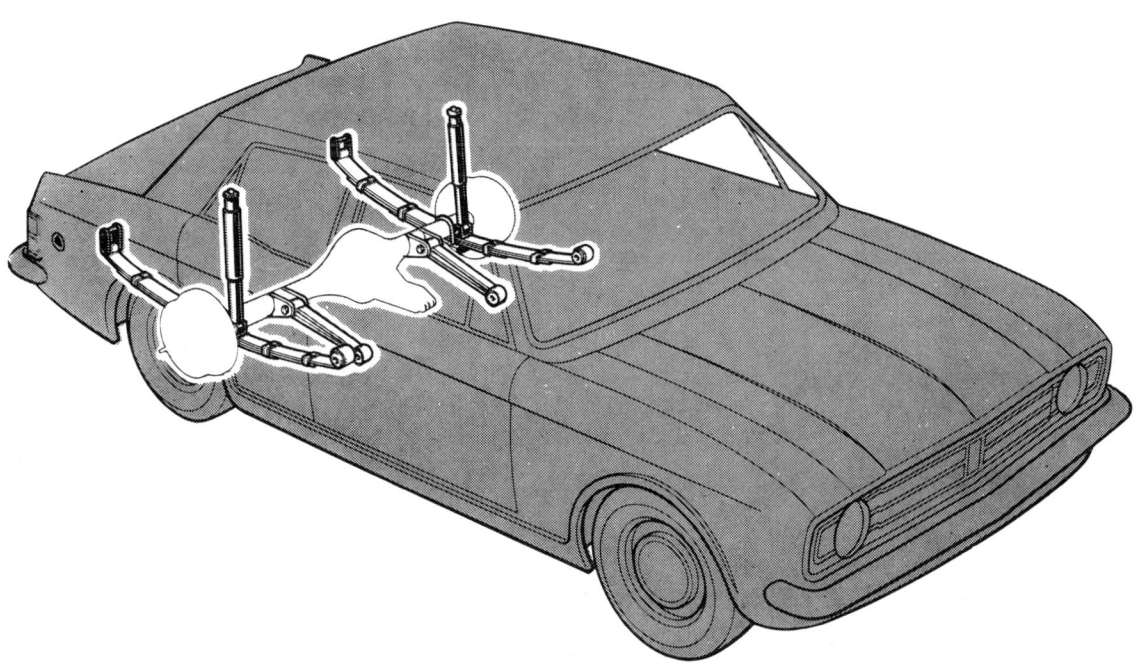

CORTINA - LOTUS

SECTION INDEX

QUICK REFERENCE DATA

SERVICE AND REPAIR OPERATIONS

OPERATION	5500-A	REAR SHOCK ABSORBER – ONE – REMOVE AND INSTALL
"	5500-A1	Extra: remaining rear shock absorber – Remove and install (Rear of car on stands)
"	5500-B	REAR SHOCK ABSORBERS – BOTH – REMOVE AND INSTALL (Includes OPS 5500-A and A1)
"	5545-A	RADIUS ARM – REMOVE AND INSTALL
"	5545-A1	Extra: front bush – renew
"	5560-A	REAR LEAF SPRING ASSEMBLY – ONE – REMOVE AND INSTALL
"	5560-A1	Extra: remaining rear leaf spring assembly – remove and install (Rear of car on stands)
"	5560-A2	Extra: rear spring leaf – one – remove and install (Rear spring assembly removed)
"	5560-B	REAR LEAF SPRING ASSEMBLIES – BOTH – REMOVE AND INSTALL (Includes OPS 5560-A and A1)
"	5560-C	REAR SPRING LEAF – ONE – RENEW (Includes OPS 5560-A and A2)
"	5560-M	REAR SPRING SHACKLE PLATES AND/OR BUSHES – ONE – SPRING – REMOVE AND INSTALL
"	5560-M1	Extra: remaining rear spring shackle plates and/or bushes – remove and install (Rear of car on stands)
"	5560-N	REAR SPRING SHACKLE PLATES AND/OR BUSHES – BOTH SPRINGS – REMOVE AND INSTALL (Includes OPS 5560-M and M1)
"	5560-P	REAR SPRING FRONT BUSH – ONE SIDE – RENEW
"	5560-P1	Extra: remaining rear spring front bush – renew
"	5560-R	REAR SPRING FRONT BUSH – BOTH SIDES – RENEW (Includes OPS 5560-P and P1)
"	5560-S	REAR SPRING FRONT AND REAR BUSHES – ALL – RENEW (Includes OPS 5560-M, M1, P and P1)
"	5707-A	REAR SPRING 'U' BOLT – ONE SPRING – REMOVE AND INSTALL
"	5707-A1	Extra: remaining rear spring 'U' bolts – remove and install (Rear of car on stands)
"	5707-B	REAR SPRING 'U' BOLTS – BOTH SPRINGS – REMOVE AND INSTALL (Includes OPS 5707-A and A1)

GENERAL DESCRIPTION

The rear suspension is of the semi-elliptic leaf spring type. The rear axle is located asymmetrically on the springs, i.e. it is closer to the forward mount than it is to the rear one. This assists in reducing axle tramp and spring wind-up during acceleration or when driving over rough surfaces. To further reduce these effects, two trailing links (radius arms) are utilised, these being mounted on rubber bushes at both ends.

Telescopic, hydraulic, double acting shock absorbers are fitted between the rear axle and a reinforced mounting inside the top of the wheel arch. They are sealed units and therefore do not require topping-up.

The leaf springs and shock absorbers are mounted on fibre bushes at each end to minimise the transmission of noise and vibration to the car body and interior.

QUICK REFERENCE DATA

PERIODIC SERVICE ATTENTION

At the first 600 miles (1,000 km.)
 Check torque of rear spring "U" bolts

Every 6,000 miles (10,000 km.) or six months
 Check torque of rear spring "U" bolts

DATA

Tightening Torques

	lb. ft.	kg.m.
*Radius arm to axle	22 – 27	3.04 – 3.73
*Radius arm to body	45 – 50	6.22 – 6.93
*Shock absorber to body	15 – 20	2.0 – 2.75
Shock absorber to axle	40 – 45	5.54 – 6.22
*Rear spring "U" bolts	18 – 26	2.50 – 3.60
*Rear spring front securing nut	25 – 30	3.46 – 4.15
*Rear spring rear shackle nuts	8 – 10	1.10 – 1.40

*These items to be tightened with the component in the kerb weight position, ie. the car must be resting on its wheels.

CORTINA - LOTUS

OP 5500-A REAR SHOCK ABSORBER – ONE – REMOVE AND INSTALL

To Remove

1. Chock the front wheels, jack up the rear of the car and fit chassis stands under the body jacking points.
2. From inside the luggage compartment remove the two nuts from the top of the shock absorber, using a second spanner on the flats on the piston rod to prevent it turning.
3. Lift off the top steel washer, the rubber bush and the lower steel washer.
4. Detach the lower end of the shock absorber from the bracket on the axle by removing the nut, lockwasher and bolt. Remove the shock absorber from the car.
5. Remove the rubber bush and steel washer from the top of the shock absorber.

To Install

6. Assemble the large steel washer (lipped edge upwards) and the rubber bush to the top of the shock absorber.
7. Fit the lower end of the shock absorber into the bracket on the axle and line up the holes in the bracket with the hole in the shock absorber. Fit the bolt, lockwasher and nut, tighten the nut to a torque of 40 to 45 lb. ft. (5.54 to 6.22 kg.m.).
8. Extend the shock absorber upwards and pass it through the mounting aperture in the body. Position the cupped washer (large opening upwards) followed by the rubber bush and dished washer over the piston rod.
9. Fit the first nut and tighten to a torque of 15 to 20 lb. ft. (2.0 to 2.75 kg.m.), remembering to prevent the piston rod rotating. If the torque wrench permits, this can be done by holding the top of the piston rod, using a ¼ in. A/F open-ended spanner across the flats or, alternatively, if the torque wrench prevents access, grip the upper part of the shock absorber body (this requires the assistance of a second mechanic).

10. Fit the locknut and tighten it securely.
11. Jack up the rear of the car, remove the chassis stands and lower the car to the ground. Remove the chocks from the front wheels.

OP 5500-A1 EXTRA: REMAINING REAR SHOCK ABSORBER – REMOVE AND INSTALL

Repeat sub-operations 2 to 10 above.

OP 5500-B REAR SHOCK ABSORBERS – REMOVE AND INSTALL
(Includes OPS 5500-A and A1)

OP 5545-A RADIUS ARM – REMOVE AND INSTALL

To Remove

1. Chock the front wheels, jack up the rear of the car and fit chassis stands.
2. Detach the rear end of the radius arm from the rear axle by unscrewing the nut and withdrawing the through bolt.
3. Detach the forward end of the radius arm from the body mounting.

To Install

4. Attach the forward end of the radius arm to the body mounting.
5. Attach the rear end of the radius arm to the mounting bracket on the axle casing.
6. Remove the chassis stands and lower the car so that the rear wheels are either on the ground or on blocks. Torque the radius arm securing bolts to the following torque figures: forward end (body mounting) – 45 to 50 lb. ft. (6.22 to 6.91 kg.m.), rear end (axle mounting) – 22 to 27 lb. ft. (3.04 to 3.73 kg.m.). **These bolts MUST be tightened with the weight of the car on its wheels.**
7. If the rear wheels are on blocks, jack up the car and remove them.

OP 5545-A1 EXTRA: FRONT BUSH – RENEW

1. Press the bush out of the end of the rod using a suitable adaptor.
2. Press the new bush into the rod, ensuring that it is square in the aperture.

OP 5560-A REAR LEAF SPRING ASSEMBLY – ONE – REMOVE AND INSTALL

Tools Required

P 5029   Spring bush remover and replacer

To Remove

1. Chock the front wheels, jack up the rear of the car and fit chassis stands under the body jacking points.
2. Position a trolley jack under the rear axle and extend it sufficiently to support the axle.

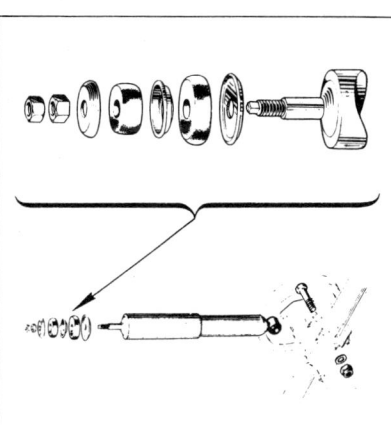

Shock Absorber Mounting

CORTINA - LOTUS

3. Remove the rear shackle nuts and detach the combined shackle bolt and plate assemblies. Remove the four loose rubber bushes.

   When replacing the left rear spring, remove the spare wheel and the rubber plug in the wheel well adjacent to the upper shackle nut.

4. Unscrew the nut from the front mounting bracket and withdraw the through bolt.
5. Remove the "U" bolts and the attachment plate.
6. Remove the spring assembly.
7. Using Tool No. P.5029, pull the bush out of the spring front eye.

To Replace

8. Replace the spring front mounting bush using Tool No. P.5029 to ensure that the bush is fitted squarely in its aperture.
9. Locate the front of the spring in its body mounting bracket. Fit the through bolt and loosely assemble the nut. Do not tighten it at this stage.
10. Fit the spring to the axle using the "U" bolts, the plate and the lockwashers and nuts. Do not tighten the nuts.
11. Fit new loose rubber bushes to the rear spring "eye" and the apertures in the body for the rear shackle.
12. Locate the spring in position and assemble the rear shackle bolt and plate assemblies. Fit the nuts, but again do not tighten.
13. Remove the trolley jack supporting the rear axle.
14. Lower the car so that the rear wheels are on the ground or on blocks.
15. Tighten the spring front and rear attachment points and the "U" bolts to the following torques:

"U" bolts	18 to 26 lb. ft. (2.50 to 3.60 kg.m.)
Spring front hanger	25 to 30 lb. ft. (3.46 to 4.15 kg.m.)
Spring rear shackles	8 to 10 lb. ft. (1.10 to 1.40 kg.m.)

    NOTE – These nuts must be tightened with the car on its wheels.

OP 5560-A1 EXTRA: REMAINING REAR LEAF SPRING ASSEMBLY – REMOVE AND INSTALL

Repeat sub-operations 3 to 12 above.

OP 5560-A2 EXTRA: REAR SPRING LEAF – ONE – REMOVE AND INSTALL

To Dismantle

1. Remove the spring centre bolt and nut.
2. Bend back the ends of the leaf cups and remove rubbers.
3. Remove the appropriate leaf.

Rear Suspension — Exploded

CORTINA - LOTUS

To Install

4. Check that the fibre pads are correctly located between the ends of the spring leaves and rebuild the spring.
5. Fit the centre bolt and nut.
6. Refit rubbers and bend leaf clips back over top leaf.

OP 5560-B REAR LEAF SPRING ASSEMBLIES – BOTH – REMOVE AND INSTALL
(Includes OPS 5560-A and A1)

OP 5560-C REAR SPRING LEAF – ONE – RENEW
(Includes OPS 5560-A and A2)

OP 5560-M REAR SPRING SHACKLE PLATES AND/OR BUSHES – ONE SPRING – REMOVE AND INSTALL

To Remove

1. Chock the front wheels, jack up the rear of the car and fit chassis stands under the body jacking points.
2. Position a trolley jack under the rear axle and extend it sufficiently to support the axle.
3. Remove the rear shackle nuts and detach the combined shackle bolt and plate assemblies. Remove the four rubber bushes.

To Install

4. Fit new rubber bushes to the rear spring "eye" and the apertures in the body for the rear shackle.
5. Locate the rear end of the spring in position and assemble the rear shackle bolt and plate assemblies. Fit the nuts but do not tighten at this stage.
6. Remove the trolley jack supporting the axle.
7. Jack up, remove the stands and lower the car to the ground. Remove the chocks from the front wheels.
8. Tighten the spring rear attachment nuts to the following torques:
   Spring rear shackles ... ... 8 to 10 lb. ft. (1.10 to 1.40 kg.m.)
   NOTE – These nuts must be tightened with the car on its wheels.

OP 5560-M1 EXTRA: REMAINING REAR SPRING SHACKLE PLATES AND/OR BUSHES – REMOVE AND INSTALL

Repeat sub-operations 3 to 5 above.

OP 5560-N REAR SPRING SHACKLE PLATES AND/OR BUSHES – BOTH SPRINGS – REMOVE AND INSTALL
(Includes OPS 5560-M and M1)

OP 5560-P REAR SPRING FRONT BUSH – ONE SPRING – RENEW

Tools Required

P.5029   Spring bush remover and replacer

To Remove

1. Chock the front wheels, jack up the rear of the car and fit chassis stands under the body jacking points.
2. Unscrew the nut from the front mounting bracket and withdraw the through bolt.
3. Pull the front end of the spring downwards and, using Tool No. P.5029, remove the bush from the front "eye".

To Install

4. Using Tool No. P.5029, fit a new bush into the front spring "eye".
5. Position the front of the spring in its body mounting bracket. Fit the through bolt and assemble the nut, but do not tighten at this stage.
6. Jack up, remove the stands and lower the car to the ground. Remove the chocks from the front wheels.
7. Tighten the spring front securing nut to a torque of 25 to 30 lb. ft. (3.46 to 4.15 kg.m.). This nut must be tightened with the car on the ground.

● OP 5560-P1 EXTRA: REMAINING REAR SPRING FRONT BUSH – RENEW

Repeat sub-operations 2 to 5 above.

OP 5560-R REAR SPRING FRONT BUSHES – BOTH SIDES – RENEW
(Includes OPS 5560 P and P1)

OP 5560-S REAR SPRING FRONT AND REAR BUSHES – ALL – RENEW
(Includes OPS 5560-M, M1, P and P1)

Rear Spring 'U' Bolts

CORTINA - LOTUS

OP 5707-A   REAR SPRING 'U' BOLTS – ONE SPRING – REMOVE AND INSTALL

To Remove

1. Chock the front wheels, jack up the rear of the car and fit chassis stands under the body jacking points.
2. Unscrew the "U" bolt nuts and detach the "U" bolts.

To Install

3. Locate the "U" bolts in position and loosely fit the securing nuts.
4. Jack up, remove the stands and lower the car to the ground. Remove the chocks from the front wheels.
5. Tighten the "U" bolt nuts to a torque of 18 to 26 lb. ft. (2.50 to 3.60 kg.m.). These nuts must be tightened with the car on the ground.

OP 5707-A1   EXTRA: REMAINING REAR SPRING 'U' BOLTS – REMOVE AND INSTALL

Repeat sub-operations 2 and 3 above.

OP 5707-B   REAR SPRING 'U' BOLTS – BOTH SPRINGS – REMOVE AND INSTALL
(Includes OPS 5707-A and A1)

CORTINA - LOTUS

6/1
ENGINE

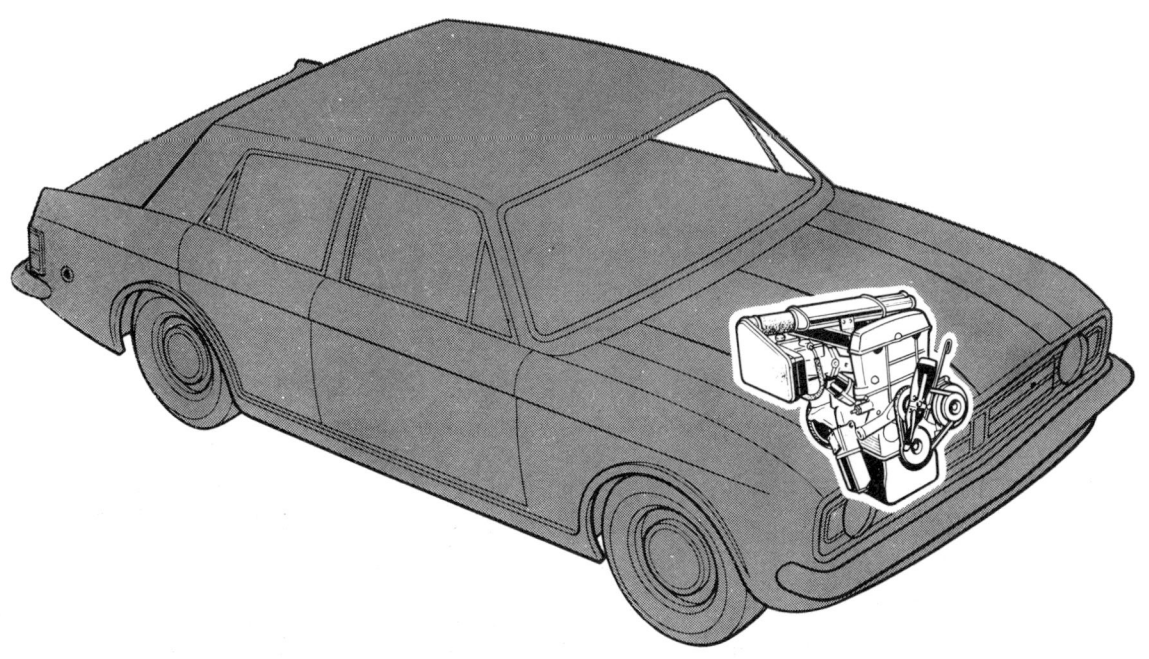

CORTINA - LOTUS

SECTION INDEX

GENERAL DESCRIPTION

QUICK REFERENCE DATA

SERVICE AND REPAIR OPERATIONS

OPERATION	6000-A	ENGINE – TUNE
,,	6000-B	ENGINE – CHECK COMPRESSION
,,	6000-C	ENGINE ASSEMBLY – REMOVE AND INSTALL
,,	6000-C1	Extra: ancillaries – remove and install
,,	6000-C2	Extra: clutch disc and/or pressure plate – remove and install
,,	6000-C3	Extra: clutch pilot spigot bearing – remove and install
,,	6000-C4	Extra: flywheel – remove and install
,,	6000-C5	Extra: sump and/or gaskets – remove and refit
,,	6000-C6	Extra: engine rear oil seal carrier – remove and install
,,	6000-C7	Extra: engine rear oil seal – renew
,,	6000-C8	Extra: front cover and/or gasket – remove and install
,,	6000-C9	Extra: front cover oil seal – remove and install
,,	6000-C10	Extra: timing chain – remove and install
,,	6000-C11	Extra: crankshaft sprocket – remove and install
,,	6000-C12	Extra: connecting rod bearings – remove and install
,,	6000-C13	Extra: main bearing clearances – check
,,	6000-C14	Extra: main bearing liners and thrust washers – remove and install
,,	6000-C15	Extra: all main bearing liners and thrust washers – check clearances and renew
,,	6000-C16	Extra: auxiliary shaft – remove and install
,,	6000-C17	Extra: auxiliary shaft bearings – renew
,,	6000-C18	Extra: crankshaft – remove and install
,,	6000-C19	Extra: cylinder head and pistons – decarbonise
,,	6000-C20	Extra: valves – all – reface, reseat and grind-in
,,	6000-C21	Extra: cylinder assembly – remove and install
,,	6000-C22	Extra: cylinder block – remove and install
,,	6000-C23	Extra: cylinder block – rebore
,,	6000-C24	Extra: cylinder block – rebore and fit liners
,,	6010-A	CYLINDER ASSEMBLY – REPLACE (Includes 6000-C, C1 and C21)
OPERATION	6010-B	CYLINDER BLOCK – REPLACE (Includes 6000-C, C1 and C22)
,,	6010-C	CYLINDER BLOCK – REBORE (Includes 6000-C, C1, C21, C22 and C23)
,,	6010-D	CYLINDER BLOCK – SLEEVE (Includes 6000-C, C1, C21, C22, C23 and C24)
,,	6015-A	ENGINE ASSEMBLY – FIT NEW, SERVICE, OR RECONDITIONED UNIT
,,	6015-A1	Extra: engine compartment – clean
,,	6015-A2	Extra: clutch disc and/or pressure plate – remove and install
,,	6015-A3	Extra: clutch release bearing – renew
,,	6015-A4	Extra: clutch fork – remove and install
,,	6015-A5	Extra: main drive gear bearing retainer gasket and/or main drive gear oil seal – renew
,,	6015-B	ENGINE AND GEARBOX ASSEMBLY – REMOVE AND INSTALL
,,	6015-B1	Extra: engine and gearbox – separate and reconnect
,,	6038-A	ENGINE FRONT MOUNTING – ONE – RENEW
,,	6038-A1	Extra: remaining front engine mounting – renew
,,	6038-B	ENGINE FRONT MOUNTINGS – BOTH – RENEW (Includes 6038-A and A1)
,,	6038-C	ENGINE FRONT MOUNTINGS – BOTH – CHECK TORQUE OF BOLTS
,,	6051-A	CYLINDER HEAD GASKET – RENEW
,,	6051-A1	Extra: cylinder head – decarbonise
,,	6051-A2	Extra: pistons – decarbonise
,,	6051-A3	Extra: one valve – remove and install
,,	6051-A4	Extra: each additional valve – remove and install
,,	6051-A5	Extra: each valve seat – re-cut
,,	6051-A6	Extra: each valve – grind-in
,,	6051-A8	Extra: each valve guide – renew
,,	6051-A9	Extra: cylinder head – renew
,,	6051-B	CYLINDER HEAD AND PISTONS – DECARBONISE (Includes 6051-A, A1 and A2)
,,	6051-C	VALVE – ONE – REMOVE AND INSTALL (Includes 6051-A, A1 and A3)
,,	6051-D	VALVE – ONE – REMOVE AND INSTALL (Includes 6051-A, A1, A3, A5 and A6)

CORTINA - LOTUS

OPERATION 6051-E	**VALVES – ALL – REMOVE AND INSTALL** (Includes 6051-A, A1, A3, A4, A5 and A6)
" 6051-F	**DECARBONISE CYLINDER HEAD AND PISTONS AND RESEAT AND GRIND-IN ALL VALVES** (Includes 6051-A, A1, A2, A3, A4, A5 and A6)
" 6051-I	**VALVE GUIDE – ONE – RENEW** (Includes 6051-A, A1, A3, A5 and A8)
" 6051-J	**VALVE GUIDES – ALL – RENEW** (Includes 6051-A, A1, A3, A4, A5 and A8)
" 6051-K	**CYLINDER HEAD – REMOVE AND INSTALL** (Includes 6051-A, A2, A3, A4 and A9)
" 6051-M	CYLINDER HEAD BOLTS – TORQUE
" 6068-A	ENGINE REAR MOUNTING – RENEW
" 6250-A	CAMSHAFT – ONE – AND/OR BEARINGS – AND/OR TAPPETS – REMOVE AND INSTALL
" 6250-A1	Extra: remaining camshaft and/or bearings and/or tappets – remove and install
" 6261-A	AUXILIARY SHAFT BEARINGS – RENEW
" 6270-A	TIMING CHAIN TENSION – ADJUST
" 6271-A	CAMSHAFT COVER GASKET – RENEW
" 6271-A1	Extra: valve clearances – all – adjust
" 6271-A2	Extra: timing chain tension – adjust
" 6271-A3	Extra: camshaft sprockets – remove and install
" 6271-B	**VALVE CLEARANCES – ADJUST** (Includes 6271-A and A1)
" 6279-A	FRONT COVER OIL SEAL, GASKETS AND/OR TIMING CHAIN – RENEW
" 6279-A1	Extra: auxiliary shaft – remove and install
" 6279-A2	Extra: crankshaft sprocket – remove and install
" 6280-A	**AUXILIARY SHAFT – REMOVE AND INSTALL** (Includes 6279-A and A1)
" 6303-A	CRANKSHAFT – REMOVE AND INSTALL
" 6312-A	CRANKSHAFT PULLEY – REMOVE AND INSTALL
" 6335-A	CRANKSHAFT REAR OIL SEAL CARRIER – REMOVE AND INSTALL
" 6335-A1	Extra: rear oil seal and/or carrier – renew
" 6335-B	**CRANKSHAFT REAR OIL SEAL AND/OR CARRIER – RENEW** (Includes 6335-A and A1)

OPERATION 6375-A	FLYWHEEL ASSEMBLY – REMOVE AND INSTALL
" 6375-A1	Extra: flywheel ring gear – renew
" 6375-B	**FLYWHEEL RING GEAR – RENEW** (Includes 6375-A and A1)
" 6450-A	CAMSHAFT COVER AND/OR GASKET – REMOVE AND INSTALL
" 6450-A1	Extra: all valve clearances – adjust
" 6450-B	**VALVE CLEARANCES – ADJUST** (Includes 6450-A and A1)
" 6600-A	OIL PUMP ASSEMBLY – REMOVE AND INSTALL
" 6600-A1	Extra: oil pump – overhaul
" 6600-B	**OIL PUMP – OVERHAUL** (Includes 6600-A and A1)
" 6675-A	SUMP AND/OR GASKETS – REMOVE AND INSTALL
" 6675-A3	Extra: all main bearing clearances – check
" 6675-A4	Extra: all main bearing liners and/or thrust washers – remove and install
" 6675-A5	Extra: all main bearing liners and thrust washers – check clearances and renew
" 6675-A6	Extra: all connecting rod liners – remove and install
" 6675-A7	Extra: one piston, connecting rod and ring assembly – remove and install
" 6675-A8	Extra: each additional piston, connecting rod and rings assembly – remove and install
" 6675-A9	Extra: each set of piston rings – renew
" 6675-A10	Extra: each connecting rod and/or piston pin – renew
" 6675-A11	Extra: each piston, pin and rings assembly – renew
" 6675-C	**MAIN BEARING CLEARANCES – ALL – CHECK** (Includes 6675-A and A3)
" 6675-D	**MAIN BEARING LINERS – ALL – RENEW** (Includes 6675-A and A4)
" 6675-E	**MAIN BEARING LINERS – ALL – CHECK CLEARANCES AND RENEW** (Includes 6675-A and A5)
" 6675-F	**CONNECTING ROD LINERS – ALL – RENEW** (Includes 6675-A and A6)
" 6675-G	**PISTON RINGS – ONE PISTON – REMOVE AND INSTALL** (Includes 6675-A and A9)

CORTINA - LOTUS

OPERATION	6675-H	PISTON RINGS – ALL – REMOVE AND INSTALL
		(Includes 6675-A, A7, A8 and A9)
,,	6675-I	PISTON – ONE – REMOVE AND INSTALL
		(Includes 6675-A, A7 and A10)
,,	6675-J	PISTONS – ALL – REMOVE AND INSTALL
		(Includes 6675-A, A7, A8 and A10)
,,	6675-K	CONNECTING ROD AND/OR PISTON PIN – ONE – REMOVE AND INSTALL
		(Includes 6675-A, A7 and A11)
,,	6675-L	CONNECTING RODS AND PISTON PINS – ALL – REMOVE AND INSTALL
		(Includes 6675-A, A7, A8 and A11)
,,	6675-M	SUMP BOLTS – TORQUE
,,	6731-A	OIL FILTER ELEMENT – RENEW
,,	6731-B	OIL FILTER ELEMENT AND ENGINE OIL – RENEW
,,	6900-B	TEMPERATURE GAUGE SENDER UNIT – REMOVE AND INSTALL

GENERAL DESCRIPTION

The engine is a four cylinder, twin overhead camshaft unit with a bore of 3.2506 in. (82.565 mm.) and a stroke of 2.867 in. (72.746 mm.). The capacity is 95.2 cu. in. (1,560 c.c.) and the compression ratio is 9.5 : 1.

The cylinder bores are machined directly in the cast iron cylinder block, which is cast integral with the upper half of the crankcase, and are provided with full length water jacketing.

The cast iron crankshaft runs in five large diameter main bearings fitted with steel-backed copper/lead bearing liners. End-float and thrust are controlled by half-thrust washers located in the cylinder block on either side of the centre main bearing.

Seals pressed in the front cover and the rear oil seal carrier prevent oil leaks from the front and rear of the crankshaft. The front seal runs on the pulley hub whilst the rear seal runs on the crankshaft flange itself. A sintered bronze spigot or needle roller bearing is pressed into the end of the crankshaft to support the gearbox first motion shaft.

The connecting rods are 'H' section forgings having separate big end caps retained by two bolts and located by spring dowel pins. Big end bearing liners are, again, steel-backed copper/lead. The small ends have steel-backed bronze bushes.

Solid skirt aluminium alloy pistons with two compression and one oil control ring situated above the piston pin bore are used. The piston pins are fully floating and are retained in position by circlips installed in grooves at each end of the piston pin bore.

The cylinder head is an aluminium casting with fully machined hemispherical combustion chambers and separate ports for each valve. The valves, which have replaceable guides and valve seat inserts, are inclined to each other and the inlets are larger than the exhausts. The valves are operated by two overhead camshafts, one for inlet valves and one for exhaust, the valves being opened directly by the cams acting on piston-type tappets.

The camshafts are driven at half engine speed by a single row timing chain from a sprocket on the crankshaft via a sprocket on an auxiliary shaft and an idler sprocket on an adjustable tensioner. The camshafts each run in five bearings which have steel-backed white metal liners. A shoulder at the front of the camshaft locates it axially in the cylinder head and controls end-float.

The auxiliary shaft is a modification of the camshaft normally used in the push rod overhead valve unit and is retained to drive the oil pump, distributor and fuel pump. The auxiliary shaft runs in three steel-backed white metal bushes and is located by a sintered metal thrust plate bolted to the cylinder block front face. A skew gear, integral with the auxiliary shaft drives the distributor and oil pump, which are both mounted on the right-hand side of the engine. An eccentric, on the auxiliary shaft, operates the fuel lift pump also situated on the right-hand side of the engine towards the rear. The front journal of the auxiliary shaft has four slots machined in its periphery to regulate the supply of oil to the camshafts and tappet gear.

A cast iron flywheel is mounted on the crankshaft flange and ensures a smooth-running engine. The drive for the starter motor is provided by a steel ring gear shrunk onto the flywheel periphery.

The sump is a steel pressing and has a front well for the lubricating oil. The engine lubrication system is the force feed type incorporating a full flow oil filter. The oil pump, which is mounted externally on the engine, is of the eccentric bi-rotor type incorporating a non-adjustable plunger type relief valve.

An oil filler cap is located in the camshaft cover. Crankcase ventilation is by a closed system, crankcase fumes being discharged directly into the carburettor air intake cover.

A three point mounting for the engine and gearbox assembly is provided on sandwich type rubber insulators.

CYLINDER BLOCK

The cylinder block is cast iron and is cast integral with the upper half of the crankcase and is basically a 1,500 block bored out. This cylinder block can be identified by the number, 681F-6015-GA cast on the left-hand side of the block in place of the cast number prefix.

Internally the crankcase incorporates five main bearings with removable caps retained by bolts fitted without lockwashers. The intermediate and rear caps are identical, but their positions must not be interchanged. When dismantling these caps ensure that their positions are marked, this is normally done, in production, by a number 2 stamped on the front intermediate cap and 4 on the rear intermediate cap. The rear cap is not marked. All caps must be fitted with the cast arrows pointing forwards.

The cylinder bores are machined directly into the cylinder block and, in production, are graded for size, the grade number, of each bore, being stamped on the push rod side of the cylinder block adjacent to the top face.

CYLINDER HEAD

The cylinder head is an aluminium casting with fully machined hemispherical combustion chambers giving a compression ratio of 9.5 : 1, and having separate ports for each valve.

The cylinder head is secured to the cylinder block by ten $\frac{7}{16}$ in. (11.11 mm.) diameter bolts, 5.7 in. (144.78 mm.) long, with plain washers. The washers for the front left-hand bolts and the rear right-hand bolts have flats to provide clearance for the exhaust camshaft collar and the inlet camshaft rear bearing cap respectively. Three additional bolts, one each side of the front cover and one at the front of the cylinder head, in the centre, secure the front of the cylinder head to the timing case.

The tightness of the cylinder head bolts should be checked, with the engine cold, at the first 600 miles (1,000 km.) service. The tightness should also be checked 600 miles (1,000 km.) after replacing the cylinder head should it have been removed for any reason.

The cylinder head gasket is made of asbestos covered with copper on the top and tinplate underneath.

The valves are in two banks, the inlets on the right and the exhausts on the left, and operate in replaceable guides pressed into the cylinder head. These guides are located by circlips which are retained by the valve spring seats. The inlet and exhaust valve guides are not identical, the inlet guide being 1.520 in. (38.61 mm.) long and the exhaust 1.480 in. (37.59 mm.). The two guides may be readily identified by the fact that the inlet has a taper at the port end, which is 0.4 in. (10.16 mm.) long, whereas the exhaust guide is only slightly tapered, 0.2 in. (5.08 mm.) long. The valve guides may be removed and replaced, using a suitable remover and replacer tool. When replacing the valve guides heat the cylinder head to 100 to 150 C. (212 to

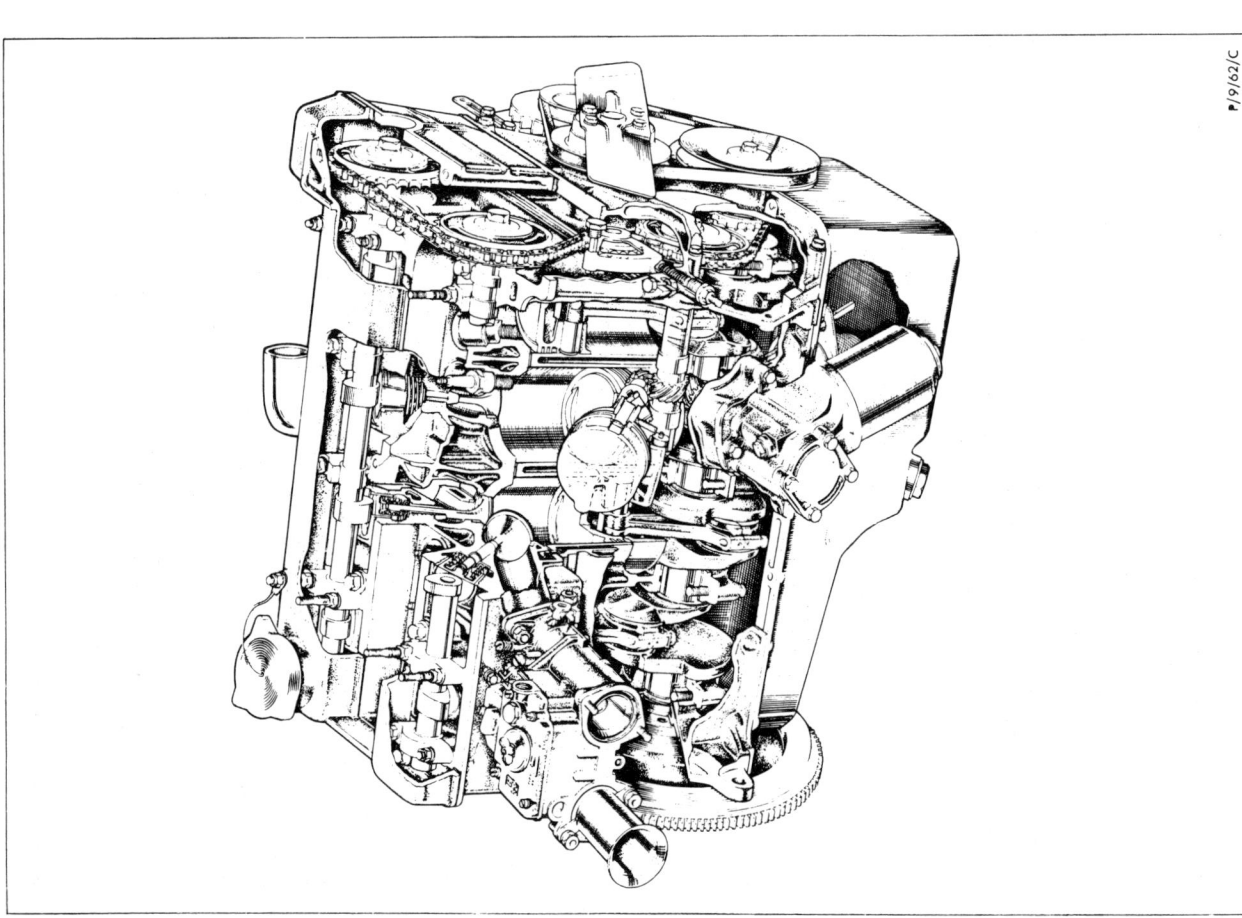

The Engine Assembly

CORTINA - LOTUS

302°F.). Valve guides are also available in 0.001 in. (0.03 mm.), 0.005 in. (0.13 mm.) and 0.006 in. (0.15 mm.) oversizes on the outside diameter, and it is advisable to check the size against the originals when replacing guides.

After fitting, ream the valve guide bore 0.3113 to 0.3123 in. (7.907 to 7.932 mm.) using a suitable reamer. The valve seats must then be recut, with the appropriate cutters, and the valves lapped in to ensure that the seal is concentric with the valve stem bore.

Recut the seats with the appropriate cutters fitted to pilot Tool No. 316-10 in handle Tool No. 316X. Where necessary the seats may be narrowed with top face and port cutters. If there is a hard glazed carbon deposit on the seat this may be removed with a glaze breaker. The valve seats should also be recut when they show signs of pitting or burning.

Seat 45	Top Face-15°	Cutter Tool Number Port-75°	Glaze Breaker-45°
317-25	317T-25	317P-22	317G-25

The valve seats have inserts which can be replaced when they become pocketed and can be obtained in 0.005 in. (0.127 mm.), 0.010 in. (0.254 mm.), 0.015 in. (0.381 mm.) oversizes.

Remove the inserts by machining two grooves 180° apart and using a small chisel remove the remaining metal in the grooves, when the insert can be prised out of its location. Care must be taken during this operation to avoid damage to the sides and bottom of the recess. Machine the recess to the required dimension appropriate to the selected replacement insert (see table below). When fitting new inserts the cylinder head should be heated to a temperature of 200°C (392°F) maximum and immersing the valve seat inserts to a temperature not lower than −80°C (−112°F). Press the insert into place, using a suitable replacer tool and allow the cylinder head to cool slowly and evenly in air.

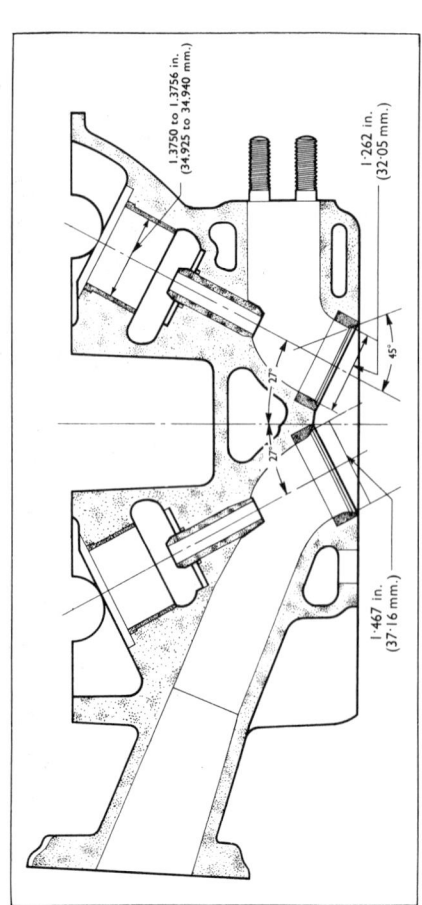

Tappet Sleeve Bore and Valve Seat Insert Dimensions

VALVE SEAT INSERTS

INLET

Size	Dimensions	Recess diameter
Standard	1.6245 in.–1.6235 in.	1.621 in.–1.620 in.
+0.005 in.	1.6295 in.–1.6285 in.	1.626 in.–1.625 in.
+0.010 in.	1.6345 in.–1.6335 in.	1.631 in.–1.630 in.
+0.015 in.	1.6395 in.–1.6385 in.	1.636 in.–1.635 in.

EXHAUST

Size	Dimensions	Recess diameter
Standard	1.4995 in.–1.4985 in.	1.496 in.–1.495 in.
+0.005 in.	1.5045 in.–1.5035 in.	1.501 in.–1.500 in.
+0.010 in.	1.5095 in.–1.5085 in.	1.506 in.–1.505 in.
+0.015 in.	1.5145 in.–1.5135 in.	1.511 in.–1.510 in.

The tappet bores are sleeved, these being available in 0.001 in. (0.025 mm.) and 0.015 in. (0.38 mm.) oversize. Should it be necessary to fit new sleeves, the cylinder head must be machined to obtain an interference fit of 0.0035 in. (0.089 mm.) to 0.0045 in. (0.114 mm.) between the head and the new tappet sleeve. Heat the cylinder head to 150°C (302°F) before fitting the sleeves. When cool machine the sleeve bore to the dimensions shown. After machining, recut the scallops to give clearance to the cams.

EXHAUST MANIFOLDS

The free flow multi-branch exhaust manifold is fabricated from welded steel tubes and has a separate pipe for each cylinder. These pipes are paired together, Nos. 1 and 4 forming one pair and Nos. 2 and 3, which pass inside of Nos. 1 and 4, the other, the pipes of each pair being joined to form a common outlet. The outlet pipes from each pair then enter the exhaust system. The exhaust manifold assembly is secured to the cylinder head by flanges welded to the pipe ends and retained with brass nuts.

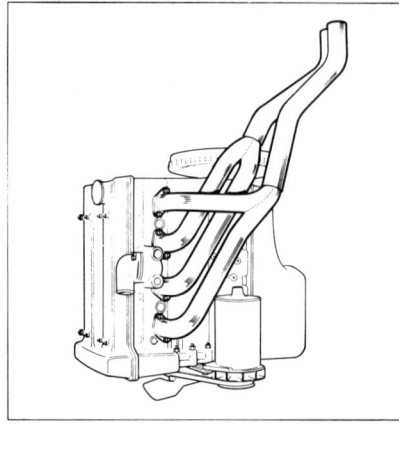

Exhaust Manifolds

TIMING CASE

The timing case is made in two pieces, a back plate on the cylinder block and a front cover which also incorporates the water pump and the oil level dipstick tube. Four timing marks are also incorporated on the front cover which, when a timing mark on the rear flange of the crankshaft pulley is aligned with one of them gives 30 B.T.D.C., 20 B.T.D.C., 10 B.T.D.C. and T.D.C. respectively.

The back plate is located on the cylinder block front face by a single bolt situated immediately below the water pump and is retained by the front cover, the front cover bolts passing through the back plate into the cylinder block. Where the back plate is not directly attached to the cylinder block, the back plate and front cover are secured together with nuts and bolts having fine threads. All bolts screwing into the cylinder block have coarse threads. As the bolts have different lengths and threads it is important that they are correctly located, as shown.

(1) Bolt ¼ in. 20 U.N.C. × 2¼ in. long
(2) Bolt ¼ in. 28 U.N.F. × 2¼ in. long and nut
(3) Bolt 5/16 in. 18 U.N.C. × 2¼ in. long
(4) Bolt 5/16 in. 18 U.N.C. × 1 in. long
(5) Bolt ¼ in. 20 U.N.C. × 3 in. long
(6) Bolt 5/16 in. 18 U.N.C. × 1¼ in. long
(7) Bolt 5/16 in. 24 U.N.F. × 1¾ in. long and nut

The back plate gasket is made of brown waxed paper and must be fitted without sealer as this may clog the timing chain oil feed. Before assembling the back plate to the cylinder block, check that it is flat. Ensure that the back plate lies flat on the cylinder block front face and is not cocked up by the main oil gallery plug.

The front cover is fitted without a gasket, an oil-tight joint being made by using ESEE-M4G-1008A jointing compound. A gasket on top of the front cover and back plate seals the joint between the cylinder head and the timing case. Three bolts are on each side of the front cover and one at the front of the cylinder head in the centre under the camshaft cover, secure the timing case to the cylinder head.

To prevent oil leaks from around the crankshaft pulley boss an oil seal is pressed in the front cover.

The oil seal can be removed, after first removing the front cover, by supporting the cover around the seal and driving the seal out from the rear with remover Tool No. P.6161 fitted to a 550 handle. Invert the tool and drive a new seal into the housing, again from the rear, and supporting the cover around the seal. The use of Tool No. P.6161 ensures that the seal protrudes 3/64 in. (1.19 mm.) inside the cover.

When fitting the cover it is important that the oil seal is aligned concentrically with the crankshaft and pulley boss. To facilitate this a centraliser Tool No. P.6150 is inserted into the seal while fitting the cover.

The oil level dipstick tube is pressed into the front cover and its upper end must be 4.90 in. (124.5 mm.) vertically above the front cover bottom face if the correct oil level is to be attained.

REAR OIL SEAL CARRIER

The crankshaft rear oil seal is pressed into the aluminium carrier, bolted to the cylinder block rear face, and runs on the periphery of the flywheel mounting flange. After removing the carrier, the oil seal can be easily replaced. Support the carrier, close to the seal, and drive the seal out, using a remover/replacer Tool No. P.6165 fitted to a 550 handle. Reverse the carrier, the remover/replacer tool and fit a new seal.

When fitting the carrier the seal must be aligned concentrically with the crankshaft if oil leaks are to be avoided. Locate a centraliser Tool No. CP.6173 in the seal and over the crankshaft while tightening the seal carrier retaining bolts.

SUMP

The pressed steel sump has a front well, incorporating a longitudinal baffle, for the lubricating oil and is bolted to the base of the cylinder block. A drain plug is located in the right-hand side. The gaskets are made of cork with aluminium foil lamination.

When fitting the sump apply sealer ESEE-M4G-1008A to the front cover and rear oil seal carrier to cylinder block joints, and also to the ends of the grooves in the front cover and rear oil seal carrier. Then fit new sump gasket, followed by the cork packing strips. Refit the sump and tighten the bolts following the procedure described on page 33.

CAMSHAFT COVER

The camshafts and valves are enclosed by an aluminium cover, which also covers the camshaft sprockets and chain. This cover is retained by eight self-locking nuts and flat washers which screw onto four of the camshaft bearing cap studs. The camshaft cover gasket is made of cork. Jointing compound is not normally used, although it can be applied to the cover to hold the gasket in place.

The oil filler cap is situated on the camshaft cover at the rear of the left-hand side.

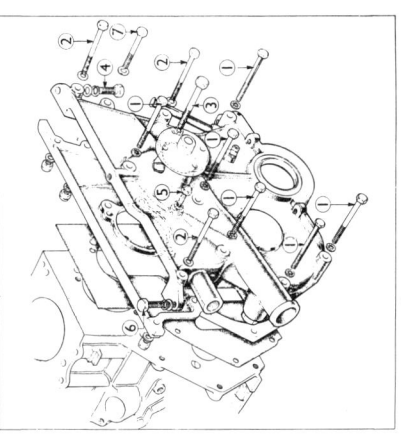

Crankshaft Rear Oil Seal Alignment

Front Cover Bolt Locations

CORTINA - LOTUS

VALVES AND SPRINGS

The valves, which have 45° seats, are inclined at 27° to the vertical, the inlet valve head being larger than the exhaust. Their respective diameters are $1\frac{17}{32}$ in. (38.89 mm.) and $1\frac{5}{16}$ in. (33.34 mm.).

The valves are fitted with double valve springs, the inner spring has right-hand coils and the outer left-hand. The springs sit on a pressed steel seat located around the valve guide and retained by the valve guide circlips. The springs are attached to the valve stems by spring retainers with tapered split collets. The valve springs may be fitted either way round.

CAMSHAFTS AND TAPPETS

The valves are operated by two identical overhead camshafts, one for inlet valves and one for exhaust valves, the valves being opened directly by the cams acting on piston type tappets. The tappets are located in the cylinder head immediately above the valves. All the tappet bores in the cylinder head have sleeves. The valve clearances are adjusted by shims located underneath the tappets and resting in the valve spring retainer on top of the valve stem. The shims are available in a number of sizes from 0.065 in. (1.651 mm.) to 0.120 in. (3.048 mm.) in increments of 0.001 in. (0.025 mm.). The shim's thickness is etched around the periphery although originally it was etched on one side. When fitting this type the etched side must be adjacent to the valve stem. Any roughness caused by the etching should be removed before fitting the shim by rubbing on fine emery cloth.

The two camshafts, which are identical, give a lift of 0.35 in. (8.89 mm.) and the timing shown. The camshafts each run in five bearings with steel-backed white metal liners, four of them $\frac{3}{4}$ in. (19.05 mm.) wide and the rear one $\frac{1}{2}$ in. (12.7 mm.).

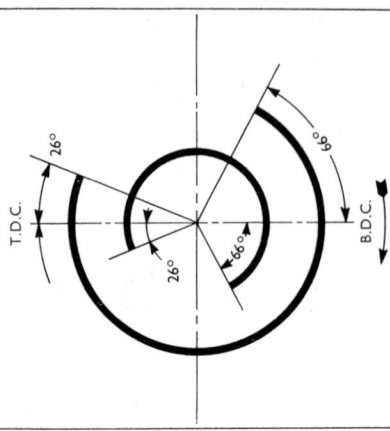

Nominal Valve Timing Diagram

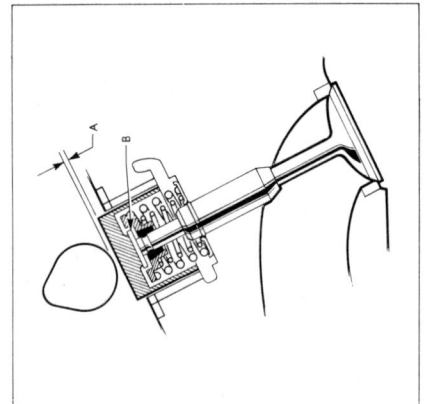

Valve and Tappet
('A' = Clearance 'B' = Shims)

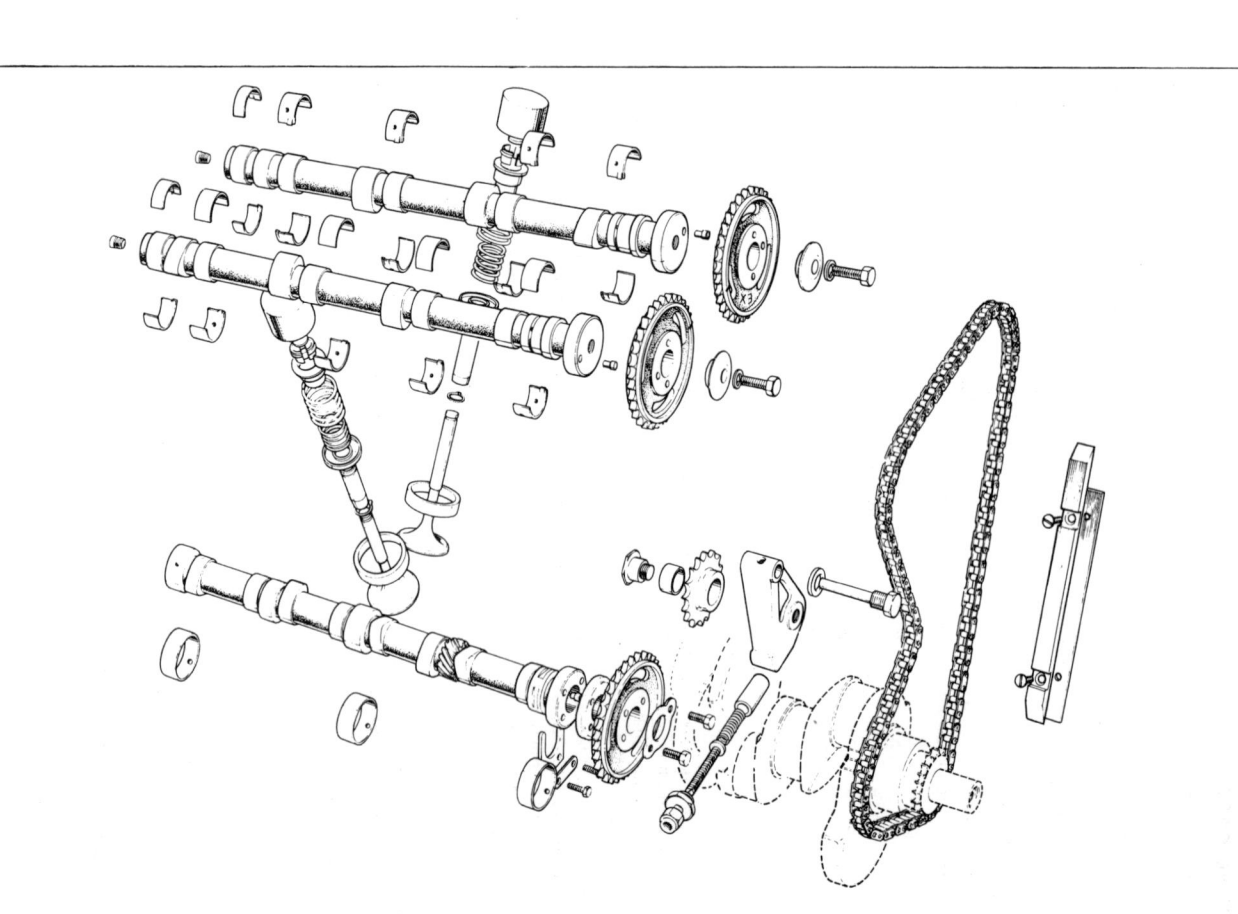

Valve Operating Mechanism

The camshaft bearing caps are located by studs screwed into the cylinder head and retained by self-locking nuts and flat washers. These caps must always be kept in their respective positions and are numbered to facilitate this. The numbers are also stamped on the cylinder head adjacent to the bearing caps.

Adjusting Valve Clearances

The valve clearances should be, inlet 0.005 to 0.007 in. (0.13 to 0.18 mm.) and exhaust 0.006 to 0.008 in. (0.15 to 0.20 mm.) Later exhaust valves with Part No. B26E020 rolled on the stem have a clearance of 0.009 in. (0.23 mm.) to 0.011 in. (0.28 mm.).

In order to check the clearance of a valve, turn the camshaft until the toe of the cam is at 180 to the tappet being checked. Then select a feeler blade or blades that can just be inserted between the tappet and the heel of the cam to measure the clearance. Note the clearance and repeat the procedure for the other valves.

Should the clearances require adjusting, remove the appropriate camshaft by unscrewing the bearing cap nuts evenly and lift the camshaft out.

NOTE – If this operation is being done on the engine it will be necessary to remove the camshaft sprocket first. To facilitate replacement it is advisable to set the valve timing marks and to slacken the timing chain tensioner before removing the sprocket.

Remove the tappets by lifting them out with a valve grinding sucker and remove the shims, keeping them in their correct order. The correct clearance may be obtained by fitting a different size shim between the valve stem and tappet. A thinner shim will be required to increase the valve clearance and a thicker one to reduce the clearance. The shim's thickness is etched around the periphery or on the side, but if this has worn off or if the shim appears to be worn, measure the thickness accurately with a micrometer. Select a shim to give the correct size from the following formula:

Shim thickness required = A—C.C.

Where A. is the actual valve clearance

C.C. is the correct valve clearance

Fit the shims into the recess in the valve spring retainer, where applicable, with the etched number on the underside. Do not use more than one shim for each valve. Any roughness caused by the etching should be removed before fitting the shim by rubbing on fine emery cloth. Fit the tappets and the camshaft. Ensure that the camshaft bearing caps are fitted in their correct positions and that the nuts are tightened gradually and evenly, starting with the centre pair and working outwards, to a torque of 9 lb. ft. (1.2 kg.m.). Refit the sprockets and timing chain (see "Valve Timing") commencing with the exhaust and then reset the ignition timing.

AUXILIARY SHAFT

The auxiliary shaft runs in three steel-backed white metal bushes located in the right-hand side of the cylinder block. The front journal of the auxiliary shaft has four flats machined in its periphery to regulate the oil supply to the camshafts.

The auxiliary shaft front and rear bushes are both approximately $\frac{3}{4}$ in. (19.05 mm.) wide, the front one having an additional oil hole for the camshaft oil feed, and the centre bush approximately $\frac{5}{8}$ in. (15.88 mm.) wide.

The bushes available in service are pre-sized and require no machining after fitting. When one bush requires replacement it is advisable to replace all three bushes as auxiliary shaft alignment may be affected if only one bush is changed.

Remove the auxiliary shaft bearing bushes, using camshaft bearing remover Tool No. P.6031 with adaptor set P.6031-3. Locate the remover and guide detail "-3a" adjacent to the collar and with the spigot in the bearing. If the centre bush is being removed use the centraliser detail "-3d". Remove the bush by screwing down the wing nut.

Replace the auxiliary shaft bushes in a similar manner (using Tool No. P.6031 with adaptor set P.6031-3).

These bushes must be assembled with the split upwards and the oil holes in line with the corresponding holes in the cylinder block. A line scribed on the remover and guide detail "-3a" can be used as a guide to facilitate this.

The auxiliary shaft is retained by a sintered iron thrust plate bolted to the cylinder block front face and located in a groove behind the auxiliary shaft flange.

TIMING CHAIN AND SPROCKETS

The camshafts and auxiliary shaft are driven at half engine speed by a single row timing chain from a sprocket on the crankshaft. Chain tension is controlled by an adjustable tensioner located between the auxiliary shaft and the inlet camshaft.

The chain tensioner consists of an idler sprocket mounted on a lever pivoted in the cylinder head and a spring-loaded adjustable plunger, acting on the lever against chain tension, located in the right-hand side of the front cover.

When correctly adjusted there should be $\frac{1}{2}$ in. (12.7 mm.) free movement in the chain between the two camshaft sprockets. This can be checked with the camshaft cover removed. To adjust the chain tension slacken the adjuster locknut and screw the adjuster in or out until the correct free movement has been obtained. Then tighten the locknut. Alternatively, the chain tension may be adjusted dynamically. To adjust, screw the adjuster in or out (as for static adjustment) to achieve the minimum noise level. A tight chain will whine and a slack chain will rattle.

Excessive chain movement may also result in the chain knocking against the damper pad, located down the left-hand side of the front cover. When replacing the retaining screws, ensure that they are thoroughly clean and apply a thin line of EM-4G-52 plastic sealer to the screw threads to prevent any possibility of oil leaks or loosening in service.

The inlet camshaft and the auxiliary shaft sprocket are not interchangeable. The exhaust camshaft sprocket, although similar, has the timing mark in a different position and must only be used on the exhaust camshaft if the correct timing is to be readily obtained. To identify it from the other sprockets it is marked "EX".

CORTINA - LOTUS

The camshaft sprockets, which are located by dowels, are retained by centre bolts and large flanged washers. The bolts are locked by spring washers and should be tightened to 25 to 30 lb.ft. (3.46 to 4.15 kg.m.). The auxiliary shaft sprocket is also located by a dowel but is retained by two bolts locked with a locking plate. An adaptor positioned between the sprocket and the auxiliary shaft gives correct timing chain alignment and is retained by the dowel.

Valve Timing

Maximum performance can only be obtained if the valve timing is correctly set. To facilitate this, timing marks are incorporated on the camshaft sprockets and the crankshaft pulley.

The timing is correct when the timing mark on the pulley is in line with the T.D.C. timing mark on the front cover and the timing marks on the inlet and exhaust camshaft sprockets are inwards and level with the camshaft cover mounting face.

CRANKSHAFT AND BEARINGS

The cast iron dynamically balanced crankshaft runs in five main bearings having steel-backed copper/lead liners with a lead/indium overlay.

The crankshaft main bearing journals may be ground 0.010 in. (0.25 mm.) or 0.020 in. (0.51 mm.) undersize. When grinding crankshafts undersize it is important to maintain the correct fillet radii at all times. The centre main bearing journal has a double radius of 0.070 in. (1.78 mm.) and 0.080 in. (2.03 mm.) and the rear main bearing has a double fillet radius, the inner radius of which must be maintained at 0.100 to 0.110 in. (2.54 to 2.79 mm.) when regrinding. The remaining main journal fillet radii are 0.110 to 0.096 in. (2.79 to 2.44 mm.). The crankpin journal fillet radii are 0.080 to 0.094 in. (2.03 to 2.39 mm.). Grind the bearing journals with the crankshaft and grinding wheel revolving in an anti-clockwise direction as viewed from the front of the shaft. Ensure that the fillet radii are smooth and free from visual chatter marks. The main bearing journal length between the thrust faces can be increased by up to 0.020 in. (0.51 mm.) providing an equal amount is machined from each face and the corresponding oversize thrust washers fitted. When grinding the rear main journal, ensure that the width of the machining wheel is 1.30 to 1.33 in. (33.02 to 33.78 mm.), that it has a fillet radii of 0.100 to 0.110 in. (2.54 to 2.79 mm.), and that the wheel is positioned 0.074 in. (1.879 mm.) from the rear face. The crankpin length must not exceed 0.010 in. (0.25 mm.) oversize.

Main bearing journal ovality should not exceed 0.0004 in. (0.010 mm.) T.I.R. and taper 0.0005 in. (0.013 mm.). The centre main bearing run-out relative to the front and rear journals should not exceed 0.002 in. (0.05 mm.) T.I.R. The thrust faces should be smooth and square to the bearing journal within 0.0005 in. (0.013 mm.) T.I.R. After grinding, the journals should be polished with a fine lapping paper and the crankshaft revolving clockwise to produce a good surface finish.

Crankshaft end-float is controlled between 0.003 in. (0.076 mm.) 0.008 in. (0.203 mm.) by half thrust washers located in the cylinder block at the centre main bearing. Standard size washers and 0.0025 in. (0.064 mm.), 0.005 in. (0.127 mm.), 0.0075 in. (0.191 mm.) and 0.010 in. (0.254 mm.) oversize washers are available.

A sprocket is located, by a woodruff key, on the front end of the crankshaft, adjacent to the front main bearing journal to drive the auxiliary shaft and camshafts. This sprocket can be removed by using a puller Tool No. P.6116, and replaced with Tool No. P.6032. When replacing the sprocket ensure that the key is pressed squarely into the keyway and that the sprocket is fitted with the boss to the rear.

A cast iron pulley for driving the water pump, fan and generator is also located by the same key as the sprocket and is retained by a centre bolt fitted with a spring and a flat washer.

Oil leaks around the crankshaft pulley boss are prevented by an oil seal pressed into the front cover. To further assist sealing, the crankshaft is fitted with an oil slinger which reduces the quantity of oil around the seal.

The clutch pilot spigot bearing is located in a bore machined in the centre of the crankshaft flange. The spigot bearing can be removed with adaptor Tool No. CP.7600-7 fitted to remover Tool No. CP.7600A or B and replaced with Tool No. P.7137.

CONNECTING RODS

The connecting rods are H section steel forgings with detachable big end caps. The caps are located by two hollow dowel pins pressed into the connecting rod and retained by two bolts fitted without lockwashers. The connecting rod can be identified by the number forged on the web. An oil squirt hole machined in the connecting rod feeds oil from the crankpin to the non-thrust side of the cylinder bore.

The steel-backed big end bearing liners have copper/lead bearing surfaces with a lead/indium overlay. The upper liner which locates in the connecting rod incorporates an oil hole, the lower liner being plain.

The connecting rod small end bearing is a steel-backed lead/bronze bush, which is not available in service, the connecting rod being serviced with the bush already fitted.

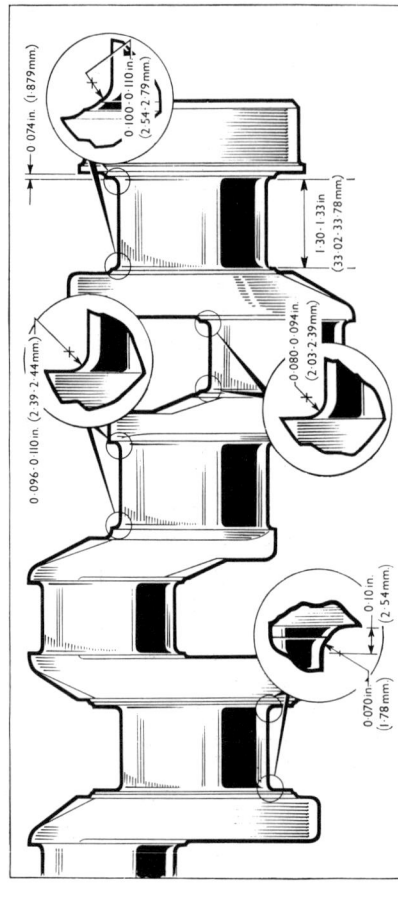

Crankshaft Fillet Radii

When dismantling an engine examine the piston markings to check the connecting rods for straightness. A heavy marking on the piston skirt above the pin on one side together with a correspondingly heavy marking below the pin on the other side indicates a bent connecting rod which should be replaced. The connecting rod small and big ends should be parallel and square to the longitudinal centre line within 0.0005 in. per inch (cm.) length.

When assembling a connecting rod to the piston ensure that it is fitted the correct way round. The marking "125E FRONT" is embossed on the web to facilitate this.

Connecting Rod Numbering

Connecting rods are numbered when installed in the engine during manufacture, to facilitate correct reassembly should they be dismantled.

The number is etched on the side of the big end so that a cap replaced with the numbers together must be in its original position. Never reassemble a bearing cap to another connecting rod.

It is advisable when removing connecting rods from an engine to check that the connecting rods have been numbered correctly. Where the connecting rods are unmarked they should be suitably stamped unless the connecting rods are being scrapped.

PISTONS, PISTON PINS AND RINGS

The aluminium alloy pistons are of the solid skirt type and have recesses machined in the crown to give clearance for the inclined valves. Each piston has three rings, two compression and one oil control ring, situated above the piston pin bore. The top compression rings are cargraph plated (red) to provide initial lubrication and the two lower rings are copper plated for identification purposes.

The lower compression ring is stepped externally on the bottom face and the upper ring is chrome plated and tapered. Both rings are marked "TOP" and must be fitted this way round. The "micro-land" oil control rings may be fitted either way round. This type of oil control ring can be identified by the narrow ring lands.

The piston pin is offset in the piston 0.04 in. (1.016 mm.) towards the thrust side of the engine, to minimise piston slap and uneven loading of the skirt thrust face during the power stroke. Therefore, it is important that the piston is fitted the correct way round and to facilitate this the piston crown is marked "FRONT" and must face forwards when the piston is fitted to the engine.

The tubular steel piston pins are fully floating and are retained in position by circlips installed in grooves at each end of the piston pin bore.

On assembling the piston, pin and connecting rods the assemblies are weighed. The maximum variation of weight between the piston and connecting rod assemblies fitted in an engine is 6 grams. When changing pistons or connecting rods in service, it is good practice to check the weights of the piston, pin and connecting rod assemblies and, if necessary, select parts to ensure that the weight variation between the respective assemblies does not exceed 6 grams. Oversize pistons and rings are available in 0.015 in. (0.38 mm.) sizes.

Piston Selection

During engine manufacture the cylinder bores and pistons are graded. The piston grade number is stamped on the piston crown and, in production, each cylinder bore grade number is stamped on the push rod side of the cylinder block, adjacent to the top face, during engine assembly. These grade numbers ensure that when the piston is fitted there is a clearance of 0.0030 to 0.0036 in. (0.076 to 0.091 mm.) between the piston and the cylinder bore.

When selecting standard size pistons, measure each cylinder bore at a point $1\tfrac{9}{16}$ in. (39.69 mm.) from the cylinder block top face, across the axis of the crankshaft, and refer to the table in the Specification, Servicing and Repair Data section, to determine the grade of piston required. Select a piston with a grade number corresponding to that found for the cylinder bore.

FLYWHEEL AND RING GEAR

The cast iron flywheel is located concentrically on the crankshaft flange and retained by six bolts fitted without lockwashers.

The flywheel ring gear is shrunk onto the flywheel and locates in a retention groove. The ring gear can be removed by cutting between two adjacent teeth with a hack saw and splitting the gear with a chisel. In no circumstances should pressure be applied in an attempt to dismantle the ring gear for repositioning on the flywheel.

When replacing the ring gear it must be heated evenly to a temperature not exceeding 600°F (316°C) or the ring gear wear-resistant properties will be destroyed. If the ring gear is heated by a naked flame place the ring gear on a bed of fire bricks and then play the flame in a circular motion onto the bricks about $\tfrac{1}{4}$ to $\tfrac{1}{2}$ in. (6.35 to 12.7 mm.) from the inside of the gear until it reaches the required temperature. The correct temperature can be detected by using a special type of temperature sensitive crayon, or alternatively by polishing a section of the ring gear and the ring gear for repositioning on the flywheel.

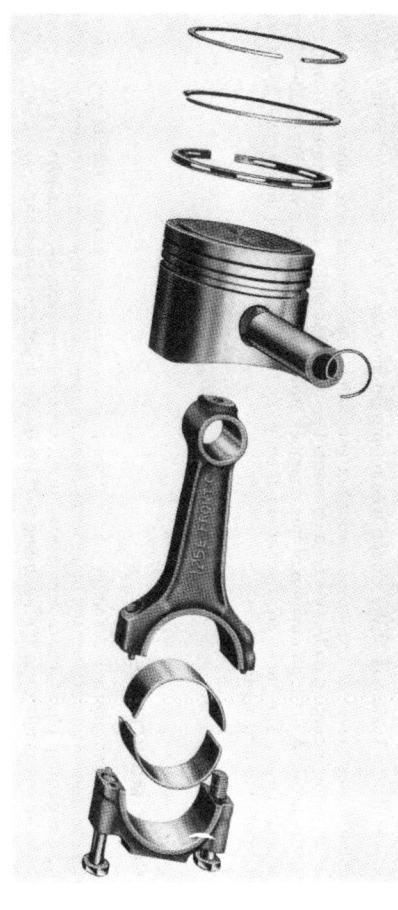

Connecting Rod and Piston Assembly

heating until it turns blue. Fit the ring gear with the chamfers on the leading faces of the gear teeth relative to the direction of rotation. Allow the ring gear to cool naturally in air, DO NOT QUENCH.

The flywheel and ring gear assembly are dynamically balanced to close limits. The clutch is located on the flywheel by three dowels and is retained by six bolts with spring washers.

THE OIL PUMP

The oil pump and filter assembly is bolted to the right-hand side of the cylinder block and can be removed with the engine in place. The oil pump, which is driven by a skew gear on the engine auxiliary shaft, is of the eccentric bi-rotor type and has the full flow element type filter bolted to a mounting flange integral with the oil pump body.

Oil enters the pump through a tube pressed into the cylinder block sump face. A spring-loaded filter gauze located on the end of this tube provides primary filtration. The gauze can be removed by bending back the retaining lug and sliding the gauze out sideways. A pressure relief valve oil return pipe is also pressed into the cylinder block sump face parallel to the inlet tube.

THE OIL FILTER

The full flow type oil filter is bolted to a mounting flange integral with the oil pump body.

To remove the filter unscrew the securing bolt and withdraw the filter body and element. Remove the sealing ring from the groove in the filter body mounting flange, then locate the new ring (supplied with the replacement element) in the groove at four diametrically opposite points. Do not fit the ring at one point and then work it round the groove as the rubber may stretch, thus leaving a surplus which may cause an oil leak. Thoroughly clean the filter body and insert the new element. Locate a new washer (again supplied with the replacement element) on the securing bolt and refit the filter assembly to the oil pump body.

VENTILATION SYSTEM

The ventilation system is of the closed type relying on blow-by for displacement of the gases.

Engine fumes leave the crankcase through a ventilation tube located vertically on top of the fuel lift pump mounting pad. This tube connects the crankcase with a chamber integral with the cylinder head. A drilling in the top of this chamber provides a connection with the inside of the camshaft cover. The fumes are discharged directly into the carburettor air intake cover by a tube incorporating a flame trap.

ENGINE MOUNTINGS

The engine and gearbox assembly has a three-point mounting on bonded rubber insulators. The two front insulators consist of a rubber block bonded between the engine mounting bracket and a mounting plate, incorporating a weld nut on the inner face. A single bolt is thus used to secure the insulator to the mounting bracket on the front crossmember.

The single rear insulator is secured to a pad on the gearbox extension housing and consists of rubber blocks bonded between steel channel section plates. A retainer is located beneath and in front of the insulator when fitted to the crossmember, which is located by a spacer at each end and is bolted to the car underbody.

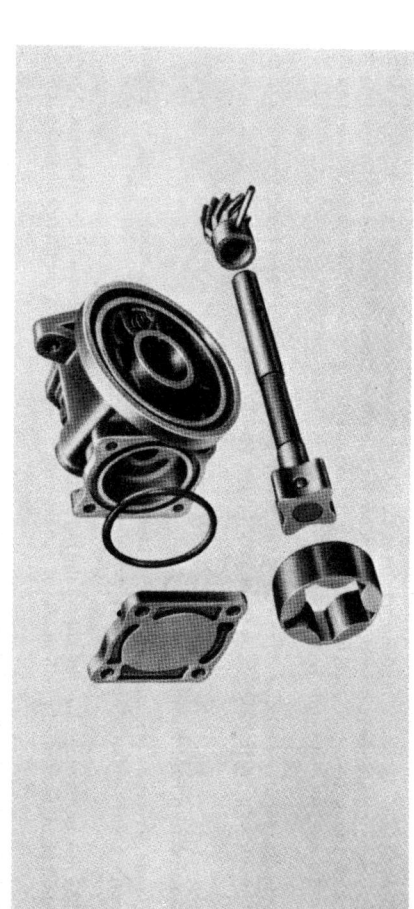

Oil Pump

CORTINA - LOTUS

QUICK REFERENCE DATA

PERIODIC SERVICE ATTENTION

At the first 600 miles (1,000 km.)

- (a) Tighten the cylinder head bolts (when cold).
- (b) Check the valve clearances and adjust if necessary.
- (c) Check the timing chain tension and adjust if necessary.
- (d) Tighten manifold and sump bolts to correct torque.
- (e) Adjust ignition timing.
- (f) Check and adjust if necessary fan belt tension.

At first 3,000 miles (5,000 km.) or three months (whichever occurs first)

- (a) Change the engine oil and renew the filter.
- (b) Clean sparking plugs and set gaps.
- (c) Check and adjust valve clearances.
- (d) Check timing chain tension.
- (e) Adjust carburettor idling and ignition timing.
- (f) Check and adjust, if necessary, fan belt tension.
- (g) Tighten generator mounting bolts to correct torque.
- (h) Check engine and cooling system for oil and water leaks as applicable.

Every 3,000 miles (5,000 km.) or three months (whichever occurs first)

- (a) Check and top-up engine oil level.
- (b) Carry out operations (b) to (h) detailed above.

Every 6,000 miles (10,000 km.) or six months (whichever occurs first)

- (a) Change the engine oil and renew the filter.
- (b) Grease distributor cam.
- (c) Lubricate distributor, examine points and adjust.
- (d) Tighten manifold bolts to correct torque.
- (e) Carry out operations (b) to (h) detailed above.

Every 12,000 miles (20,000 km.) or twelve months (whichever occurs first)

- (a) Clean crankcase emission flame trap.

DATA

Compression pressures (hot)	180 to 200 lb./sq. in. (12.66 to 14.06 kg./sq. cm.) at cranking speed
Idling speed	800 rev./min.
Firing order	1, 3, 4, 2
Location of No. 1 cylinder	Next to radiator

Auxiliary shaft end-float	0.002 to 0.007 in. (0.05 to 0.178 mm.)
Connecting rod end-float on crankpin	0.004 to 0.010 in. (0.10 to 0.25 mm.)
Crankshaft end-float	0.003 to 0.008 in. (0.08 to 0.20 mm.)
Valve seat angle	45°
Inner and outer rotor clearance	0.006 in. (0.15 mm.) maximum
Outer rotor and housing clearance	0.010 in. (0.25 mm.) maximum
Inner and outer rotor end-float	0.005 in. (0.13 mm.) maximum
Oil pressure	35 to 40 lb./sq. in. (2.46 to 2.81 kg./sq. cm.)
Sump capacity (including oil filter)	7½ Imp. pints (9 U.S. pints, 4.26 litres)
Oil filter capacity	⅔ Imp. pint (0.8 U.S. pint, 0.38 litre)

Grade of oil:

SAE Grade	Use Between
5W-20, 5W-30	−40°F to + 32°F
10W-30	−10°F to + 70°F
10W-40	−10°F to + 90°F
10W-50	−10°F to + 120°F
20W-40	+25°F to + 90°F
20W-50	+32°F to + 120°F
20W-50 "All Season"	+25°F to + 120°F
10W	−10°F to + 32°F
20W-20	+25°F to + 70°F
30	+32°F to + 90°F
40	+50°F to + 120°F

Piston clearance in bore	0.0030 to 0.0036 in. (0.076 to 0.091 mm.)
Compression ring to groove clearance — upper	0.0016 to 0.0031 in. (0.041 to 0.079 mm.)
— lower	0.0016 to 0.0036 in. (0.041 to 0.091 mm.)
Oil control ring to groove clearance	0.0015 to 0.0030 in. (0.040 to 0.076 mm.)
Compression ring gap	0.009 to 0.014 in. (0.23 to 0.36 mm.)
Oil control ring gap	0.010 to 0.020 in. (0.25 to 0.51 mm.)
Valve clearance — (cold) — Inlet	0.005 to 0.007 in. (0.13 to 0.18 mm.)
— Exhaust (Part No. A26E020)	0.006 to 0.008 in. (0.15 to 0.20 mm.)
— Exhaust (Part No. B26E020)	0.009 to 0.011 in. (0.23 to 0.28 mm.)

Tightening Torques

Cylinder head	7/16 in. − 14 UNC	60 to 65 lb. ft. (8.29 to 8.98 kg.m.)
Main bearing cap	7/16 in. − 14 UNC	55 to 60 lb. ft. (7.60 to 8.29 kg.m.)
Connecting rod big end	3/8 in. − 24 UNF	44 to 46 lb. ft. (6.08 to 6.36 kg.m.)
Flywheel (hexagonal)	3/8 in. − 24 UNF	45 to 50 lb. ft. (6.22 to 6.91 kg.m.)
(bi-hexagonal)	3/8 in. − 24 UNF	50 to 55 lb. ft. (6.91 to 7.60 kg.m.)
Oil filter centre bolt	3/8 in. − 24 UNF	12 to 15 lb. ft. (1.66 to 2.07 kg.m.)

Manifold nuts	$\frac{5}{16}$ in. – 24 UNF	12 to 15 lb. ft. (1.66 to 2.07 kg.m.)
Front cover	$\frac{1}{4}$ in. – 20 UNC	5 to 7 lb. ft. (0.69 to 0.97 kg.m.)
	$\frac{1}{4}$ in. – 28 UNF	5 to 7 lb. ft. (0.69 to 0.97 kg.m.)
	$\frac{5}{16}$ in. – 18 UNC	10 to 15 lb. ft. (1.38 to 2.07 kg.m.)
	$\frac{5}{16}$ in. – 24 UNF	10 to 15 lb. ft. (1.38 to 2.07 kg.m.)
Sump	$\frac{1}{4}$ in. – 20 UNC	7 to 9 lb. ft. (0.97 to 1.24 kg.m.)
Rear oil seal retainer	$\frac{5}{16}$ in. – 18 UNC	12 to 15 lb. ft. (1.66 to 2.07 kg.m.)
Crankshaft pulley	$\frac{7}{16}$ in. – 20 UNF	24 to 28 lb. ft. (3.32 to 3.87 kg.m.)
Oil pump	$\frac{5}{16}$ in. – 18 UNC	12 to 15 lb. ft. (1.66 to 2.07 kg.m.)
Auxiliary shaft thrust plate	$\frac{1}{4}$ in. – 20 UNC	5 to 7 lb. ft. (0.69 to 0.97 kg.m.)
Auxiliary shaft sprocket	$\frac{5}{16}$ in. – 18 UNC	12 to 15 lb. ft. (1.66 to 2.07 kg.m.)
Sump drain plug	$\frac{1}{2}$ in. – 20 UNF	20 to 25 lb. ft. (2.76 to 3.46 kg.m.)
Camshaft bearing cap nuts	$\frac{5}{16}$ in. – 24 UNF	9 lb. ft. (1.24 kg.m.)
Camshaft sprocket bolts	$\frac{7}{16}$ in. – 20 UNF	25 to 30 lb. ft. (3.46 to 4.15 kg.m.)
Chain tensioner sprocket pin	$\frac{5}{8}$ in. – 18 UNF	40 to 45 lb. ft. (5.53 to 6.22 kg.m.)
Chain tensioner retaining bolt	$\frac{5}{8}$ in. – 11 UNC	45 to 50 lb. ft. (6.22 to 6.91 kg.m.)
Chain tensioner pivot pin	$\frac{1}{4}$ in. – BSP	40 to 45 lb. ft. (5.53 to 6.22 kg.m.)

SERVICE AND REPAIR OPERATIONS

When working around the engine compartment care must be taken to prevent damage to the paintwork. Immediately after opening the bonnet, wing covers must be fitted. This is assumed in all operations where required.

Before reassembling, during any operation, all components should be thoroughly cleaned, paying particular attention to joint faces and bearing surfaces. Any local high spots or burrs on the joint faces should be carefully removed with a fine oil stone. Ensure that any piece of gasket material or dirt which enters a blind tapped hole, during cleaning, is removed as the bolt may bottom on the resulting plug of dirt before the bolt head clamps the mating part. When tightening a bolt which bottoms, a characteristic springiness may be felt through the spanner or torque wrench. If this occurs, the bolt should be removed and the hole cleaned out.

Inspect all moving parts and bearing surfaces for wear. Check the dimensions of worn parts against the "Specification" and select new parts where necessary.

OP 6000-A TUNE ENGINE

(Includes Cleaning and Adjusting Spark Plugs, Cleaning or Replacing Air Cleaner Element, Adjusting or Replacing Contact Breaker Points, Checking Timing, Cleaning and Adjusting Carburettor and Fuel Pump and Adjusting Valves)

Tools Required

Spark plug cleaner
Dwell meter
Timing light

1. Remove the air cleaner and the air intake cover.
2. Pull off the spark plug leads and remove the spark plugs. Clean the spark plugs and reset the gaps to 0.023 in. (0.58 mm.).
3. Replace the spark plugs and reconnect the spark plug leads.
4. Remove the distributor cap and examine the contact breaker points. Replace the points if badly burnt or excessive metal transfer has occurred.
5. Adjust the dwell angle to 57° to 63° at 1,000 rev./min. Alternatively adjust the point gap to 0.014 to 0.016 in. (0.36 to 0.41 mm.) and refit the distributor cap.
6. Unscrew the fuel pump sediment bowl retainer clamp and lift off the bowl and filter screen.
7. Carefully wash the screen in petrol and flush all traces of sediment from the sediment chamber and bowl.
8. Refit the screen to the fuel pump body, ensure that the gasket is in good condition, refit the sediment bowl and tighten the clamp.
9. Remove the air cleaner element and clean by shaking or blowing it through. If unserviceable, replace the element. Wash the air cleaner body with petrol.
10. Refit the air cleaner element.

OP 6000-B ENGINE ASSEMBLY – CHECK COMPRESSION

Tools Required

500X   Gang Gauge Set

1. Warm up the engine to the normal operating temperature.
2. Remove the air cleaner.
3. Remove all the spark plugs.
4. Set the throttle to the wide open position.
5. Place the gang gauge in a convenient position, insert the expanding rubber plug into the No. 1 spark plug orifice and expand the plug by pulling up on the handle.
6. Crank the engine with the starter motor until full pressure is recorded on the gauge. The normal compression pressure is 180 to 200 lb./sq. in. (12.66 to 14.06 kg./sq. cm.) but at altitudes appreciably above sea level proportionally lower pressures will be obtained.
7. Test the remaining cylinders in a similar manner.
8. Replace the spark plugs and connect the plug leads.
9. Refit the air cleaner.

OP 6000-C ENGINE ASSEMBLY – REMOVE AND INSTALL

To Remove

1. Remove the bonnet.
2. Disconnect the battery.
3. Drain the cooling system by opening the drain plugs on the radiator and engine cylinder block. Store the coolant if it is to be used again.
4. Disconnect the radiator hoses at the engine and then remove the radiator assembly.
5. Remove the air cleaner and the carburettor air intake cover.
6. Disconnect both heater hoses, one from the front cover and one from cylinder head.
7. Disconnect the accelerator linkage from the carburettor.
8. Disconnect the choke control from the carburettor.
9. Disconnect the temperature gauge sender unit lead and the two generator leads.
10. Disconnect the exhaust manifolds and pull to one side.
11. Disconnect the fuel inlet pipe at the fuel pump.

---

11. Disconnect the fuel feed pipes at the carburettors.
12. Remove the carburettor float chamber cover.
13. Withdraw the float arm pivot pin and remove the float and gasket.
14. Unscrew all the jets and blow them clear with an air line.
15. Remove the needle valve and the needle valve body and blow it clear with an air line.
16. Clean the float, float chamber and filter gauze using clean petrol.
17. Replace all the jets and the needle valve body and needle valve.
18. Locate the gasket on the upper body and fit the float assembly, sliding the pivot pin into position. Check the float setting.
19. Refit the float chamber cover.
20. Reconnect the fuel feed pipes.
21. Adjust the valve clearances see Operation No. 6271-B.
22. Connect the leads of a timing light using the clips provided in accordance with the manufacturer's instructions.
23. Check that the mark on the crankshaft pulley is visible and mark with chalk or paint if necessary.
24. Start the engine and point the timing light at the crankshaft pulley adjacent to the timing scale.
25. Progressively increase the engine speed to 3,500 rev./min. observing the timing mark, with the aid of the timing light, to check that the distributor advances the ignition timing.
26. At 3,500 rev./min. adjust the ignition timing to 26° B.T.D.C., if necessary, by slackening the distributor clamp and turning the distributor body as required.
27. After making an adjustment, tighten the clamp sufficiently to hold the distributor in position. DO NOT OVER-TIGHTEN.
28. Remove the timing light.
29. Adjust the slow-running as follows with the engine at the normal operating temperature:
    (a) Adjust all four volume control screws three-quarters of a turn out after first screwing right in.
    (b) Adjust the throttle stop screw to give a speed of 1,000 to 1,200 rev./min.
    (c) Synchronise the two carburettors using a proprietary synchronising tool or alternatively improvise a stethoscope from a piece of tube. Balance the two carburettors by adjusting the coupling screw until the air flow is the same through each, with a stethoscope this is denoted by the same "hiss" level in each carburettor throat.
    (d) Adjust each volume control screw in turn to give the maximum speed possible, readjust the throttle stop screw, as necessary, to maintain a speed of 1,000 to 1,200 rev./min.
    (e) Adjust the idling speed to 800 to 1,000 rev./min.
30. Refit the air cleaner and the air intake cover.

CORTINA - LOTUS

12. Disconnect the distributor high and low tension leads from the coil.
13. Disconnect the brake servo vacuum pipe.
14. Disconnect the oil pressure gauge line at the adaptor.
15. Jack up the front of the car and fit stands. Remove the sump shield where fitted.
16. Disconnect the starter motor lead and remove the starter motor.
17. Unscrew the lower clutch housing bolts and remove the cover.
18. Remove the stands and jack from beneath the car.
19. Unscrew the clutch housing to engine bolts.
20. Fit a sling and support the engine on suitable tackle.
21. Disconnect the engine mountings from the front crossmember.
22. Suitably support the gearbox.
23. Pull the engine unit forward off the main drive gear and lift the assembly from the engine compartment.
24. Drain the engine oil.

To Install

25. Position the engine assembly in the engine compartment and engage the unit on the main drive gear. Ensure that the upper flywheel cover is located on the dowels.
26. Secure the engine front mountings to the front crossmember.
27. Fit the engine to clutch housing bolts. Note that the engine earth strap is secured by the top left-hand bolt.
28. Remove the sling and lifting tackle.
29. Reconnect the fuel pipe to the fuel pump.
30. Replace the distributor cap and reconnect the high and low tension leads to the coil.
31. Reconnect the brake servo vacuum pipe.
32. Replace the two generator leads and the temperature gauge sender unit lead.
33. Refit the exhaust manifolds, using new gaskets, tighten the nuts to a torque of 12 to 15 lb. ft. (1.66 to 2.07 kg.m.).
34. Reconnect the choke control to the carburettors.
35. Reconnect the throttle linkage to the carburettors.
36. Reconnect the heater hoses.
37. Replace the radiator assembly and reconnect the radiator hoses.

38. Make sure the two coolant drain plugs are closed. Refill the cooling system with long life anti-freeze.
39. Fit the air intake cover and the air cleaner to the carburettors.
40. Jack up the front of the car and fit stands.
41. Replace the starter motor and reconnect the starter motor lead.
42. Replace the sump shield if fitted. Fit the lower clutch housing cover and all the lower clutch housing bolts.
43. Jack up the car, remove the stands, lower car to the ground and remove the jack.
44. Reconnect the battery.
45. Reconnect the oil pressure gauge line at the adaptor.
46. Refill the engine sump with approved oil.
47. Run up and check engine for leaks.
48. Check and readjust the ignition if necessary to give 26° advance at 3,500 rev./min.
49. Adjust the slow-running.
50. Refit the bonnet.

OP 6000-C1 EXTRA: ANCILLARIES – REMOVE AND INSTALL
(Includes, carburettor, distributor, fuel pump, generator, engine mountings, oil pressure gauge adaptor).

To Remove

1. Remove the engine ancillaries. Slacken the generator mounting and remove the fan belt. Remove the fan, water pump pulley, generator and the front engine mountings. Disconnect the fuel pipe and remove the carburettors and air intake back plate, accelerator return spring bracket, fuel pump, distributor, oil pressure gauge adaptor, thermostat housing, thermostat and temperature gauge sender unit.

To Replace

2. Fit the thermostat, thermostat housing, temperature gauge sender unit, front engine mountings, oil pressure gauge adaptor, fuel pump, generator, water pump pulley and fan. Fit the fan belt and adjust the tension to give ½ in. (12.7 mm.) total free play. Tighten the generator mounting bolts to a torque of 15 to 18 lb. ft. (2.08 to 2.49 kg.m.).

3. Fit and time the distributor. Turn the engine until the crankshaft pulley timing mark is adjacent to the 12° mark on the front cover with number 1 cylinder on compression. With the low tension terminal adjacent to the cylinder block, turn the rotor until it points towards the distributor cap rear clip. Insert the distributor into its bore in this position and secure the clamp bracket to the cylinder block. Slacken the clamp and turn the distributor body until the points just open while holding the rotor against the direction of rotation to take up lost motion. Tighten the clamp sufficiently to hold the distributor in this position. DO NOT OVER-TIGHTEN.

CORTINA - LOTUS

4. Fit the carburettors and back plate assembly. Locate new "O" rings in the spacers and tighten the retaining nuts until there is 0.040 in. (1.02 mm.) clearance between the coils of the double coil spring washers. Connect the fuel pipe and refit the accelerator return spring bracket.

OP 6000-C2 EXTRA: CLUTCH DISC AND/OR PRESSURE PLATE – REMOVE AND INSTALL

See Operation No. 7000-A2 of Section 7/1

OP 6000-C3 EXTRA: CLUTCH PILOT SPIGOT BEARING – REMOVE AND INSTALL

See Operation No. 7000-A3 of Section 7/1

OP 6000-C4 EXTRA: FLYWHEEL – REMOVE AND INSTALL
(engine, clutch disc, pressure plate and spigot bearing removed)

Tools Required

P.4008   Crown wheel and pinion backlash gauge

To Remove

1. Remove the flywheel.

To Install

2. Locate the flywheel squarely on the crankshaft flange and tighten the retaining bolts to a torque of:—
   (a) 45 to 50 lb. ft. (6.22 to 6.91 kg.m.) for hexagonal bolts.
   (b) 50 to 55 lb. ft. (6.91 to 7.60 kg.m.) for bi-hexagonal bolts.

3. Check the flywheel run-out using the gauge Tool No. P.4008 at the rim. The flywheel run-out should not exceed 0.004 in. (0.10 mm.) total indicator reading.

OP 6000-C5 EXTRA: SUMP AND/OR GASKETS – REMOVE AND REFIT

Tools Required

200A or B   Engine stand
P.6107   Universal stand adaptor

To Remove

1. Fit a universal stand adaptor Tool No. P.6107 and mount the engine on a stand Tool No. 200A or B.
2. Remove the sump and gasket.
3. Clean the sump and cylinder block faces and remove the cork packing strips.

To Replace

4. Apply sealer ESEE-M4G-1008A to the front cover and rear oil seal carrier to cylinder block joints, and also to the ends of the grooves in the front cover and rear oil seal carrier. Then refit the new sump gaskets, followed by the cork packing strips.
5. Refit the sump and engage the corner bolts and tighten all bolts sufficiently to clamp gasket. Engage remaining bolts and tighten all bolts sufficiently to clamp gasket.
6. Engage remaining bolts and tighten all bolts sufficiently to clamp gasket.
7. Tighten all bolts to 7 to 9 lb. ft. (0.97 to 1.24 kg.m.) following the sequence given in alphabetical order.
8. Retighten all bolts to 7 to 9 lb. ft. (0.97 to 1.24 kg.m.) following the sequence given in numerical order.
9. Remove the engine from stand and adaptor.

OP 6000-C6 EXTRA: ENGINE REAR OIL SEAL CARRIER AND/OR GASKET – REMOVE AND INSTALL (with flywheel and sump removed)

Tools Required

CP.6173   Crankshaft rear oil seal aligner

To Remove

1. Remove the rear oil seal carrier.

To Replace

2. Locate a new gasket on the rear oil seal carrier, using ESEE-M4G-1008A jointing compound at the ends, and fit the carrier to the block rear face using an aligner Tool No. CP.6173. Tighten the bolts evenly to a torque of 12 to 15 lb. ft. (1.66 to 2.07 kg.m.) and remove the aligner.

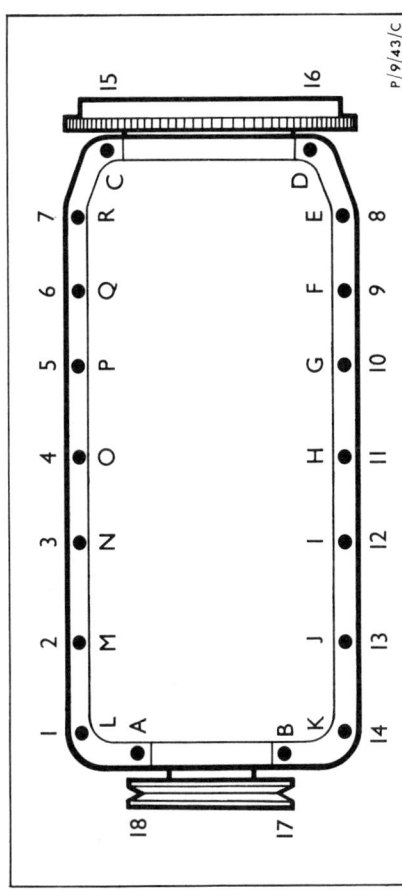

Sump Bolt Tightening Sequence

1 . 1969

CORTINA - LOTUS

OP 6000-C7   EXTRA: ENGINE REAR OIL SEAL – RENEW
(with rear oil seal carrier removed)

Tools Required

550   Driver handle
P.6165   Crankshaft rear oil seal remover/replacer

To Remove

1. Suitably support the rear oil seal carrier and remove the oil seal using a remover/replacer Tool No. P.6165 fitted to a 550 handle.

To Replace

2. Invert the remover/replacer Tool No. P.6165 on the 550 handle and with the rear oil seal carrier suitably supported, drive a new seal into the housing.

OP 6000-C8   EXTRA: FRONT COVER AND/OR GASKET – REMOVE AND INSTALL
(with sump removed)

Tools Required

550   Driver handle
P.6150   Crankshaft front cover oil seal aligner

To Remove

1. Remove the camshaft cover.
2. Remove the fan belt and then remove the fan and the water pump pulley.
3. Remove the crankshaft pulley, using suitable levers.
4. Remove the timing chain tension adjuster.
5. Remove the front cover.
6. Remove the crankshaft oil slinger.
7. Disconnect the timing chain taking care not to rotate the camshafts or crankshaft and alter the valve timing.
8. Remove the auxiliary shaft sprocket.
9. Remove the front cover back plate and gaskets.

To Install

10. Locate a new gasket on the cylinder head to the timing cover joint and on the cylinder block to front cover back plate joint.
11. Fit the front cover back plate. Align the water pump aperture with the front cover and seal aligner Tool No. P.6150 before tightening the single bolt to 5 to 7 lb. ft. (0.69 to 0.97 kg.m.) torque.
12. Fit the auxiliary shaft sprocket. Tighten the retaining bolts to 12 to 15 lb. ft. (1.66 to 2.07 kg.m.) torque and turn up the locking plate tabs.

Cylinder Block, Sump and Timing Case Assembly

CORTINA - LOTUS

13. Locate the timing chain around the sprockets ensuring that there is minimum slack between the exhaust camshaft sprocket and the crankshaft sprocket.

14. Locate the crankshaft oil slinger in place.

15. Coat the front cover joint faces with ESEE-M4G-1008A jointing compound. Align the cover in position with Tool No. P.6150. Tighten the $\frac{1}{4}$ in. retaining nuts and bolts evenly to a torque of 5 to 7 lb. ft. (0.69 to 0.97 kg.m.) and the $\frac{5}{16}$ in. to 10 to 15 lb. ft. (1.38 to 2.07 kg.m.) and remove the aligner tool. Refer to page 13 for bolt locations.

16. Fit the timing chain tension adjuster and tighten to 45 to 50 lb. ft. (6.22 to 6.91 kg.m.) torque.

17. Fit the crankshaft pulley aligning the pulley slot with the crankshaft key. Tighten the retaining bolt to 24 to 28 lb. ft. (3.32 to 3.87 kg.m.).

18. Replace the water pump pulley and the fan. Fit the fan belt and adjust the tension so that there is $\frac{1}{2}$ in. (12.7 mm.) total movement.

19. Adjust the timing chain tension to give $\frac{1}{2}$ in. (12.7 mm.) free movement between the camshaft sprockets. Ensure that there are no tight spots by turning the engine through several revolutions.

20. Fit the camshaft cover, using a new gasket and tighten the retaining nuts evenly.

OP 6000-C9 EXTRA: FRONT COVER OIL SEAL – REMOVE AND INSTALL
(with front cover removed)

Tools Required

550   Driver handle
P.6161   Crankshaft front oil seal remover/replacer

To Remove

1. Suitably support the front cover and remove the oil seal from the rear, using remover/replacer Tool No. P.6161 fitted to a 550 handle.

To Replace

2. Insert the remover/replacer Tool No. P.6161 on the 550 handle and with the front cover suitably supported drive a new seal into the housing.

OP 6000-C10 EXTRA: TIMING CHAIN – REMOVE AND INSTALL
(front cover removed)

To Remove

1. Remove the camshaft sprockets and extract the timing chain.

To Install

2. Locate the timing chain in position.

3. Fit the camshaft sprockets and timing chain ensuring there is a minimum of slack between exhaust sprocket and the camshaft sprocket.

4. Tighten the retaining bolts to a torque of 25 to 30 lb. ft. (3.46 to 4.15 kg.m.).

OP 6000-C11 EXTRA: CRANKSHAFT SPROCKET – REMOVE AND INSTALL
(with front cover and timing chain removed)

Tools Required

P.6032-B   Crankshaft sprocket replacer
P.6116   Crankshaft sprocket remover

To Remove

1. Remove the oil slinger then the crankshaft sprocket using Tool No. P.6116.

To Replace

2. Replace the crankshaft sprocket using replacer Tool No. P.6032-B ensuring that the long boss is towards the main bearing journal.

3. Refit the oil slinger.

OP 6000-C12 EXTRA: CONNECTING ROD BEARINGS – REMOVE AND INSTALL
(with sump removed)

To Remove

1. Turn the crankshaft to facilitate removal of number one big end cap. Unscrew the big end bolts two or three turns and tap them to release the cap. Completely unscrew the bolts and remove the cap.

2. Remove the upper connecting rod liner from the connecting rod and the lower from the connecting rod cap.

Timing Chain Tension Adjuster

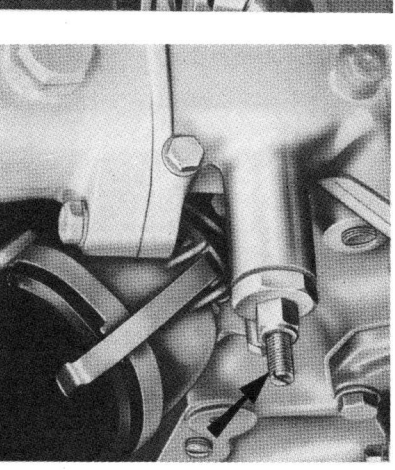

Timing Chain Tension Adjustment

CORTINA - LOTUS

To Replace

3. Replace the upper and lower bearing liners in their appropriate locations.
4. Locate the big end caps on the connecting rod and tighten the bolts to a torque of 44 to 46 lb. ft. (6.08 to 6.36 kg.m.).
5. Repeat sub-operations 1 to 4 for the other three connecting rods.

OP 6000-C13 EXTRA: MAIN BEARING CLEARANCES – CHECK
(with sump removed)

Tools Required

Micrometer, or means such as "Plastigage" to measure bearing clearance.

1. Remove the bearing cap and wipe the bearing and journal clean.
2. Place a piece of "Plastigage" on the bearing surface the full width of the bearing cap and about $\frac{1}{4}$ in. (6.4 mm.) off centre.
3. Install the cap and tighten the bolts to a torque of 65 to 70 lb. ft. (8.98 to 9.67 kg.m.). DO NOT TURN THE CRANKSHAFT WHILE THE "PLASTIGAGE" IS IN PLACE.
4. Remove the cap and using the "Plastigage" scale check the width of the now compressed "Plastigage" strip.
   Check at the widest point to get the minimum clearance.
   Check at the narrowest point to get the maximum clearance.
   The difference between the two readings is the taper.
5. Clean the bearing liner and refit the cap.
6. Check the remaining bearing clearances using the same procedure.

OP 6000-C14 EXTRA: MAIN BEARING LINERS AND THRUST WASHERS – REMOVE AND INSTALL
(with sump removed)

Tools Required

P.6110   Main bearing liner remover/replacer

To Remove

1. Remove number one main bearing cap.
2. Remove the upper bearing liner from the cylinder block using Tool No. P.6110 and the lower bearing liner from the cap.

To Install

3. Fit the new upper bearing liner in the cylinder block using Tool No. P.6110 and the new lower bearing liner in the cap.
4. Refit the main bearing cap and tighten the bolts to a torque of 55 to 60 lb. ft. (7.60 to 8.29 kg.m.).

Change the remaining main bearing liners in sequence

Intermediate and Rear Main Bearing

5. Complete sub-operations 1, 2, 3 and 4.

Centre Main Bearing

6. Complete sub-operation 1.
7. Remove the thrust washers.
8. Complete sub-operations 2 and 3.

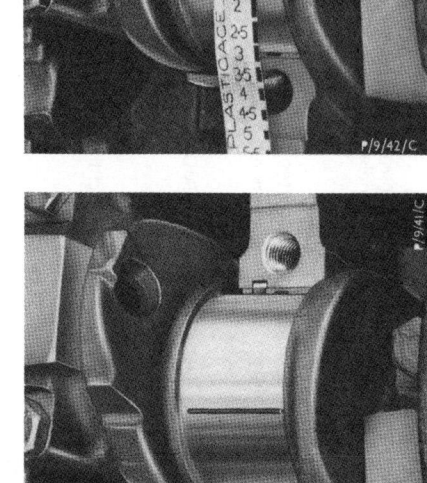

Measuring Bearing Clearance with Plastigage

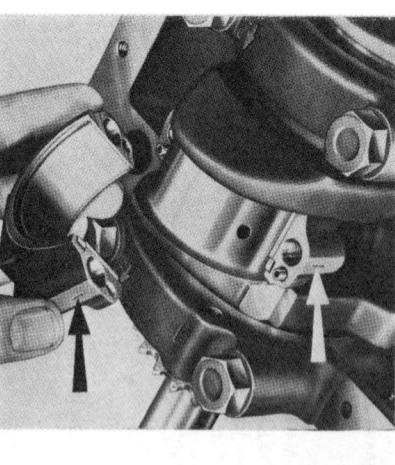

Fitting a Big End Cap

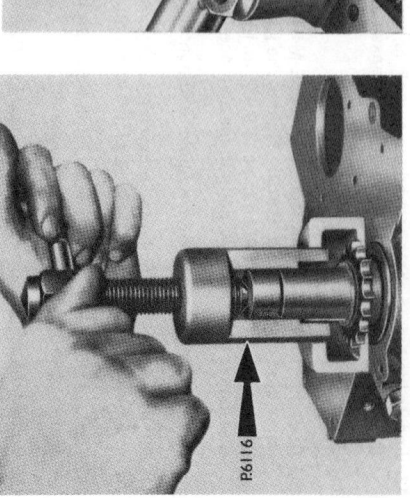

Removing the Crankshaft Sprocket

CORTINA - LOTUS

9. Fit new thrust washers.
10. Complete sub-operation 4.

OP 6000-C15 EXTRA: ALL MAIN BEARING LINER AND THRUST WASHERS – CHECK CLEARANCES AND RENEW (with sump removed)

1. Check the clearances, see Operation No. 6000-C13.
2. Remove the bearing liner, see Operation No. 6000-C14 sub-operations 1 and 2.
3. Select bearing liners to give the correct clearance.
4. Fit the bearing liners, see Operation No. 6000-C14 sub-operation 3.
5. Recheck the bearing clearance, see Operation No. 6000-C13 sub-operations 2 to 7.
6. Repeat the procedure for the remaining bearings.

OP 6000-C16 EXTRA: AUXILIARY SHAFT – REMOVE AND INSTALL (engine, timing chain and sprocket removed)

To Remove

1. Remove the distributor (see operation No. 12100-A).
2. Remove the oil pump and filter assembly.
3. Disconnect the fuel pipes and remove the fuel pump.
4. Remove the auxiliary shaft sprocket adaptor.
5. Remove the auxiliary shaft thrust plate.
6. Withdraw the auxiliary shaft.

To Install

7. Fit a new dowel to the new shaft.
8. Slide the auxiliary shaft into position.
9. Fit the thrust plate in the auxiliary shaft groove. Tighten the retaining bolts to 5 to 7 lb. ft. (0.69 to 0.97 kg.m.) and bend up the locking tabs.
10. Check the auxiliary shaft end-float with feeler blades between the thrust plate and the auxiliary shaft flange. This should be between 0.002 and 0.007 in. (0.050 and 0.178 mm.)
11. Refit the auxiliary shaft sprocket adaptor.
12. Fit the fuel pump and connect the fuel pipes.
13. Refit the oil pump and filter assembly, tighten the retaining bolts to 12 to 15 lb. ft. (1.66 to 2.07 kg.m.) torque.
14. Time the distributor (see operation No. 12100-A as part of operation No. 6279-A).

OP 6000-C17 EXTRA: AUXILIARY SHAFT BEARINGS – RENEW (engine, auxiliary shaft and crankshaft removed)

Tools Required

P.6031   Auxiliary shaft bearing bush remover/replacer
P.6031-3   Auxiliary shaft bearing bush remover/replacer adaptors.

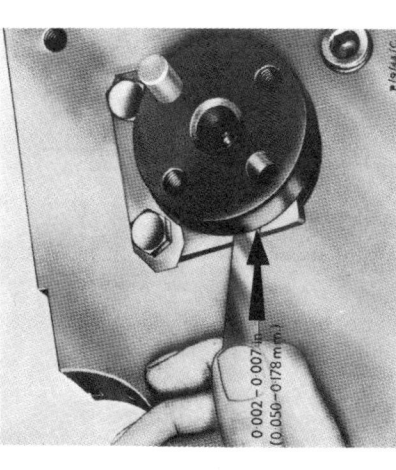

Checking the Auxiliary Shaft End-float

Fitting the Auxiliary Shaft Thrust Plate

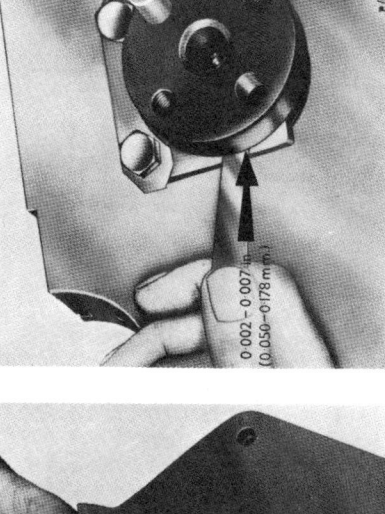

Remove the Oil Pump

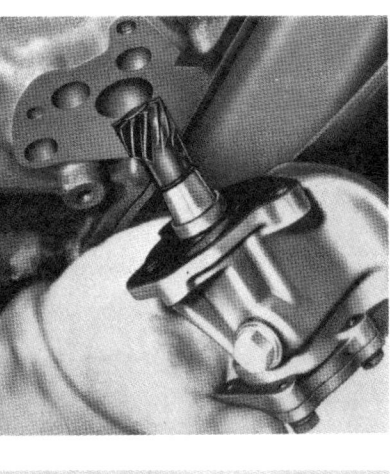

Fitting a Main Bearing Liner

CORTINA - LOTUS

To Remove

1. Remove the auxiliary shaft bearing bushes using Tool No. P.6031-3.
2. Check all the oilways to ensure that they are clear, apply EM-4G-52 sealing compound to the oil gallery plugs prior to refitting.

To Replace

3. Fit new auxiliary shaft bearing bushes again using Tool No. P.6031 and P.6031-3. Ensure that the oil holes in the bushes and cylinder block are aligned. The splits in the bushes should be upwards and outwards at 45° to the vertical.

OP 6000-C18 EXTRA: CRANKSHAFT – REMOVE AND INSTALL
(with sump, front cover, rear oil seal carrier, connecting rods and main bearings removed)

To Remove

1. Lift out the crankshaft and remove the bearing liners and thrust washers.

To Replace

2. Fit a new spigot bearing. Refer to operation 7000-A3 of Section 7/1.
3. Replace the crankshaft sprocket using replacer Tool No. P.6032A or B.
4. Fit the main bearing liners and replace the crankshaft. Fit the crankshaft thrust washers with the oil grooves facing the crankshaft flange. Refit the main bearing caps and tighten the bolts to 55 to 60 lb. ft. (7.60 to 8.29 kg.m.) torque.
5. Check the crankshaft end-float with feeler blades between the crankshaft and the thrust washers. This should be between 0.003 and 0.008 in. (0.08 and 0.20 mm.).

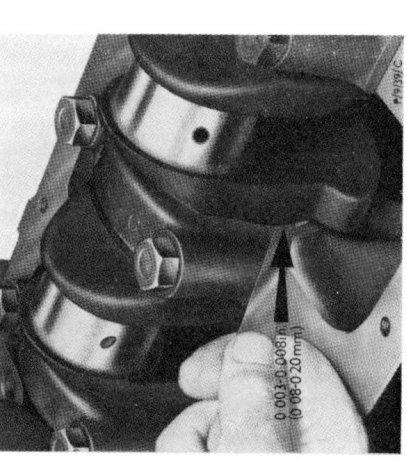

Fitting the Auxiliary Shaft Bushes

OP 6000-C19 EXTRA: CYLINDER HEAD AND PISTONS – DECARBONISE
(with engine removed)

See Operation No. 6051-A, sub-operations 11 to 18 and Operation Nos. A1 and A2.

OP 6000-C20 EXTRA: VALVES – ALL – REFACE, RESEAT AND GRIND-IN
(with engine removed)
(Includes initial setting of valves, running-up engine and final setting of valve clearances)

See Operation No. 6051-A, sub-operations 11 to 18 and Operation Nos. A3, A4, A5 and A6

OP 6000-C21 EXTRA: CYLINDER ASSEMBLY – REMOVE AND INSTALL
(with engine and ancillaries removed)

Tools Required

200 A or B	Engine stand
P.4008	Crown wheel and pinion backlash gauge
PT.4063A	Cylinder head gasket locating studs
CP.6032B	Crankshaft sprocket replacer
P.6041	Crankshaft pulley remover
P.6107	Universal stand adaptor
P.6116	Crankshaft sprocket remover
P.6150	Front oil seal aligner
P.6161	Front oil seal remover and replacer
P.7137	Spigot bearing replacer and clutch disc locator

To Remove

1. Fit a universal stand adaptor Tool No. P.6107 and mount the engine on a stand Tool No. 200 A or B.
2. Unscrew the pressure plate bolts evenly and detach the pressure plate and clutch disc.
3. Remove the flywheel.
4. Remove the crankshaft pulley, using remover Tool No. P.6041.
5. Remove the oil pump and filter assembly.
6. Remove the camshaft cover and gasket.
7. Slacken the timing chain tensioner.
8. Remove the camshaft sprockets and disconnect the timing chain.
9. Unscrew the cylinder head bolts evenly and lift off the cylinder head gasket.

NOTE – DO NOT lay the cylinder head flat on its face as damage to the valves can occur while the camshafts are still fitted.

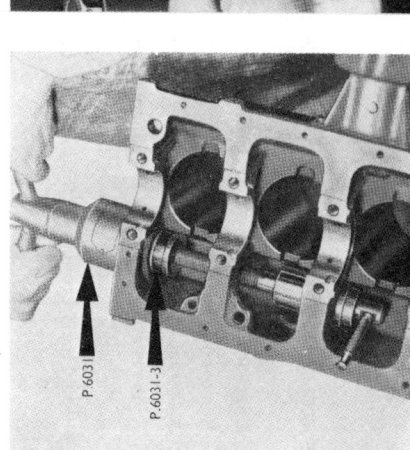

Checking the Crankshaft End-float

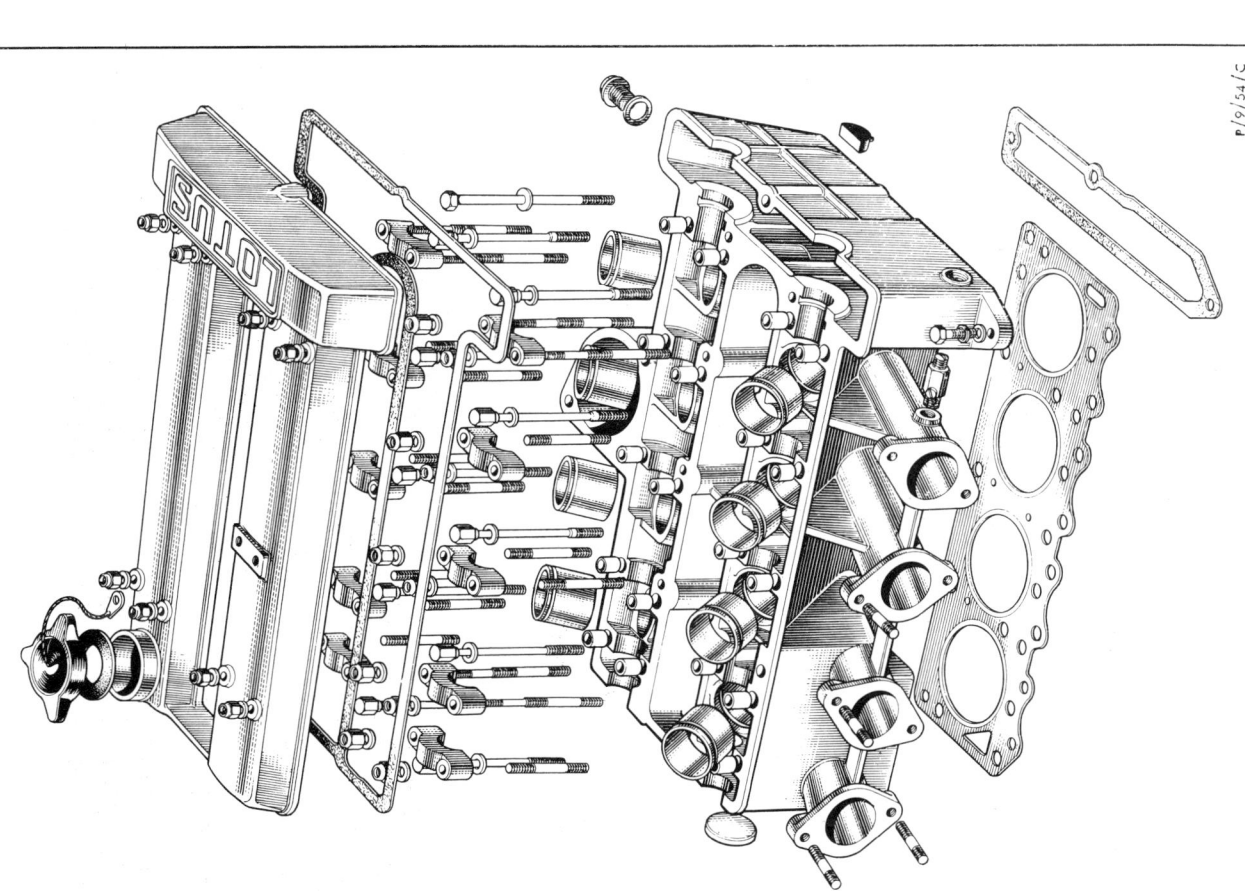

Cylinder Head and Camshaft Cover

10. Remove the breather pipe.
11. Invert the engine on the stand and remove the sump and gaskets.
12. Remove the front cover.
13. Remove the timing chain.
14. Remove the crankshaft oil slinger.
15. Remove the auxiliary shaft sprocket and adaptor.
16. Remove the front cover back plate.
17. Remove the crankshaft sprocket using Tool No. P.6116 and then extract the key.
18. Remove the oil pump filter gauze.
19. Remove the oil pump inlet tube and oil return pipe.
20. Remove the cylinder assembly from the stand and bolt the adaptor Tool No. P.6107 to the new cylinder assembly. Mount the new cylinder assembly on the universal stand.

To Replace

21. Replace the oil pump inlet tube and oil return pipe. Press the pipe fully home to the full depth of the counterbored hole.
22. Fit the filter gauze to the oil inlet tube.
23. Fit the crankshaft key and sprocket to the front end of the crankshaft. Press the sprocket home using replacer Tool No. CP.6032B. Ensure that the long boss is adjacent to the main bearing journal.

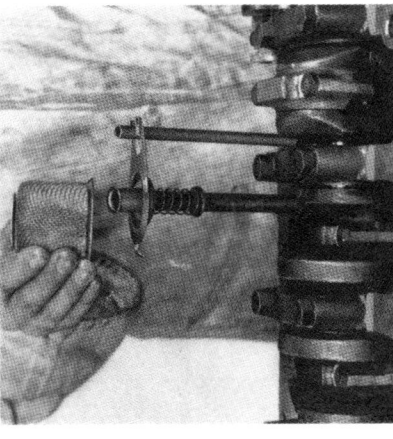

Fitting the Oil Pump Filter Gauze

CORTINA - LOTUS

24. Fit the front cover back plate. Align the water pump aperture with the front cover and seal aligner Tool No. P.6150 before tightening the single bolt to 5 to 7 lb. ft. (0.69 to 0.97 kg.m.) torque.

25. Fit the auxiliary shaft sprocket adaptor followed by the sprocket after first fitting a new dowel, if not already fitted. Tighten the retaining bolts to 12 to 15 lb. ft. (1.66 to 2.07 kg.m.) and turn up the locking plate tabs.

26. Locate the timing chain in position around the crankshaft and auxiliary shaft sprockets and around the water pump aperture in the back plate.

27. Fit the oil slinger to the crankshaft with the dished face away from the sprocket.

28. Fit a new oil seal to the front cover using Tool No. P.6161 and a 550 handle.

29. Coat the front cover joint faces with ESEE-M4G-1008A jointing compound and fit the front cover, aligning the seal with Tool No. P.6150. Tighten the $\frac{1}{4}$ in. nuts and bolts evenly to 5 to 7 lb. ft. (0.69 to 0.97 kg.m.) torque and the $\frac{5}{16}$ in. to 10 to 15 lb. ft. (1.38 to 2.07 kg.m.) and remove the aligner.

30. Fit new gaskets on the block flange using ESEE-M4G-1008A jointing compound at each end on the front cover and rear oil seal carrier. Fit the cork strips again using EM-4G-47. Refit the sump and then tighten the bolts evenly to a torque of 7 to 9 lb. ft. (0.97 to 1.24 kg.m.) in the sequence given on page 33.

31. Fit the crankshaft pulley aligning the pulley slot with the crankshaft key. Tighten the retaining bolt to 24 to 28 lb. ft. (3.32 to 3.87 kg.m.).

32. Fit a new clutch pilot spigot bearing using replacer Tool No. P.7137.

33. Locate the flywheel squarely on the crankshaft flange. Tighten the retaining bolts evenly to a torque of 45 to 50 lb. ft. (6.22 to 6.91 kg.m.) for hexagonal bolts and 50 to 55 lb. ft. (6.91 to 7.60 kg.m.) for bi-hexagonal bolts. Check the flywheel run-out using gauge Tool No. P.4008. This should not exceed 0.004 in. (0.10 mm.) total indicator reading.

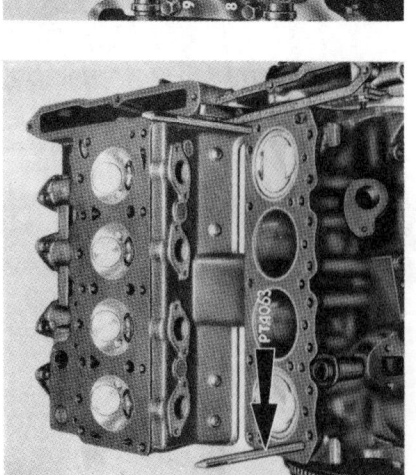

Fitting the Cylinder Head

Cylinder Head Bolt Tightening Sequence

34. Fit the clutch assembly to the flywheel. Centralise the clutch disc with the hub assembly away from the flywheel using the locator Tool No. P.7137. Tighten the bolts evenly to a torque of 12 to 15 lb. ft. (1.66 to 2.07 kg.m.), then remove the clutch disc locator.

35. Locate the breather pipe in its bore in the cylinder block and new gasket on top of the front cover.

36. Fit the cylinder head assembly. Locate the cylinder head gasket on the cylinder block using the locating studs Tool No. PT.4063A screwed into diagonally opposite bolt holes in the block face. Fit the cylinder head assembly engaging the breather pipe in its bore. Screw the cylinder head bolts home before removing the locating studs and then tighten in sequence to 60 to 65 lb. ft. (8.29 to 8.98 kg.m.) torque. Tighten the three front cover bolts to 10 to 15 lb. ft. (1.38 to 2.07 kg.m.) torque.

37. Fit the camshaft sprockets and timing chain. Align the timing mark on the crankshaft pulley with the lower mark on the front cover and the timing marks on the sprockets adjacent to each other and level with the camshaft cover mounting face. Fit the exhaust sprocket first. Tighten the retaining bolts to 25 to 30 lb. ft. (3.46 to 4.15 kg.m.) torque.

38. Adjust the timing chain tension to give $\frac{1}{2}$ in. (12.7 mm.) free movement between the camshaft sprockets. Ensure that there are no tight spots by turning the engine through several revolutions.

39. Fit the camshaft cover and tighten the retaining nuts evenly.

40. Fit the oil pump and filter assembly.

41. Remove the engine from the stand and the adaptor.

Camshaft Sprocket Valve Timing Marks

Crankshaft Valve Timing Mark

OP 6000-C22 EXTRA: CYLINDER BLOCK – REMOVE AND INSTALL
(cylinder assembly removed)

Tools Required

550	Drive handle
38 U 3	Piston ring squeezer
CP.6173	Crankshaft rear oil seal aligner
P.6165	Crankshaft rear oil seal remover and replacer

To Remove

1. Remove the rear oil seal carrier.

2. Unscrew the bolts several turns and tap them to release the big end caps. Unscrew the bolts completely and remove the caps. Push the pistons out of the bores and withdraw the assemblies.

3. Remove the auxiliary shaft thrust plate and withdraw the auxiliary shaft.

4. Unscrew the main bearing cap bolts evenly and lift off each cap. Lift out the crankshaft and remove the main bearing liners and thrust washers.

5. Dismantle the piston and connecting rod assemblies. Remove the piston rings and remove the two piston pin circlips. Push the piston pin out of each piston.

NOTE – It is permissible to heat the piston to a temperature of 120°C (248°F) to assist piston pin removal.

To Install

6. Locate the auxiliary shaft thrust plate in the shaft groove and tighten the retaining bolts to 5 to 7 lb. ft. (0.69 to 0.97 kg.m.) torque. Check the end-float which should be between 0.002 and 0.007 in. (0.051 to 0.178 mm.). Turn over the locking tabs.

7. Fit the crankshaft main bearing liners and thrust washers. Install crankshaft in cylinder block and refit main bearing caps. Tighten main bearing cap bolts evenly to 55 to 60 lb. ft. (7.60 to 8.29 kg.m.) torque and check crankshaft rotation.

8. Check the crankshaft end-float. Take up the end-float in one direction and insert a feeler blade between the crankshaft and the thrust washer to measure the clearance. The end-float should be between 0.003 and 0.008 in. (0.08 and 0.20 mm.).

9. Select pistons to suit the new cylinder block. Refer to page 21.

10. Locate the piston rings in the cylinder bore and check the ring gaps which should be between 0.009 and 0.014 in. (0.23 and 0.36 mm.) for the compression rings and 0.010 to 0.020 in. (0.25 to 0.51 mm.) for the oil control rings.

11. Check piston ring to groove clearances which should be as follows:—

Upper compression ring	0.0016 to 0.0031 in. (0.041 to 0.081 mm.)
Lower compression ring	0.0016 to 0.0036 in. (0.041 to 0.091 mm.)
Oil control ring	0.0015 to 0.0030 in. (0.04 to 0.076 mm.)

12. Fit the piston rings, fitting the oil control ring first, followed by the lower and then the upper compression rings. Ensure that the compression rings are fitted the correct way up.

13. Assemble the piston to the connecting rod. Ensure that the "FRONT" marking on the connecting rod is on the same side of the assembly as the mark on the piston crown. Heat the piston in water or oil prior to inserting the piston pin. Retain the piston with the circlips.

14. Fit the piston and connecting rod assemblies into the appropriate bores. Position the oil control ring gap to the rear and the compression ring gaps to 150° on either side of this. Compress the rings using Tool No. 38 U 3. Turn the crankshaft as necessary to fit the connecting rod big end caps to the crankpins. Tighten the bolts to 44 to 46 lb. ft. (6.08 to 6.36 kg.m.) torque. Check the big end float on the crankpin, this should be 0.004 to 0.010 in. (0.10 to 0.25 mm.).

15. Fit a new oil seal to the rear oil seal carrier using Tool No. P.6165 and a 550 handle.

16. Fit a new gasket to the rear oil seal carrier using ESEE-M4G-1008A jointing compound at the ends. Secure the carrier to the cylinder block, aligning it with Tool No. CP.6173. Tighten the bolts evenly to 12 to 15 lb. ft. (1.66 to 2.07 kg.m.) torque and remove the aligner.

OP 6000-C23 EXTRA: CYLINDER BLOCK – REBORE
(with cylinder block removed)

Tools Required
Boring bar

1. Rebore cylinder block using proprietary boring equipment and adhering to the manufacturer's instructions.

OP 6000-C24 EXTRA: CYLINDER BLOCK – REBORE AND FIT LINERS
(with cylinder block removed)

Tools Required
Boring bar
Locally manufactured remover and replacer ring

1. Bore the cylinder block to the specified size for the cylinder liner using a proprietary boring bar.

2. Place the remover adaptor in the replacer ring and locate in the cylinder liner which should be lubricated on the outside with tallow. Press the liner into the bore from the top on a suitable press.

3. Cut the connecting rod clearance slots in the base of the liner and machine the bore to give the correct clearance for the piston being fitted.

OP 6010-A   CYLINDER ASSEMBLY – REMOVE AND INSTALL
(Includes 6000-C, C1 and C21)

OP 6010-B   CYLINDER BLOCK – REMOVE AND INSTALL
(Includes 6000-C, C21 and C22)

OP 6010-C   CYLINDER BLOCK – REBORE
(Includes 6000-C, C1, C21, C22 and C23)

OP 6010-D   CYLINDER BLOCK – SLEEVE
(Includes 6000-C, C1, C21, C22, C23 and C24)

OP 6015-A   ENGINE ASSEMBLY – FIT NEW, SERVICE, OR RECONDITIONED UNIT
(Includes transferring, but not overhauling, ancillaries, adjusting carburettor and ignition timing, cleaning exterior of ancillaries and checking for water, oil or fuel leaks. Does not include valve adjustment)

1. Remove and install the engine assembly, see OP 6000-C and C1.

OP 6015-A1   EXTRA: ENGINE COMPARTMENT – CLEAN
(with steam cleaner with engine removed)

Tools Required

Steam cleaner

Clean engine compartment using proprietary steam cleaner, following manufacturer's instructions.

OP 6015-A2   EXTRA: CLUTCH DISC AND/OR PRESSURE PLATE – REMOVE AND INSTALL
(with engine removed)

To Remove

1. Unscrew the pressure plate bolts evenly and remove the pressure plate and disc.

To Replace

2. Centralise the clutch disc, with the hub assembly away from the flywheel, using Tool No. P.7137. Fit the pressure plate and cover assembly and tighten the bolts evenly to a torque of 12 to 15 lb. ft. (1.66 to 2.07 kg.m.) and remove the clutch disc locator.

OP 6015-A3   EXTRA: CLUTCH RELEASE BEARING – RENEW
(with engine removed)

To Remove

1. Remove the rubber gaiter.
2. Withdraw the release arm and bearing assembly from the clutch housing.
3. Remove the release arm from the hub and bearing assembly.

To Replace

4. Apply a smear of molybdenum based grease to the hub and release arm then engage the arm in the slots in the hub and release bearing assembly.
5. Pass the release arm through the aperture in the clutch housing and slide the release bearing onto the main drive gear bearing retainer.
6. Replace the rubber gaiter.

OP 6015-A4   EXTRA: CLUTCH FORK – REMOVE AND INSTALL
(with clutch release bearing removed)

To Remove

1. Remove the rubber gaiter.
2. Withdraw the release arm from the clutch housing.

To Install

3. Pass the release arm through the aperture in the clutch housing and refit the rubber gaiter.

OP 6015-A5   EXTRA: MAIN DRIVE GEAR BEARING RETAINER GASKET AND/OR MAIN DRIVE GEAR OIL SEAL – REMOVE AND INSTALL
(with engine and clutch release bearing removed)

See Operation No. 7000-A6 of Section 7.

OP 6015-B   ENGINE AND GEARBOX ASSEMBLY – REMOVE AND INSTALL
(Includes remove assembly, position on blocks on floor and refit assembly)

CORTINA - LOTUS

To Remove

1. Disconnect the engine components prior to removal, see Operation No. 6000-C sub-operations 1 to 24, but do not disconnect the gearbox or remove the starter motor.
2. Remove the driveshaft (see Section 4, Operation No. 4602-A sub-operations 1 to 4) and fit a dummy sliding spline.
3. Remove the gear lever.
4. Disconnect the speedometer cable.
5. Remove the clutch operating cylinder circlip and withdraw the cylinder to one side.
6. Remove the bolts securing the gearbox crossmember to the body and the bolt securing the crossmember to the gearbox.
7. Remove the engine and gearbox assembly and place on blocks.

To Install

8. Install the engine and gearbox in the engine compartment.
9. Fit the gearbox crossmember to the body and gearbox.
10. Connect the speedometer cable.
11. Fit the clutch operating cylinder.
12. Fit the gearlever.
13. Fit the driveshaft. See Section 4, Operation No. 4602-A, sub-operations 5 to 8.
14. Reconnect the engine components. See Operation 6000-C, sub-operations 29 to 49.

OP 6015-B1   EXTRA: ENGINE AND GEARBOX — SEPARATE AND RECONNECT

To Separate

1. Remove the starter motor.
2. Remove the bolts securing the clutch housing to the engine. Note that a top bolt secures an earth strap.
3. Remove the bolts securing the lower dust cover and remove the cover.
4. Separate the gearbox and engine.

To Reconnect

5. Offer the gearbox up to the engine so that the main drive gear spigot enters the crankshaft bearing and the splines engage with the clutch disc splines. Turn the mainshaft with the box in gear, if necessary, to engage the splines. Push the gearbox fully home.
6. Refit the clutch housing to engine bolts.
7. Position the lower dust cover and secure with bolts.
8. Refit the starter motor.

OP 6038-A   ENGINE FRONT MOUNTING — ONE — RENEW

To Remove

1. With handbrake applied, jack up the front of car and fit stands.
2. Support the engine on a jack. (Remove the engine sump shield if fitted.)
3. Remove one of the engine mountings.

To Replace

4. Replace the engine mounting.
5. Remove the jack from under the engine. (Replace the sump shield if fitted.)
6. Remove the stands and lower the car to ground.

OP 6038-A1   EXTRA: REMAINING FRONT ENGINE MOUNTING — RENEW

To Remove

1. Remove other engine mounting.

To Replace

2. Replace other engine mounting.

OP 6038-B   ENGINE FRONT MOUNTINGS — BOTH — RENEW
(Includes 6038-A and A1)

OP 6038-C   ENGINE FRONT MOUNTINGS — BOTH — CHECK TORQUE OF BOLTS

OP 6051-A   CYLINDER HEAD GASKET — RENEW
(Includes remove and install cylinder head assembly, clean cylinder head and block mating faces. Does not include decarbonise cylinder head or pistons)

Tools Required

PT.4063   Cylinder head gasket locating studs.

To Remove

1. Drain the engine coolant.
2. Remove the carburettor air cleaner and air intake cover.
3. Disconnect the radiator top hose and the heater hose at the cylinder head.
4. Disconnect the brake servo vacuum hose.
5. Disconnect the temperature gauge sender unit.
6. Disconnect the throttle and choke cables and the fuel pipe from the carburettors.

CORTINA - LOTUS

7. Detach the exhaust manifolds and move them clear of the cylinder head.
8. Remove the camshaft cover and gasket.
9. Slacken the timing chain tensioner.
10. Remove the camshaft sprockets and disconnect the timing chain.
11. Pull the plug leads off the sparking plugs.
12. Unscrew the cylinder head bolts evenly and lift off the cylinder head and gasket.

NOTE – DO NOT lay the cylinder head flat on its face as damage to the valves can occur.

To Replace

13. Fit the cylinder head assembly. Locate the cylinder head gasket, copper side uppermost, on the cylinder block. Screw the studs Tool No. PT.4063A into diagonally opposite bolt holes on the block face to locate the gasket. Locate a new gasket on top of the front cover. Install the cylinder head assembly, engaging the breather pipe in its bore.

14. Refit the cylinder head bolts before removing the locating studs and then tighten in the sequence shown on page 47 to 60 to 65 lb. ft. (8.29 to 8.98 kg.m.) torque. Tighten the three front cover bolts to 10 to 15 lb. ft. (1.38 to 2.07 kg.m.) torque.

15. Fit the camshaft sprockets and timing chain. Align the timing mark on the crankshaft pulley with the lower mark on the front cover and the timing marks on the sprockets adjacent to each other and level with the camshaft cover mounting face. Fit the exhaust sprocket first. Tighten the retaining bolts to 25 to 30 lb. ft. (3.41 to 4.15 kg.m.) torque. Rotate the engine one revolution in its normal direction of rotation and re-check valve timing.

16. Adjust the timing chain tension to give ½ in. (12.7 mm.) free movement between the camshaft sprockets. Ensure that there are no tight spots by turning the engine through several revolutions.

17. Check that the plug caps are secure on their respective leads then couple the plug leads to their respective plugs.

18. Fit the camshaft cover plugs and gasket and refit the cover.

19. Position the manifold gaskets on the head and fit the manifolds. Tighten the retaining nuts to 12 to 15 lb. ft. (1.66 to 2.07 kg.m.) torque.

20. Reconnect the fuel pipe and the throttle and choke cables to the carburettors.

21. Reconnect the temperature gauge sender unit.

22. Reconnect the brake servo vacuum hose.

23. Reconnect the radiator top hose and the heater hose.

24. Replace the carburettor air intake cover and fit the air cleaner.

25. Refill the cooling system with long life anti-freeze.

26. Retime the ignition (see operation No. 12100-A).

27. If necessary, adjust the timing chain tension dynamically. Refer to OP 6270-A.

OP 6051-A1 EXTRA: CYLINDER HEAD – DECARBONISE
(with cylinder head removed)

1. Remove the carburettors and air intake back plate as an assembly.
2. Remove the sparking plugs.
3. Using a suitable implement, remove all carbon deposits from cylinder head faces, cylinder head ports, valve heads and piston crowns. Care must be taken to ensure that carbon scraped off is prevented from contaminating any parts of the engine.
4. Refit the sparking plugs.
5. Fit the carburettors and back plate assembly. Locate new 'O' ring in the spacers and tighten the retaining nuts until there is 0.040 in. (1.02 mm.) clearance between the coils of the double coil spring washers. Connect the fuel pipe and refit the accelerator return spring bracket.

OP 6051-A2 EXTRA: PISTONS – DECARBONISE
(with cylinder head removed)

Using a suitable implement, remove all carbon deposits from the combustion bowl and piston crown. Care must be taken to ensure that carbon scraped off is prevented from contaminating any part of the engine, and that no damage is done to the piston by the scraping tool.

OP 6051-A3 EXTRA: ONE VALVE – REMOVE AND INSTALL
(does not include refacing seat or grind-in valve)

To Remove

1. Extract the tappet and adjustment shim.
2. Compress the valve springs and extract the collets. Remove the valve spring retainer and valve spring.
3. Remove the valve.

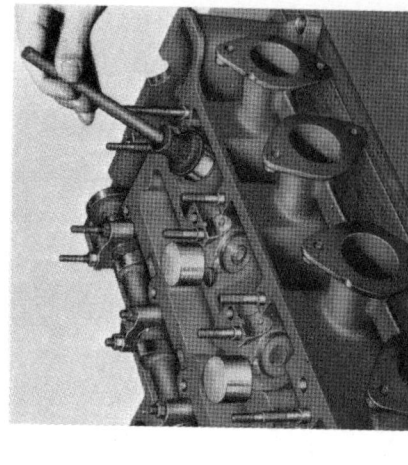

Removing the Tappets

To Replace

4. Fit the valve springs and retainer and compress to fit the collets.
5. Place the adjustment shim in the retainer recess on top of the valve stem and insert the tappet into its bore.
6. Refit the camshaft ensuring that the bearing liners are correctly located. Tighten the bearing cap nuts evenly to 9 lb. ft. (1.24 kg.m.) torque.
7. Check and adjust the valve clearance (see operation No. 6051-A8).

OP 6051-A4 EXTRA: EACH ADDITIONAL VALVE – REMOVE AND INSTALL

Using the procedure in Operation No. 6051-A3 remove and install each additional valve.

OP 6051-A5 EXTRA: EACH VALVE SEAT – RECUT
(with valve removed)

Tools Required

316X   Valve seat cutter
316-10   Valve seat cutter pilot
317   Valve seat cutter

1. Recut the valve seat to ensure that the seat is concentric with the valve stem bore. Use cutter Tool No. 317-25 fitted to pilot Tool No. 316-10 in handle Tool No. 316X.

OP 6051-A6 EXTRA: EACH VALVE – GRIND-IN
(with valve removed)

Tools Required

Valve lapping dolly

1. Grind or lap in the valves using a suitable grade of grinding paste.

OP 6051-A8 EXTRA: EACH VALVE GUIDE – RENEW
(with valve removed)

Tools Required

P.6054   Valve guide remover/replacer
Tap wrench
Valve guide reamer (0.3113 to 0.3123 in. (7.907 to 7.932 mm.) diameter).

To Remove

1. Remove the valve spring lower seat.
2. Remove the valve guide using Tool No. P.6054.

To Replace

3. Heat the cylinder head to 100 to 150°C (212 to 303°F) locate a circlip on the new valve guide and press a new guide into the cylinder head up to the circlip.
4. Ream the valve guide bore to 0.3113 to 0.3123 in. (7.907 to 7.932 mm.) using a suitable reamer and tap wrench.
5. Refit the valve spring lower seat.

OP 6051-A9 EXTRA: CYLINDER HEAD – RENEW
(with valves removed)

1. Remove the temperature gauge sender unit and adaptor.
2. Remove the heater water adaptor.
3. Remove the thermostat housing and thermostat and transfer to the new cylinder head.
4. Fit the heater water adaptor to the new cylinder head.
5. Replace the temperature gauge sender unit and adaptor on the new cylinder head. Apply EM-4G-52 sealer to the adaptor prior to fitting it to the cylinder head.
6. Remove the manifold studs and accelerator return spring bracket from the head being replaced and fit them to the new cylinder head. Alternatively fit new studs in the new cylinder head.

OP 6051-B CYLINDER HEAD AND PISTONS – DECARBONISE
(Includes 6051-A, A1 and A2)

OP 6051-C VALVE – ONE – REMOVE AND INSTALL
(Includes 6051-A, A1 and A3)
(Does not include reface valve seat or grind-in valve)

OP 6051-D VALVE – ONE – REMOVE AND INSTALL
(Includes 6051-A, A1, A3, A5 and A6)
(Includes reface valve seat and grind-in valve)

OP 6051-E VALVES – ALL – REMOVE AND INSTALL
(Includes 6051-A, A1, A3, A4, A5 and A6)

OP 6041-F DECARBONISE CYLINDER HEAD AND PISTONS AND RESEAT AND GRIND-IN ALL VALVES
(Includes 6051-A, A1, A2, A3, A4, A5 and A6)

OP 6051-I VALVE GUIDE – ONE – RENEW
(Includes 6051-A, A1, A3, A5, A6 and A8)

OP 6051-J VALVE GUIDES – ALL – RENEW
(Includes 6051-A, A1, A3, A4, A5, A6 and A8)

OP 6051-K CYLINDER HEAD – REMOVE AND INSTALL
(Includes 6051-A, A2, A3, A4, and A9)

CORTINA - LOTUS

OP 6051-M CYLINDER HEAD BOLTS – TORQUE
(includes adjust valve clearances)

Tools Required

P.6129   Cylinder head bolt socket.

1. Raise the bonnet and fit wing covers.
2. Remove the air cleaner.
3. Remove the camshaft cover.
4. Tighten the bolts evenly and in the sequence shown on page 46.
5. Measure and note the clearance for each valve.
6. Slacken the timing chain tensioner.
7. Set the valve timing marks.
8. Remove the camshaft sprockets and disconnect the timing chain.
9. Remove each camshaft.
10. Remove the tappets keeping them in their respective order.
11. Remove each tappet adjustment shim in turn and substitute one giving the correct clearance.
12. Refit the tappets in their respective bores.
13. Fit the camshafts, ensuring that the bearing liners are correctly located. Tighten the bearing cap nuts evenly to 9 lb. ft. (1.24 kg.m.) torque.
14. Re-check the valve clearances and readjust if necessary.
15. Fit the camshaft sprockets and timing chain. Align the timing marks on the crankshaft pulley with the lower mark on the front cover and the timing marks on the sprockets adjacent to each other and level with the camshaft cover mounting face. Fit the exhaust sprocket first. Tighten the retaining bolts to 25 to 30 lb. ft. (3.46 to 4.15 kg.m.) torque.
16. Adjust the timing chain tension to give ½ in. (12.7 mm.) free movement between the camshaft sprockets. Ensure that there are no tight spots in the chain by turning the engine through several revolutions.
17. Re-time the ignition (see operation No. 12100-A).
18. If necessary adjust the timing chain tension dynamically. Refer to OP 6270-A.

CORTINA - LOTUS

OP 6068-A ENGINE REAR MOUNTING INSULATOR – REPLACE

To Remove

1. With the handbrake applied, jack up the car front end and fit stands.
2. Suitably support the gearbox with a jack.
3. Disconnect the crossmember and spacers from the body floor pan.
4. Unscrew the engine rear mounting centre bolt and remove the crossmember.
5. Remove the engine mounting and retainer from the crossmember.

To Replace

6. Fit the engine mounting and retainer to the crossmember.
7. Locate the crossmember on the gearbox extension housing and fit the centre bolt.
8. Fit the crossmember and spacers to the body.
9. Remove the jack from under the gearbox.
10. Jack up the car, remove the stands, and lower the car to the ground.

OP 6250-A CAMSHAFT – ONE – AND/OR BEARINGS AND/OR TAPPETS – REPLACE

To Remove

1. Remove the air cleaner.
2. Remove the camshaft cover.
3. Slacken the timing chain tensioner.
4. Set the engine in the timing position. Remove the camshaft sprocket and disconnect the timing chain.
5. Remove the camshaft.
6. Extract the bearing liners and/or the tappets.

To Replace

7. Fit new tappets and/or bearing liners.
8. Fit the camshaft and tighten the bearing cap nuts evenly to 9 lb. ft. (1.24 kg.m.) torque.
9. Check and adjust the valve clearances (see operation No. 6051-A8).
10. Fit the camshaft sprocket and timing chain aligning the timing marks, as pre-set during removal. Tighten the retaining bolt to 25 to 30 lb. ft. (3.46 to 4.15 kg.m.) torque.
11. Adjust the timing chain tension to give ½ in. (12.7 mm.) free movement between the camshaft sprockets. Ensure that there are no tight spots by turning the engine through several revolutions.

12. Refit the camshaft cover.
13. Fit the air cleaner.
14. Re-time the ignition (see operation No. 12100-A).
15. If necessary adjust the timing chain tension dynamically. Refer to OP 6270-A.

OP 6250-A1 EXTRA: REMAINING CAMSHAFT AND/OR BEARINGS AND/OR TAPPETS – REMOVE AND INSTALL
(Includes adjust valves)

To Remove

1. Remove the camshaft sprocket and disconnect the timing chain.
2. Remove the camshaft.
3. Extract the bearing liners and/or tappets.

To Install

4. Fit new tappets and/or bearings.
5. Fit the camshaft and tighten the bearing cap nuts evenly to 9 lb. ft. (1.24 kg.m.) torque.
6. Check and adjust the valve clearances (see operation No. 6051-A8).
7. Fit the camshaft sprocket and timing chain aligning the timing marks. Tighten the retaining bolt to 25 to 30 lb. ft. (3.46 to 4.15 kg.m.) torque.

OP 6261-A AUXILIARY SHAFT BEARINGS – REPLACE
(Includes remove engine)

Tools Required

200 A or B	Engine stand
550	Driver handle
P.4008	Crown wheel and pinion backlash gauge
P.6031	Auxiliary shaft bearing bush remover/replacer
P.6031-3	Auxiliary shaft bearing bush remover/replacer adaptors
CP.6041	Crankshaft pulley remover
P.6107	Engine bracket
CP.6173	Crankshaft rear oil seal aligner
P.6150	Front cover oil seal aligner
P.6161	Front cover oil seal remover/replacer
P.6165	Crankshaft rear oil seal remover/replacer
P.7137	Spigot bearing replacer and clutch disc locator

To Remove

1. Remove the engine assembly as described in OP 6000-C.
2. Fit the bracket Tool No. P.6107 and mount the engine on a stand Tool No. 200 A.
3. Remove the camshaft cover.
4. Remove the timing chain tension adjuster.
5. Remove the crankshaft pulley using Tool No. CP.6041.
6. Remove the oil pump and filter assembly.
7. Invert the engine and remove the sump, gaskets and cork packing strips.
8. Remove the front cover.
9. Remove the crankshaft oil slinger.
10. Disconnect the timing chain.
11. Remove the auxiliary shaft sprocket.
12. Remove the front cover back plate and gaskets.
13. Remove the distributor and fuel pump.
14. Remove the auxiliary shaft thrust plate. Withdraw the auxiliary shaft.
15. Unscrew the pressure plate bolts evenly and detach the pressure plate and disc.
16. Remove the flywheel.
17. Remove the crankshaft rear oil seal carrier.
18. Unscrew the bolts several turns and tap them to release the big end caps. Completely unscrew the bolts and remove the caps. Push the pistons up into the cylinder bores.
19. Unscrew the main bearing cap bolts evenly and lift off each cap. Lift out the crankshaft. Remove the thrust washers and bearing liners.
20. Remove the auxiliary shaft bearing bushes using Tool No. P.6031 with adaptors Tool No. P.6031-3.
21. Check all the oilways to ensure that they are clear, apply EM-4G-52 sealing compound to the oil gallery plugs prior to refitting.

To Replace

22. Fit new auxiliary shaft bearing bushes again using Tool No. P.6031 and P.6031-3. Ensure that the oil holes in the bushes and cylinder block are aligned. The splits in the bushes should be upwards and outwards at 45° to the vertical.
23. Fit the main bearing liners and replace the crankshaft. Locate the crankshaft thrust washers with the oil grooves facing the crankshaft flange. Refit the main bearing caps and tighten the retaining bolts to 55 to 60 lb. ft. (7.60 to 8.29 kg.m.) torque.

24. Check the crankshaft end-float with feeler blades between the crankshaft and the thrust washers. This should be between 0.003 and 0.008 in. (0.08 and 0.203 mm.).

25. Turn the crankshaft as necessary to fit the connecting rod big ends to the crankpins. Tighten the bolts to a torque of 44 to 46 lb. ft. (6.08 to 6.36 kg.m.).

26. Fit a new crankshaft rear oil seal using remover/replacer Tool No. P.6165 and a 550 handle.

27. Fit a new gasket to the rear oil seal carrier using ESEE-M4G-1008A jointing compound at the ends. Secure the carrier to the cylinder block, aligning it with Tool No. CP.6173. Tighten the bolts evenly to 12 to 15 lb. ft. (1.66 to 2.07 kg.m.) torque and remove the aligner.

28. Locate the flywheel squarely on the crankshaft flange. Tighten the retaining bolts to 45 to 50 lb. ft. (6.22 to 6.91 kg.m.) torque for hexagonal bolts and 50 to 55 lb. ft. (6.91 to 7.60 kg.m.) for bi-hexagonal bolts.

29. Check the flywheel run-out using gauge Tool No. P.4008. This should not exceed 0.004 in. (0.10 mm.) total indicator reading.

30. Centralise the clutch disc with the hub assembly away from the flywheel using Tool No. P.7137 or P.7091-A. Tighten the bolts evenly to a torque of 12 to 15 lb. ft. (1.66 to 2.07 kg.m.) and remove the clutch disc locator.

31. Slide the auxiliary shaft into position. Fit the thrust plate in the auxiliary shaft groove. Tighten the retaining bolts to 5 to 7 lb. ft. (0.69 to 0.97 kg.m.) and bend up the locking tabs.

32. Check the auxiliary shaft end-float with feeler blades between the thrust plate and the auxiliary shaft flange. This should be between 0.002 and 0.007 in. (0.050 and 0.178 mm.).

33. Locate a new gasket on the cylinder head to timing cover joint.

34. Fit the front cover back plate. Align the water pump aperture with the front cover and seal aligner Tool No. P.6150 before tightening the single bolt to 5 to 7 lb. ft. (0.69 to 0.97 kg.m.) torque.

35. Fit the auxiliary shaft sprocket. Tighten the retaining bolts to 12 to 15 lb. ft. (1.66 to 2.07 kg.m.).

36. Locate the timing chain around the sprockets after aligning the timing marks.

37. Fit the oil slinger on the crankshaft.

38. Fit a new oil seal to the front cover using Tool No. P.6161 and a 550 handle.

39. Coat the front cover joint faces with ESEE-M4G-1008A jointing compound. Align the front cover with Tool No. P.6150 and tighten the ¼ in. nuts and bolts evenly to 5 to 7 lb. ft. (0.69 to 0.97 kg.m.) torque and the 5⁄16 in. to 10 to 15 lb. ft. (1.38 to 2.07 kg.m.) before removing the aligner.

40. Fit the timing chain tension adjuster and adjust the timing chain tension to give ½ in. (12.7 mm.) free movement between the camshaft sprockets. Ensure that there are no tight spots by turning the engine through several revolutions.

41. Locate a new gasket on the oil pump mounting flange and fit the oil pump and filter assembly. Tighten the retaining bolts to 12 to 15 lb. ft. (1.66 to 2.07 kg.m.).

Crankshaft, Flywheel, Connecting Rod and Piston

42. Fit new gaskets on the block flange using ESEE-M4G-1008A jointing compound at each end. Fit the cork packing strips, chamfered ends into the groove, again using ESEE-M4G-1008A jointing compound at the ends and refit the sump. Tighten the sump bolts to 6 to 8 lb. ft. (0.83 to 1.11 kg.m.) torque in the sequence shown on page 33.
43. Fit the crankshaft pulley aligning the pulley slot with the crankshaft key. Tighten the pulley retaining bolt to 24 to 28 lb. ft. (3.32 to 3.87 kg.m.) torque.
44. Replace the camshaft cover plugs and gasket and refit the cover.
45. Refit the distributor and fuel pump.
46. Remove the engine from the work stand.
47. Refit the engine assembly in the car as in OP 6000-C.

OP 6270-A   TIMING CHAIN TENSION – ADJUST (dynamic)

With the engine running:

1. Slacken the locknut.
2. Screw the adjuster in until the noise disappears.
3. Tighten the locknut.

   NOTE – A tight chain will whine and a slack chain will rattle.

OP 6271-A   CAMSHAFT COVER GASKET – RENEW

To Remove

1. Remove the air cleaner.
2. Remove the camshaft cover.

To Replace

3. Ensure mating faces are clean.
4. Fit a new gasket to the camshaft cover and refit the cover.
5. Refit the air cleaner.

OP 6271-A1   EXTRA: VALVE CLEARANCES – ADJUST (camshaft cover removed)

1. Measure and note the clearance for each valve.
2. Slacken the timing chain tensioner.
3. Set the valve timing marks.
4. Remove the camshaft sprockets and disconnect the timing chain.
5. Remove each camshaft.

6. Remove the tappets keeping them in their respective order.
7. Remove each tappet adjustment shim in turn and substitute one giving the correct clearance.
8. Refit the tappets in their respective bores.
9. Fit the camshafts, ensuring that the bearing liners are correctly located. Tighten the bearing cap nuts evenly to 9 lb. ft. (1.24 kg.m.) torque.
10. Re-check the valve clearances and readjust if necessary.
11. Fit the camshaft sprockets and timing chain. Align the timing marks on the crankshaft pulley with the lower mark on the front cover and the timing marks on the sprockets adjacent to each other and level with the camshaft cover mounting face. Fit the exhaust sprocket first. Tighten the retaining bolts to 25 to 30 lb. ft. (3.46 to 4.15 kg.m.) torque.
12. Adjust the timing chain tension to give ½ in. (12.7 mm.) free movement between the camshaft sprockets. Ensure that there are no tight spots in the chain by turning the engine through several revolutions.
13. Re-time the ignition (see operation No. 12100-A).

OP 6271-A2   EXTRA: TIMING CHAIN TENSION – ADJUST (Static)

1. Adjust the timing chain tension to give ½ in. (12.7 mm.) free movement between the camshaft sprockets. Ensure that there are no tight spots in the chain by turning the engine through several revolutions.

OP 6271-A3   EXTRA: CAMSHAFT SPROCKETS – REMOVE AND INSTALL (camshaft cover removed)

To Remove

1. Slacken the timing chain tension adjuster.
2. Remove the camshaft sprockets and disconnect the timing chain.

To Replace

3. Fit the camshaft sprockets and timing chain. Align the timing mark on the crankshaft pulley with the lower mark on the front cover and the timing marks on the sprockets adjacent to each other and level with the camshaft cover mounting face. Fit the exhaust sprocket first. Tighten the retaining bolts to 25 to 30 lb. ft. (3.46 to 4.15 kg.m.) torque.
4. Adjust the timing chain tension to give ½ in. (12.7 mm.) free movement between the camshaft sprockets. Ensure that there are no tight spots by turning the engine through several revolutions.
5. Re-time the ignition (see operation No. 12100-A).

OP 6271-B   VALVE CLEARANCES – ADJUST (Includes 6271-A and A1)

CORTINA - LOTUS

OP 6279-A   FRONT COVER OIL SEAL, GASKETS AND/OR TIMING CHAIN – RENEW
(Includes, remove radiator, sump, cylinder head and front cover back plate)

Tools Required

550	Driver handle
P.6150	Crankshaft front cover oil seal aligner
P.6161	Crankshaft front oil seal remover/replacer

To Remove

1. Drain the engine coolant by opening the drain plugs on the radiator and cylinder block.
2. Disconnect the radiator hoses at the engine.
3. Remove the radiator assembly.
4. Remove the air cleaner.
5. Remove the camshaft cover.
6. Remove the fan belt and then remove the fan and the water pump pulley.
7. Remove the crankshaft pulley, using suitable levers.
8. Remove the sump.
9. Remove the timing chain tension adjuster.
10. Remove the front cover.
11. Remove the crankshaft oil slinger.
12. Disconnect the timing chain.
13. Remove the auxiliary shaft sprocket.
14. Remove the front cover back plate and gaskets.
15. Remove the cylinder head. See OP 6051-A.

To Replace

16. Using a new gasket fit the front cover back plate. Align the water pump aperture with the front cover and seal aligner Tool No. P.6150 before tightening the single bolt to 5 to 7 lb. ft. (0.69 to 0.97 kg.m.) torque.
17. Fit the auxiliary shaft sprocket. Tighten the retaining bolts to 12 to 15 lb. ft. (1.66 to 2.07 kg.m.) torque and turn up the locking plate tabs.
18. Locate the crankshaft oil slinger in place.
19. Fit a new oil seal, using a remover/replacer Tool No. P.6161 and 550 handle. If required fit a new timing chain vibration damper.
20. Coat the front cover joint faces with ESEE-M4G-1008A jointing compound. Align the cover in position with Tool No. P.6150. Tighten the $\frac{1}{4}$ in. retaining nuts and bolts evenly to a torque of 5 to 7 lb. ft. (0.69 to 0.97 kg.m.) and the $\frac{5}{16}$ in. to 10 to 15 lb. ft. (1.38 to 2.07 kg.m.) and remove the aligner tool.
21. Locate a new gasket on the cylinder head to the timing cover joint. Refit the cylinder head. See OP 6051-A.
22. Locate the timing chain around the sprockets ensuring that there is minimum slack between the exhaust camshaft sprocket and the crankshaft sprocket.
23. Fit the timing chain tension adjuster and tighten to 45 to 50 lb. ft. (6.22 to 6.91 kg.m.) torque.
24. Fit the sump. Fit new gaskets on the block flange using ESEE-M4G-1008A jointing compound at each end. Fit the cork packing strips with the chamfered ends into the grooves, again using ESEE-M4G-1008A jointing compound and refit the sump. Tighten the bolts to 6 to 8 lb. ft. (0.83 to 1.11 kg.m.) torque.
25. Fit the crankshaft pulley aligning the pulley slot with the crankshaft key. Tighten the retaining bolt to 24 to 28 lb. ft. (3.32 to 3.87 kg.m.).
26. Replace the water pump pulley and the fan. Fit the fan belt and adjust the tension so that there is $\frac{1}{2}$ in. (12.7 mm.) total movement.
27. Adjust the timing chain tension to give $\frac{1}{2}$ in. (12.7 mm.) free movement between the camshaft sprockets. Ensure that there are no tight spots by turning the engine through several revolutions.
28. Fit the camshaft cover and tighten the retaining nuts evenly.
29. Fit the air cleaner.
30. Replace the radiator assembly.
31. Refit the radiator top and bottom hoses and tighten the clips.
32. Refill the radiator with long life anti-freeze.
33. Re-time the ignition (see operation No. 12100-A).
34. If necessary, set the timing chain dynamically. See OP 6270-A.

OP 6279-A1   EXTRA: AUXILIARY SHAFT – REMOVE AND INSTALL
(front cover assembly - removed)
(Includes remove fuel and oil pump and radiator grille)

To Remove

1. Remove the radiator grille. See OP M 1030-A, Section 12.
2. Remove the distributor. See OP 12100-A.
3. Remove the oil pump and filter assembly.
4. Disconnect the fuel pipes and remove the fuel pump.

5. Remove the auxiliary shaft sprocket adaptor.
6. Remove the auxiliary shaft thrust plate.
7. Withdraw the auxiliary shaft.

To Replace

8. Fit a new dowel to the new shaft.
9. Slide the auxiliary shaft into position.
10. Fit the thrust plate in the auxiliary shaft groove. Tighten the retaining bolts to 5 to 7 lb. ft. (0.69 to 0.97 kg.m.) and bend up the locking tabs.
11. Check the auxiliary shaft end-float with feeler blades between the thrust plate and the auxiliary shaft flange. This should be between 0.002 and 0.007 in. (0.05 and 0.178 mm.)
12. Refit the auxiliary shaft sprocket adaptor.
13. Fit the fuel pump and connect the fuel pipes.
14. Refit the oil pump and filter assembly, tighten the retaining bolts to 12 to 15 lb. ft. (1.66 to 2.07 kg.m.) torque.
15. Time the distributor (see operation No. 12100-A as part of operation No. 6279-A).
16. Refit the radiator grille. See OP M 1030-A, Section 12.

OP 6279-A2   EXTRA: CRANKSHAFT SPROCKET – REMOVE AND INSTALL

Tools Required

P.6032   Crankshaft sprocket replacer
P.6116   Crankshaft sprocket remover

To Remove

1. Remove the crankshaft sprocket using Tool No. P.6116.

To Install

2. Fit the crankshaft sprocket. Press the sprocket home using replacer Tool No. P.6032. Ensure that the long boss is adjacent to the main bearing journal.

OP 6280-A   AUXILIARY SHAFT – REMOVE AND INSTALL
(Includes OP 6279-A and A1)

OP 6303-A   CRANKSHAFT – REMOVE AND INSTALL
(Includes remove and install engine, connecting rod and main bearing liners and transferring crankshaft sprocket)

Tools Required

200A or B	Engine stand
550	Driver handle
P.4008	Crown wheel and pinion backlash gauge
CP.6041	Crankshaft pulley remover
P.6032B	Crankshaft sprocket replacer
P.6107	Engine bracket
P.6116	Crankshaft sprocket remover
CP.6173	Crankshaft rear oil seal aligner
P.6150	Crankshaft front cover oil seal aligner
P.6161	Front cover oil seal remover/replacer
P.6165	Crankshaft rear oil seal remover/replacer
P.7137	Spigot bearing replacer and clutch disc locator

To Remove

1. Remove the engine assembly as described in OP 6000-C.
2. Fit a bracket Tool No. P.6107 and mount the engine on a stand Tool No. 200 A or B.
3. Remove the camshaft cover.
4. Set the valve timing marks in the timed position.
5. Remove the crankshaft pulley using Tool No. CP.6041.
6. Unscrew the pressure plate bolts evenly and remove the pressure plate and clutch disc.
7. Remove the flywheel.
8. Remove the sump and gaskets.
9. Remove the timing chain tensioner.
10. Remove the front cover.
11. Remove the crankshaft oil slinger.
12. Disconnect the timing chain.
13. Remove the auxiliary shaft sprocket.
14. Remove the front cover back plate and gaskets.
15. Remove the crankshaft sprocket using Tool No. P.6116.
16. Remove the rear oil seal housing.
17. Unscrew the big end bolts two or three turns and tap them to release the caps. Completely unscrew the bolts and remove the caps. Push the pistons up into the cylinder bores.
18. Unscrew the main bearing cap bolts evenly and lift off each cap. Lift out the crankshaft and remove the bearing liners and thrust washers.

CORTINA - LOTUS

To Replace

19. Fit a new spigot bearing into the crankshaft using Tool No. P.7137.

20. Replace the crankshaft sprocket using replacer Tool No. P.6032A or B.

21. Fit the main bearing liners and replace the crankshaft. Fit the crankshaft thrust washers with the oil grooves facing the crankshaft flange. Refit the main bearing caps and tighten the bolts to 55 to 60 lb. ft. (7.60 to 8.29 kg.m.) torque.

22. Check the crankshaft end-float with feeler blades between the crankshaft and the thrust washers. This should be between 0.003 and 0.008 in. (0.08 and 0.20 mm.).

23. Turn the crankshaft as necessary to fit the connecting rod big ends to the crankpins. Tighten the connecting rod bolts to 44 to 46 lb. ft. (6.08 to 6.36 kg.m.) torque. Check the end-float on the crankpin, this should be 0.004 to 0.010 in. (0.10 to 0.25 mm.).

24. Fit a new crankshaft rear oil seal using remover/replacer Tool No. P.6165 and a 550 handle.

25. Locate a new gasket on the rear oil seal carrier using ESEE-M4G-1008A jointing compound at the ends and fit the carrier to the block rear face using an aligner Tool No. CP.6173. Tighten the bolts evenly to a torque of 12 to 15 lb. ft. (1.66 to 2.07 kg.m.) and remove the aligner.

26. Locate a new gasket on the cylinder head to timing cover joint.

27. Fit the front cover back plate. Align the water pump aperture with the front cover and seal aligner Tool No. P.6150 before tightening the single bolt to 5 to 7 lb. ft. (0.69 to 0.97 kg.m.) torque.

28. Fit the auxiliary shaft sprocket. Tighten the retaining bolts to 12 to 15 lb. ft. (1.66 to 2.07 kg.m.) torque and turn up the locking plate tabs.

29. Locate the timing chain around the sprockets ensuring that the crankshaft is reset in the timed position and that there is minimum slack between the exhaust camshaft sprocket and the crankshaft sprocket.

30. Fit the oil slinger on the crankshaft.

31. Fit a new oil seal to the front cover using Tool No. P.6161 and a 550 handle.

32. Coat the front cover joint faces with ESEE-M4G-1008A jointing compound. Secure the cover in place using an aligner Tool No. P.6150. Tighten the ¼ in. bolts evenly to a torque of 5 to 7 lb. ft. (0.69 to 0.97 kg.m.) and the 5/16 in. to 10 to 15 lb. ft. (1.38 to 2.07 kg.m.), and remove the aligner.

33. Fit the timing chain tension adjuster and tighten to a torque of 45 to 50 lb. ft. (6.22 to 6.91 kg.m.).

34. Adjust the timing chain tension to give ½ in. (12.7 mm.) free movement between the camshaft sprockets. Ensure that there are no tight spots by turning the engine through several revolutions.

35. Locate the flywheel squarely upon the crankshaft flange. Tighten the bolts evenly to 45 to 50 lb. ft. (6.22 to 6.91 kg.m.) torque for hexagonal bolts and 50 to 55 lb. ft. (6.91 to 7.90 kg.m.) for bi-hexagonal bolts.

36. Check the flywheel run-out using the gauge Tool No. P.4008 at the rim. The flywheel run-out should not exceed 0.004 in. (0.10 mm.) total indicator reading.

37. Centralise the clutch disc, with the hub assembly away from the flywheel using Tool No. P.7137. Tighten the bolts evenly to a torque of 12 to 15 lb. ft. (1.68 to 2.07 kg.m.) then remove the disc locator.

38. Fit the crankshaft pulley aligning the pulley slot with the crankshaft key. Tighten the retaining bolt to 24 to 28 lb. ft. (3.32 to 3.87 kg.m.) torque.

39. Fit new gaskets on the block flange using ESEE-M4G-1008A jointing compound at each end. Fit the cork packing strips with the chamfered ends into the grooves, again using ESEE-M4G-1008A jointing compound and refit the sump. Tighten the bolts to 6 to 8 lb. ft. (0.83 to 1.11 kg.m.) torque in the sequence shown on page 33.

40. Fit the camshaft cover and tighten the retaining nuts evenly.

41. Remove the engine from the stand and remove the bracket Tool No. P.6107.

42. Refit the engine assembly in the car as described in OP 6000-C.

OP 6312-A   CRANKSHAFT PULLEY – REMOVE AND INSTALL
(engine in-situ)

To Remove

1. Slacken the generator mounting bolts and remove the fan belt.
2. Remove crankshaft pulley using suitable levers.

To Install

3. Replace crankshaft pulley aligning the pulley slot with the crankshaft key. Tighten the retaining bolt to a torque of 24 to 28 lb. ft. (3.32 to 3.87 kg.m.).

4. Replace the fan belt and adjust the tension to give ½ in. (12.7 mm.) total movement. Tighten the generator mounting bolts to a torque of 15 to 18 lb. ft. (2.08 to 2.49 kg.m.).

OP 6335-A   CRANKSHAFT REAR OIL SEAL CARRIER – REMOVE AND INSTALL
(includes remove and install engine)

Tools Required

200A or B	Engine stand
P.4008	Crown wheel and pinion backlash gauge
P.6107	Engine bracket
CP.6173	Crankshaft rear oil seal aligner
P.7137	Spigot bearing replacer and clutch disc locator

CORTINA - LOTUS

To Remove

1. Remove the engine assembly as described in OP 6000-C.
2. Fit a bracket Tool No. P.6107 and mount the engine on a stand Tool No. 200A or B.
3. Unscrew the pressure plate bolts evenly and remove the pressure plate and clutch disc.
4. Remove the flywheel.
5. Remove the sump and gaskets.
6. Remove the rear oil seal carrier.

To Install

7. Locate a new gasket on the rear oil seal carrier using ESEE-M4G-1008A jointing compound at the ends, and fit the carrier to the block rear face using an aligner Tool No. CP.6173. Tighten the bolts evenly to a torque of 12 to 15 lb. ft. (1.66 to 2.07 kg.m.) and remove the aligner.
8. Locate the flywheel squarely upon the crankshaft flange. Tighten the bolts evenly to a torque of 45 to 50 lb. ft. (6.22 to 6.91 kg.m.) for hexagonal bolts and 50 to 55 lb. ft. (6.91 to 7.90 kg.m.) for bi-hexagonal bolts.
9. Check the flywheel run-out using the gauge Tool No. P.4008 at the rim. The flywheel run-out should not exceed 0.004 in. (0.10 mm.) total indicator reading.
10. Centralise the clutch disc, with the hub assembly away from the flywheel using Tool No. P.7091A. Tighten the bolts evenly to a torque of 12 to 15 lb. ft. (1.68 to 2.07 kg.m.) then remove the disc locator.
11. Fit new gaskets on the block flange using ESEE-M4G-1008A jointing compound in the grooves. Fit the cork packing strips with the chamfered ends into the grooves, again using ESEE-M4G-1008A jointing compound and refit the sump. Tighten the bolts to a torque of 6 to 8 lb. ft. (0.83 to 1.11 kg.m.) in the sequence shown on page 33.
12. Remove the engine from the stand and remove the bracket Tool No. P.6107.
13. Refit the engine assembly in the car as described in OP 6000-C.

OP 6335-A1   EXTRA: REAR OIL SEAL CARRIER AND/OR SEAL – RENEW
(with rear oil seal carrier removed)

See Operation No. 6000-C7.

OP 6335-B   CRANKSHAFT REAR OIL SEAL AND/OR CARRIER – RENEW
(Includes OP 6335-A and A1)

OP 6375-A   FLYWHEEL ASSEMBLY – REMOVE AND INSTALL
(Includes remove and install gearbox assembly, clutch disc and pressure plate)

Tools Required

P.4008   Crown wheel and pinion backlash gauge
P.7137   Spigot bearing replacer and clutch disc locator or P.7091A clutch disc locator.

To Remove

1. Remove the gearbox as described in OP 7003-A of Section 7.
2. Unscrew the pressure plate bolts evenly and remove the pressure plate and disc.
3. Remove the flywheel.

To Install

4. Locate the flywheel squarely upon crankshaft flange and tighten the retaining bolts evenly to a torque of 45 to 50 lb. ft. (6.22 to 6.91 kg.m.) for hexagonal bolts or 50 to 55 lb. ft. (6.91 to 7.90 kg.m.) for bi-hexagonal bolts.
5. Check the flywheel run-out using the gauge Tool No. P.4008 at the rim. The flywheel run-out should not exceed 0.004 in. (0.10 mm.) total indicator reading.
6. Centralise the clutch disc, with the hub assembly away from the flywheel using Tool No. P.7137 or P.7091-A. Tighten the bolts evenly to a torque of 12 to 15 lb. ft. (1.66 to 2.07 kg.m.) and remove the clutch disc locator.
7. Replace the gearbox in the car as described in OP 7003-A of Section 7.

OP 6375-A1   EXTRA: FLYWHEEL RING GEAR – RENEW
(flywheel removed)

To Remove

1. Cut between two adjacent teeth with a hacksaw and split the flywheel ring gear with a chisel.

To Replace

2. Heat the new ring gear evenly to a temperature not exceeding 600°F (316°C) and fit it to the flywheel with the chamfers on the leading faces of the teeth in the normal direction of rotation. Allow the ring to cool naturally in air. DO NOT QUENCH.

OP 6375-B   FLYWHEEL RING GEAR – RENEW
(Includes OP 6375-A and A1)

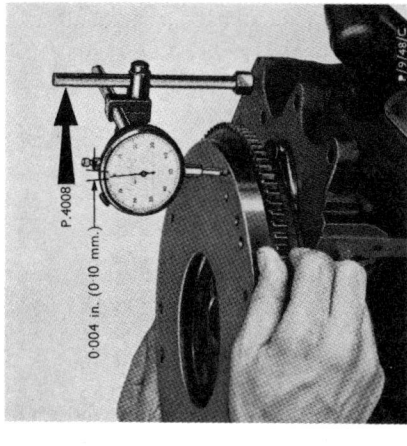

Checking the Flywheel Run-Out

CORTINA - LOTUS

OP 6450-A CAMSHAFT COVER AND/OR GASKET – REMOVE AND INSTALL
(engine in-situ)

To Remove
1. Remove the air cleaner.
2. Remove the camshaft cover.

To Install
3. Ensure mating faces are clean and free of old gasket material.
4. Fit a new gasket to the camshaft cover and refit the cover.
5. Refit the air cleaner.

OP 6450-A1 EXTRA: ALL VALVE CLEARANCES – ADJUST
(rocker cover removed)

1. Measure and note the clearance for each valve.
2. Set the valve timing marks.
3. Slacken the timing chain tensioner.
4. Remove the camshaft sprockets and disconnect the timing chain.
5. Remove each camshaft.
6. Remove the tappets keeping them in their respective order.
7. Remove each tappet adjustment shim in turn and substitute one giving the correct clearance.
8. Refit the tappets in their respective bores.
9. Fit the camshafts, ensuring that the bearing liners are correctly located. Tighten the bearing cap nuts evenly to 9 lb. ft. (1.24 kg.m.) torque.
10. Re-check the valve clearances and readjust if necessary.
11. Fit the camshaft sprockets and timing chain. Align the timing mark on the crankshaft pulley with the lower mark on the front cover and the timing marks on the sprockets adjacent to each other and level with the camshaft cover mounting face. Fit the exhaust sprocket first. Tighten the retaining bolts to 25 to 30 lb. ft. (3.46 to 4.15 kg.m.) torque.
12. Adjust the timing chain tension to give ½ in. (12.7 mm.) free movement between the camshaft sprockets. Ensure that there are no tight spots by turning the engine through several revolutions.
13. Re-time the ignition (see operation No. 12100-A).

OP 6450-B VALVE CLEARANCES – ADJUST
(Includes OP 6450-A and A1)

OP 6600-A OIL PUMP ASSEMBLY – REMOVE AND INSTALL
(engine in-situ)

To Remove
1. With the handbrake applied, jack up the front of the car and fit stands.
2. Remove the oil pump and filter assembly.

To Replace
3. Locate a new gasket on the oil pump mounting flange and fit the oil pump and filter assembly to the cylinder block. Tighten the bolts to 12 to 15 lb. ft. (1.66 to 2.07 kg.m.) torque.
4. Jack up the car, remove the stands and lower to the ground.

OP 6600-A1 EXTRA: OIL PUMP ASSEMBLY – OVERHAUL
(oil pump assembly removed)

To Dismantle
1. Remove the filter body and element and extract the sealing ring from the groove.
2. Remove the end plate and withdraw the rubber 'O' ring from the groove in the pump body.
3. Check the clearance between the lobes of the inner and outer rotors. This should not exceed 0.006 in. (0.15 mm.). The rotors are supplied as a matched pair only so that if clearance is excessive a new rotor must be fitted.
4. Check the clearance between the outer rotor and the housing, this should not exceed 0.010 in. (0.25 mm.). If clearance between the outer rotor and pump body is excessive a new rotor assembly and/or pump body should be fitted.

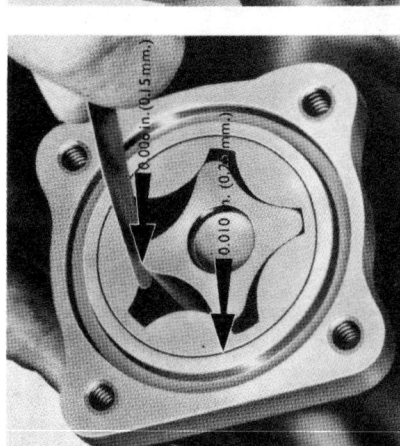

Checking the Eccentric Bi-Rotor Clearances

Checking the Eccentric Bi-Rotor Clearances

5. Place a straight edge across the face of the pump body and check the clearance between the face of the rotors and the straight edge. This should not exceed 0.005 in. (0.13 mm.). If this clearance is excessive the face of the pump body can be carefully lapped on a flat surface.

6. If it is necessary to renew the rotor or drive shaft, remove the outer rotor, then drive out the retaining pin securing the skew gear to the drive shaft and pull off the gear.

7. Withdraw the inner rotor and drive shaft.

To Reassemble

8. If the pump has been completely dismantled, fit the inner rotor and drive shaft assembly to the pump body. Press the skew gear onto the drive shaft end supporting the shaft, at the rotor end, on a suitable spacer. Replace the gear retaining pin and peen over the ends securely.

9. Install the outer rotor with its chamfered face inwards, towards the pump body.

10. Place the rubber 'O' ring in the groove in the pump body.

11. Fit the end plate with the machined face towards the rotors.

12. Locate a new filter body sealing ring in the groove and fit the filter assembly to the oil pump. Fit a new aluminium washer to the centre bolt and tighten to 12 to 15 lb. ft. (1.66 to 2.07 kg.m.).

OP 6600-B   OIL PUMP – OVERHAUL
(Includes OP 6000-A and A1)

OP 6675-A   SUMP AND/OR GASKET – REMOVE AND INSTALL
(engine in situ)

To Remove

2. Drain the cooling system.
3. Disconnect the battery.
4. Disconnect the bottom radiator hose at the engine end.
5. With the handbrake applied, jack up the front of the car and fit stands.
6. Position a jack under the gearbox.
7. Remove the front engine mounting bolts.
8. Remove the sump shield if fitted.
9. Remove the two bolts and withdraw the starter motor.
10. Remove the sump and gasket, adjusting the jack under the gearbox as necessary to raise the engine.
11. Clean the sump and cylinder block faces and remove the cork packing strips.

To Install

12. Fit new gaskets on the block flange, using ESEE-M4G-1008A jointing compound at each end. Fit the cork packing strips, with the chamfered ends into the grooves, again using ESEE-M4G-1008A jointing compound. Refit the sump and tighten the bolts evenly to a torque of 6 to 8 lb. ft. (0.83 to 1.11 kg.m.) in the sequence shown on page 33.
13. Replace the engine front mounting bolts and remove the jack.
14. Clean and replace the starter motor securing it with the two bolts.
15. Replace the sump shield if fitted.
16. Jack up the car, remove the stands and lower the car to the ground.
17. Reconnect the radiator bottom hose to the engine.
18. Reconnect the battery.
19. Refill the radiator with long life anti-freeze.
20. Refill the sump with the correct grade engine oil.
21. Run up the engine and check for oil or water leaks.
22. Check the oil level and top-up if necessary.

OP 6675-A3   EXTRA: ALL MAIN BEARING CLEARANCES – CHECK
(with sump removed)

See Operation No. 6000-C13, but support the crankshaft, with a jack adjacent to the bearing being measured, to transfer the clearance to the bottom of the journal.

OP 6675-A4   EXTRA: ALL MAIN BEARING LINERS AND/OR THRUST WASHERS – REMOVE AND INSTALL
(with sump removed)

See Operation No. 6000-C14

Checking the Eccentric Bi-Rotor End-float

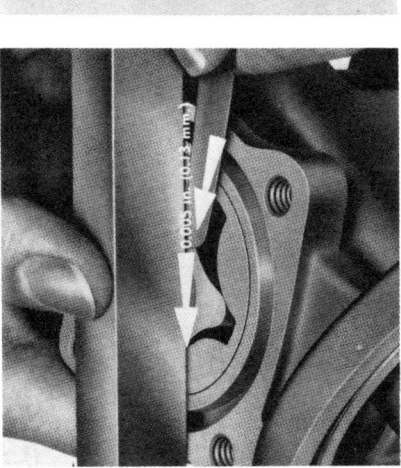

Oil Pressure Relief Valve

CORTINA - LOTUS

OP 6675-A5 EXTRA: ALL MAIN BEARING LINERS AND THRUST WASHERS – CHECK CLEARANCES AND RENEW
(with sump removed)

See Operation No. 6000-C15, but support the crankshaft, with a jack adjacent to the bearing being measured, to transfer the clearance to the bottom of the journal.

OP 6675-A6 EXTRA: ALL CONNECTING ROD LINERS – REMOVE AND INSTALL
(with sump removed)

See Operation No. 6000-C12

OP 6675-A7 EXTRA: ONE PISTON, CONNECTING ROD AND RINGS ASSEMBLY – REMOVE AND INSTALL
(Includes remove and install the cylinder head. Does not include any dismantling of the piston rod and rings assembly. Sump removed.)

Tools Required

38 U 3 Piston ring squeezer
PT.4063A Cylinder head gasket locating studs

To Remove

1. Lower the jack positioned under the gearbox.
2. Remove the cylinder head (see operation No. 6051-A).
3. Unscrew the big end bolts several turns and tap them to release the cap. Completely unscrew the bolts and remove the big end cap. Push the piston out of the bore and withdraw the assembly.
4. Remove the piston rings.
5. Extract the two piston pin circlips and push the pin out of the piston. Separate the piston and connecting rod.

Checking the Piston Ring Gap

To Install

6. Select a new piston if required to obtain a clearance of 0.003 to 0.0036 in. (0.0762 to 0.081 mm.) between the piston and piston bore.

   NOTE – Pistons are graded as listed below and the grades are stamped on the crown of the piston.

Grade	Diameter
1	3.2500 to 3.2503 in. (82.550 to 82.557 mm.)
2	3.2503 to 3.2506 in. (82.557 to 82.565 mm.)
3	3.2506 to 3.2509 in. (82.565 to 82.573 mm.)
4	3.2509 to 3.2512 in. (82.573 to 82.580 mm.)

7. Locate the piston rings in the unworn portion of the cylinder bore and check the ring gaps, which should be between 0.009 to 0.014 in. (0.23 to 0.36 mm.) for the compression rings and 0.010 to 0.020 in. (0.25 to 0.51 mm.) for the oil control rings.

8. Check piston ring to groove clearances which should be as follows:—

Upper compression ring	0.0016 to 0.0031 in. (0.041 to 0.079 mm.)
Lower compression ring	0.0016 to 0.0036 in. (0.041 to 0.091 mm.)
Oil control ring	0.0015 to 0.0030 in. (0.040 to 0.076 mm.)

9. Fit the piston rings, fitting the oil control ring first, followed by the lower and then the upper compression rings. Ensure that the compression rings are fitted the correct way up.

10. Assemble the piston to the connecting rod. Ensure that the "FRONT" marking on the connecting rod is on the same side of the assembly as the mark on the piston crown. Heat the piston in water or oil prior to inserting the piston pin. Retain the piston pin with the circlips.

Checking the Piston Ring to Groove Clearance

11. Position the oil control ring gap to the rear and the compression ring gaps to 150° on either side of this. Compress the rings using Tool No. 38 U 3 and push the piston into its cylinder bore with the arrow on the crown pointing towards the front of the engine. Turn the crankshaft as necessary to fit the connecting rod big end to the crank pin. Tighten the connecting rod bolts to a torque of 44 to 46 lb. ft. (6.08 to 6.36 kg.m.).

12. Refit the cylinder head (see operation No. 6051-A).

13. Lift the engine on the gearbox jack.

OP 6675-A8 EXTRA: EACH ADDITIONAL PISTON, CONNECTING ROD AND RINGS ASSEMBLY – REMOVE AND INSTALL
(with cylinder head removed)

Repeat sub-operations 3 to 11 of OP 6675-A7 for each additional connecting rod and piston assembly.

OP 6675-A9 EXTRA: EACH SET OF PISTON RINGS – RENEW
(with piston, connecting rod and rings assembly removed)
(Includes checking ring gaps and cleaning ring grooves)

Tools Required

Piston ring groove cleaner

1. Remove the piston rings.

2. Clean the ring grooves in the piston with a proprietary ring groove cleaner.

3. Locate the piston rings in the unworn portion of the cylinder bore and check the ring gaps, which should be between 0.009 to 0.014 in. (0.23 to 0.36 mm.).

4. Fit the piston rings, fitting the oil control ring first, followed by the lower and then the upper compression rings. Ensure that the compression rings are fitted the correct way up.

5. Check piston ring to groove clearances which should be as follows:—

   Upper compression ring ... ... 0.0016 to 0.0031 in. (0.041 to 0.079 mm.)
   Lower compression ring ... ... 0.0016 to 0.0036 in. (0.041 to 0.091 mm.)
   Oil control ring ... ... 0.0015 to 0.0030 in. (0.040 to 0.076 mm.)

OP 6675-A10 EXTRA: EACH PISTON, PIN AND RINGS ASSEMBLY – RENEW
(with piston, connecting rod and rings assembly removed)
(Includes checking ring gaps and fitting rings to piston)

1. Extract the two piston pin circlips and push out the piston pin.
   NOTE – It is permissible to heat the piston to a temperature of 120°C (248°F) to assist piston pin removal).

2. Select a new piston as detailed in OP 6675-A7.

3. Select a new piston pin from the grades listed below:—

PISTON PIN

Grade	Diameter
A	0.8121 to 0.8122 in. (20.627 to 20.630 mm.)
B	0.8122 to 0.8123 in. (20.630 to 20.632 mm.)

4. Assemble the piston to the connecting rod. Ensure that the "FRONT" marking on the connecting rod is on the same side of the assembly as the mark on the piston crown. Heat the piston in water or oil prior to inserting the piston pin. Retain the piston with the circlips.

5. Locate the piston rings in the cylinder bore and check the ring gaps which should be between 0.009 to 0.014 in. (0.23 and 0.36 mm.) for the compression rings and 0.010 to 0.020 in. (0.25 to 0.51 mm.) for the oil control rings.

6. Fit the piston rings, fitting the oil control ring first, followed by the lower and then the upper compression rings. Ensure that the compression rings are fitted the correct way up.

OP 6675-A11 EXTRA: EACH CONNECTING ROD AND/OR PISTON PIN – RENEW
(with piston, connecting rod and rings assembly removed)
(Includes measuring size of piston pin)

1. Extract the two piston pin circlips and push the pin out of the piston. Separate the piston and connecting rod.

2. Select a new connecting rod from the grades listed below, or alternatively, select a new piston pin from the grades listed in OP 6675-A10 sub-operation 3.

CONNECTING RODS

Grade	Small end bore	Colour Code
A	0.8124 to 0.81255 in. (20.635 to 20.639 mm.)	Silver
B	0.81255 to 0.8127 in. (20.639 to 20.643 mm.)	Green

3. Assemble the piston to the connecting rod. Ensure that the "FRONT" marking on the connecting rod is on the same side of the assembly as the mark on the piston crown. Heat the piston in water or oil prior to inserting the piston pin. Retain the pin with the circlips.

OP 6675-C MAIN BEARING CLEARANCES – ALL – CHECK
(Includes 6675-A, and A3)

OP 6675-D MAIN BEARING LINERS – ALL – RENEW
(Includes 6675-A and A4)

CORTINA - LOTUS

OP 6675-E   MAIN BEARING LINERS – ALL – CHECK AND RENEW
(Includes 6675-A and A5)

OP 6675-F   CONNECTING ROD LINERS – ALL – RENEW
(Includes 6675-A and A6)

OP 6675-G   PISTON RINGS – ONE PISTON – REMOVE AND INSTALL
(Includes 6675-A and A9)

OP 6675-H   PISTON RINGS – ALL – REMOVE AND INSTALL
(Includes 6675-A, A7, A8 and A9)

OP 6675-I   PISTON – ONE – REMOVE AND INSTALL
(Includes 6675-A, A7 and A10)

OP 6675-J   PISTONS – ALL – REMOVE AND INSTALL
(Includes 6675-A, A7, A8 and A10)

OP 6675-K   CONNECTING ROD AND/OR PISTON PIN – ONE – REMOVE AND INSTALL
(Includes 6675-A, A7 and A11)

OP 6675-L   CONNECTING RODS AND PISTON PINS – ALL – REMOVE AND INSTALL
(Includes 6675-A, A7, A8 and A11)

OP 6675-M   SUMP BOLTS – TORQUE
(Includes removing and installing any components necessary to gain access to all the sump bolts)

1. Remove the sump shield if fitted.
2. Tighten the sump bolts following the procedure described on page 33.
3. Refit the sump shield if fitted.

OP 6731-A   OIL FILTER ELEMENT – RENEW
(Does not include change engine oil)

1. Unscrew the securing bolt and withdraw the filter body and element.
2. Remove the sealing ring from the groove in the filter body mounting flange and the washer from the securing bolt.
3. Thoroughly clean the filter body.
4. Insert the new element into the body and fit the sealing ring, supplied with the element, to the groove in the filter body mounting flange.
   NOTE – Do not fit the ring at one point and then work it round the groove, as the rubber may stretch, thus leaving a surplus which may cause an oil leak.
5. Refit the filter assembly to the oil pump body and tighten the securing bolt to a torque of 12 to 15 lb. ft. (1.66 to 2.07 kg.m.).

OP 6731-B   OIL FILTER ELEMENT AND ENGINE OIL – RENEW

1. Remove the sump plug and drain the engine oil while it is hot.
2. Renew the oil filter, see Operation No. 6731-A.
3. Refit the sump drain plug and tighten to a torque of 20 to 25 lb. ft. (2.76 to 3.46 kg.m.).
4. Refill the sump with an approved grade of engine oil, see Lubrication and Maintenance section.

OP 6900-B   TEMPERATURE GAUGE SENDER UNIT – REMOVE AND INSTALL

To Remove
1. Raise the bonnet and fit wing covers.
2. Disconnect the lead.
3. Ensure that the radiator filler cap is secure and remove the sender unit.

To Install
4. Install the sender unit and top-up the radiator if any coolant has been lost.
5. Reconnect the lead.
6. Remove the wing covers and close the bonnet.

CORTINA - LOTUS

6/2
EXHAUST SYSTEM

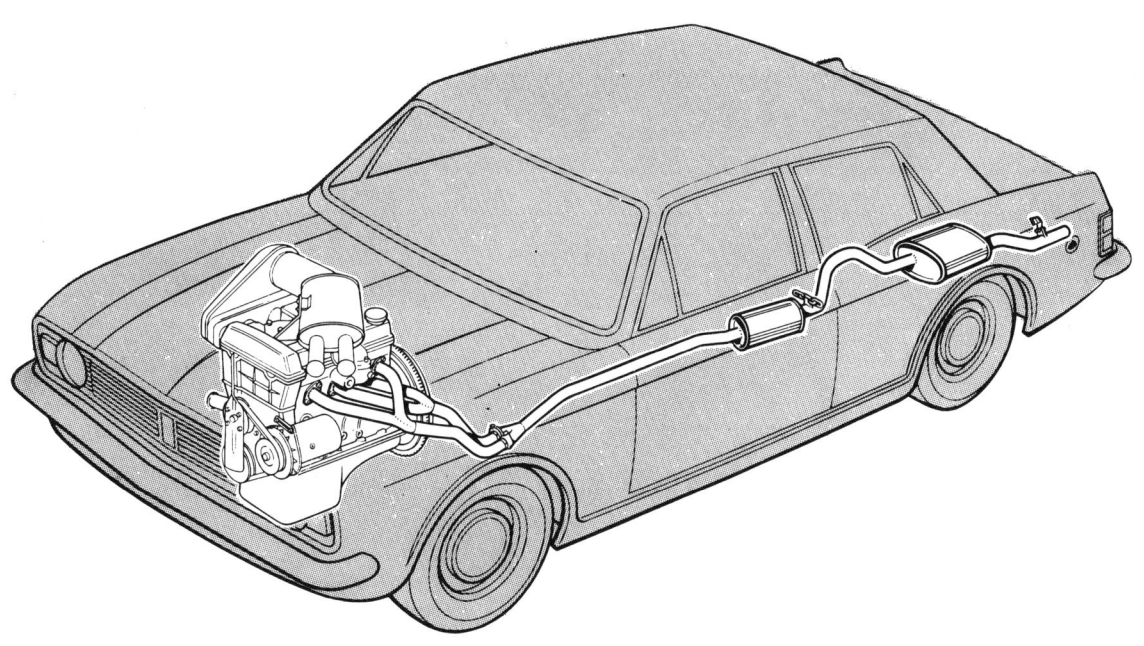

CORTINA - LOTUS

SECTION INDEX

GENERAL DESCRIPTION
QUICK REFERENCE DATA
SERVICE AND REPAIR OPERATIONS

OPERATION 5220-A   EXHAUST SYSTEM – RENEW
"   5230-A   INLET PIPE AND FRONT MUFFLER – RENEW
"   5231-A   REAR MUFFLER AND TAIL PIPE – RENEW

GENERAL DESCRIPTION

The exhaust system is in two parts consisting of an inlet pipe and front muffler assembly and a rear muffler and tail pipe assembly, the sections being joined just in front of the rear axle. The exhaust is routed along the driveshaft tunnel to the front muffler then over the rear axle to the rear muffler and a tail pipe alongside the fuel tank. Suspension is by two "O" rings hooked on brackets, welded to the floor pan and the front end of the inlet pipe to the rear muffler, and by a support strap attached to the tail pipe.

The front muffler is circular in section and is of the straight-through absorption type, whereas the rear is elliptical in shape and incorporates both a baffled and an absorption section.

QUICK REFERENCE DATA

DATA

Tightening Torques

Clamp bolts (old type)—Manifold to inlet pipe	7 to 10 lb. ft. (0.97 to 1.38 kg.m.)
—Front muffler to rear inlet pipe	7 to 10 lb. ft. (0.97 to 1.38 kg.m.)
Clamp nuts (new type)—Manifold to inlet pipe	12 to 15 lb. ft. (1.66 to 2.07 kg.m.)
—Front muffler to rear inlet pipe	12 to 15 lb. ft. (1.66 to 2.07 kg.m.)
Tail pipe support strap—bracket to body—bolts	12 to 15 lb. ft. (1.66 to 2.07 kg.m.)
—strap to tail pipe—bolts	5 to 7 lb. ft. (0.70 to 0.97 kg.m.)

CORTINA - LOTUS

OP 5220-A   EXHAUST SYSTEM – RENEW

To Remove

1. Jack up the car, support on stands and remove the jack.
2. Slacken the clamps at the inlet pipe to exhaust manifold joints and slide forward.
3. Unscrew the bolts from the tail pipe bracket to disconnect the support strap and reinforcing plate.
4. Unhook the "O" rings supporting the front end of the rear muffler inlet pipe.
5. Disconnect the exhaust system from the exhaust manifolds and remove the assembly.
6. Slacken the clamp and separate the front muffler and inlet pipe from the rear muffler and tail pipe.

To Install

7. Position the clamp on the inlet pipe to the rear muffler and connect the front muffler and inlet pipe to the rear muffler and tail pipe. Loosely clamp together in approximate alignment.
8. Position the exhaust system beneath the car and engage with the exhaust manifolds.
9. Fit the "O" rings to the brackets to support the rear muffler inlet pipe.
10. Connect the support strap.
11. Align the exhaust system and tighten all the clamp bolts to a torque of 7 to 10 lb. ft. (0.97 to 1.38 kg.m.).
12. Jack up the car, remove the stands, lower the car to the ground and remove the jack.

OP 5230-A   INLET PIPE AND FRONT MUFFLER – RENEW

To Remove

1. Jack up the car, support on stands and remove the jack.
2. Slacken the clamps at the inlet pipe to exhaust manifold joints and slide forward.
3. Slacken the clamp at the front muffler to rear inlet pipe joint and slide rearwards.
4. Withdraw the inlet pipe and front muffler from the exhaust manifold and rear inlet pipe.

To Install

5. Engage the inlet pipe and front muffler on the exhaust manifold and in the rear inlet pipe.
6. Position the clamps and tighten to a torque of 7 to 10 lb. ft. (0.97 to 1.38 kg.m.) if old type clamps are used or 12 to 15 lb. ft. (1.66 to 2.07 kg.m.) if new type clamps are used.
7. Jack up the car, remove the stands, lower the car to the ground and remove the jack.

OP 5231-A   REAR MUFFLER AND TAIL PIPE – RENEW

To Remove

1. Jack up the car, support on stands and remove the jack.
2. Slacken the clamp at the front muffler to rear inlet pipe joint.
3. Unscrew the bolts from the tail pipe bracket to disconnect the support strap and reinforcing plate.
4. Unhook the "O" rings supporting the front end of the rear muffler inlet pipe.
5. Disconnect the rear muffler and tail pipe from the front muffler and remove the assembly.

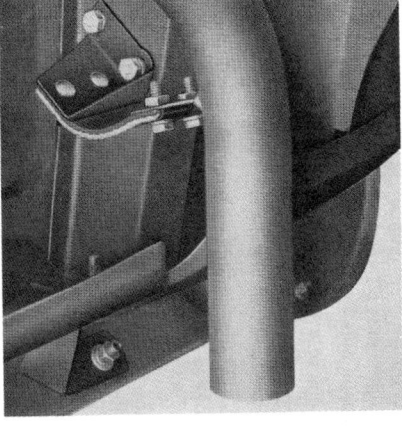

Tail Pipe Support Strap

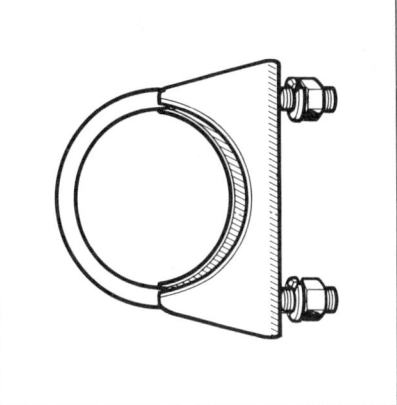

Manifold and Front Muffler to Inlet Pipe Clamp (new type)

Front Muffler to Rear Inlet Pipe Clamp (old type) and Support Bracket

Manifolds to Inlet Pipe Clamps (old type)

To Install

6. Locate the clamp on the front end of the rear muffler inlet pipe and position the rear muffler and tail pipe assembly beneath the car and engage with the front muffler.

7. Connect the support strap, and tighten the bolts to a torque of 5 to 7 lb. ft. (0.07 to 0.97 kg.m.).

8. Fit the "O" rings to the brackets to support the rear muffler inlet pipe.

9. Align the rear muffler and tail pipe and tighten the clamp bolt to a torque of 7 to 10 lb. ft. (0.97 to 1.38 kg.m.) if old type clamp is used or 12 to 15 lb. ft. (1.66 to 2.07 kg.m.) if new type clamp is used.

10. Jack up the car, remove the stands, lower the car to the ground and remove the jack.

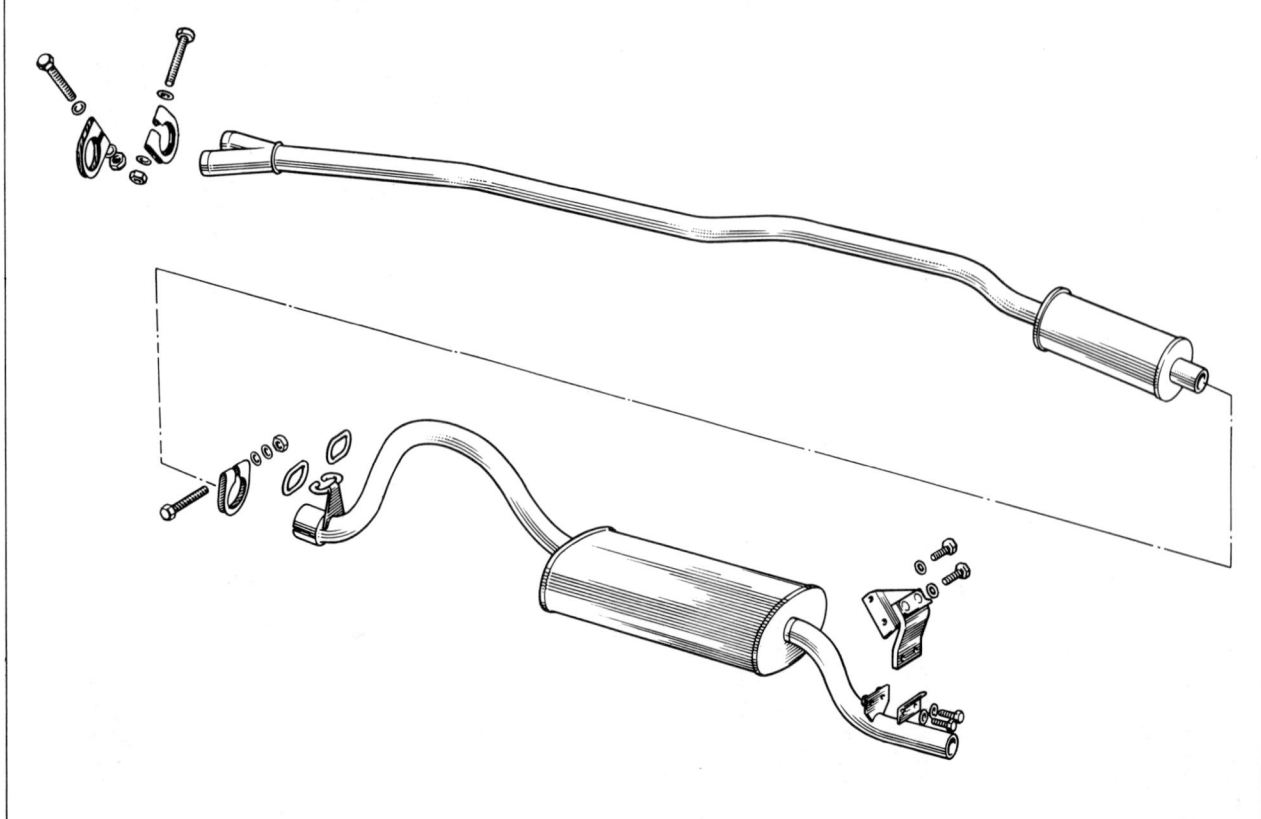

Exhaust System — Exploded

CORTINA - LOTUS

7
CLUTCH and GEARBOX

1 . 1969　　　　　　　　　　　　　　　　　　　　　　　　　　　　　　　　Section 7 — 1

CORTINA - LOTUS

SECTION INDEX

GENERAL DESCRIPTION

OPERATION	7000-A	GEARBOX AND CLUTCH HOUSING ASSEMBLY – REMOVE AND INSTALL
"	7000-A1	Extra: clutch release bearing – renew
"	7000-A2	Extra: clutch pressure plate and/or disc – remove and install
"	7000-A3	Extra: clutch pilot bearing – renew
"	7000-A6	Extra: main drive gear bearing retainer oil seal – renew
"	7000-A7	Extra: extension housing rear oil seal – renew
"	7000-A8	Extra: extension housing rear bush – renew
"	7000-B	CLUTCH RELEASE BEARING – RENEW (Includes OPS 7000-A and A1)
"	7000-C	CLUTCH DISC AND/OR PRESSURE PLATE – REMOVE AND INSTALL (Includes OPS 7000-A and A2)
"	7000-D	CLUTCH PILOT BEARING – RENEW (Includes OPS 7000-A, A2 and A3)
"	7003-A	GEARBOX ASSEMBLY – REMOVE AND INSTALL
"	7003-A1	Extra: selector mechanism – overhaul
"	7003-A2	Extra: extension housing and mainshaft assembly – remove and install
"	7003-A3	Extra: third top synchroniser assembly – overhaul
"	7003-A4	Extra: first/second synchroniser assembly – overhaul
"	7003-A5	Extra: extension housing – overhaul
"	7003-A6	Extra: main drive gear bearing retainer oil seal – renew
"	7003-A7	Extra: main drive gear – remove and install
"	7003-A8	Extra: countershaft gear – remove and install
"	7003-A9	Extra: reverse idler gear – remove and install
"	7003-A10	Extra: reverse selector fork – remove and install
"	7003-B	SELECTOR MECHANISM – OVERHAUL (Includes OPS 7003-A and A1)
"	7003-C	THIRD/TOP SYNCHRONISER ASSEMBLY – OVERHAUL (Includes OPS 7003-A, A1, A2 and A3)
"	7003-D	FIRST/SECOND SYNCHRONISER ASSEMBLY – OVERHAUL (Includes OPS 7003-A, A1, A2 and A4)
"	7003-E	MAIN DRIVE GEAR BEARING RETAINER OIL SEAL – RENEW (Includes OPS 7003-A and A6)
OPERATION	7003-F	MAIN DRIVE GEAR – REMOVE AND INSTALL (Includes OPS 7003-A, A1, A6 and A7)
"	7003-G	COUNTERSHAFT GEAR – REMOVE AND INSTALL (Includes OPS 7003-A, A1, A2, A6, A7 and A8)
"	7003-H	REVERSE IDLER GEAR – REMOVE AND INSTALL (Includes OPS 7003-A, A1, A2 and A9)
"	7003-J	REVERSE SELECTOR FORK – REMOVE AND INSTALL (Includes OPS 7003-A, A1, A2, A9 and A10)
"	7003-K	GEARBOX ASSEMBLY – OVERHAUL (Includes OPS 7003-A, A1, A2, A3, A4, A5, A6, A7, A8, A9 and A10 where applicable)
"	7003-L	EXTENSION HOUSING – REMOVE AND INSTALL (Includes OPS 7003-A, A1 and A2)
"	7202-A	GEAR LEVER – REMOVE AND INSTALL
"	7500-A	CLUTCH HYDRAULIC SYSTEM – BLEED
"	7501-B	CLUTCH HYDRAULIC SLAVE CYLINDER – REMOVE AND INSTALL
"	7501-B1	Extra: clutch hydraulic slave cylinder – overhaul
"	7501-C	CLUTCH HYDRAULIC SLAVE CYLINDER – OVERHAUL (Includes OPS 7501-B and B1)
"	7534-A	CLUTCH HYDRAULIC MASTER CYLINDER – REMOVE AND INSTALL
"	7534-A1	Extra: clutch hydraulic master cylinder – overhaul
"	7534-B	CLUTCH HYDRAULIC MASTER CYLINDER – OVERHAUL (Includes OPS 7534-A and A1)
"	7657-A	EXTENSION HOUSING OIL SEAL – RENEW
"	7657-A1	Extra: extension housing bush – renew
"	7657-B	EXTENSION HOUSING BUSH – RENEW (Includes OPS 7657-A and A1)

CORTINA - LOTUS

GENERAL DESCRIPTION

CLUTCH

The clutch mechanism consists of a single dry plate disc with a diaphragm spring pressure plate bolted to the engine flywheel. The clutch release system is self-adjusting and is hydraulically actuated. A flexible pipe line connects the operating cylinder with the master cylinder bolted to the engine rear bulkhead.

The diaphragm spring is pivoted on specially shouldered pins and retained to these pins by two fulcrum rings. The spring is retained to the pressure plate by three spring steel clips which are riveted to the pressure plate. When the diaphragm's centre is moved towards the flywheel by the release bearing the diaphragm's outer edge deflects towards the gearbox causing the clutch to disengage.

GEARBOX

The gearbox has four forward ratios and one reverse. All the forward gears are engaged through forged blocker ring synchromesh units.

The constant mesh gears, i.e. the main drive gear, the three other forward gears on the mainshaft and the corresponding gears on the countershaft, are helically cut to promote quiet running. The reverse gear on the countershaft incorporates straight cut spur teeth which mesh, through an idler gear, with teeth machined on the outside of the first/second gear synchroniser sleeve. Gear selection is by means of a remote floor change lever. The gear lever operates directly on the single selector rail which has pinned to it a selector lever. Movement of the gear lever, and thus the selector rail, causes the selector lever to "pick up" in the appropriate selector fork and move it to the required position. Movement of the first/second or third/top selector fork causes the appropriate synchroniser sleeve to move as necessary so that it engages with the dog teeth on the required gear. When reverse gear is required, movement of the reverse selector relay lever draws the reverse idler into mesh with both the reverse gear on the countershaft and on the mainshaft.

Engagement of two gears at once is prevented by a guard cam pivoted in the gearbox top cover on the right-hand sice. This engages with the selector forks which are not in use and holds them positively in the disengaged position. It is emphasised that the selector forks are not attached to the selector rail. The first/second and third/top forks, while they are mounted on the rail, are free to slide. Gear engagement only takes place when the selector lever locates in the appropriate fork and moves it into the required position.

Two selective circlips compensate for the tolerances which must be allowed in manufacture. Excessive end-float or backlash can be eliminated without resorting to unnecessarily close, and thus expensive, limits and fits. Whenever overhauling the gearbox, or a part of the gearbox that involves removing a circlip, always fit new circlips, never replace circlips that have been used. Of course, every care should be taken to ensure that the correct size circlip is selected and properly fitted.

Routine maintenance is confined to topping-up the gearbox oil at 6,000 mile (10,000 km.) intervals. At the first 3,000 miles (5,000 km.) the gearbox oil must be drained and renewed.

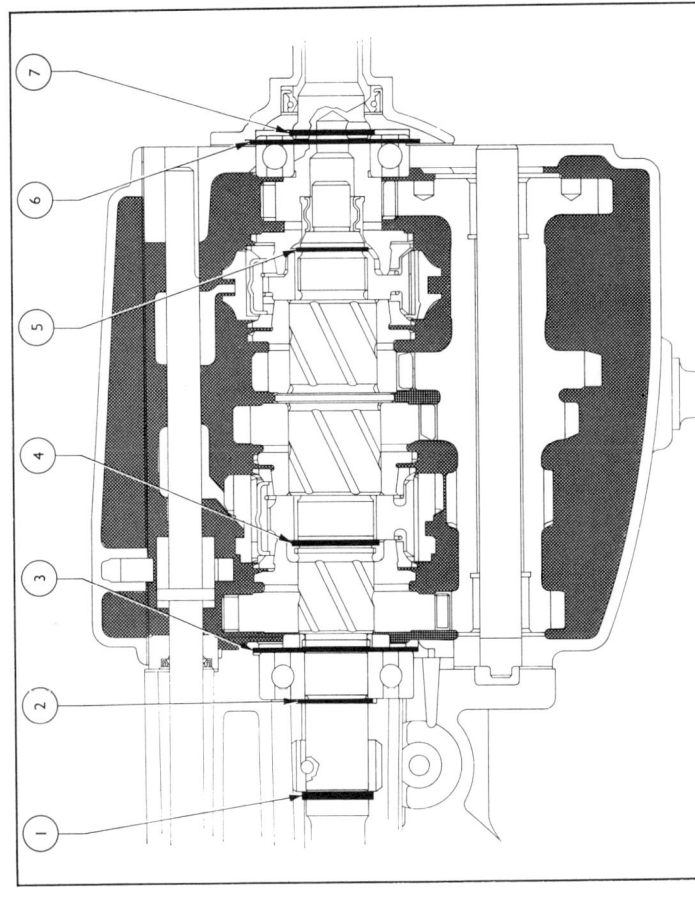

Circlip Locations (2 and 3 are selective)

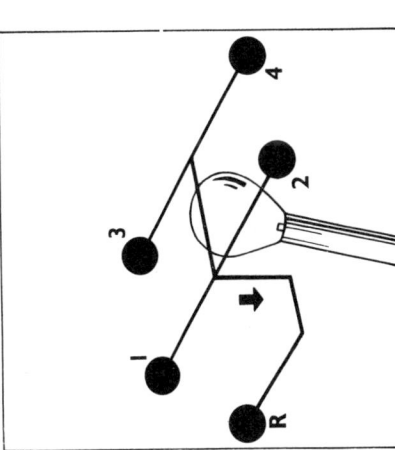

Gear Lever Positions

QUICK REFERENCE DATA

PERIODIC SERVICE ATTENTION

At first 600 miles (1,000 km.)

Check clutch reservoir fluid level

Top-up to the correct level, if necessary, approximately ½ in. (12.7 mm.) below the top face, with approved fluid, Part No. ME-3833-F. The cap and the area surrounding it should be wiped with a clean rag before removing the cap, to prevent dirt entering when it is removed.

Ensure that the air vent in the filler cap is clear before replacing the cap.

At first 3,000 miles (5,000 km.)
   (a) Check clutch reservoir fluid level
   (b) Change the gearbox oil

Every 6,000 miles (10,000 km.)
   (a) Check clutch reservoir fluid level
   (b) Top-up gearbox oil

Every 36,000 miles (60,000 km.)
   (a) Renew all clutch seals and fluid

DATA

Clutch

Type	Single dry plate
Actuation	Hydraulic
Hydraulic fluid	ME-3833-F
Master cylinder diameter	0.70 in. (1.778 cm.)
Operating cylinder diameter	0.875 in. (2.22 cm.)
Clutch disc lining – outside diameter	8 in. (20.4 cm.)
Pressure plate – diameter	8.8 in. (22.43 cm.)
– to flywheel bolt tightening torque	12 to 15 lb. ft. (1.66 to 2.07 kg.m.)

Gearbox

Grade of lubricant	S.A.E. 80 E.P.
Gearbox capacity	2⅜ Imp. pints (2.85 U.S. pints, 1.35 litres)
Amount required to refill gearbox in service	1¾ Imp. pints (2.00 U.S. pints, 0.992 litres)
Speedometer driven gear	23
Speedometer driving gear	7
Ratios—First	2.972:1 Third 1.397:1
—Second	2.010:1 Top 1.000:1
	Reverse 3.324:1

Tightening Torques, lb. ft. (kg.m.)

Clutch pressure plate to flywheel	12 to 15 (1.66 to 2.07)
Clutch housing to transmission case	40 to 45 (5.53 to 6.22)
Transmission case drain and filler plugs	25 to 30 (3.46 to 4.15)
Transmission extension to transmission case	30 to 35 (4.15 to 4.84)

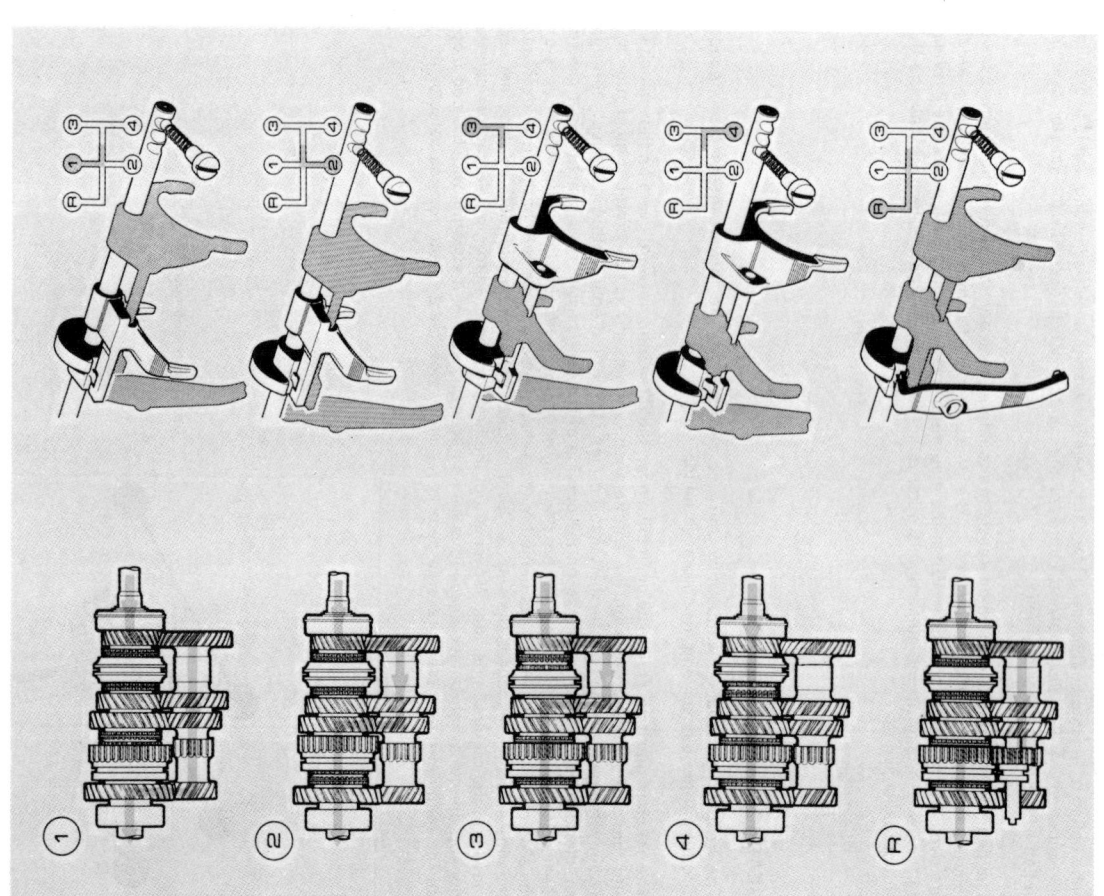

Power Train and Selector Positions in each gear

CORTINA - LOTUS

SERVICE AND REPAIR OPERATIONS

OP 7000-A GEARBOX AND CLUTCH HOUSING ASSEMBLY – REMOVE AND INSTALL

To Remove

1. Open the luggage compartment lid and disconnect the battery. Raise the bonnet.
2. Disconnect the throttle linkages at the carburettors.
3. Remove the centre console after removing four crosshead securing screws and the gear lever knob. Bend up lock tab, unscrew the plastic dome nut and withdraw the gear lever.
4. Jack up the car and fit stands all round.
5. Remove the four bolts joining the flanges of the two halves of the drive shaft. Lower the rear half of the drive shaft. Remove the bolts securing the centre bearing carrier to the floor pan, lower the carrier and bearing assembly and slide the front yoke from the gearbox. Fit a dummy yoke in the gearbox to prevent oil loss.
6. Remove plastic clip and handbrake cable from drive shaft support bracket.
7. Unscrew the forked retainer or remove the circlip, as applicable, and disconnect the speedometer cable and driven gear.
8. Remove the clutch operating cylinder circlip and move the cylinder to one side.
9. Remove the two starter motor securing bolts and move the starter to one side.
10. Remove the bolts securing the clutch housing to the engine. Note that a top bolt also secures an earth strap.
11. Remove the bolts securing the lower dust cover and detach the cover.
12. Place a support jack beneath the rear of the engine.
13. Remove the bolts securing the gearbox crossmember to the body. Slide the gearbox rearwards whilst supporting its weight and detach it from the engine.
14. Remove the crossmember centre bolt and detach it from the gearbox.

To Install

15. Place the crossmember on the rubber mounting and fit the centre bolt to attach it to the gearbox.
16. Ensure that the adaptor plate is positioned on the rear of the engine and that the main drive gear spline is lightly smeared with Molygrease. Offer the gearbox up so that the main drive gear spigot enters the crankshaft pilot bearing. Push the gearbox fully home.
17. Replace the four crossmember-to-body bolts and spring washers. Remove the engine support jack.
18. Replace the bolts securing the clutch housing to the engine. The uppermost pair are plain bolts and one also secures an earth strap. The remainder are dowel bolts.
19. Replace the lower dust cover.
20. Refit the starter motor.
21. Replace the clutch operating cylinder and retain it with a circlip.
22. Replace the speedometer driven gear and retain it with the forked retainer or circlip, as applicable.
23. Refit the handbrake cable and plastic clip to the drive shaft bracket.
24. Refit the drive shaft yoke on the gearbox and bolt the bearing carrier to the floor pan. Replace the four bolts that secure the front and rear halves of the drive shaft together.
25. Jack up the car and remove the body stands.
26. Refit the gear lever lower gaiter and retain it beneath the carpet to the floor pan with the metal ring. Fit the gear lever and turret. Replace the centre console and upper gaiter and fit the gear lever knob.
27. Reconnect the throttle linkage at the carburettors.
28. Reconnect the battery earth lead.
29. Top-up the gearbox with oil as required, ensuring that the car is level.

OP 7000-A1 EXTRA: CLUTCH RELEASE BEARING – RENEW

To Remove

1. Remove the release arm rubber gaiter.
2. Withdraw the release arm and bearing assembly from the clutch housing.

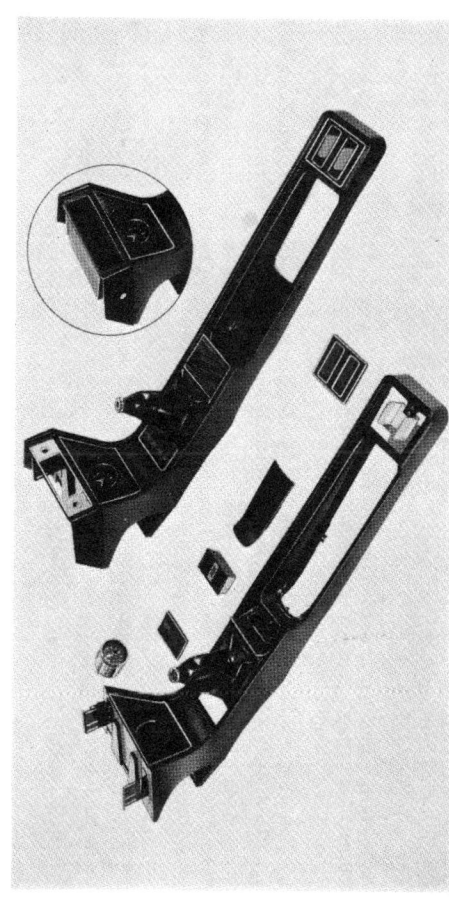

The Centre Console

CORTINA - LOTUS

3. Unhook the release arm from the bearing.

4. Engage the release arm in the slots of the release bearing hub. Apply molygrease between fork and hub.

5. Pass the release arm through the aperture in the clutch housing and slide the release bearing onto the main drive gear bearing retainer. Previously apply a smear of molygrease to bearing retainer.

6. Replace the rubber gaiter.

OP 7000-A2   EXTRA: CLUTCH PRESSURE PLATE AND/OR DISC – REMOVE AND INSTALL

Tools Required

P.7091-A   Clutch disc locator (modified) or Tool P.7137

1. Slacken the six clutch retaining bolts evenly, working diagonally across the clutch.

2. Remove the clutch disc and pressure plate.

3. Apply a light smear of molygrease to the main drive spline then place the clutch disc in position on the flywheel with the hub towards the flywheel (the flywheel side of the disc is marked near the centre). Align the clutch disc with the locator Tool No. P.7091-A (modified) or Tool No. P.7137.

4. Refit the pressure plate assembly, locating on the dowels. Fit the six securing bolts and spring washers; torque to 12 to 15 lb. ft. (1.7 to 2.1 kg.m.).

5. Remove the locator tool.

Clutch Components – Exploded View

OP 7000-A3   EXTRA: CLUTCH PILOT BEARING – RENEW

Tools Required

7600-A or B   Clutch pilot bearing remover (main tool)
CP.7600-7   Clutch pilot bearing remover (adaptor)
P.7137   Spigot bearing replacer and clutch disc locator

1. Push the adaptor, Tool No. CP.7600-7, behind the bearing and screw the main tool, 7600-A or B, into the adaptor. Tighten the wing nut to extract the bearing.

2. Position the new bearing on Tool No. P.7137 with the integral grease seal away from the crankshaft. Tap it into place in the crankshaft flange, ensuring that the bearing is 0.156 to 0.175 in. (3.96 to 4.44 mm.) below the crankshaft flange.

OP 7000-A6   EXTRA: MAIN DRIVE GEAR BEARING RETAINER OIL SEAL – RENEW

Tools Required

P.7136   Main drive gear oil seal replacer

To Dismantle

1. Remove clutch release arm and bearing.

2. Unscrew the three bolts and spring washers securing the main drive gear retainer to the gearbox case.

3. Withdraw the retainer and paper gasket.

4. Remove the retainer oil seal with suitable tool.

To Reassemble

5. Place a new oil seal on the replacer (Tool No. P.7136) so that when fitted the lips of the seal face the gearbox. Drive the seal into the retainer.

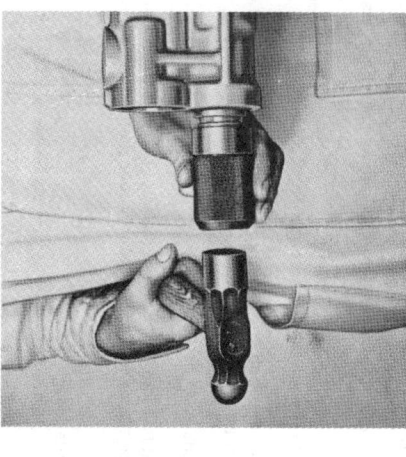

Replacing Extension Housing Oil Seal

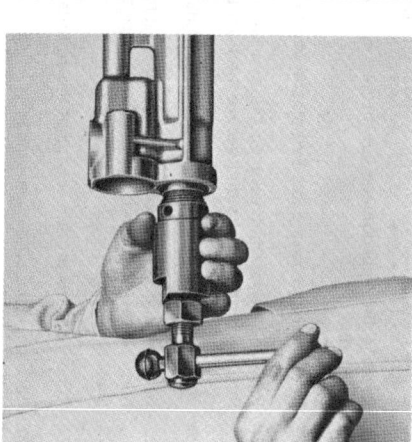

Removing Extension Housing Oil Seal

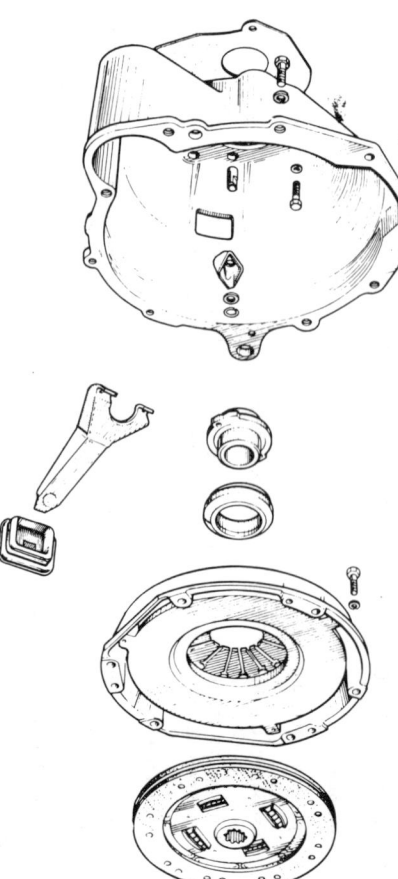

6. Apply a smear of molygrease to the main drive gear bearing retainer then fit the retainer to the gearbox. Cover the main drive gear splines before fitting the retainer and oil seal to prevent damage to the seal lip on assembly. First fit a new gasket on the gearbox front face. Ensure that the oil groove in the retainer is in line with the oil passage in the gearbox casing and that the gasket does not cover this passage. Coat the three retaining bolts with sealer (Part No. ESEE-M4G-1008A) and fit them, complete with spring washers. Tighten securely.

7. Replace clutch release arm and bearing.

OP 7000-A7 EXTRA: EXTENSION HOUSING REAR OIL SEAL – RENEW

Tools Required

7657   Extension housing oil seal remover
P.7657-4 Extension housing oil seal remover (adaptor)
P.7095   Extension housing oil seal replacer

1. Extract the extension housing oil seal by screwing Tool No. P.7657-4 into Tool No. 7657 and screw the assembly into the seal. Tighten the tool centre bolt to withdraw the seal.

2. Fit new extension housing oil seal by driving it squarely into position using Tool No. P.7095.

OP 7000-A8 EXTRA: EXTENSION HOUSING REAR BUSH – RENEW

Tools Required

7657   Extension housing oil seal remover
P.7657-4 Extension housing oil seal remover (adaptor)
P.7095   Extension housing oil seal replacer
P.7149   Extension housing bush remover
P.7150   Extension housing bush replacer

1. Extract the extension housing oil seal by screwing Tool No. P.7657-4 into Tool No. 7657 and screw the assembly into the seal. Tighten the tool centre bolt to withdraw the seal.

2. Extract extension housing bush using Tool No. P.7149.

3. Assemble a new bush to the replacing Tool No. P.7150 and drive into position.

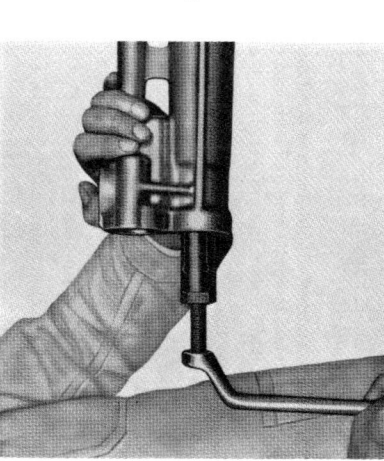

Removing Extension Housing Bush

4. Fit new extension housing oil seal by driving it squarely into position using Tool No. P.7095.

OP 7000-B CLUTCH RELEASE BEARING – RENEW
(Includes OPS 7000-A and A1)

OP 7000-C CLUTCH DISC AND/OR PRESSURE PLATE – REMOVE AND INSTALL
(Includes OPS 7000-A and A2)

OP 7000-D CLUTCH PILOT BEARING – RENEW
(Includes OPS 7000-A, A2 and A3)

OP 7003-A GEARBOX ASSEMBLY – REMOVE AND INSTALL

To Remove

1. Open the luggage compartment lid and disconnect the battery. Raise the bonnet.

2. Disconnect the throttle linkages at the carburettors.

3. Remove the centre console after removing four crosshead securing screws and the gear lever knob.

   Beneath the carpet is a rubber gaiter secured by a metal ring to the floor pan. Remove the ring and gaiter. Unscrew the gear lever turret.

4. Jack up the car and fit stands all round.

5. Remove plastic clip and handbrake cable from drive shaft support bracket.

6. Remove the four bolts joining the flanges of the two halves of the drive shaft. Lower the rear half of the drive shaft. Remove the bolts securing the centre bearing carrier to the floor pan, lower the carrier and bearing assembly and slide the front yoke from the gearbox. Fit a dummy yoke in the gearbox to prevent oil loss.

7. Unscrew the forked retainer or circlip, as applicable, and disconnect the speedometer cable and driven gear.

8. Remove the clutch operating cylinder circlip and move the cylinder to one side.

9. Remove the two starter motor securing bolts and move the starter to one side.

10. Remove the bolts securing the clutch housing to the engine. Note that a top bolt also secures an earth strap.

11. Remove the bolts securing the lower dust cover and detach the cover.

12. Place a support jack beneath the rear of the engine.

13. Remove the bolts securing the gearbox crossmember to the body. Slide the gearbox rearwards whilst supporting its weight and detach it from the engine.

14. Remove the clutch release arm and bearing.

15. Unscrew the four bolts securing the clutch housing to the gearbox and remove it.

16. Remove the crossmember centre bolt and detach it from the gearbox.

To Install

17. Place the crossmember on the rubber mounting and fit the centre bolt to attach it to the gearbox.

18. Fit the clutch housing to the gearbox and tighten the four bolts. Fit the clutch release arm and bearing.
19. Ensure that the adaptor plate is positioned on the rear of the engine and that the main drive gear spline is lightly smeared with molygrease. Offer the gearbox up so that the main drive gear spigot enters the crankshaft pilot bearing. Push the gearbox fully home.
20. Replace the four crossmember-to-body bolts and spring washers. Remove the engine support jack.
21. Replace the bolts securing the clutch housing to the engine. The uppermost pair are plain bolts and one also secures an earth strap. The remainder are dowel bolts.
22. Replace the lower dust cover.
23. Refit the starter motor.
24. Replace the clutch operating cylinder and retain it with a circlip.
25. Replace the speedometer driven gear and retain it with the forked retainer or circlip, as applicable.
26. Refit the handbrake cable and plastic clip to the drive shaft bracket.
27. Refit the drive shaft yoke on the gearbox and bolt the bearing carrier to the floor pan. Replace the four bolts that secure the front and rear halves of the drive shaft together.
28. Jack up the car and remove the body stands.
29. Refit the gear lever lower gaiter and retain it beneath the carpet to the floor pan with the metal ring. Fit the gear lever and turret. Replace the centre console and upper gaiter and fit the gear lever knob.
30. Reconnect the throttle linkage at the carburettors.
31. Reconnect the battery earth strap.
32. Top-up the gearbox with oil as required, ensuring that the car is level.

OP 7003-A1 EXTRA.: SELECTOR MECHANISM – OVERHAUL
(Includes removing selector lever and selector forks)

Tools Required

200A   Engine stand
P.7089   Gearbox mounting bracket

To Remove

1. Using the gearbox mounting bracket Tool No. P.7089, mount the gearbox on the engine stand.
2. Remove the four bolts securing the top cover plate to the gearbox. Lift off the plate.
3. Using a suitable tool, prise off the blanking plug from the rear of the extension housing.
4. Remove the plunger screw from the side of the gearbox case.
5. Using suitable punch, remove the spring pin securing the selector boss to the rail. Ensure that the spring pin can be punched through clear of any mainshaft components. If necessary, position the synchroniser hub on the mainshaft to suit.
6. Withdraw the selector rail rearwards taking care not to let the selector boss and 'C' cam drop into the gearbox casing.
7. To remove the selector forks, move the first/second gear and third/top gear synchromesh hubs to their foremost positions towards the main drive gear bearing. Lift out the forks.
8. If necessary, remove the selector rail plunger spring from its bore in the gearbox casing.
9. Remove the spring pin securing the third/fourth selector fork to the relay arm and remove both selector forks.

To Install

10. Replace both selector forks on the relay arm and secure the third and top fork to the arm with a new spring pin.
11. Position the assembled selector forks on their synchromesh sleeves and move the synchromesh hubs into the neutral position so that the selector fork extension arms locate beneath the reverse idler selector arm mounted on the side of the gearbox casing.
12. Grease the selector rail oil seal in the rear of the gearbox casing and slide the rail through the extension housing. Position the selector boss and the 'C' Cam so that the cam locates in the cutouts in the selector fork extension arms. Pass the selector rail through the boss and selector forks until the spring pin holes in the boss and selector rail align. During this operation take care not to damage the selector rail oil seal.
13. Assemble the plunger ball and spring to their bore and fit securing screw using sealer.
14. Fit the spring pin to retain the selector boss to the selector rail.
15. Apply sealer to the blanking plug and tap it in the extension housing behind the selector rail.
16. Fit a new gasket, using a suitable sealing compound to the top of the gearbox. Locate the cover plate on its dowels to the gearbox casing and secure with the four bolts.
17. Remove the gearbox from the engine stand and detach the mounting bracket.

OP 7003-A2 EXTRA.: EXTENSION HOUSING AND MAINSHAFT ASSEMBLY – REMOVE AND INSTALL
(Selector mechanism removed)

Tools Required

P.7113   Dummy countershaft
P.7152   Selector rail oil seal replacer
575   Light driver handle

To Remove

1. Unscrew the four bolts securing the extension housing to the gearbox casing.
2. Using a hide mallet, tap the extension housing slightly rearwards until it is possible to rotate it so that the countershaft aligns with the cut-away in the extension housing flange.
3. Tap the countershaft rearwards using a drift until it is just clear of the front of the gearbox casing. Push the countershaft out using a dummy countershaft (Tool No. P.7113). The cluster gear will now drop to the bottom of the gearbox.
4. Remove extension housing and mainshaft assembly. It is necessary to push the third/top synchroniser sleeve slightly forward to provide clearance between the synchroniser and the cluster gear.

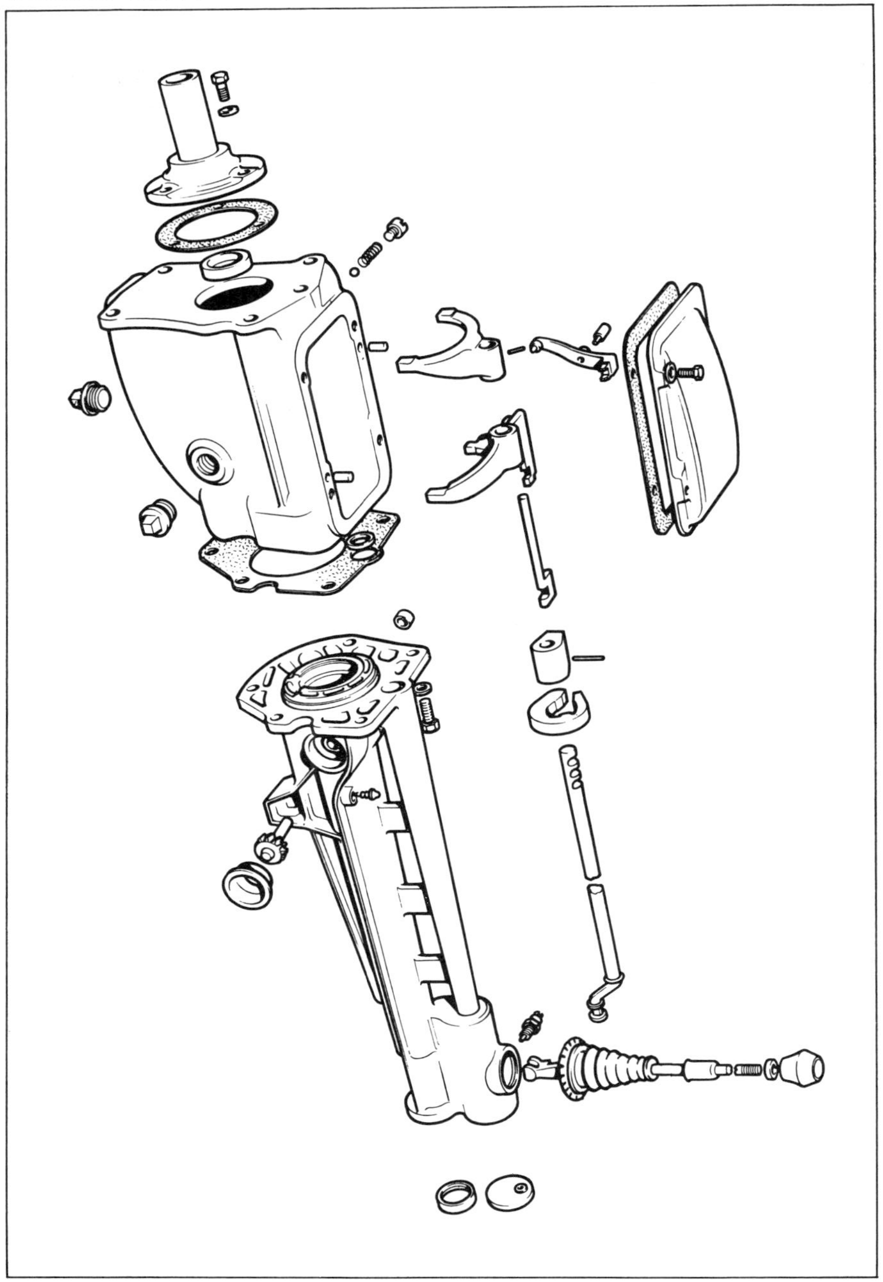

CORTINA - LOTUS

To Install

5. Fit a new oil seal to the selector rail aperture in the rear face of the gearbox case using P.7152 oil seal replacer and the drive handle 575.

6. Thread cord or suitable plastic covered wire under the cluster gear at each end to facilitate lifting it into position later. Take care that the thrust washers in the case at each end of the countershaft gear are not displaced.

7. Fit a new gasket to the extension housing using a jointing compound.

8. (i) Slide the extension housing and mainshaft into position after pulling the third/top synchroniser sleeve forward to clear the cluster gear. Ensure the top gear blocker ring locates correctly. It is important that care is taken when locating the spigot on the main drive gear bearing as the needle rollers could be dislodged.

   (ii) Align the cut-away on the extension housing with the countershaft aperture in the rear face of the gearbox.

9. Carefully, with pieces of string at each end, lift the countershaft gear into mesh with the mainshaft and main drive gears. Take care that the thrust washers in the case at each end of the countershaft gear are not displaced.

10. Check that the cluster gear bore aligns with the apertures of the countershaft. Push the dummy countershaft out by inserting the countershaft from the rear. Finally tap the countershaft into place using a hide or copper mallet. Ensure that the lug on the rear of the countershaft is positioned so that it will fit in the recess on the extension housing flange.

11. Rotate the extension housing so that the bolt holes align. Push the extension housing fully home onto the gearbox case.

12. Using a suitable sealer on the bolts, secure the extension housing to the gearbox case. Tighten the bolts to the correct torque.

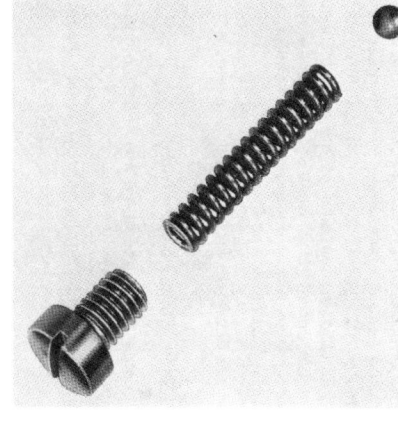

Selector Ball and Spring

Removing Selector Ball and Spring Retaining Screw

Gearbox Externals — Exploded View

OP 7003-A3 EXTRA: THIRD TOP SYNCHRONISER ASSEMBLY – OVERHAUL
(Mainshaft and extension housing assembly removed)

Tools Required

P.4090-9	Split rings
P.4090-7A/b	Replacer adaptor
370	Universal taper base
STN.7245	Thrust button

To Dismantle

1. Lift off top gear blocker ring from the main drive gear side of the third/top gear synchroniser assembly.

2. Remove circlip at forward end of mainshaft and discard it. Locate the split rings Tool No. P.4090-9 around the rear face of the third gear and in the base plate of the press. Press the mainshaft out of the third/top gear synchroniser and the third gear whilst supporting the mainshaft from beneath to prevent it dropping.

3. Dismantle the synchroniser assembly by pulling the sleeve off the hub and withdrawing the blocker bars and springs.

4. Check all parts for wear. If the synchroniser hub or sleeve show signs of damage or wear, both must be renewed.

To Reassemble

5. Slide the synchroniser sleeve over the hub and locate a blocker bar in each of the three slots cut in the hub.

NOTE – If a new synchroniser unit is being installed, slide the sleeve off the hub and clean all traces of preservative from the hub, sleeve, blocker bars and springs. Lightly oil them.

6. Install a blocker bar spring to run around, clockwise or anti-clockwise, inside the synchroniser sleeve beneath the blocker bars. The tagged end of the spring must be located in the "U" section of a blocker bar. Fit the other spring to the opposite face of the synchroniser unit ensuring that the spring tag locates in the same blocker bar as the spring just fitted and runs in the same rotational direction. View direct onto one side of the synchroniser assembly and note the direction of rotation of the spring (clockwise or anti-clockwise). View direct onto the other side of the synchroniser assembly — the direction of rotation of the spring should be the same as for the first spring, when viewed face on.

7. Position the third gear on the mainshaft so that the dog teeth face forward. Assemble the blocker ring on the third gear cone.

8. Position the synchroniser assembly on the mainshaft with the boss forward.

9. Support the hub on Tool No. P.4090-7A/b and locate it in the bed of a press. Press the hub fully home and then fit a new circlip on the mainshaft in front of the hub.

10. Prior to fitting the mainshaft and extension housing assembly to the gearbox, locate the top gear blocker ring on the main drive gear cone.

OP 7003-A4 EXTRA: FIRST/SECOND SYNCHRONISER ASSEMBLY – OVERHAUL
(Mainshaft and extension housing assembly removed)

Tools Required

P.4090-7A/a	Split rings — 1st gear
P.4090-9	Split rings — 2nd gear
P.4090-6	Split rings — 1st gear and mainshaft bearing
P.4000-31A/a	Replacer adaptor — 1st gear and mainshaft bearing
P.4090-7A/b	Replacer adaptor — 1st/2nd synchro hub assembly
P.7154	Circlip assessment gauge
370	Universal taper base

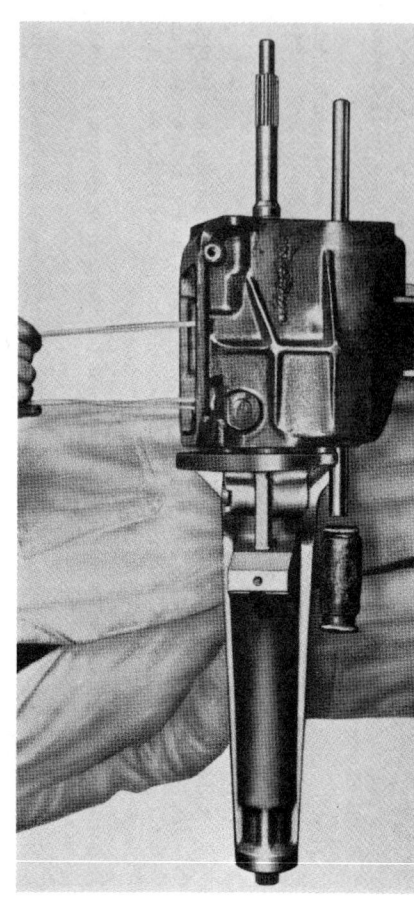

Replacing the Countershaft

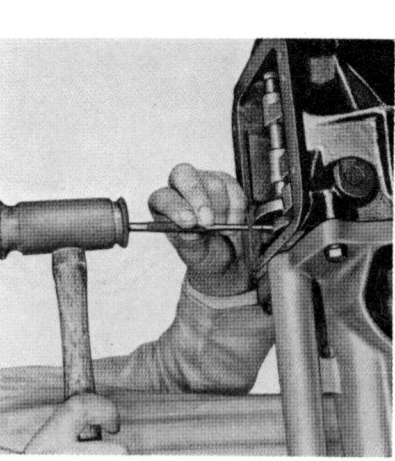

Replacing Selector Boss Spring Pin

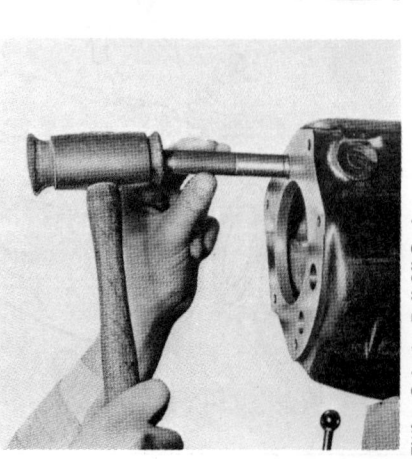

Fitting Selector Rail Oil Seal

Gearbox Internals – Exploded

To Dismantle

1. Withdraw speedometer drive gear after removing the plug on the extension housing.

2. Remove the circlip securing the mainshaft bearing to the extension housing from its groove. Tap mainshaft assembly out of extension housing using a hide mallet. Remove the circlip retaining speedometer drive gear to the mainshaft and pull off the gear. Remove the circlip retaining the mainshaft bearing.

3. Position the split rings Tool No. P.4090-7A/a behind the first gear and place assembly in the press base Tool No. 370. Press the first gear, spacer and mainshaft bearing from the mainshaft.

4. Remove circlip securing first and second synchroniser assembly to mainshaft.

5. Position split rings Tool No. P.4090-9 behind the second gear, and place the assembly in the press base Tool No. 370. Press the second gear and first and second synchroniser assembly complete with blocker rings off the mainshaft. Dismantle the synchroniser assembly.

6. Check all parts for wear. If the synchroniser hub or sleeve show signs of damage or wear, both must be renewed. The mainshaft bearing must be discarded.

To Reassemble

7. Assemble the second speed gear to the mainshaft so that the cone and dog teeth are rearwards.

8. Slide the synchroniser sleeve over the hub and locate a blocker bar in each of the three slots cut in the hub.

NOTE – If a new synchroniser unit is being installed, slide the sleeve off the hub and clean all traces of preservative from the hub, sleeve, blocker bars and springs. Lightly oil them.

Removing Third/Fourth Synchroniser Assembly

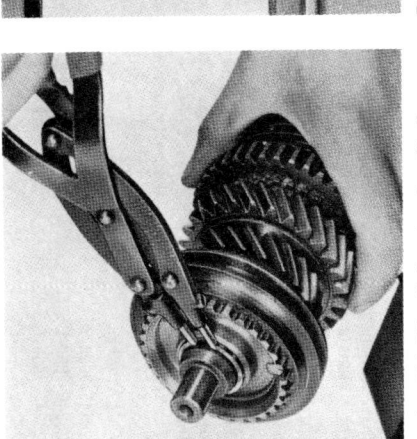

Removing Third/Fourth Synchroniser Circlip

9. Install a blocker bar spring to run around, clockwise or anti-clockwise, inside the synchroniser sleeve beneath the blocker bars. The tagged end of the spring must locate in the "U" section of a blocker bar. Fit the other spring to the opposite face of the synchroniser unit ensuring that the spring tag locates in the same blocker bar as the spring just fitted and runs in the same rotational direction. View direct onto one side of the synchroniser assembly and note the direction of rotation of the spring (clockwise or anti-clockwise). View direct onto the other side of the synchroniser assembly — the direction of rotation of the spring should be the same as for the first spring, when viewed face on.

10. Assemble a blocker ring to the cone on second gear.

11. Fit the synchroniser assembly onto the mainshaft so that the reverse gear teeth on the periphery of the synchroniser sleeve are forwards.

12. Locate the replacer, P.4090-7A/b, so that the plain side abuts the rear of the synchroniser hub.

13. Locate the mainshaft in the press base plate, 370.

14. Press the synchroniser assembly onto the mainshaft as far as possible. Secure the synchroniser assembly in position with the circlip.

15. Assemble a blocker ring to the first gear side of the first/second synchroniser assembly on the mainshaft. Fit the first speed gear, cone side forwards, to the mainshaft.

16. (i) Position the spacer (or oil slinger) on the mainshaft so that the larger diameter is adjacent to the first speed gear.

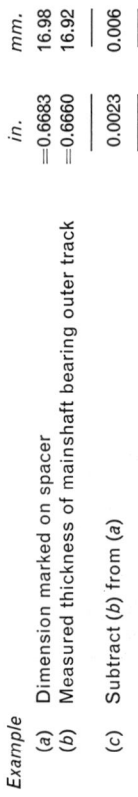

Speedometer Driven Gear Retaining Plug

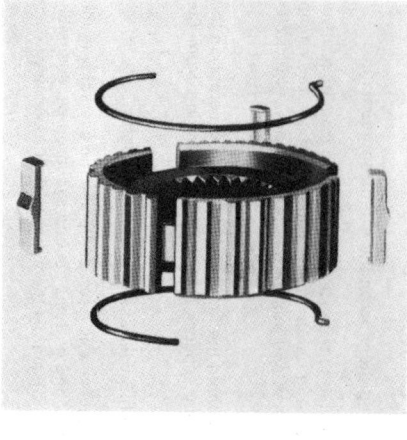

Synchroniser Spring Rotation

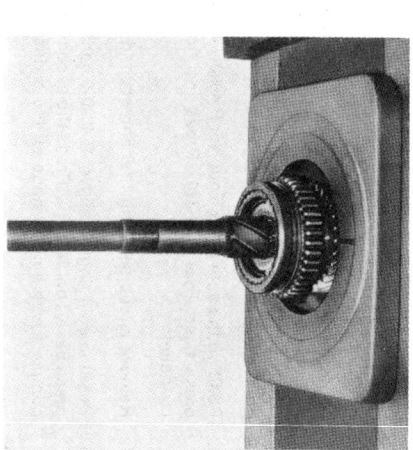

Removing First/Second Gear Synchroniser

(ii) Assemble the circlip assessment gauge, P.7154, to the mainshaft bearing recess in the extension housing so that the flat side is visible. Assess which circlip would be required to clamp the spacer into the bearing recess so that there is no end-float. Then measure the width of the bearing outer track, using a micrometer. If this dimension is different to the larger dimension marked on the spacer, it will be necessary to vary the circlip to compensate.

Example

		in.	mm.
(a)	Dimension marked on spacer	=0.6683	16.98
(b)	Measured thickness of mainshaft bearing outer track	=0.6660	16.92
(c)	Subtract (b) from (a)	0.0023	0.006

Therefore the circlip required must be approximately 0.0023 in. (0.006 mm.) thicker than the one selected when using the spacer. The circlip selected must not exceed the required increase in thickness, if it does it may be too large and will not go into the circlip groove. The following table shows the part numbers, dimensions and colour codings of the circlip:—

Part Number	Size—in. (mm.)	Colour Code
2824E-7030-A	0.0731 (1.860)	Yellow
2824E-7030-B	0.0720 (1.830)	Red
2824E-7030-C	0.0708 (1.800)	Blue
2824E-7030-D	0.0696 (1.770)	Violet
2824E-7030-E	0.0684 (1.737)	Green
2824E-7030-F	0.0682 (1.732)	Magenta
2824E-7030-G	0.0670 (1.702)	Plain

(iii) Position the selected circlip loosely on the mainshaft adjacent to the spacer (or oil slinger) fitted in sub-operation 19.

Locate the replacer, P.4000-31A/a, on the bearing so that the recessed side abuts the inner race.

With the split rings, P.4090-6, in the press base plate, 370, locate the bearing and replacer so that they fit into the split rings. Press the bearing into position on the mainshaft.

With the thickest circlip which fits the groove, secure the bearing to the mainshaft using one of the following selective range:—

Circlip Securing Mainshaft Bearing to Extension Housing

CORTINA - LOTUS

Part Number		Size—in. (mm.)	Colour Code
2824E-7669-A	...	0.0707 (1.795)	Plain
2824E-7669-B	...	0.0719 (1.825)	Pink
2824E-7669-C	...	0.0731 (1.860)	Magenta
2824E-7669-D	...	0.0743 (1.890)	Violet
2824E-7669-E	...	0.0755 (1.920)	Green
2824E-7669-F	...	0.0767 (1.950)	Blue
2824E-7669-G	...	0.0779 (1.980)	Red
2824E-7669-H	...	0.0791 (2.010)	Yellow

Locate ball bearing in mainshaft indent, and push speedometer drive gear onto mainshaft so that it just clears circlip groove in mainshaft. Fit new circlip to mainshaft to retain gear in position.

Heat the front end of the extension housing using a suitable hot plate or placing in hot water. This will expand the extension housing so that the mainshaft can easily be fitted. DO NOT USE A WELDING TORCH.

Fit the circlip to secure the mainshaft bearing to the extension housing.

Refit the speedometer driven gear and fit a new plug, using a suitable sealer, to the extension housing.

OP 7003-A5 EXTRA: EXTENSION HOUSING – OVERHAUL
(Identical to operation 7000-A8)

OP 7003-A6 EXTRA: MAIN DRIVE GEAR BEARING RETAINER OIL SEAL – RENEW
(Identical to operation 7000-A6)

OP 7003-A7 EXTRA: MAIN DRIVE GEAR – REMOVE AND INSTALL
(Main drive gear bearing retainer and mainshaft assembly removed)

Replacing First/Second Gear Synchroniser Assembly

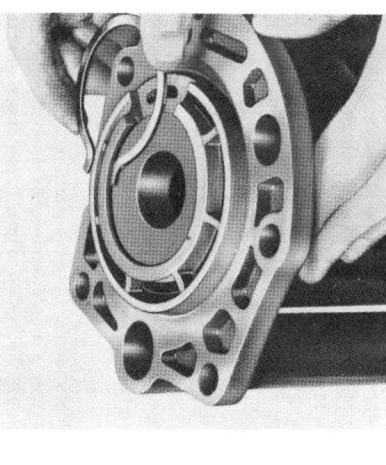

Assessing Circlip Retaining Mainshaft Bearing to Extension Housing

1 . 1969

Section 7 — 24

Tools Required

P.4090-7A	Split rings – removing
P.4090-6	Split rings – replacing
P.4000-39a	Replacer adaptor
P.4000-39b	Thrust pad
370	Press base plate

To Dismantle

1. Remove the spigot bearing from the recess in the end of the main drive gear.
2. Remove the circlip around the main drive gear bearing. Using a suitable copper drift, tap the main drive gear and bearing assembly out of the gearbox.
3. Remove the circlip from the main drive gear. Discard the circlip.
4. With the split rings, P.4090-7A, located round the bearing, press the bearing off the main drive gear. Use the 370 base plate to position the split rings in the press.

To Reassemble

5. Position the recessed adaptor, Tool No. P.4000-39a, so that the centre boss locates on the bearing inner track. Fit the thrust button, Tool No. P.4000-39b, to the cone end of the main drive gear.
6. With the split rings, Tool No. P.4090-6, in the base plate, 370, locate the main drive gear so that the adaptor, Tool No. P.4000-39a, is downwards and press the bearing fully home.

NOTE – The tools used to replace the bearing, which is an interference fit, ensure that all the load is taken through the inner bearing track. If the bearing is fitted so that the press load is taken through the ball bearings, then the bearing will be damaged. Therefore, ensure that the correct tools are used.

7. Fit the circlip to secure the bearing to the main drive gear.

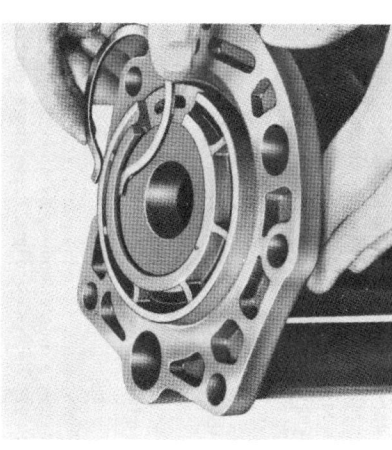

Removing Main Drive Gear Bearing

Replacing Main Drive Gear Bearing

1 . 1969

Section 7 — 25

CORTINA - LOTUS

8. Assemble the main drive gear to the gearbox and, using a copper drift on the bearing outer race, tap it into place until the circlip groove appears on the outside of the gearbox case. Take care that the dog teeth on the main drive gear are not damaged by the cluster gear.

NOTE – The bearing is an interference fit in the casing. Therefore, when fitting the bearing, it is important that the outer race is tapped. Do not tap on the main drive gear, this will result in a load passing through the ball bearings, consequently the bearing will be damaged.

9. Fit the circlip to the periphery of the bearing.
10. Assemble the spigot bearing to the recess in the end of the main drive gear.

OP 7003-A8   EXTRA: COUNTERSHAFT GEAR – REMOVE AND INSTALL
(Main drive gear and mainshaft assembly removed)

To Dismantle

1. Remove the countershaft gear and the two thrust washers from the gearbox. In both ends of the countershaft gear there are twenty-one needle rollers retained by a small retaining washer on each side of each set of rollers. Remove the needle rollers, the washers and the dummy countershaft (P.7113).

To Reassemble

2. Slide the dummy countershaft into the countershaft gear. Fit a retainer washer over the dummy countershaft and push it into the gear bore. Grease the needle rollers and assemble twenty-one into the recess, fit the second retaining washer. Repeat this procedure for the rollers at the other end of the gear. Grease the thrust washers and locate them so that their convex side fits into the recess.
3. Position the countershaft gear in the bottom of the gearbox case, taking care not to displace the thrust washers. Position the thrust washers so that the "ears" are located on each side of the location boss and with the flat on the washer uppermost.

OP 7003-A9   EXTRA: REVERSE IDLER GEAR – REMOVE AND INSTALL
(Mainshaft and extension housing assembly removed)

Tools Required

P.7043   Reverse idler shaft remover

To Remove

1. Withdraw the reverse idler shaft with Tool No. P.7043. Should this tool not be available, locate a nut, a flat washer and a sleeve on a $\frac{5}{16}$ in. 24 UNF threaded bolt. Screw the bolt into the reverse idler shaft and tighten the nut to withdraw the shaft.

To Install

2. Push the idler shaft into the gearbox, fit the reverse idler gear on the shaft and locate the reverse selector relay lever in the groove in the reverse idler gear.
3. Tap the reverse idler shaft into position using a copper mallet.

OP 7003-A10   EXTRA: REVERSE SELECTOR FORK – REMOVE AND INSTALL
(Reverse idler gear removed)

To Remove

1. Slide the reverse relay lever from the fulcrum pin on the gearbox casing. *Do not remove the pin.*

CORTINA - LOTUS

OP 7003-B   SELECTOR MECHANISM – OVERHAUL
(Includes OPS 7003-A and A1)

OP 7003-C   THIRD/TOP SYNCHRONISER ASSEMBLY – OVERHAUL
(Includes OPS 7003-A, A1, A2 and A3)

OP 7003-D   FIRST/SECOND SYNCHRONISER ASSEMBLY – OVERHAUL
(Includes OPS 7003-A, A1, A2 and A4)

OP 7003-E   MAIN DRIVE GEAR BEARING RETAINER OIL SEAL – RENEW
(Includes OPS 7003-A and A6)

OP 7003-F   MAIN DRIVE GEAR – REMOVE AND INSTALL
(Includes OPS 7003-A, A1, A2, A6 and A7)

OP 7003-G   COUNTERSHAFT GEAR – REMOVE AND INSTALL
(Includes OPS 7003-A, A1, A2, A6, A7 and A8)

OP 7003-H   REVERSE IDLER GEAR – REMOVE AND INSTALL
(Includes OPS 7003-A, A1, A2 and A9)

OP 7003-J   REVERSE SELECTOR FORK – REMOVE AND INSTALL
(Includes OPS 7003-A, A1, A2, A9 and A10)

OP 7003-K   GEARBOX ASSEMBLY – OVERHAUL
(Includes OPS 7003-A, A1, A2, A3, A4, A5, A6, A7, A8, A9 and A10)

OP 7003-L   EXTENSION HOUSING – REMOVE AND INSTALL
(Includes OPS 7003-A, A1 and A2)

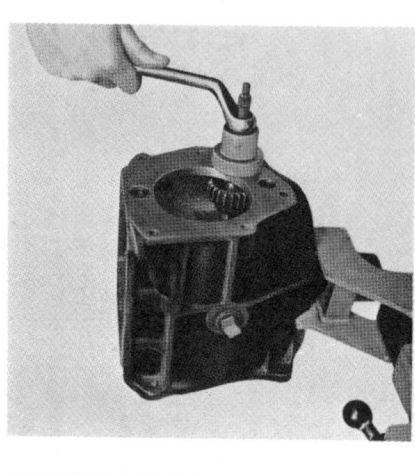

Removing Reverse Idler Shaft

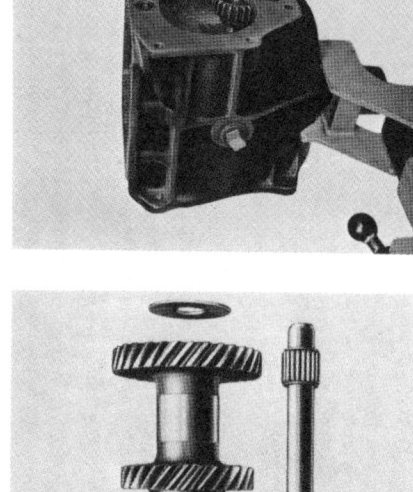

Countershaft Assembly – Exploded

OP 7202-A GEAR LEVER – REMOVE AND INSTALL

To Remove

1. Lift the gear lever gaiter and remove the circlip holding the rubber spring in compression. Bend up the lock tab and unscrew the plastic dome nut and withdraw the gear lever.

To Install

2. Locate the gear lever in the extension housing so that the forked end engages correctly with the selector rail. Secure in position by screwing the plastic dome nut into the extension housing and lock by bending down a tab on the integral locking plate. Depress the spring and fit the circlip. Locate the gear lever gaiter and check that all gears are obtainable. If necessary move the gear lever knob so that the gearshift pattern is correctly shown, lock the adjustment with the threaded ferrule beneath the knob.

OP 7500-A CLUTCH HYDRAULIC SYSTEM – BLEED

Tools Required

P.2006   Bleed tube

1. Clean the area around the bleed valve on the operating cylinder and remove the dust excluding rubber cap.
2. Fit a bleed tube, Part No. P.2006, and place the other end of the tube in a jar containing fluid, Part No. ME-3833-F, keeping the end of the tube beneath the surface of the fluid throughout the bleeding operation.
3. Open the bleed valve by turning it anti-clockwise and slowly depress and release the clutch pedal several times. For each stroke some fluid and/or air should be pumped out of the tube. If neither fluid nor air is pumped out, the bleed valve is not properly opened or there is a blockage in the pipe line.
   NOTE – Where air in the system is suspected, remember that initial application of the clutch pedal will cause air trapped in the bleed tube to be forced into the fluid container.
4. Continue depressing and releasing the clutch pedal slowly until no more air bubbles emerge from the tube, ensuring that the fluid level in the reservoir is maintained during the bleeding operation.
   Do not replenish the reservoir with fluid obtained from the system as it may be aerated or contaminated.
5. Close the bleed valve tightly with the pedal fully depressed, when fluid alone comes out of the bleed tube with each stroke of the clutch pedal. Refit the dust cap on the valve.
6. Refill the reservoir to the correct level and refit the cap.

OP 7501-B CLUTCH HYDRAULIC SLAVE CYLINDER – REMOVE AND INSTALL

To Remove

1. Jack up the front of the car and fit chassis stands.
2. Detach the fluid pipe by unscrewing the union nut, using a blanking plug to prevent dirt entering the pipe.
3. Remove the retaining circlip from around the cylinder body after slipping the rubber boot off the operating cylinder.
4. Push the cylinder forwards out of its location, removing the boot and the push rod simultaneously.

To Install

5. Slide the cylinder into its location in the clutch housing flange from the front. Push the push-rod through the rubber boot and insert the push-rod with the boot hanging loose, into the operating cylinder and clutch release arm.
6. Fit a circlip, ensuring that it is correctly located in its groove and fit the rubber boot on the operating cylinder.
7. Reconnect the fluid pipe, tighten the union nut.
8. Bleed the system as described in OP 7500-A.

OP 7501-B1 EXTRA: CLUTCH HYDRAULIC SLAVE CYLINDER – OVERHAUL

To Dismantle

1. Remove the piston and seal by extracting the circlip from the cylinder body, and then removing the piston and spring assembly from the cylinder.
2. Unscrew the bleed valve on the side of the operating cylinder.
3. Pull the spring and then the rubber piston seal off the spigot at the front of the piston.
4. Wash all parts in hydraulic fluid, Part No. ME-3833-F, methylated spirit or commercial alcohol and examine the rubber piston seal carefully. Renew the seal if there is any sign of damage to the sealing lip.

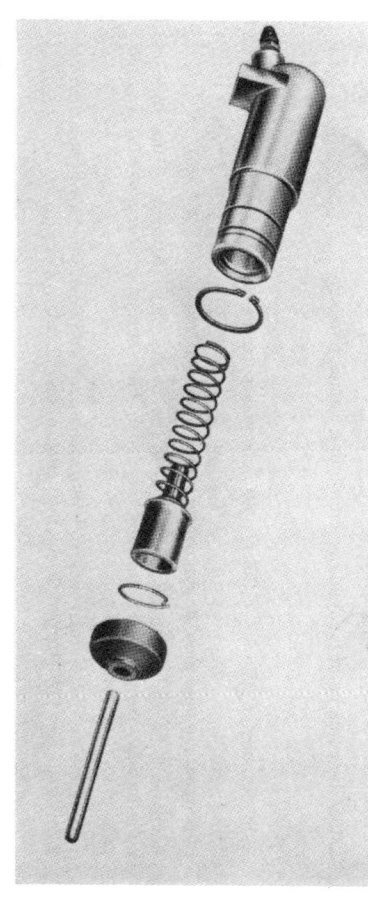

Clutch Operating Cylinder – Exploded View

CORTINA - LOTUS

To Reassemble

5. Locate the piston seal on the spigot at the front end of the piston with the recess in the seal away from the piston. Locate the spring on the piston spigot.
6. Dip the piston and seal in hydraulic fluid and carefully insert, spring first, into the cylinder.
7. Replace the bleed valve but do not tighten.

OP 7501-C CLUTCH HYDRAULIC SLAVE CYLINDER – OVERHAUL
(Includes 7501-B and B1)

OP 7534-A CLUTCH HYDRAULIC MASTER CYLINDER – REMOVE AND INSTALL

To Remove

1. Disconnect the clutch master cylinder push-rod from the pedal by unscrewing the nut and withdrawing the spring washer and concentric bolt.
2. Detach the fluid pipe by unscrewing the union nut, using a blanking plug to prevent dirt entering the line.
3. Withdraw the master cylinder after removing two spring washers and nuts securing the master cylinder to the bulkhead.
4. Empty the contents of the fluid reservoir into a waste container.

To Install

5. Refit the master cylinder to the engine bulkhead, securing with two spring washers and nuts.
6. Reconnect the fluid pipe, do not overtighten the union.
7. Reconnect the clutch master cylinder push-rod to the pedal by passing the concentric bolts through the push-rod and then the pedal. Fit a spring washer and nut. Tighten to 12 to 15 lb. ft. (1.66 to 2.07 kg.m.) torque.
8. Top-up the master cylinder reservoir with clean approved fluid, ME-3833-F, and then bleed the system; see OP 7500-A. Check the action of the clutch.

OP 7534-A1 EXTRA: CLUTCH HYDRAULIC MASTER CYLINDER – OVERHAUL

To Dismantle

1. Remove the rubber boot. Withdraw the circlip and remove the push-rod.
2. Withdraw the piston and valve assembly from the cylinder.
3. Remove the piston from the valve assembly. The spring retainer is held in position on the spigot end of the piston by a tab which engages under a shoulder on the front of the piston. Lift up the tab and remove the spring retainer, spring and valve assembly from the piston.
4. Dismantle the valve assembly by compressing the spring and moving the valve stem to one side in the retainer, so releasing the end of the valve stem from the key-hole slot in the retainer. Slide the valve spacer and shim off the valve stem.

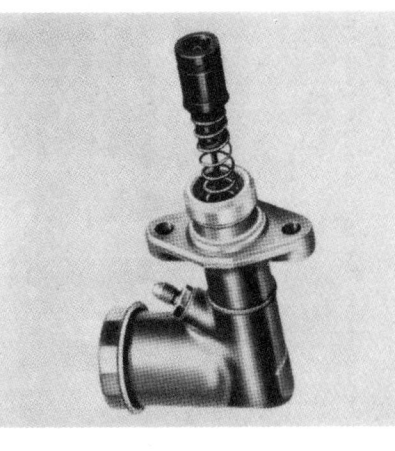

Refitting the Piston Valve Assembly

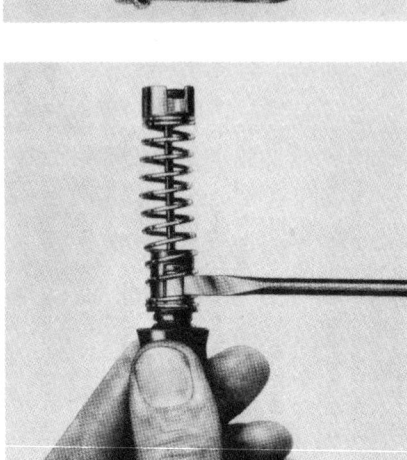

Removing the Piston Valve

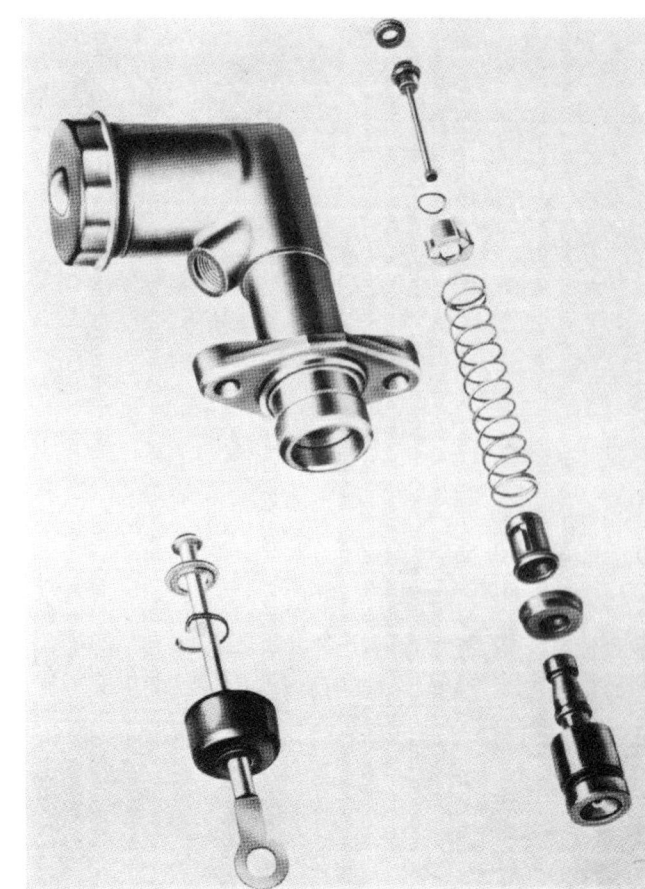

Clutch Master Cylinder – Exploded View

5. Remove the rubber valve seal and piston seal if necessary.
6. Wash the parts in methylated spirits, approved fluid ME-3833-F, or commercial alcohol. Carefully inspect the piston rubber seal and renew if there is any sign of damage to the sealing lip.

To Reassemble

7. Replace the piston seal with the lip away from the large diameter of the piston.
8. Fit the valve seal to the valve stem with the lip outwards and away from the spring. Slide the shim, valve spacer, with legs over the valve seal, and return spring over the valve stem. Ensure that the convex face of the shim abuts the valve stem flange.
9. Fit the spring retainer in the rear end of the return spring, compress the spring and locate the valve stem in the key-hole slot in the end of the spring retainer.
10. Insert the front of the piston in the spring retainer, and secure it by locating the spring retainer tab under the front shoulder of the piston.
11. Dip the piston and seal in hydraulic fluid and insert the piston assembly in the cylinder, valve seal end first.
12. Install the push-rod in the master cylinder. Locate the washer and fit the retaining circlip.
13. Refit the rubber boot to the clutch master cylinder.

OP 7534-B   CLUTCH MASTER CYLINDER – OVERHAUL
(Includes 7534-A and A1)

OP 7657-A   EXTENSION HOUSING OIL SEAL – RENEW
(Gearbox in situ)

Tools Required

7657   Oil seal remover (main tool)
P.7657-4   Adaptor for 7657
P.7095   Oil seal replacer

To Remove

1. Chock the front wheels and jack up the rear of the car. Fit stands.
2. Mark the drive shaft pinion flanges and remove the four bolts and self-locking nuts.
3. Remove the bolts securing the centre bearing carrier to the floor pan. Lower the carrier and bearing assembly and slide the front yoke from the gearbox. Fit a dummy yoke to the gearbox to prevent oil loss.
4. Extract the oil seal from the rear of the extension housing using Tool No's. 7657 and P.7657-4.

To Install

5. Locate a new extension housing oil seal on replacer Tool No. P.7095, so that the lip on the seal faces into the extension housing and then drive the seal into position in the housing.

6. Replace the drive shaft by sliding the front universal joint yoke onto the splines of the mainshaft, taking care not to damage the extension housing oil seal. Secure the carrier and bearing assembly to the floor pan with bolts. Align the mating marks on the drive shaft and pinion flanges, fit the retaining bolts and secure with four new self-locking nuts.
7. Remove the stands, lower the car to the ground and remove the chocks from the front wheels.
8. With the car on level ground, check gearbox oil level and top-up if necessary.
(See also OP 7000-A7).

OP 7657-A1   EXTRA: EXTENSION HOUSING REAR BUSH – RENEW
(Extension housing oil seal removed)

Tools Required

P.7149   Gearbox extension housing bush remover
P.7150   Gearbox extension housing bush replacer

1. Insert the bush removing Tool No. P.7149 into the rear of the extension housing so that it locates on the forward end of the bush. Withdraw the bush.
2. Assemble the new bush to the replacing Tool No. P.7150, and drive it into position.

OP 7657-B   EXTENSION HOUSING REAR BUSH – RENEW
(Includes OPS 7657-A and A1)

CORTINA - LOTUS
Supplement

CLUTCH and GEARBOX

(1967-1968)

CORTINA - LOTUS

GEARBOX (1967-68)

While the gearbox internals remain basically the same, the three-rail system of gear selection is utilized as opposed to the single rail described in Section 7.

Reverse gear selection is opposite to that shown for '69 model.

SERVICE AND REPAIR OPERATIONS

With the exception of the following procedures, operations on '67-'68 vehicles may be carried out as detailed in Section 7.

SERVICE AND REPAIR OPERATIONS

OP 7003-A1 EXTRA : SELECTOR MECHANISM – OVERHAUL
(Includes removing selector lever and selector forks)

Tools Required

200A   Engine stand
P.7089 Gearbox mounting bracket

To Remove

1. Using the gearbox mounting bracket mount the gearbox on the engine stand.
2. Remove the gearbox top and inspection covers, springs and detent balls.
3. Remove the tapered screw securing the selector forks to the selector rails.
4. Remove the central first/second gear rail.
5. Remove the left-hand side third/top gear rail and slide off the over-run stop tube.
6. Unscrew the bolt securing the angled selector arm to the extension rod in the extension housing after removing the locking wire. Slide off the arm.
7. Remove the extension rod forward bearing bridge.
8. Unscrew the three gear lever turret bolts and withdraw it complete with extension rod.
9. Disengage the reverse rail from the reverse relay lever and remove the rail.
10. Lift the lock tabs and remove the two retaining bolts and extract the plunger ball and springs.

To Install

11. Fit the reverse gear selector rail (identified by two detent grooves) through the right-hand bore in the extension housing.

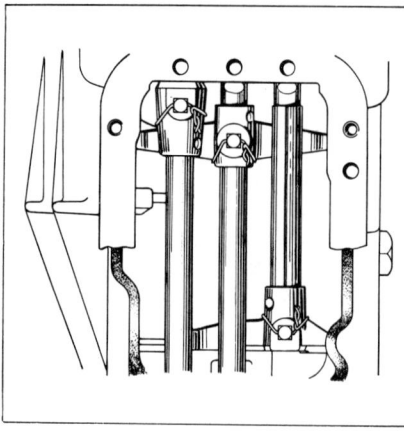

Selector Fork and Gate Positions

CORTINA - LOTUS

12. Locate the selector fork on the rail then engage the rail in the gearbox case and on the reverse relay lever.

13. Slide the extension rod through the bore in the turret housing and fit the turret to the extension housing. Secure with three bolts and spring washers.

14. Refit the forward bearing bridge over the extension rod, with the cut-out forwards, and secure it with two bolts.

15. Replace the angled selector arm on the extension rod with the arms uppermost. Secure it with a bolt and lock with wire.

16. Fit the centre rail (identified by a cross drilling with a floating pin in one end). Ensure that the uppermost arm of the angled selector arm is to the left of the rail. Engage the first/second gear selector fork as the rail is pushed in.

17. Slide in the third/top gear rail, passing it through the over-run stop tube and the third/top selector fork.

18. Secure the selector forks to the rails, and fit the gear lever.

19. Fit the spring-loaded plunger and the detent ball and spring in the bores beneath the gear lever turret. Do not bend up the lock tabs at this stage.

20. Select either third or fourth gear and check that the angled selector arm clears the first/second gear selector rail flat (C). If a fouling condition occurs, fit the next sized plunger down. Conversely, if fouling occurs between the angled selector arm and the reverse gear face (B) the next sized plunger up should be fitted, see following table.

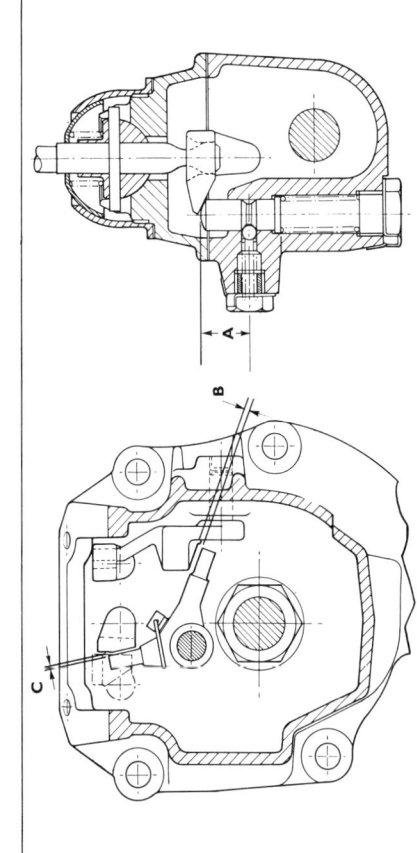

Reverse Plunger Selection

Gearbox Externals – Exploded View

CORTINA - LOTUS

The plungers available are as follows:—

Part Number	Size & Dimension A
2821E-7K187-H	0.870 in. (2.21 cm.)
2821E-7K187-G	0.840 in. (2.10 cm.)
2821E-7K187-F	0.810 in. (2.06 cm.)
2821E-7K187-E	0.780 in. (1.98 cm.)
2821E-7K187-D	0.750 in. (1.91 cm.)

21. When the correct plunger has been fitted, bend up the lock tabs.

22. Replace the top cover, detent balls and springs, securing with four bolts and spring washers.

23. Replace the inspection cover, and secure with bolts and spring washers. The breather hole should be towards the rear of the gearbox.

24. Check engagement of all gears.

CORTINA - LOTUS

8
COOLING SYSTEM

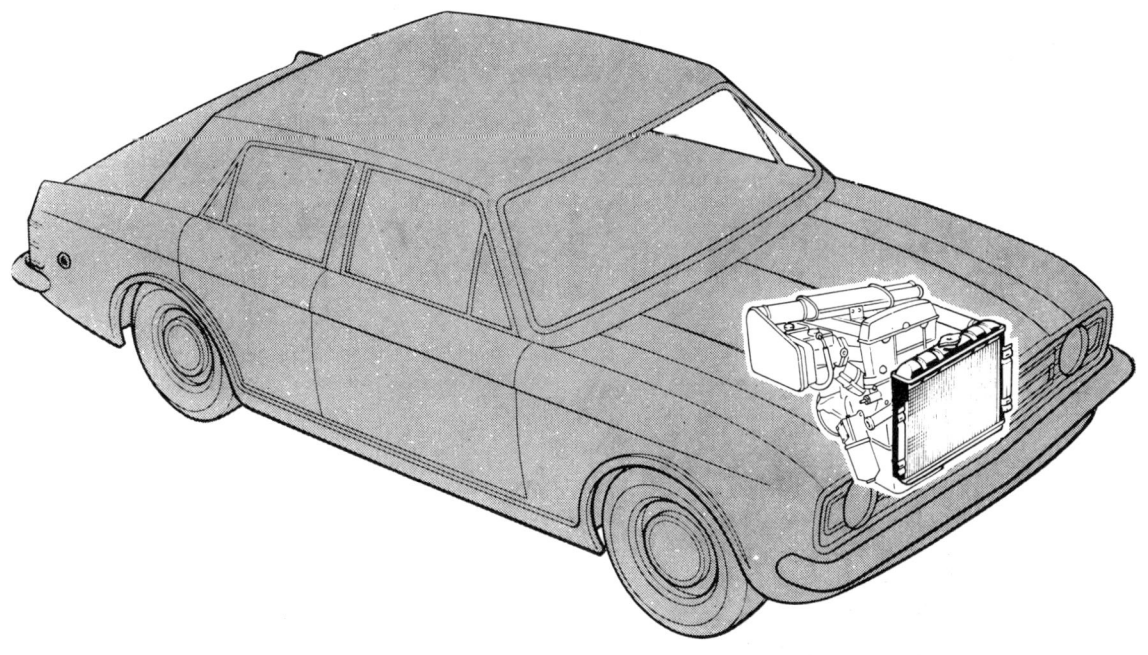

CORTINA - LOTUS

SECTION INDEX

GENERAL DESCRIPTION
QUICK REFERENCE DATA
SERVICE AND REPAIR OPERATIONS

OPERATION	8005-A	RADIATOR ASSEMBLY – REMOVE AND INSTALL
"	8005-A1	Extra: radiator hoses – renew
"	8115-A	DRAIN PLUG – REMOVE AND INSTALL
"	8150-A	RADIATOR GRILLE – REMOVE AND INSTALL
"	8250-A	RADIATOR TOP HOSE – RENEW
"	8250-A1	Extra: thermostat or water outlet – renew
"	8250-A2	Extra: thermostat – test
"	8286-A	RADIATOR BOTTOM HOSE – RENEW
"	8501-B	**WATER PUMP – OVERHAUL**
"	8606-A	FAN – REMOVE AND INSTALL
"	8620-A	FAN BELT – RENEW
"	8620-B	FAN BELT – ADJUST

CORTINA - LOTUS

GENERAL DESCRIPTION

The pressurised cooling system is of the impeller-assisted thermo-syphon type with the water pump integral with the engine front cover.

Coolant is circulated from the base of the radiator up through the pump and into the cylinder block. The coolant then circulates through the cylinder block and cylinder head to the thermostat, located on the cylinder head. At operating temperature hot coolant is returned to the radiator top tank. The coolant then flows down the radiator tubes and is cooled by air passing through the radiator, induced by the fan. The fan is belt-driven from the crankshaft pulley in tandem with the generator.

The thermostat in conjunction with a by-pass tube, cast in the cylinder head, assists in the rapid warming up of the engine and controls the normal engine running temperature.

A coolant temperature gauge on the facia warns the driver of any cooling irregularities.

The corrugated fin high efficiency radiator is located in the front of the engine compartment and incorporates a pressure cap and drain plug. It is connected to the cylinder head and water pump by reinforced neoprene hoses.

In production the system is normally filled with a 50% solution of FoMoCo long life anti-freeze. (A tag is usually fitted on the radiator filler neck or a sticker on the windscreen when long life anti-freeze has been used.) This anti-freeze offers adequate cooling system and engine protection for three years or 36,000 miles (whichever occurs first) provided that its strength is not allowed to fall below 30% solution.

The coolant is used as a source of heat for the interior heater.

QUICK REFERENCE DATA

PERIODIC SERVICE ATTENTION

Daily
    Top-up radiator coolant level
    Check for water leaks

Every 3,000 miles (5,000 km.) or three months (whichever occurs first)
    Top-up radiator coolant level
    Adjust fan belt tension

At 36,000 miles (60,000 km.) or four years (whicever occurs first)
    Change long life anti-freeze and hoses

DATA

Cooling system capacity	12.50 pints (7.10 litres, 15.2 U.S. pints)
Radiator cap release pressure	13 p.s.i. (0.914 kg./sq. cm.)
Fan belt tension	½ in. (12.7 mm.) free movement
Thermostat – starts to open	85° to 89°C (185° to 192°F)
– fully open	99° to 102°C (210° to 216°F)

CORTINA - LOTUS

SERVICE AND REPAIR OPERATIONS

OP 8005-A RADIATOR ASSEMBLY – REMOVE AND INSTALL

To Remove

1. Position a drain tray under the radiator, open the drain plug and remove the radiator cap. Leave the radiator to drain. If long life anti-freeze is used it should be retained.
2. Disconnect the radiator top and bottom hoses.
3. Remove the four retaining bolts and lift out the radiator.
4. Remove the drain tap and the flexible overflow pipe from the base of the filler neck.

To Install

5. Replace the drain tap and the flexible overflow pipe.
6. Fit the radiator and secure it with the four bolts.
7. Reconnect the radiator top and bottom hoses and tighten the two hose clips.
8. Refill the cooling system with the anti-freeze solution and check for leaks.

OP 8005-A1 EXTRA: RADIATOR HOSES – RENEW

1. Unscrew the two hose clips and remove the hoses.
2. Refit the hoses, and secure firmly with the hose clips.

OP 8115-A DRAIN PLUG – REMOVE AND INSTALL (Radiator or Cylinder Block)

To Remove

1. Drain the radiator coolant.
2. Remove the radiator or cylinder block drain plug.

To Install

3. Replace the radiator or cylinder block drain plug.
4. Refill the cooling system with the anti-freeze solution and check for leaks.

OP 8150-A RADIATOR GRILLE – REMOVE AND INSTALL

To Remove

1. Open the bonnet.
2. Remove the five screws securing the headlamp bezels and remove the bezels.
3. Remove the two screws and drill out the six pop rivets securing the radiator grille. Pull out the grille.

To Install

4. Locate the grille in its surround and secure it with the two screws at its upper edge. Pop rivet the grille in the six locations around the grille surround. (As an alternative six self-tapping screws may be used to secure the grille.)
5. Replace the headlamp bezels and secure them with the five screws.
6. Close the bonnet.

OP 8250-A RADIATOR TOP HOSE – RENEW

To Remove

1. Partially drain the radiator to below the top hose level and store the anti-freeze.
2. Remove the top hose.

To Install

3. Replace the top hose and tighten the clips.
4. Top-up the radiator with the anti-freeze solution.

OP 8250-A1 EXTRA: THERMOSTAT OR WATER OUTLET – RENEW

To Remove

1. Remove the water outlet and gasket.
2. Extract the thermostat from the recess in the cylinder head.

To Install

3. Locate the thermostat in the recess in the cylinder head, fit a new gasket and fit the water outlet connection.

OP 8250-A2 EXTRA: THERMOSTAT – TEST

1. Suspend the thermostat in water in a suitable container so that it does not touch the sides of the container.
2. Gradually heat the water, frequently checking the temperature with an accurate thermometer. The thermometer must not touch the container.
3. The thermostat should start to open at 85°C to 89°C (185°F to 192°F) and be fully opened at 99°C to 102°C (210°F to 216°F).
4. If the thermostat does not function properly, do not attempt any adjustment, but replace with a new unit.

OP 8286-A RADIATOR BOTTOM HOSE – RENEW

To Remove

1. Drain the cooling system and store the anti-freeze solution.
2. Remove the bottom hose.

To Install

3. Replace the bottom hose and tighten the clips.
4. Refill the cooling system with the long life anti-freeze.

OP 8501-B WATER PUMP – OVERHAUL

Tools Required

370   Taper base
CP.4111 Hub remover
P.8000-4B Water pump overhaul kit

P.8008-A   Slave ring
550   Driver handle
P.6150   Crankshaft front cover oil seal aligner
P.6161   Crankshaft front cover oil seal remover/replacer

To Remove

1. Drain the engine coolant by opening the drain plugs on the radiator and cylinder block.
2. Disconnect the radiator hoses at the engine.
3. Remove the radiator assembly.
4. Remove the air cleaner.
5. Remove the camshaft cover.
6. Remove the fan belt and then remove the fan and the water pump pulley.
7. Remove the crankshaft pulley, using suitable levers.
8. Remove the sump.
9. Remove the timing chain tension adjuster.
10. Remove the front cover.
11. Remove the crankshaft oil slinger.
12. Disconnect the timing chain taking care not to rotate the camshafts or crankshaft and alter the valve timing.
13. Remove the auxiliary shaft sprocket.
14. Remove the front cover back plate and gaskets.
15. Withdraw the bearing retainer clip from the slot in the housing.
16. Remove the pump pulley hub from the shaft using Tool No. CP.4111.
17. Press the impeller, seal, slinger, shaft and bearing assembly out of the housing, using the ring and thrust block adaptors, details 'd' and 'e' of Tool No. P.4000-4B together with P.8008A and 370 base on a suitable press. Adaptor detail 'e' is hollow and fits over the shaft and bears against the outer diameter of the bearing.
18. Press the impeller off the end of the shaft, using adaptors, details 'a' and 'b', ensuring that the vanes avoid the slots.
19. Remove the pump seal from the shaft.
20. Carefully split the slinger bush with a chisel to detach it from the shaft.
21. Remove the insert from the front cover.

To Install

22. Press the shaft and bearing assembly into the housing (short end of shaft to the front of the housing) until the groove in the shaft is in line with the groove inside the housing, again using the ring and thrust block adaptors, details 'd' and 'e'.
23. Refit the bearing retainer clip in the groove of the bearing and housing.
24. Press the pump pulley hub on to the front end of the shaft until the end of the shaft is flush with the end of the hub, using the split ring (detail 'a') and thrust block adaptor (detail 'e').

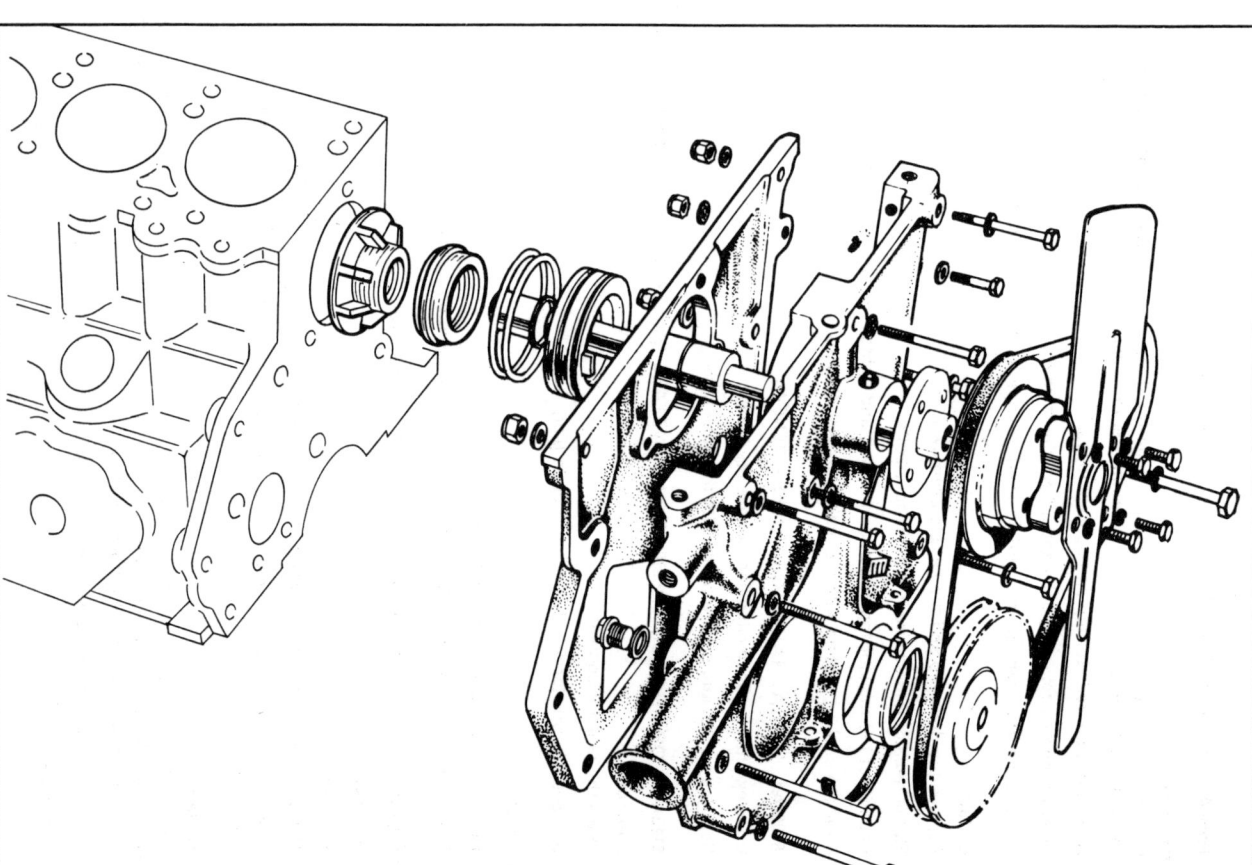

Timing Case and Water Pump Assembly

25. Fit a new slinger bush (flanged end first) on the rear end of the shaft, using driver (detail 'f').

26. Fit the pump seal on the slinger bush with the carbon thrust face towards the impeller and press it into the housing using the replacer (detail 'k').

27. Fit the "O" ring to the insert and fit the insert to the front cover.

28. Press the impeller on to the shaft, using the adaptor details 'a' and 'b' until a clearance of 0.030 in. (0.762 mm.) is obtained between the impeller blades and the insert face. Ensure that load is applied to the shaft.

29. Locate a new gasket on the cylinder head to the timing cover joint.

30. Fit the front cover back plate. Align the water pump aperture with the front cover and seal aligner Tool No. P.6150 before tightening the single bolt to 5 to 7 lb. ft. (0.69 to 0.97 kg.m.) torque.

31. Fit the auxiliary shaft sprocket. Tighten the retaining bolts to 12 to 15 lb. ft. (1.66 to 2.07 kg.m.) torque and turn up the locking plate tabs.

32. Locate the timing chain around the sprockets ensuring that there is a minimum slack between the exhaust camshaft sprocket and the crankshaft sprocket.

33. Locate the crankshaft oil slinger in place.

34. Fit a new oil seal, using a remover/replacer Tool No. P.6161 and 550 handle. If required fit a new timing chain vibration damper.

35. Coat the front cover joint faces with EM-4G-47 jointing compound. Align the cover in position with Tool No. P.6150. Tighten the $\frac{3}{8}$ in. retaining nuts and bolts evenly to a torque of 5 to 7 lb. ft. (0.69 to 0.97 kg.m.) and the $\frac{5}{16}$ in. to 10 to 15 lb. ft. (1.38 to 2.07 kg.m.) and remove the aligner tool.

36. Fit the timing chain tension adjuster and tighten to 45 to 50 lb. ft. (6.22 to 6.91 kg.m.) torque.

37. Fit the sump. Fit new gaskets on the block flange using EM-4G-47 jointing compound at each end. Fit the cork packing strips with the chamfered ends into the grooves, again using EM-4G-47 jointing compound and refit the sump. Tighten the bolts to 6 to 8 lb. ft. (0.83 to 1.11 kg.m.) torque.

38. Fit the crankshaft pulley aligning the pulley slot with the crankshaft key. Tighten the retaining bolt to 24 to 28 lb. ft. (3.32 to 3.87 kg.m.).

39. Replace the water pump pulley and the fan. Fit the fan belt and adjust the tension so that there is $\frac{1}{2}$ in. (12.7 mm.) total movement.

40. Adjust the timing chain tension to give $\frac{1}{2}$ in. (12.7 mm.) free movement between the camshaft sprockets. Ensure that there are no tight spots by turning the engine through several revolutions.

41. Fit the camshaft cover and tighten the retaining nuts evenly.

42. Fit the air cleaner.

43. Replace the radiator assembly.

44. Refit the radiator top and bottom hoses and tighten the clips.

45. Refill the radiator with long life anti-freeze.

46. Retime the ignition (see operation No. 12100-A).

OP 8606-A   FAN – REMOVE AND INSTALL

To Remove

1. Unscrew the four bolts, fitted with spring washers, and remove the metal fan blades, if fitted.

2. Unscrew the four bolts, fitted with heavy duty flat washers, and remove the plastic fan blades, if fitted.

To Install

3. Reposition the fan blades and secure with the four bolts and spring washers or heavy duty flat washer, as applicable. Tighten bolts to a torque of 7 to 9 lb. ft. (0.968 to 1.244 kg.m.).

OP 8620-A   FAN BELT – RENEW

To Remove

1. Slacken the generator mounting bolts and move the generator towards the engine.

2. Slip the belt over the edge of the generator pulley, taking care not to damage the pulley. The belt may then be detached from the crankshaft and water pump pulleys.

To Install

3. Pass the fan belt around the water pump and crankshaft pulleys and engage it in the generator pulley.

4. Adjust the fan belt tension to give $\frac{1}{2}$ in. (12.7 mm.) total movement between the water pump and generator pulleys.

5. Tighten the generator mounting bolts.

OP 8620-B   FAN BELT – ADJUST

1. Slacken the generator adjustment locking bolt and the generator mounting bolts.

2. Move the generator as necessary until there is $\frac{1}{2}$ in. (13 mm.) free movement at a point mid-way between the generator and fan pulleys.

3. Tighten the adjusting bolt and generator mounting bolts.

Fan Belt Tension

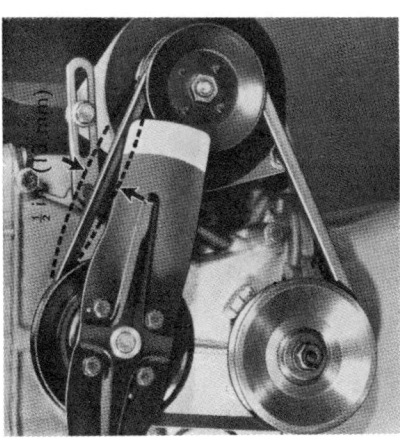

Radiator Filler Cap

CORTINA - LOTUS

9
FUEL SYSTEM

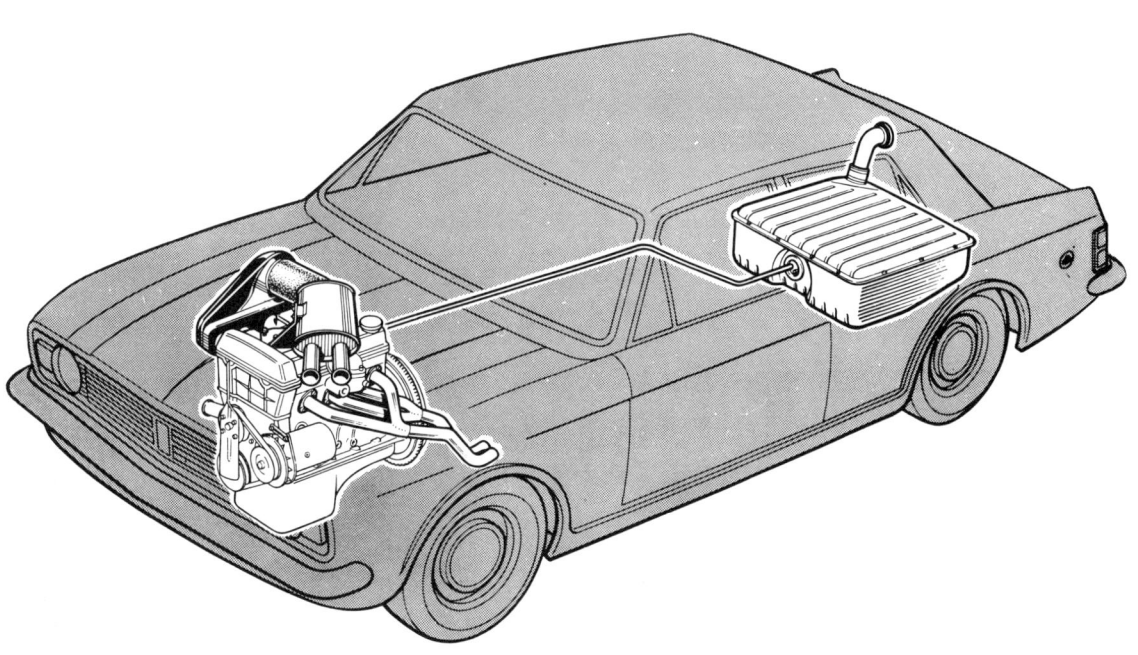

1 . 1969 Section 9 — 1

CORTINA - LOTUS

SECTION INDEX

GENERAL DESCRIPTION
QUICK REFERENCE DATA
SERVICE AND REPAIR OPERATIONS

OPERATION	9000-A	FUEL SYSTEM AND CARBURETTOR – CLEAN
,,	9002-A	FUEL TANK – REMOVE AND INSTALL
,,	9273-A	OIL PRESSURE GAUGE – REMOVE AND INSTALL
,,	9275-A	FUEL GAUGE TANK UNIT – REMOVE AND INSTALL
,,	9280-A	FUEL GAUGE – REMOVE AND INSTALL
,,	9350-A	FUEL PUMP – REMOVE AND INSTALL
,,	9350-A1	Extra: diaphragm – renew
,,	9350-A2	Extra: fuel pump mechanism – overhaul
,,	9350-A3	Extra: valves – renew
,,	9350-B	**FUEL PUMP DIAPHRAGM – RENEW** (Includes OPS 9350-A and A1)
,,	9350-C	**FUEL PUMP MECHANISM – OVERHAUL** (Includes OPS 9350-A, A1 and A2)
,,	9350-D	**FUEL PUMP VALVES – RENEW** (Includes OPS 9350-A and A3)
,,	9350-E	**FUEL PUMP DIAPHRAGM AND VALVES – RENEW** (Includes OPS 9350-A, A1 and A3)
,,	9350-F	**FUEL PUMP – OVERHAUL** (Includes OPS 9350-A, A1 A2 and A3)
,,	93509	FUEL PUMP – TEST
,,	9364-A	FUEL FILTER BOWL AND/OR SEAL – REMOVE AND INSTALL
,,	9428-A	EXHAUST MANIFOLD – REMOVE AND INSTALL
,,	9448-A	EXHAUST MANIFOLD GASKET – RENEW
,,	9510-A	CARBURETTOR AND/OR GASKET – REMOVE AND INSTALL
,,	9510-A1	Extra: carburettor gaskets – renew
,,	9510-A2	Extra: carburettor – overhaul
,,	9510-A3	Extra: studs – carburettor to manifold – renew
,,	9510-B	CARBURETTOR GASKETS – RENEW (Includes 9510-A and A1)
,,	9510-C	**CARBURETTOR – ONE – OVERHAUL** (Includes OPS 9510-A and A2)
,,	9510-D	**CARBURETTOR – BOTH – OVERHAUL** (Includes 9510-A and A2 × 2)
OPERATION	9510-M	CARBURETTOR – ADJUST
,,	9533-A	CARBURETTOR JETS AND FLOAT CHAMBER – CLEAN
,,	9600-A	AIR CLEANER – REMOVE AND INSTALL
,,	9700-A	CHOKE CONTROL CABLE – RENEW
,,	9725-A	THROTTLE CONTROL ROD – REMOVE AND INSTALL
,,	9735-A	THROTTLE PEDAL ASSEMBLY – REMOVE AND INSTALL
,,	9735-A1	Extra: accelerator pedal and shaft bushes – renew
,,	9735-B	**ACCELERATOR PEDAL AND SHAFT BUSHES – RENEW** (Includes 9735-A and A1)

CORTINA - LOTUS

GENERAL DESCRIPTION

The fuel system has a 10 Imp. gallon (12 U.S. gallon, 45.4 litre) fuel tank located in the luggage compartment floor pan. A flush fitting filler cap is located in the luggage compartment rear panel, the filler pipe being connected to the tank by a short length of hose retained with clips. Fuel tank ventilation is by a groove in the filler cap sealing flange. The fuel tank gauge unit and fuel pipe are located in the front face of the tank.

A nylon fuel line connects the fuel tank to a diaphragm type mechanical fuel pump, mounted on the right-hand side of the engine and operated by the auxiliary shaft. The fuel pump incorporates a gauze screen and an inverted sediment bowl. From the fuel pump a branched pipe delivers fuel to the carburettors.

Two side draught Weber type 40 DCOE 31 carburettors are fitted, these being of the dual barrel type with two venturis in each barrel. A single accelerator pump, which discharges fuel into the barrels, is fitted in each carburettor. Also, a progressive starting device is mounted on top of each carburettor to supply a rich mixture to the barrels for easy cold starting.

A separate inlet manifold is not provided, the carburettors feeding the cylinders by means of inlet tracts cast with the cylinder head. These tracts are not interconnected by a bridge pipe, etc. at any point. It will be appreciated that, since the inlet tracts are completely separate from one another, each carburettor is composed of and functions as, in effect, two carburettors with a separate set of identical jets for each barrel. The layout is, therefore, similar to a system comprised of four single barrel carburettors. Apart from the throttle linkage, the carburettor assemblies are identical.

Between each carburettor barrel and inlet tract there is a metal plate with a rubber 'O' ring in each face. These 'O' rings act as a seal to prevent the ingress of air and, together with the double coil spring washers fitted between the carburettor retaining nuts and carburettor mounting flanges, absorb vibration and thereby prevents frothing in the float chambers.

The air cleaner is of the replaceable paper element type, mounted on top of the engine. A rubber moulded hose connects the air cleaner to an air box mounted on the carburettors.

The fuel gauge, on the instrument panel, registers the quantity of fuel in the tank with the ignition switched on. The gauge is designed to eliminate needle fluctuation whilst the car is in motion, the gauge taking about thirty seconds to reach a true reading after switching the ignition on.

QUICK REFERENCE DATA

PERIODIC SERVICE ATTENTION

At first 600 miles (1,000 km.)
   Adjust the carburettor slow-running

At first 3000 miles (5000 km.) or three months
   (whichever occurs first)
   Adjust the carburettor slow-running

Every 3000 miles (5000 km.) or three months
   Adjust the carburettor slow-running

Every 6000 miles (10,000 km.) or six months
   (whichever occurs first)
   Adjust the carburettor slow-running

Every 12,000 miles (20,000 km.) or twelve months
   (whichever occurs first)
   Adjust the carburettor slow-running
   Clean sediment from fuel pump filter and bowl

Every 18,000 miles (30,000 km.) or eighteen months
   (whichever occurs first)
   Adjust slow-running
   Renew air cleaner element
   Renew fuel line filter (where fitted)

DATA

Fuel tank capacity ... 10 Imp. gallons (12 U.S. gallons, 45.4 litres)

Carburettors

Type ...	Two, dual barrel, horizontal Weber type 40 DCOE 31
Main venturi	30
Auxiliary venturi	4.5
Main jet	110
Idling jet	45/F9
Accelerator pump jet	35
Accelerator pump inlet valve bleed	40
Accelerator pump spring length	1.00 in. with 10.75 oz. load
Progression holes	1×120, 2×100
Starting jet	100/F5
Emulsion tube	F11
Air corrector jet	155
Needle valve	1.75
Starting air jet	100
Float weight	26 gms.
Float level	8.5 mm. (including gasket)
Float stroke	6.5 mm.
Petrol level	29 mm.

Fuel Pump

Type ...	Mechanical
Delivery pressure	$1\frac{1}{4}$ to $2\frac{1}{2}$ lb./sq. in. (0.088 to 0.175 kg./sq. cm.)
Inlet depression	$8\frac{1}{2}$ in. mercury (21.59 cm.)
Diaphragm spring test length	0.468 in. (11.883 mm.)
Diaphragm spring test pressure	$3\frac{1}{4}$ to $3\frac{1}{2}$ lb. (1.474 to 1.588 kg.)
Rocker arm spring test length	0.44 in. (11.18 mm.)
Rocker arm spring test pressure	5 to $5\frac{1}{2}$ lb. (2.268 to 2.495 kg.)
Colour code	Red

Tightening Torques

Exhaust manifold to exhaust pipe clamp bolts	7 to 10 lb. ft. (0.97 to 1.38 kg.m.)
Exhaust manifold to exhaust pipe clamp nuts	12 to 15 lb. ft. (1.66 to 2.07 kg.m.)
Fuel pump retaining bolts	12 to 15 lb. ft. (1.66 to 2.07 kg.m.)
Exhaust manifold nuts	12 to 15 lb. ft. (1.66 to 2.07 kg.m.)

CORTINA - LOTUS

SERVICE AND REPAIR OPERATIONS

OP 9000-A FUEL SYSTEM AND CARBURETTOR – CLEAN

1. Raise the bonnet and fit wing covers.
2. Slacken the clip securing the air cleaner to air box hose and remove the hose from the air cleaner.
3. Slacken the four nuts and lift off the air cleaner assembly.
4. Release the air cleaner clips and pull out the end plate complete with the air cleaner element.
5. Unscrew the wing nut and lift off the retainer plate and air cleaner element.
6. Shake the element clean or blow it out with compressed air and replace it on the air cleaner end plate. Replace the retainer plate, washer and wing nut.
7. Push the end plate into the air cleaner body and secure it with the two clips.
8. Disconnect the fuel supply pipes from the carburettors.
9. Unscrew the wing nuts and lift off the two main and idling jet covers.
10. Progressively unscrew the five screws, spring washers and flat brass washers securing the carburettor covers and remove the two covers.
11. Remove the jets and blow them clean with an air line (see OP 9533-A sub-Ops 6 – 19 for jet removal and replacement).
12. Clean the floats and float chamber with petrol and blow clean with an air line.
13. Refit the carburettor covers, ensuring that the floats are free to move in the bodies. Secure them with five screws, flat washers and spring washers, tightening them evenly. Refit the small circular main and idling jet covers, ensuring that the cork seal is in place, retain them with the wing nuts.
14. Remove the fuel pump to carburettors fuel pipe and blow it out with an air line. Replace the pipe, at the pump.
15. Disconnect the fuel tank to fuel pump pipe and blow it out with an air line. Replace the pipe, at the pump end.
16. Remove the fuel pump sediment bowl and filter, blow them clean or wash them in petrol and replace them on the pump.
17. Reconnect the fuel supply pipes at the carburettors.
18. Replace the air cleaner on the two camshaft cover mounting brackets and secure it with the four nuts and flat washers. (Ensure that a flat washer is fitted on either side of each slot in the brackets.)
19. Replace the hose on the air cleaner and tighten the retaining clip.
20. Drain the fuel from the tank by syphoning the fuel into a suitable container.
21. Flush the fuel tank and reconnect the pipe.
22. Strain the fuel and return to the fuel tank.
23. Remove the wing covers and close the bonnet.

OP 9002-A FUEL TANK – REMOVE AND INSTALL

Tools Required
P. 9082   Fuel tank sender unit lock ring wrench
           Bostik model D hand gun

To Remove

1. Drain the fuel from the tank by syphoning the fuel into a suitable container.
2. Disconnect the fuel gauge wire and the fuel feed pipe from the sender unit.
3. Open the boot and remove the floor covering if fitted, from the luggage compartment floor.
4. Slacken the two hose clips on the flexible pipe which connects the filler pipe with the fuel tank neck and then slide the filler pipe away from the fuel tank.
5. Remove the fuel tank shield, where fitted.
6. Lift out the fuel tank after removing the self-tapping bolts securing the fuel tank to the floor pan.
7. The fuel tank sender unit can be removed, if necessary, using Tool No. P.9082. Remove the sealing ring from the fuel tank.

To Install

8. If the fuel tank gauge unit has been removed, fit a new sealing ring in the recess provided in the tank, then fit the gauge unit. Using Tool No. P.9082, tighten the unit to provide an efficient seal.

Air Cleaner and Air Box

9. Apply a bead of Bostik sealing compound No. 1222 F on the floor pan where it mates with the fuel tank flange, using a suitable sealer gun. Alternatively Bostik No. 6 may be used. Fit the fuel tank, retaining it with the self-tapping bolts.

10. Fit the fuel tank shield where applicable.

11. Slide the flexible pipe which connects the filler pipe with the fuel tank neck over the tank neck and secure with a hose clip.

12. Refit the floor covering in the luggage compartment, where applicable.

13. Fit the fuel feed pipe and fuel gauge unit wire to the sender unit.

14. Refill the tank.

15. Check the operation of the fuel gauge, after replacing the filler cap, by switching on the ignition and observing the needle, bearing in mind that the gauge takes about thirty seconds to reach a true reading.

OP 9273-A OIL PRESSURE GAUGE – REMOVE AND INSTALL

To Remove

1. Turn the gauge approximately $\frac{1}{8}$ of a turn anti-clockwise and ease the instrument out of the facia sufficiently to obtain access to the pipe union.

2. Disconnect the pipe union and illumination bulb wire.

To Install

3. Reconnect the pipe union and illumination bulb wire.

4. Push the gauge into the facia and rotate approximately $\frac{1}{8}$ of a turn clockwise.

OP 9275-A FUEL TANK GAUGE UNIT – REMOVE AND INSTALL

Tools Required

P.9082   Fuel tank sender unit lock ring wrench

To Remove

1. Drain the fuel tank by syphoning the fuel into a suitable container.

2. Disconnect the fuel tank gauge unit wire and feed pipe.

3. Unscrew the sencer unit lock ring with Tool No. P.9082. Remove the sealing ring.

To Install

4. Fit a new sealing ring in the recess in the tank and fit the sender unit tightening the lock ring with Tool No. P.9082.

5. Reconnect the wire and feed pipe to the unit.

6. Refill the tank.

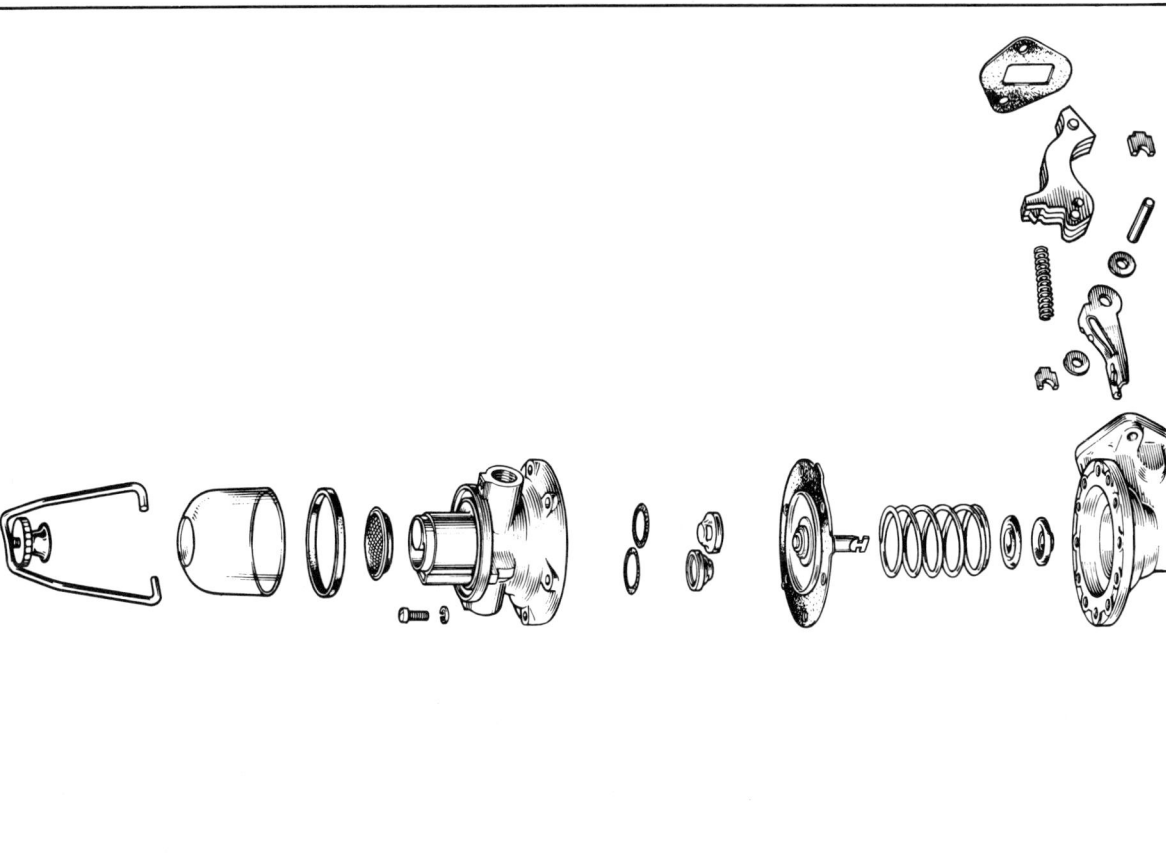

Fuel Pump - Exploded

CORTINA - LOTUS

OP 9280-A FUEL GAUGE – REMOVE AND INSTALL

To Remove

1. Turn the fuel gauge approximately $\frac{1}{8}$ of a turn anti-clockwise and pull the instrument out of the facia.
2. Make a note of the wiring positions then disconnect the wires and illumination bulb from the fuel gauge and remove.

To Install

3. Reconnect the wires and illumination bulb to the gauge.
4. Push the gauge into the facia and rotate approximately $\frac{1}{8}$ of a turn clockwise.

OP 9350-A FUEL PUMP ASSEMBLY – REMOVE AND INSTALL

To Remove

1. Raise the bonnet and fit wing covers.
2. Disconnect the fuel lines at the pump. The line should be suitably plugged to prevent loss of fuel or the ingress of foreign matter.
3. Unscrew and remove the two bolts and spring washers securing the fuel pump to the cylinder block and detach the fuel pump, lifting the operating lever to clear the eccentric and the slotted hole in the block. Remove the gasket.

To Install

4. Clean the mounting face on the cylinder block, removing any trace of gasket which may be adhering to the face. Fit a new gasket to the cylinder block flange, holding it in place with a smear of grease.
5. Insert the rocker arm through the slot in the block wall so that the arm lies on the camshaft eccentric.
6. Secure the fuel pump to the cylinder block with two spring washers and bolts, tightening the bolts evenly to a torque of 12 to 15 lb. ft. (1.66 to 2.07 kg.m.).
7. Ensure that the pipe joints are clean and refit the fuel pipes.
8. Run the engine and check for leaks at the joints.
9. Remove the wing covers and lower the bonnet.

OP 9350-A1 EXTRA: DIAPHRAGM – RENEW
(Fuel pump removed)

To Remove

1. Slacken the clamp nut and remove the sediment bowl and gasket.
2. Remove the filter and clean.
3. Mark the position of the diaphragm tab on both halves of the pump body, remove the screws and separate the two halves of the pump body.

4. Turn the diaphragm, approximately a quarter turn (in either direction), to free the diaphragm rod from the rocker arm link, and detach the diaphragm.
5. Remove the diaphragm spring, oil seal retaining washer and rubber oil seal.

To Install

6. Replace the oil seal, oil seal retainer and the diaphragm spring.
7. Insert the end of the rod in the slotted end of the link. Engage the grooves in the pull rod end by turning the diaphragm a quarter of a turn, so that the **Smaller Tab** on the diaphragm aligns with the mating mark on the lower body flange.
8. Replace the pump upper body, align the mating marks and loosely secure with six screws.
   NOTE – The centre line of the ports should be at 30° to the mounting flange.
9. Operate the rocker arm several times to centralise the diaphragm and fully tighten the screws with the rocker arm raised.
10. Replace the filter.
11. Replace the sediment bowl and gasket and secure with the clamp.

OP 9350-A2 EXTRA: FUEL PUMP MECHANISM – OVERHAUL

To Dismantle

1. Relieve the staking over the two pin retainers and withdraw the retainers. The pin, rocker arm, spring, link and two washers may then be removed as an assembly.

To Reassemble

2. Position the rocker arm, with the boss between the flanges of the link, ensuring that the central web of the link and the spring seat location on the rocker arm are uppermost. Align the holes in the link and rocker arm and insert the pivot pin.

Refitting the Upper Body

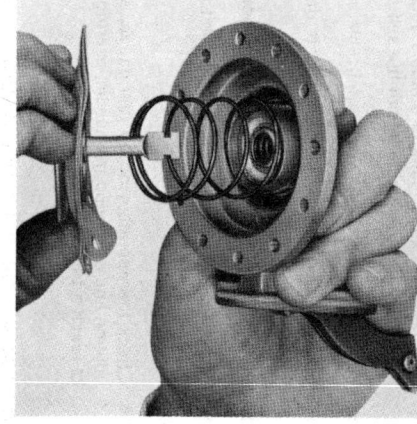

Refitting the Diaphragm

CORTINA - LOTUS

3. Fit one thrust washer to each end of the pin, next to the link, and carefully insert the assembly into the lower pump body casting, with the spring seat on the rocker arm uppermost. Place the rocker arm spring in position so that the ends are located by the registers on the body and the rocker arm.

4. Insert two new pin retainers, one at each end of the pin, ensuring that these positively locate the pin in the casting. Stake over the casting to the pin retainers in two locations each side. Check the operation of the rocker arm and link.

NOTE – New pin retainers should always be fitted after dismantling the lower pump body, as service replacement parts are supplied oversize with a shorter shoulder to enable the staking to be carried out satisfactorily. No attempt should be made to refit the old pin retainers.

OP 9350-A3   EXTRA: VALVES – RENEW
(Fuel pump removed and upper and lower bodies separated.)

To Remove

1. Carefully relieve the staking and remove the valves from the upper body.

To Install

2. Fit the gaskets in the upper body, then fit the two valve assemblies.
   NOTE – The valves will only seat properly when in their correct locations and the right way up.

3. Ensure that the valves are pressed fully home and retain each valve securely by staking at six points around its housing.

OP 9350-B   FUEL PUMP DIAPHRAGM – RENEW
(Includes OPS 9350-A and A1)

OP 9350-C   FUEL PUMP MECHANISM – OVERHAUL
(Includes OPS 9350-A, A1 and A2)

OP 9350-D   FUEL PUMP VALVES – RENEW
(Includes OPS 9350-A and A3)

OP 9350-E   FUEL PUMP DIAPHRAGM AND VALVES – RENEW
(Includes OPS 9350-A, A1 and A3)
(Using diaphragm and valve kit)

OP 9350-F   FUEL PUMP – OVERHAUL
(Includes OPS 9350-A, A1, A2 and A3)
(Using fuel pump repair kit)

OP 9350-G   FUEL PUMP – TEST

Tools Required

500 X   Gang gauge

1. Raise the bonnet and fit wing covers.

Fuel Pump Inlet Depression Test

2. Fill the carburettor float chamber with petrol. If necessary, a separate gravity feed fuel supply can be connected to the carburettor.

3. Disconnect the fuel line from the tank fuel at the pump inlet, suitably plugging the end of the pipe to prevent loss of fuel from the tank or the ingress of foreign matter.

4. Connect the vacuum gauge to the inlet connection.

5. Start the engine and allow it to run at idling speed, when a vacuum reading of at least 8½ in. (21.59 cms.) mercury should be obtained.

6. Stop the engine, when the gauge needle should take at least one minute to return to zero.

7. Disconnect the vacuum gauge from the pump and reconnect the fuel line to the pump inlet.

Fuel Pump Delivery Pressure Test

8. Fill the carburettor float chamber with petrol. If necessary a separate gravity feed fuel supply can be connected to the carburettor.

9. Disconnect the fuel line from the pump to the carburettor.

10. Connect the pressure gauge to the pump outlet.

11. Start the engine and observe the pressure when running at idling speed. Momentarily race the engine and observe the pressure. This should be 1¼ to 2½ lb./sq. in. (0.087 to 0.176 kg./sq. cm.).

12. Stop the engine.

13. Disconnect the pressure gauge from the pump and reconnect the fuel pump to carburettor pipe.

14. Remove the wing covers and close the bonnet.

OP 9364-A   FUEL FILTER SEAL AND/OR BOWL – RENEW

To Remove

1. Slacken the clamp nut and remove the sediment bowl and gasket.
2. Remove the filter and clean.

To Install

3. Carefully refit the filter.
4. Replace the sediment bowl and gasket and secure with the clamp.

OP 9428-A   EXHAUST MANIFOLDS – REMOVE AND INSTALL

To Remove

1. Raise the bonnet and fit wing covers.
2. Remove the eight nuts securing the exhaust manifolds to the cylinder head.
3. Slacken the two clamps securing the exhaust manifolds to the exhaust pipe.

4. Pull the manifolds away from the cylinder head to clear the mounting studs and pull the manifolds forward to release them from the exhaust pipe.

To Install

5. Clean the manifold and cylinder head mating surfaces and position four new gaskets on the cylinder head.
6. Enter the manifolds into the exhaust pipes and position the flanges on the cylinder head mounting studs.
7. Secure the manifolds with the eight brass locknuts, tightening the nuts to a torque of 12 to 15 lb. ft. (1.66 to 2.07 kg.m.) and tighten the manifolds to exhaust pipe clamps.
8. Remove the wing covers and close the bonnet.

OP 9448-A   EXHAUST MANIFOLD GASKET – RENEW
(Includes OP 9428-A)

OP 9510-A   CARBURETTORS AND/OR GASKETS – REMOVE AND INSTALL

To Remove

1. Raise the bonnet and fit wing covers.
2. Remove the clip securing the hose, from the air cleaner to the air intake, at the air box end and remove the air intake cover by unscrewing the three bolts.
3. Remove the air box backplate assembly by unscrewing the eight nuts, withdrawing the spring washers, air horn retaining plates and air horns.
4. Detach the throttle cable from the carburettors and unlatch the return spring.
5. Disconnect the fuel supply pipes from the fuel pump at the carburettors.
6. Detach the choke cable from the starting devices.

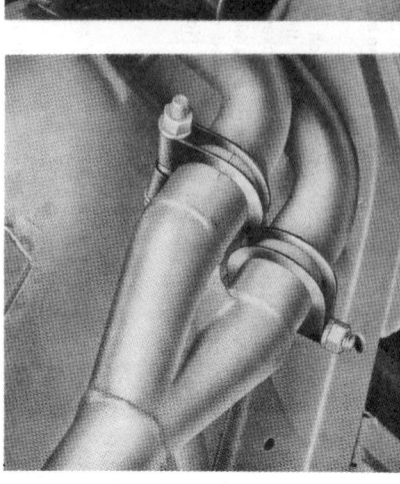

Exhaust to Manifold Clamps

Air Horns

7. Progressively unscrew the eight nuts retaining the carburettors to the inlet tracts. Four can readily be seen from above, the other four are below the carburettors. Withdraw from each stud a flat washer and a double coil spring washer. Carefully remove the carburettors, ensuring that the synchronising linkage between the two is not distorted. Remove the gaskets and 'O' rings from the inlet tract studs.

To Install

8. Carefully examine each carburettor to inlet tract gasket, ensuring that the metal mounting plate is not damaged and that the 'O' rings in the faces of the plate are in position. Fit the gaskets to the inlet tracts.
9. Fit the carburettors, ensuring that the synchronising linkage is correctly positioned so that the lug on the rear carburettor throttle linkage is between the spring-loaded plunger and adjusting screw on the front carburettor. To each stud fit a double coil spring washer, a flat washer and nut. Tighten the eight nuts progressively until a 0.040 in. (1.016 mm.) clearance exists between the coils of the double coil spring washers. This clearance should be checked with feeler blades. Do not overtighten the nuts otherwise the 'O' rings will be flattened into the recesses in the plate.
10. Refit the fuel supply pipes to the carburettors.
11. Reconnect the choke control by securing the cable casing in the cast arm of each starting device cover with the clamp screw. Ensure that the choke control on the instrument panel is pushed fully home and that the starting device operating levers are in the off position. Retain the inner cable in the operating levers with the clamp screws.

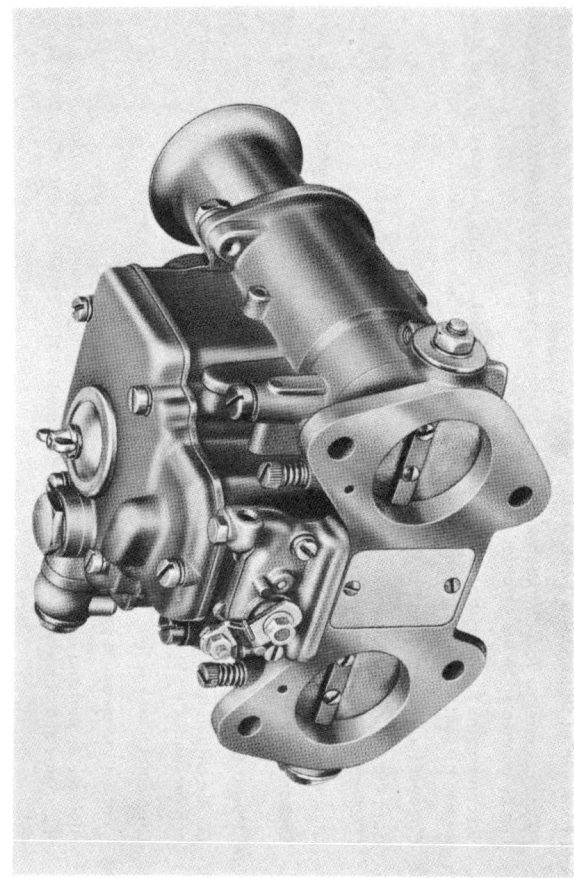

Weber Carburettor

CORTINA - LOTUS

12. Fit the backplate assembly of the air box to the carburettors, ensuring that the cork gasket is not distorted, then an air horn in each barrel and retain with eight air horn retaining plates, spring washers and nuts.
13. Refit the throttle cable to the throttle linkage and refit the return spring.
14. Adjust the carburettor slow-running as in OP 9510-C.
15. Replace the air box cover and secure it with the three bolts. Reposition the hose on the air box and tighten the clip.
16. Remove the wing covers and lower the bonnet.

OP 9510-A1 EXTRA: CARBURETTOR GASKET – RENEW

A gasket kit is not available for this carburettor. However, the following instructions detail the procedure for the replacement of all gaskets which may be obtained individually.

Before fitting a new gasket, ensure that the mating faces are clean and free of old gasket material.

1. Unscrew the fuel filter retaining plug from the top cover.
2. Fit a new washer to the plug then replace the plug.
3. Unscrew the wing nut, retaining the main and idling jet cover, and remove the five screws to release the top cover. Lift off the top cover and gasket.
4. Carefully push out the float fulcrum pin and remove the float assembly, needle valve and cover gasket.
5. Unscrew the needle valve seat from the cover, fit a new sealing washer and refit the valve seat.
6. Locate a new gasket on the cover, the needle valve within its seat and refit the float assembly.
7. Remove the four screws retaining the jet inspection cover to the base of the carburettor.
8. Locate a new gasket on the cover, then refit the cover.
9. Remove the two screws retaining the body plate, located between the carburettor mounting flanges, and withdraw the plate and gasket.
10. Locate a new gasket on the plate and secure the plate and gasket to the body.
11. Remove the accelerator pump jet retaining screw, and withdraw the pump jet.
12. Fit a new washer to the pump jet then relocate the jet within the body.
13. Fit a new 'O' ring to the jet retaining screw and refit the screw.
14. Carefully position the cover on the body and secure with screws.
15. Locate a new gasket within the main and idling jet cover then secure to the top cover with the wing nut.

OP 9510-A2 EXTRA TO OVERHAUL EACH CARBURETTOR

To Dismantle

1. Remove the auxiliary venturi followed by the main venturi from each barrel. Auxiliary venturi size:–4.5, Main venturi size:–30.

1. Volume Control Screw
2. Starting Jet
3. Starting Device Control Lever
4. Starting Device Piston Spring Guide/Retainer
5. Progression Hole Inspection Plug
6. Accelerator Pump Jet Retaining Screw

Jet Positions

7. Accelerator Pump Delivery Valve Screw
8. Air Horn
9. Accelerator Pump Inlet Valve
10. Idling Jet Holder
11. Air Horn Retaining Plate
12. Emulsion Tube, Air Correction and Main Jet Holder
13. Accelerator Pump

2. To remove the fuel filter first unscrew the hexagon-headed retainer from the carburettor cover, noting that there is a sealing washer beneath the retainer head. Withdraw the gauze filter from the cover, taking care not to lose the brass seat in the top of the filter.

3. The carburettor cover can be withdrawn after removing the small circular main and idling jet cover retained by a wing nut and then the five screws (slacken evenly) with their spring washers and flat brass washers.

4. Remove the floats and needle valve from the cover. Gently push out the float fulcrum pin from the cover, after which the needle valve may be removed from its seat. Withdraw the cover gasket and unscrew the needle valve seat from the cover, a sealing washer being fitted between the needle valve seat and cover. Float weight:—26 grams. Needle valve size:—1.75.

5. Remove the accelerator pump from the carburettor body. Pull out the inverted 'U' shaped control rod which will withdraw the split retainer, spring and piston. To dismantle the assembly, first compress the spring, slightly rotate the piston and withdraw from the hooked end of the control rod, followed by the spring and split retainer.

6. Unscrew the accelerator pump inlet valve from the base of the float chamber. Shake the inlet valve to ensure that the ball inside the valve body slides freely. Inlet valve size:—35.

7. Remove the idling jet holders (two per carburettor) and pull the idling jets from the holders. Idling jet size:—45F9.

8. Unscrew the emulsion tube holders (two per carburettor). Pull each emulsion tube from its holder and then the main jet from one end of the emulsion tube and the air corrector jet from the other. Jet sizes:—Main—110, Air corrector—155, Emulsion tube—F11.

9. Remove each accelerator pump delivery valve retaining screws (two per carburettor) and invert the carburettor to extract the balls and weights.

10. Unscrew the pump jet retaining screws (two per carburettor). Extract each pump jet from the body. Pump jet size:—100F5.

11. Remove the starting jets (two per carburettor). Jet size:—35.

12. The progression holes in each barrel may be inspected by removing the two progression hole inspection plugs. Size:—1×120, 2×100.

13. Unscrew the volume control screws and throttle stop screw, if fitted. Examine the springs.

14. The starting device cover on the side of the carburettor may be removed after unscrewing the two retaining screws which have spring and flat washers beneath their heads.

15. Carefully prise out the combined starting device piston spring guide/retainer circlips (two per carburettor). Withdraw the guides and springs and invert the carburettor to extract the starting device pistons.

16. Remove the throttle plates, one from each barrel, by unscrewing the two screws securing each plate in its shaft.

17. Remove the throttle spindle. A new carburettor body is supplied complete with the throttle spindle, pump operating arm, bearings, etc. They can be dismantled, if required, as follows. Remove the nuts, after bending back the tab washers, each end of the throttle spindle. Remove the flat washer from one end and the throttle linkage from the other. Withdraw the plate and gasket, secured by two screws, from the engine side of the carburettor to gain access to the accelerator pump control arm. Tap out the pin retaining this arm to the spindle.

Ensuring that the threads are not damaged, knock out the spindle which will also remove, from one end of the spring retainer, the spring, dust cover and bearing. After carefully prising out the spring retainer from the other end the spring and dust cover can be extracted.

Cleaning and Inspection

After dismantling and prior to reassembling, the carburettor, filter and the jets should be cleaned and checked for size, see Specification. The seating of the idling jets, starting jets, starter pistons and main jets in the carburettor body should also be examined, together with the seating faces of the jets and tubes themselves.

It is physically impossible to inter-change air correction jets with main jets, starting jets with idling jets, etc., and similarly their positions in the carburettor.

Inspect the various gaskets, also the sealing rings fitted to the accelerator pump jet screws, needle valve seat and filter retainer. Clean the filter and ensure that the gauze is undamaged.

Shake the accelerator pump inlet valve to check that the ball is free to slide.

If it is suspected that any of the ducts in the carburettor are blocked then the lead plugs which close the ducts should be removed, the ducts blown out and new lead plugs fitted. These plugs must be a good fit to prevent fuel leakage and the ingress of air which might affect mixture strength. After fitting, the plugs must not stand proud on mating faces, since if they are proud air leaks may result due to surfaces not seating.

To Reassemble

18. If removed, refit the throttle spindle across the barrels, ensuring that accelerator pump operating arm is fitted on the shaft so that the shouldered end is by a barrel and the curved end of the arm is uppermost. Retain the arm to the spindle with a tension pin. Fit a dust cover, spring and spring retainer at each end of the shaft and, depending on the end of the shaft, fit a flat washer on the throttle linkage. Secure with new tab washers and nuts, operating the throttle spindle whilst tightening the nuts to ensure ease of movement. It may be necessary to lightly tap either end of the shaft to obtain this condition. Refit the accelerator arm access plate and gasket, securing with two screws.

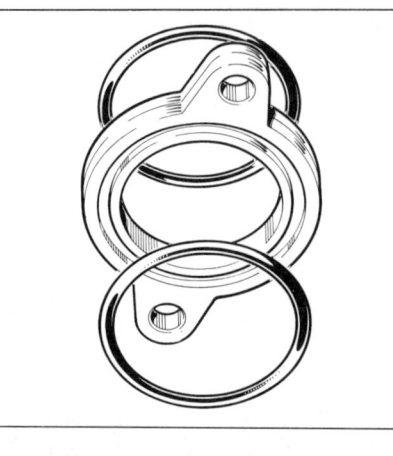

Carburettor Gasket

Double Coil Spring Washer Clearance

19. Fit the throttle plates, noting that the edges of the plates are chamfered, the plates being fitted so that they can completely close the barrels. If a new throttle plate is being fitted, check that the angle stamped on the plate is 79° 30′. If a throttle plate with a different angle is fitted then the carburettor's low speed progression will be affected, since the distance between the throttle plate's edge and adjacent progression hole will be altered with the result that vacuum at the progression hole will be either too strong or too weak with consequential irregular running. Retain the plates in the shaft with new screws, the shaft holes being countersunk for the heads; do not tighten these screws at this time. Close the throttle shaft to centralise the plates with the barrels, tighten the screws.

20. Using a 0.002 in. (5.08 mm.) thick feeler gauge in the gap between the throttle body and the throttle plate (on the centre line of the throttle plate and at right-angles to the throttle spindle) at the progression hole side of the throttle barrel, hold the throttle control lever firmly against the stop screw, and adjust the screw until a light pull is required to withdraw the feeler blade. Next, trap the feeler blade on the opposite side of the throttle plate and if the concentricity is correct then the same effort will be required to withdraw the blade with the stop screw in the same position. If the concentricity is incorrect the clamping screws must be backed off and the plate moved as required. Repeat the above procedure on the second throttle plate until concentricity is obtained on this also. Peen the threaded ends of the screws to retain them.

21. Having checked (and set if necessary) the concentricity of the two throttle plates, they should now be checked for synchronisation. Using a 0.002 in. (5.08 mm.) feeler blade positioned between the throttle body and the plate on the progression hole side of the barrel and holding the throttle control lever hard against the stop screw, adjust the screw until a light pull is required to withdraw the feeler from between the second plate and throttle barrel. Without disturbing the stop screw, the same effort should be required to withdraw the feeler from between the two throttle plates. In cases where the concentricity of the two throttle plates has been set but synchronisation is incorrect, check first that all the throttle plates are identical with respect to the degree number stamped on them (i.e. 79° 30 mins). If these are correct this normally indicates a twisted throttle spindle and this is corrected by holding one end of the spindle and turning the other end in the required direction.

22. Fit the starting device pistons in the carburettor body tapered ends first, followed by the coil springs and combined spring guide/retainers, the latter being held in position by circlips.

23. Secure the starting device cover to the carburettor, dowels being provided on the cover for locating purposes, and retain with two screws, flat washers and spring washers, tightening them evenly to avoid distorting the cover. By looking through the spring guide/retainers and operating the lever a check can be made to ensure that the pistons are raised by the operating lever and lowered by the springs when the lever is released.

24. Fit the progression hole inspection plugs, one in each barrel.

25. Fit the starting jets, two per carburettor. Jet size:—100F5.

26. Replace the accelerator pump jets (two per carburettor), noting that the smaller diameter enters first and the flat on the large diameter is to the engine side of the carburettor. Check the condition of the retaining screw rubber seals and refit together with these screws. Pump jet size:—35.

27. Refit the accelerator pump delivery valves (two per carburettor) by fitting a ball first, then a weight, concave face to the ball, and finally the retaining screws.

28. Fit the emulsion tube holders. Push a hexagonal head main jet (size 110) into the large diameter end of an emulsion tube (F11) and a circular-headed air corrector jet (size 155) in

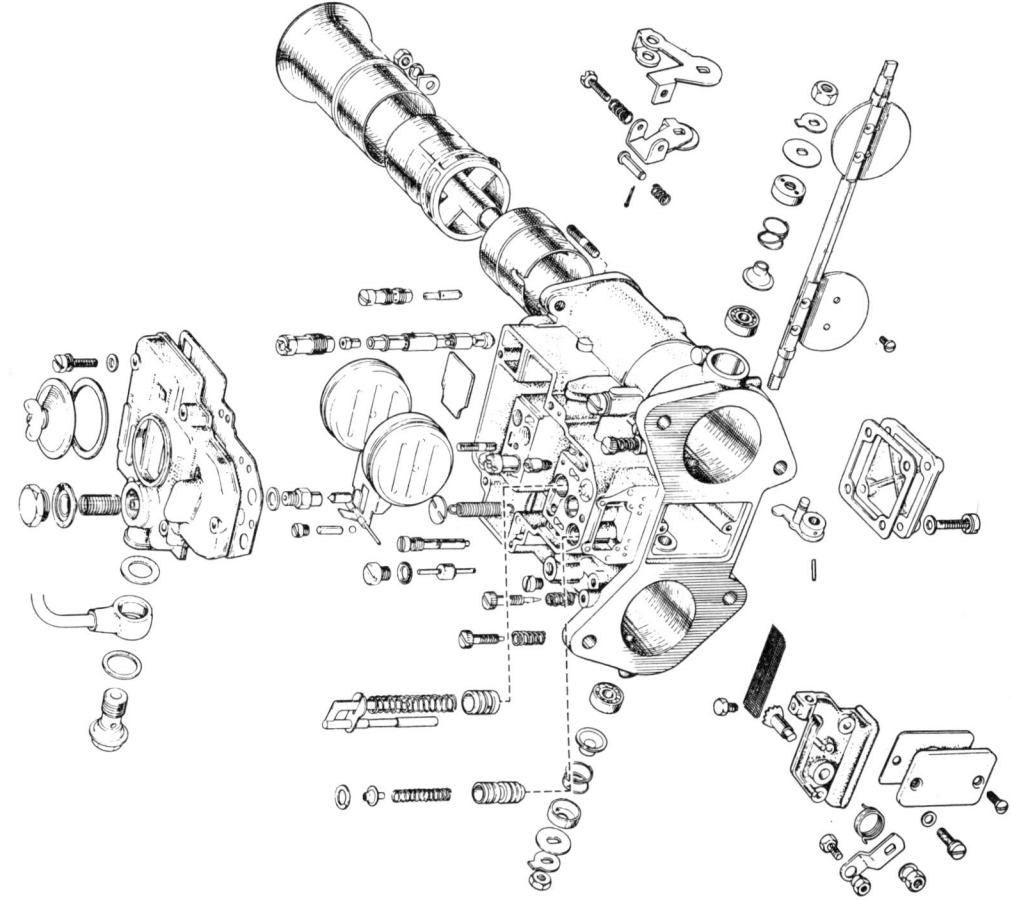

Carburettor - Exploded

the other end. Insert the assembly, air corrector jet first, into the emulsion tube holder and screw into position.

29. Replace the idling jets (two per carburettor) in their holders and fit them in the carburettor. (Jet size:—45F9).

30. Check that the accelerator pump inlet valve nylon ball moves freely and screw the valve into the base of the float chamber. Inlet valve size:—40.

31. Reassemble and fit the accelerator pump. Slide the split retainer on the accelerator pump control rod, with the dishing toward the hook end of the control rod. Pass the spring over the hooked end of the control rod, compress the spring and then locate the piston on the control rod. Position the assembly in the carburettor and press the split retainer into position. Check the operation of the pump by actuating the throttles.

32. Screw the needle valve seat into the cover, after ensuring that the seat sealing washer fitted between the seat and cover is in good condition. This sealing washer also affects the float level.

33. Refit the needle valve and floats. Check the needle valve damping ball for free operation and then place the needle valve in its seat. Place a new gasket on the cover and then push the float fulcrum pin through the cover 'legs' and float hinge. Needle valve size:—1.75. Float weight:—26 grams.

34. Check the float level. Hold the carburettor cover in the vertical position with the floats hanging down and with the tab which abuts the needle valve in light contact with the ball and perpendicular. The distance between both floats and the cover, including gasket, should be 8.5 mm. If necessary, bend the needle valve tab to obtain this measurement. After levelling the floats check that the stroke is 6.5 mm., i.e. 15 mm. from the cover. If necessary, adjust the position of the other tab, which abuts the needle valve seat, to obtain this movement.

The float level and stroke should be checked whenever the floats, needle valve, needle valve seat or sealing washer are renewed.

35. Refit the carburettor cover, ensuring that the floats are free to move in the body. Secure with five screws, flat washers and spring washers, tightening them evenly. Refit the small circular main and idling jet cover and cork seal retained by a wing nut.

36. Place the gauze fuel filter in the top cover, then the brass seat in the gauze filter and finally screw the retainer, with a sealing washer beneath its head, into the cover.

37. Carefully screw the volume control screws (two per carburettor) into position until each just contacts its seat and then unscrew one turn.

38. Fit the throttle stop screw, if fitted, until it just contacts the throttle stop lever and then screw in a further half-turn.

39. Fit the main venturis (size 30), smaller external diameter first, so that the brass pin in the larger external diameter slides in the barrel's channel.

40. Replace the auxiliary venturis (size 4.5), larger external diameter first, engaging the venturi spring tongue in the barrel channel.

OP 9510-A3 EXTRA: CARBURETTOR TO MANIFOLD STUDS – RENEW

Tools Required

Stud Remover and Replacer

To Remove

1. Using the stud remover and replacer in the prescribed manner, remove the studs.

To Install

2. Fit the new studs, again using the stud remover and replacer.

OP 9510-B CARBURETTOR GASKETS – RENEW
(Includes OPS 9510-A and A1)

OP 9510-C CARBURETTOR – ONE – OVERHAUL
(Includes OPS 9510-A and A2)

OP 9510-D CARBURETTOR – BOTH – OVERHAUL
(Includes OPS 9510-A and A2 × 2)

OP 9510-M CARBURETTORS – ADJUST

1. Remove the outer part of the air box which is secured by three bolts and spring washers.

Float Level and Float Stroke Setting.

2. Disconnect the fuel pipe to each carburettor.

3. Remove the jet cover wing nut and the five screws to release the top cover.

4. Lift off the top cover.

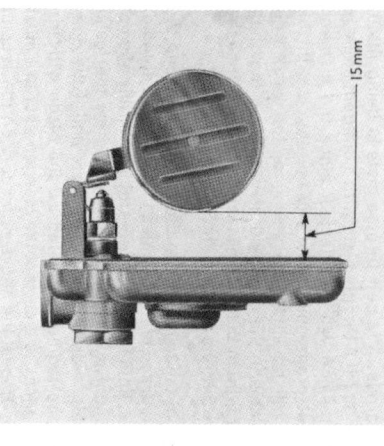

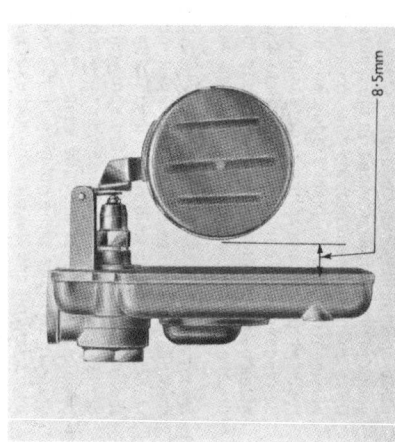

Float Level   Float Stroke

CORTINA - LOTUS

5. Carry out the following checks:—
   (a) Floats are not punctured.
   (b) Floats swing freely about its pivot.
   (c) Ensure that the needle valve housing is screwed fully home.
   (d) The ball pin moves freely when the float is raised and lowered.
   (e) The contact face of the tab located between the float arms is not pitted.

6. With the top cover held in a vertical position check that the dimension between the cover, with the gasket in position, and each float is 8.5 mm. Adjust the float arms and/or the tab, located between the float arms, to obtain this dimension.

7. With the float level correctly set and the float moved fully away from the cover, check that the dimension between the cover and float is 15 mm. Adjust the tab which contacts the needle valve housing to obtain this dimension.

8. Refit the top cover, jet cover and reconnect the fuel line to each carburettor.

Slow Running Adjustments

9. Carefully screw in each volume screw until it just contacts its seat then unscrew $\frac{3}{4}$ of a turn.

10. Start the engine and set the throttle stop screw of the rear carburettor so that the slow-running speed is slightly higher than normal (1,000 to 1,200 rev./min.).

11. Adjust the synchronising screw of the throttle spindle connecting linkage to match the throttle openings. This adjustment, which is most important, can be checked by placing one end of a piece of flexible tube against the ear and the other on a common spindle, it is only necessary to place the tube in the intake of one barrel of each carburettor. If the hiss of one carburettor is louder than the other, adjust the rear carburettor throttle stop screw and synchronising screw to equalise the intensity of the hiss and therefore the intake of air. *Once the front and rear carburettor throttle plates are synchronised, do not touch the synchronising screw of the throttle spindle connecting linkage.*

Throttle Stop Screw

12. The four volume control screws should then be adjusted together an eighth of a turn at a time, allowing time for each adjustment to take effect, until the engine runs evenly.

    Whilst one volume control screw should not normally be adjusted without adjusting the other three a similar amount, if necessary, on final adjustment, a slight variation between each volume control screw is permissible, to achieve a satisfactory idling condition. However, this variation should not exceed one quarter of a turn between each screw.

    If, after setting the volume control screws, the engine idling speed is too fast (above 800 – 1,000 rev./min.), slacken the rear stop screw until the idling is satisfactory and, if necessary, readjust the volume control screws and ignition setting.

13. Refit the air box cover.

OP 9533-A CARBURETTOR JETS AND FLOAT CHAMBERS – CLEAN

To Remove

1. Slacken the two clips and remove the air cleaner to air box hose.
2. Unscrew the three bolts and remove the air box cover.
3. Disconnect the fuel supply pipes from the carburettors.
4. Unscrew the wing nuts and lift off the two main and idling jet covers.
5. Progressively unscrew the five screws, spring washers and flat brass washers securing the carburettor covers and remove the two covers.
6. Unscrew the accelerator pump inlet valve from the base of the float chamber.
7. Remove the idling jet holders (two per carburettor) and pull the idling jets from the holders.
8. Unscrew the emulsion tube holders (two per carburettor). Pull each emulsion tube from its holder and then the main jet from one end of the emulsion tube and the air corrector jet from the other.
9. Remove each accelerator pump delivery valve retaining screws (two per carburettor).

Volume Control Screws

Throttle Synchronising Screw

10. Unscrew the pump jet retaining screws (two per carburettor) and examine the rubber seal around each screw. Extract each pump jet from the body.
11. Remove the starting jets (two per carburettor).
12. Remove the accelerator pump jets from the carburettor body. Pull out the inverted 'U' shaped control rod which will withdraw the split retainer, spring and piston.
13. Blow the accelerator pump, the jets and their housings clean. Wash the floats and the float chambers in clean petrol and blow clean.

To Install

14. Fit the starting jets (two per carburettor).
15. Replace the accelerator pump jets (two per carburettor), noting that the smaller diameter enters first and the flat on the large diameter is to the engine side of the carburettor. Check the condition of the retaining screw rubber seals and refit together with these screws.
16. Refit the accelerator pump delivery valves (two per carburettor) by fitting a ball first, then a weight, concave face to the ball, and finally the retaining screws.
17. Fit the emulsion tube holders. Push a hexagonal head main jet (size 110) into the large diameter end of an emulsion tube (F11) and a circular-headed air corrector jet (size 155) in the other end. Insert the assembly, air corrector jet first, into the emulsion tube holder and screw into position.
18. Replace the idling jets (two per carburettor) in their holders and fit them in the carburettor.
19. Check that the accelerator pump inlet valve ball moves freely and screw the valve into the base of the float chamber.
20. Reassemble and fit the accelerator pump. Slide the split retainer on the accelerator pump control rod, with the dishing toward the hook end of the control rod. Pass the spring over the hooked end of the control rod, compress the spring and then locate the piston on the control rod. Position the assembly in the carburettor and press the split retainer into position.
21. Refit the carburettor covers, ensuring that the floats are free to move in the bodies. Secure them with five screws, flat washers and spring washers, tightening them evenly. Refit the small circular main and idling jet covers retained by wing nuts.
22. Reconnect the fuel supply pipes to the carburettors.
23. Replace the air box cover and secure it with the three bolts and spring washers.
24. Position the air cleaner to air box hose on the two components and tighten the two retaining clips.

OP 9600-A   AIR CLEANER ASSEMBLY – REMOVE AND INSTALL

To Remove

1. Slacken the four nuts securing the air cleaner body to the rocker cover brackets and the hose clip adjacent to the body. Lift off the body.
2. Release the two clips at the end opposite to the air intake tubes, separate the body and remove the cleaner element.
3. Remove the air intake cover by unscrewing the three bolts.
4. Locate the air intake cover on the air box backplate and secure with three screws, plain and spring washers.
5. Position the cleaner element within the air cleaner body and secure the two parts of the body with the clips.
6. Locate the air cleaner body on its supporting brackets and tighten the four retaining nuts.
7. Position the hose on the cleaner body and tighten the clip.

OP 9700-A   CHOKE CONTROL – RENEW

To Remove

1. Raise the bonnet and fit wing covers.
2. Disconnect the choke inner cable from the operating levers by unscrewing the clamp screws. Release the outer cable from the cast arms of each starting device cover.
3. Remove the choke control from the facia by pulling out the inner cable sufficiently to unscrew the chrome bezel and remove the outer cable and inner cable assembly complete.

To Install

4. Feed the cable assembly through the facia aperture and tighten the chrome bezel to retain it.
5. Secure the cable casing in the cast arm of each starting device cover with the clamp screw. Ensure that the choke control on the facia is pushed fully 'home' and that the starting device operating levers are in the off position. Retain the inner cable in the operating levers with the clamp screws.
6. Remove the wing covers and close the bonnet.

OP 9725-A   THROTTLE CABLE – REMOVE AND INSTALL

To Remove

1. Raise the bonnet and fit wing covers.
2. Unlatch the throttle return spring from the bracket on the inlet tracts.
3. Disconnect the throttle cable from the throttle control lever on the carburettor and release the inner cable from the anchor bracket.
4. Release the cable from the throttle pedal upright by pulling it out of the slot.
5. Feed the cable through the grommet in the bulkhead and remove it from the passenger compartment side.

To Install

6. Feed a new cable through the grommet in the bulkhead from the passenger compartment side.
7. Connect the cable to the throttle pedal by locating it in the slot.
8. Connect the cable to the carburettor throttle control lever and reconnect the return spring. Locate the inner cable in the anchor bracket.
9. Remove the wing covers and lower the bonnet.

CORTINA - LOTUS

OP 9735-A ACCELERATOR PEDAL AND SHAFT ASSEMBLY – REMOVE AND INSTALL

To Remove

1. Release the cable from the throttle pedal upright by pulling it out of the slot.
2. Remove the bolts securing the bushes to their respective brackets.
3. Withdraw the throttle shaft and bearings.

To Install

4. Locate the throttle shaft and bearings on the brackets and secure with the two bolts.
5. Engage the throttle cable within the slot in the upright.

OP 9735-A1 EXTRA: ACCELERATOR PEDAL AND SHAFT BUSHES – RENEW

To Remove

1. Slide the bush, remote from the throttle pedal upright, off the shaft.
2. Rotate the bush, adjacent to the throttle pedal upright, through 90° and slide off the shaft.

To Install

3. Slide a bush onto the shaft and against the throttle pedal upright. Rotate through 90° to locate.
4. Slide the remaining bush onto the shaft for approximately one inch.

OP 9735-B ACCELERATOR PEDAL AND SHAFT BUSHES – RENEW
(Includes OPS 9735-A and A1)

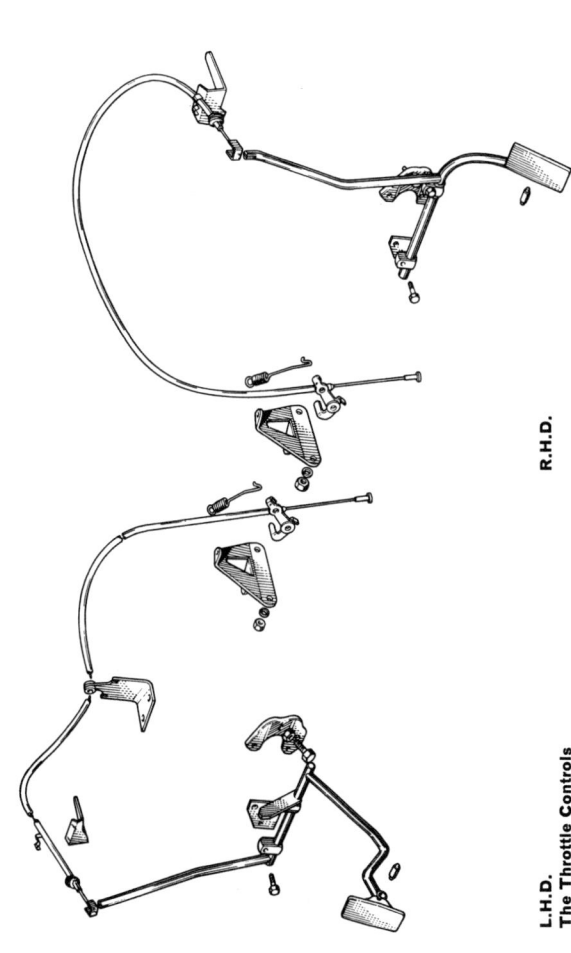

L.H.D. R.H.D.

The Throttle Controls

CORTINA - LOTUS

10
ELECTRICAL SYSTEM

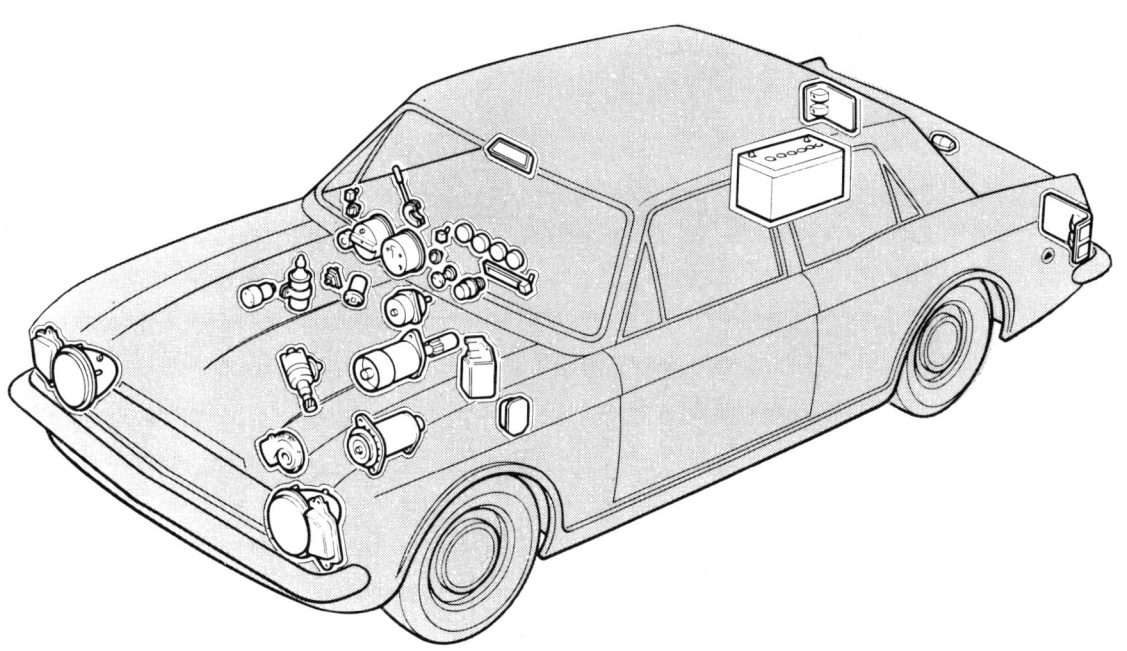

CORTINA - LOTUS

SECTION INDEX

GENERAL DESCRIPTION

CHARGING SYSTEM

GENERAL DESCRIPTION

QUICK REFERENCE DATA

SERVICE AND REPAIR OPERATIONS

OPERATION	10001-A	GENERATOR ASSEMBLY – REMOVE AND INSTALL
"	10001-A1	Extra: generator pulley – remove and install (Generator assembly removed)
"	10001-A2	Extra: generator front bearing – renew (Generator pulley removed)
"	10001-A3	Extra: brushes – renew and commutator – clean (Generator pulley removed)
"	10001-A4	Extra: generator – overhaul (Generator brushes and commutator removed)
"	10001-B	**GENERATOR PULLEY – REMOVE AND INSTALL** (Includes OPS 10001-A and A1)
"	10001-C	**GENERATOR BRUSHES – RENEW AND COMMUTATOR – CLEAN** (Includes OPS 10001-A, A1 and A3)
"	10001-D	**GENERATOR ASSEMBLY – OVERHAUL** (Includes OPS 10001-A, A1, A3 and A4)
"	10001-E	**GENERATOR FRONT BEARING – RENEW** (Includes OPS 10001-A, A1 and A2)
"	10001-M	CHARGING CIRCUIT – TEST AND ADJUST
"	10505-A	VOLTAGE REGULATOR UNIT – REMOVE AND INSTALL

STARTING SYSTEM

GENERAL DESCRIPTION

QUICK REFERENCE DATA

SERVICE AND REPAIR OPERATIONS

OPERATION	11001-A	STARTER MOTOR ASSEMBLY – REMOVE AND INSTALL
"	11001-A1	Extra: drive components – remove and install (starter motor removed)
"	11001-A2	Extra: brushes – renew and commutator – clean (drive components removed)
"	11001-A3	Extra: starter motor – overhaul (drive components, brushes and commutator removed)
"	11001-B	**STARTER MOTOR DRIVE COMPONENTS – REMOVE AND INSTALL** (Includes OPS 11001-A and A1)
"	11001-C	**STARTER MOTOR BRUSHES – RENEW AND COMMUTATOR – CLEAN** (Includes OPS 11001-A, A1 and A2)
"	11001-D	**STARTER MOTOR – OVERHAUL** (Includes OPS 11001-A, A1 A2 and A3)
"	11450-A	STARTER SOLENOID – REMOVE AND INSTALL

IGNITION SYSTEM

GENERAL DESCRIPTION

QUICK REFERENCE DATA

SERVICE AND REPAIR OPERATIONS

OPERATION	12024-A	IGNITION COIL – REMOVE AND INSTALL
"	12100-A	DISTRIBUTOR ASSEMBLY – REMOVE AND INSTALL
"	12100-A1	Extra: distributor cap – remove and install (distributor removed)
"	12100-A2	Extra: condenser – renew (distributor cap removed)
"	12100-A3	Extra: points – renew (distributor cap removed)
"	12100-A4	Extra: distributor breaker plate assembly – remove and install (distributor cap removed)
"	12100-A5	Extra: distributor breaker plate assembly – overhaul (breaker plate removed)
"	12100-A6	Extra: governor weights and springs – renew (breaker plate removed)
"	12100-A7	Extra: vacuum unit – renew (breaker plate removed)
"	12100-A8	Extra: distributor – overhaul (governor weights and vacuum unit removed)
"	12100-B	**GOVERNOR WEIGHTS AND SPRINGS – RENEW** (Includes OPS 12100-A, A1, A4 and A6)
"	12100-C	**VACUUM UNIT – RENEW** (Includes OPS 12100-A, A1, A4 and A7)
"	12100-D	**DISTRIBUTOR – OVERHAUL** (Includes OPS 12100-A, A1, A2, A3, A4, A5, A6, A7 and A8)
"	12100-E	DISTRIBUTOR ASSEMBLY – TEST AND ADJUST
"	12199-A	CONTACT BREAKER POINTS – RENEW (DISTRIBUTOR IN SITU)
"	12300-A	CONDENSER – RENEW (DISTRIBUTOR IN SITU)

CORTINA - LOTUS

OPERATION 12405-A	SPARK PLUGS – REMOVE AND INSTALL
,, 12405-A1	Extra: spark plugs – clean and reset
,, 12405-B	**SPARK PLUGS – CLEAN AND RESET** (Includes OPS 12405-A and A1)

INSTRUMENTS, CONTROLS AND ANCILLARIES

GENERAL DESCRIPTION

QUICK REFERENCE DATA

SERVICE AND REPAIR OPERATIONS

OPERATION 9280-A	FUEL GAUGE – REMOVE AND INSTALL
,, 10505-B	INSTRUMENT VOLTAGE REGULATOR – REMOVE AND INSTALL
,, 10670-A	AMMETER – REMOVE AND INSTALL
,, 10880-A	OIL PRESSURE GAUGE – REMOVE AND INSTALL
,, 10883-A	TEMPERATURE GAUGE – REMOVE AND INSTALL
,, 11572-A	IGNITION SWITCH AND/OR LOCK BARREL – REMOVE AND INSTALL
,, 11654-A	LIGHT SWITCH ASSEMBLY – REMOVE AND INSTALL
,, 13000-A	HEADLAMPS – ALL – ALIGN
,, 13005-A	HEADLAMP SEALED BEAM UNIT – REMOVE AND INSTALL
,, 13200-A	FRONT INDICATOR AND/OR SIDE LAMP ASSEMBLY – REMOVE AND INSTALL
,, 13335-A	DIRECTION INDICATOR, HEADLAMP FLASHER AND HORN SWITCH ASSEMBLY – REMOVE AND INSTALL
,, 13350-A	FLASHER UNIT – REMOVE AND INSTALL
,, 13404-A	REAR DIRECTION INDICATOR AND/OR STOP/TAIL LAMP BULBS – REMOVE AND INSTALL
,, 13404-A1	Extra: remaining rear direction indicator and/or stop/tail lamp bulb – remove and install
,, 13404-A2	Extra: rear lamp assembly – one – remove and install
,, 13404-A3	Extra: rear lamp bezel and/or lens – remove and install
,, 13404-B	**REAR DIRECTION INDICATOR AND/OR STOP/TAIL LAMP BULBS – ALL – REMOVE AND INSTALL** (Includes OPS 13404-A and A1)

OPERATION 13404-C	**REAR LAMP ASSEMBLY – ONE – REMOVE AND INSTALL** (Includes OPS 13404-A and A2)
,, 13404-D	**REAR LAMP ASSEMBLIES – BOTH – REMOVE AND INSTALL** (Includes OPS 13404-A, A1 and A2 × 2)
,, 13404-E	**REAR LAMP BEZEL AND/OR LENS – ONE SIDE – REMOVE AND INSTALL** (Includes OPS 13404-A, A2 and A3)
,, 13404-F	**REAR LAMP BEZELS AND/OR LENS – BOTH SIDES – REMOVE AND INSTALL** (Includes OPS 13404-A, A1, A2 × 2 and A3 × 2)
,, 13480-A	STOP LAMP SWITCH – REMOVE AND INSTALL
,, 13543-B	REAR LICENCE PLATE LAMP – REMOVE AND INSTALL
,, 13740-A	INSTRUMENT PANEL LIGHT SWITCH – REMOVE AND INSTALL
,, 13776-A	INTERIOR LIGHT – REMOVE AND INSTALL
,, 13801-A	HORN – REMOVE AND INSTALL
,, 13990-A	WARNING LIGHT BULB – RENEW
,, 14067-A	FUSE BLOCK ASSEMBLY – REMOVE AND INSTALL
,, 15000-A	CLOCK AND/OR BULB – REMOVE AND INSTALL
,, 15055-A	CIGARETTE LIGHTER ASSEMBLY – REMOVE AND INSTALL
,, 17255-A	SPEEDOMETER HEAD – REMOVE AND INSTALL
,, 17262-A	SPEEDOMETER INNER CABLE – REMOVE AND INSTALL
,, 17262-B	SPEEDOMETER INNER AND OUTER CABLES – REMOVE AND INSTALL
,, 17360-A	TACHOMETER – REMOVE AND INSTALL
,, 17508-A	WINDSCREEN WIPER ASSEMBLY – REMOVE AND INSTALL
,, 17535-A	COMBINED WINDSCREEN WIPER AND WASHER SWITCH ASSEMBLY – REMOVE AND INSTALL
,, 18309-A	HEATER CONTROLS ASSEMBLY – REMOVE AND INSTALL
,, 18309-M	HEATER CONTROL CABLES – BOTH – ADJUST
,, 18467-A	HEATER ASSEMBLY – REMOVE AND INSTALL
,, 18467-A2	Extra: heater radiator – remove and install
,, 18467-A3	Extra: heater motor – remove and install
,, 18467-C	**HEATER RADIATOR – REMOVE AND INSTALL** (Includes OPS 18467-A and A2)
,, 18467-D	**HEATER MOTOR – REMOVE AND INSTALL** (Includes OPS 18467-A, A2 and A3)

OPERATION 18470-A   FACE LEVEL VENT – REMOVE AND INSTALL
"   18503-A   HEATER CONTROLS BEZEL – REMOVE AND INSTALL
"   18503-B   HEATER CONTROLS MOUNTING PANEL – REMOVE AND INSTALL

WIRING SYSTEM

ILLUSTRATIONS OF LOOM CONNECTIONS

WIRING DIAGRAMS

GENERAL DESCRIPTION

The electrical system is of 12 volts with a negatively earthed battery, in line with latest practice. The lead-acid battery, of either 38 or 57 amp. hour capacity is mounted on a tray in the luggage compartment. A conventional generator, of either 22 amp. or 25 amp. maximum output, is used to charge the battery. This is mounted on a bracket at the left-hand side of the engine, and is driven at $1\frac{1}{2}$ times engine speed by the fan belt. A three-bobbin regulator controls the generator output by inserting a resistance in the field coil circuit. The three bobbins are the cut-out, the voltage regulator and the current regulator.

Two types of starter motor are available and both are fitted to the lower right-hand side of the engine. The first type is the conventional inertia starter motor, with the brushes on either side of the commutator. The second starter motor is a pre-engaged type, which, as the name implies, engages with the ring gear before beginning to rotate, and does not disengage until the starter control is released.

Both starter motors engage with the ring gear, shrunk onto the engine flywheel. Current is supplied to the motor by a solenoid switch which is controlled by the ignition switch on the facia.

The ignition system consists of a Lucas distributor, an oil filled coil and Autolite Powertip spark plugs. The distributor is mounted on the right-hand side of the engine, and driven by a skew gear from the camshaft.

The ignition advance is controlled according to engine speed by governor weights within the distributor body. The oil filled coil is used in conjunction with a special starter solenoid, and a ballast resistor wire. This arrangement ensures that during starting, full battery voltage is applied to the coil to facilitate engine firing. All high tension leads are of the suppressor type.

The two main instruments comprise a speedometer, incorporating a main beam warning light, and a tachometer, incorporating a generator and oil pressure warning light. Separate fuel, coolant temperature gauges, ammeter and an oil pressure gauge are mounted in the centre of the facia above the heater controls.

Headlamp dipping and flashing, together with direction indicator and horn controls are combined in a single steering column lever. Switches for the driving lights and instrument panel lights, the ignition switch and direction indicator warning lamps are mounted on the facia.

The rotary type windscreen wiper switch is combined with a windscreen washer plunger, and is located to the right of the heater controls.

The wiring loom is in four sections. The main loom connects all the instruments, controls and warning lights, stop light switch and interior light. The left-hand loom connects the left-hand front lights, horn, temperature sender unit, generator and regulator. The right-hand front lights, battery, oil pressure warning light switch, ignition and starting systems are connected by the right-hand loom. The rear lights, number plate light and fuel tank gauge unit are connected by the rear loom.

CHARGING SYSTEM

GENERAL DESCRIPTION

Battery The 12 volt negatively earthed battery of 38 amp hour (57 amp hour optional) capacity is mounted in the luggage compartment.

Generator The generator, of either 22 or 25 amp output, is mounted at the left-hand side of the engine. It is driven by the fan belt at 1.5 times engine speed.

Regulator A regulator, consisting of a cut-out, a voltage regulator and a current regulator controls the generator output by inserting a resistance into the field coil circuit.

QUICK REFERENCE DATA

PERIODIC SERVICE ATTENTION

Weekly

Check battery electrolyte level, check connections

At first 600 miles (1,000 km.)
Adjust fan belt tension. Tighten generator mounting bolts to a torque of 15 to 18 lb. ft. (2.08 to 2.49 kg.m.).

Every 3,000 miles (5,000 km.) or three months (whichever occurs first)
Check fan belt tension and adjust. Tighten generator mounting bolts to a torque of 15 to 18 lb. ft. (2.08 to 2.49 kg.m.).

Every 6,000 miles (10,000 km.) or six months (whichever occurs first)
Check battery condition, check connections and top-up
Lubricate generator rear bearing

DATA

Table showing relationship between Regulator, Generator and Battery

Equipment	Regulator			Generator			Battery	
	Part No.	Make	Identification No.	Part No.	Identification No.	Rated	Part No.	Capacity
Standard	3004E-10505-A / 3004E-10505-C	Autolite / Lucas	†GR.5000 *37344	2701E-10002-A	C.40	22 amp	113E-10658-A	38 A/H
Optional	3004E-10505-B / 3004E-10505-D	Autolite / Lucas	†GR.5001 *37342	2730E-10002-A	C.40L	25 amp	113E-10658-D	57 A/H

† On regulator cover
* Stamped on regulator base

Battery

Type	Lead acid
Voltage	12
Capacity (amp hr.)—Standard equipment	38 at 20 hr. rate
—Cold climate	57 at 20 hr. rate
Specific gravity charged	1.275 to 1.290
Low limit while discharging at 20 hr. rate	1.105

Generator

Type	12 volt, two brush
Speed (ratio to engine)	1.5 : 1
Maximum charge	22 amps. (standard) 25 amps. (optional)
Fan belt tension (total free movement)	½ in. (13 mm.)
Mounting bolts (tightening torque)	15 to 18 lb. ft. (2.08 to 2.49 kg.m.)
Adjusting strap (tightening torque)	12 to 15 lb. ft. (1.66 to 2.08 kg.m.)

Regulator

Lucas

Cut-out—Cut-in voltage	12.6 to 13.4 volts
—Drop-off voltage	9.25 to 11.25 volts
—Armature to core air gap	0.035 to 0.045 in. (0.9 to 1.1 mm.)
—'Follow-through' of moving contact	0.010 to 0.020 in. (0.3 to 0.5 mm.)
Current regulator on-load setting	Maximum rated generator output + 1½ amps
Armature to core air gap	0.045 to 0.049 in. (1.14 to 1.24 mm.)
Voltage regulator open circuit setting	14.4 to 15.6 volts at 20°C (68°F)
Armature to core air gap	0.045 to 0.049 in. (1.14 to 1.24 mm.)

Regulator

Autolite

Cut-out—Cut-in voltage	12.6 to 13.4 volts
—Drop-off voltage	9.25 to 11.25 volts
—Armature to core air gap	0.025 to 0.037 in. (0.64 to 0.94 mm.)
—'Follow-through' of moving contact	0.015 to 0.025 in. (0.38 to 0.64 mm.)
Current regulator on-load setting	Maximum rated generator output + 1½ amps
Armature to core air gap	0.014 to 0.019 in. (0.36 to 0.48 mm.)
Voltage regulator open circuit setting	14.4 to 15.6 volts at 20°C (68°F)
Armature to core air gap	0.024 to 0.028 in. (0.61 to 0.71 mm.)

CORTINA - LOTUS

SERVICE AND REPAIR OPERATIONS

OP 10001-A GENERATOR ASSEMBLY – REMOVE AND INSTALL

To Remove

1. Disconnect the battery.
2. Disconnect the leads from D and F terminals on the rear of the generator.
3. Slacken the three generator securing bolts and tilt the generator towards the engine.
4. Remove the fan belt.
5. Remove the generator securing bolts and detach the generator.

To Install

6. Fit the generator and retain with three bolts.
7. Replace the fan belt and tighten the generator bolts to the specified torque so that the fan belt has ½ in. (13 mm.) free movement at a point mid-way between the generator and water pump pulleys.
8. Reconnect the D and F leads to the generator and reconnect the battery.

OP 10001-A1 EXTRA: GENERATOR PULLEY – REMOVE AND INSTALL (GENERATOR ASSEMBLY REMOVED)

To Remove

1. Hold the generator pulley and slacken the pulley securing nut.
2. Remove the nut, lockwasher, generator pulley, Woodruff key and spacer.

To Install

3. Fit the spacer and pulley to the generator, taking care to first correctly position the key in its key-way.
4. Secure the pulley with a lockwasher and nut.

OP 10001-A2 EXTRA: GENERATOR FRONT BEARING – RENEW (GENERATOR PULLEY REMOVED)

To Remove

1. Raise the brushes clear of the commutator.
2. Unscrew and remove the two through bolts.
3. Lift the drive-end bracket and armature from the yoke. Remove the fibre washers from the armature shaft at the commutator end.
4. Remove the Woodruff key and press the armature from the end bracket.
5. Remove the circlip securing the bearing retaining plate.
6. Extract the bearing and withdraw the corrugated washer from the end bracket.
7. Remove the commutator end-plate from the yoke.

To Install

8. Locate the felt ring and corrugated washer in the end bracket.
9. Pack the bearing with a suitable grease and press into position.
10. Locate the bearing retaining plate in position and secure with circlip.
11. Fit the armature to the end bracket, and install the assembly in the yoke locating the peg on the bracket within the radial groove in the yoke.
12. Locate the two fibre washers on the armature shaft at the commutator end.
13. Replace the commutator end-plate and secure with the two through bolts.
14. Lower the brushes onto the commutator.

OP 10001-A3 EXTRA: GENERATOR BRUSHES – RENEW AND COMMUTATOR – CLEAN (GENERATOR PULLEY REMOVED)

To Remove

1. Unscrew and remove the two generator through bolts.
2. Withdraw the commutator end bracket from the yoke.
3. Remove the generator drive end bracket and armature.
4. Unscrew the brush securing screws and remove the brushes.

To Reassemble

5. Clean the brush holders and fit new brushes.
6. Check the circuit to earth with a simple bulb circuit.
7. Clean the armature. Test insulation between commutator and armature, and continuity of coil windings to commutator. Undercut the commutator segments if necessary (fabricated type only) and test armature casing for earth.
8. Refit the drive end bracket and armature to the yoke.
9. Refit the commutator end bracket and replace the through bolts.
10. Check the position and free movement of brush springs.

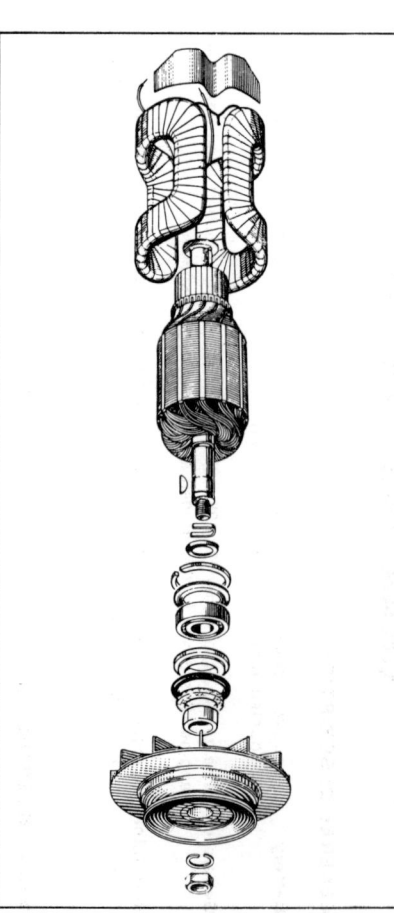

Generator - Exploded View

CORTINA - LOTUS

OP 10001-A4   EXTRA: GENERATOR – OVERHAUL
(GENERATOR BRUSHES AND COMMUTATOR REMOVED)

Tools Required

CPT.9504   Pole piece screwdriver
CPT.9507   Endplate bush remover and replacer
CPT.9509   Pole expander

1. Remove the terminal post from the yoke (secured by a rivet).
2. Using the pole piece screwdriver remove the pole piece securing screws.
3. Remove the pole pieces and field coils after marking the pole pieces and yoke so that they can be refitted in their original positions.
4. Disconnect the field coil wires at the field posts (note connections).
5. Relocate the field coils to the field posts.
6. Rivet the terminal post to the yoke.
7. Replace the pole pieces and field coils to the yoke.
8. Remove the commutator end bracket bush.
9. Fit a new bush (bush must have been soaked in oil for 24 hours).
10. Remove the circlip securing the bearing retainer plate to the drive end bracket and remove the plate.
11. Press out the bearing assembly.
12. Clean the bearing housing and fit a new bearing.
13. Test the field coils for continuity and earth.

OP 10001-B   GENERATOR PULLEY – REMOVE AND INSTALL
(Includes OPS 10001-A and A1)

OP 10001-C   GENERATOR BRUSHES – RENEW AND COMMUTATOR – CLEAN
(Includes OPS 10001-A, A1 and A3)

OP 10001-D   GENERATOR ASSEMBLY – OVERHAUL
(Includes OPS 10001-A, A1, A3 and A4)

OP 10001-E   GENERATOR FRONT BEARING – RENEW
(Includes OPS 10001-A, A1 and A2)

OP 10001-M   CHARGING CIRCUIT – TEST AND ADJUST

To isolate the source of any charging system fault, the following checks should be carried out:—

The following gives all the information required to check and adjust all parts of the charging system. Normally, the area of the fault will be pin-pointed with diagnosis equipment, and only that part of the system will require attention. For this reason, this operation should not be considered as a basic repair operation.

1. **Fan belt tension**

   Check, and if necessary adjust, the fan belt to give $\frac{1}{2}$ in. (13 mm.) total free movement at a point mid-way between the generator and water pump pulleys. Tighten the mounting bolts and adjusting strap screw to the specified torque.

2. **Battery condition**

   (a) Ensure that the battery exterior is clean and free from cracks and corrosion particularly around the terminals.

Checking Battery Specific Gravity

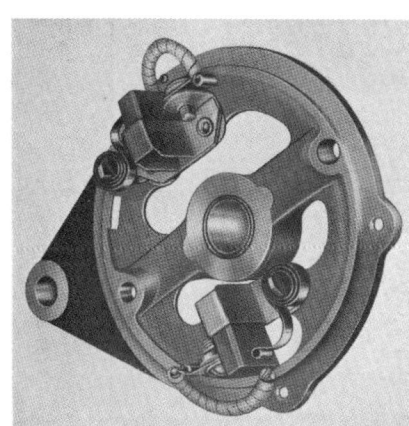

Commutator End Plate

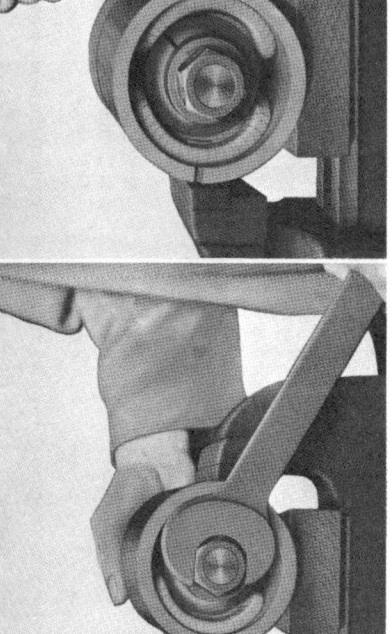

Pole Shoe Expander

Section 10 – 12

Section 10 – 13

1 . 1969

(b) Check the specific gravity with a hydrometer. If the electrolyte level is less than ¼ in. (7 mm.) above the plates, distilled water should be added and the battery bench charged for at least one hour before carrying out the check. Draw enough electrolyte into the hydrometer to make the scale float. Repeat the test for each cell.

(c) The following table relates the specific gravity to the battery condition at 16°C (60°F):—

Hydrometer Reading	Battery Condition
1.280	Fully charged
1.240	75% charged
1.200	50% charged
1.160	25% charged
1.120	Discharged

If the electrolyte temperature varies from 16°C (60°F), adjust the reading obtained as follows:—

Add   0.004 for every 5½°C (10°F) above 16°C (60°F)

Subtract 0.004 for every 5½°C (10°F) below 16°C (60°F)

For example:—

1.272 specific gravity at 27°C (80°F)

= 1.272 + 0.008

= 1.280 at 16°C (60°F), i.e. battery fully charged.

1.204 specific gravity at 10°C (50°F)

= 1.204 − 0.004

= 1.200 at 16°C (60°F), i.e. battery 50% charged.

(d) If one cell is about 0.003 lower than the rest is it possibly failing. An extended bench charge may revive it.

If the readings are irregular, with one or more cells 0.050 lower than the rest, the battery is not fit for further use.

If the readings are reasonably uniform, the battery is probably healthy, although low readings indicate a bench charge is required.

(e) Take a high rate discharge test across the battery terminals.

If a hand test instrument is used, push the probes onto the battery terminals and hold for 10 seconds. Note the voltmeter reading for the 10 seconds.

If a test set (such as Crypton, Ford and Sun) is used, connect the ammeter and voltmeter to the battery terminals: negative to negative and positive to positive. Turn the control knob to give an ammeter reading of 150 amps. Note the voltmeter reading for 10 seconds. In both tests there should be virtually no voltage fall-off and the reading should be approximately 7.2 to 9.5 volts.

Any appreciable drop-off or a voltage of less than 5 volts indicates that the battery has reached the end of its useful life.

DIAGNOSIS CHART OF TEST RESULTS

Specific Gravity Readings	High Rate Discharge Test Readings	Battery Condition
Readings uniform and within Range 1.260-1.280	Readings High and Steady	Healthy and in reasonable State of Charge
Readings uniform but lower than 1.260	Readings Low and Steady	Healthy but requires Charging
One cell about 0.030 lower than remainder	Reading shows Falling Voltage	Probable Failing Cell
Irregular Readings more than one cell 0.050 lower than remainder	Reading Low and showing rapid fall	Battery at End of Life
Very Low Readings	Very Low Voltage	Battery has internal fault or is in deeply sulphated condition

The High Rate Discharge Test and the Specific Gravity Test are complementary, no advantage will be gained by performing one test and not the other.

3. Generator Output Test

(a) Disconnect the wires from the "D" and "F" regulator terminals and join them together.

(b) Connect a 0—30 voltmeter between this junction and earth.

(c) Run the engine at approximately 1,000 rev./min.

NOTE – Do not exceed this speed or the generator may be damaged.

(d) The voltmeter reading should rise rapidly without fluctuation to more than 24 volts.

(e) Should the reading be incorrect, connect a jumper wire between the "D" and "F" terminals on the generator, and connect the voltmeter between this wire and earth.

(f) If the reading is now more than 20 volts, the continuity of the "D" or "F" leads is suspect. If the reading is still incorrect, there is a fault in the generator.

4. Test and Reset if Necessary the Regulator Open Circuit

(a) Remove the connecting block from the regulator.

(b) Insert the blade of a small screwdriver between the block and the terminal, depress the spring clip, and remove each terminal in turn.

(c) Replace each wire onto the regulator, with the exception of the wire(s) on the terminal "B".

(d) If two wires have been removed from the "B" terminal, these should be connected together.

(e) Connect a voltmeter between the "WL" terminal and the regulator base.

CORTINA - LOTUS

(f) Start the engine and increase its speed to 2,000 rev./min.

(g) Note the voltmeter reading which should be steady and between the following limits:—

Ambient Temperature	Voltage Checking Limits
10°C (50°F)	14.5 to 15.8
20°C (68°F)	**14.4 to 15.6**
30°C (86°F)	14.3 to 15.3
40°C (104°F)	14.2 to 15.1

(h) If the reading is steady but outside these limits, drill out the plastic rivets and remove the cover.

NOTE — The top cover must not be removed during Warranty Period.

(i) With the engine speed as quoted above, turn the adjustment cam with the Lucas Tool No. 54381742; clockwise to raise the voltage setting and anti-clockwise to lower the voltage setting. The figures to set to are as follows:—

Ambient Temperature	Voltage Setting Limits
10°C (50°F)	14.9 to 15.5
20°C (68°F)	**14.7 to 15.3**
30°C (86°F)	14.5 to 15.1
40°C (104°F)	14.3 to 14.9

(j) Reduce the engine speed and then raise it to the previously stated speed to check the setting.

(k) If the voltage continues to rise with engine speed, check the regulator to body earth, then replace the regulator.

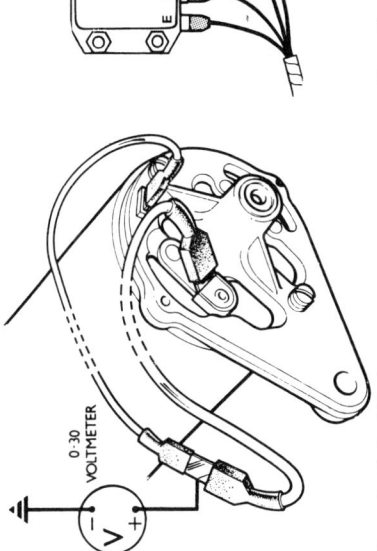

Generator Output Test

(l) Refit the cover using spire clips and self-tapping screws.

(m) Remove the voltmeter, refit the terminals into the connecting block and replace the block.

5. **Test and Reset if Necessary the Current Regulator Bobbin**

(a) The voltage regulator bobbin must be rendered inoperative by holding the points together with a spring clip (if the vehicle warranty has expired), or by earthing the terminal "WL", through a $\frac{1}{2}$ ohm resistor capable of carrying 30 amps, to the regulator base.

(b) Remove the connecting block from the regulator.

(c) Insert the blade of a small screwdriver between the block and the terminal, depress the spring clip, and remove each terminal in turn.

(d) Replace each wire onto the regulator, with the exception of the wire(s) on the terminal "B".

(e) Connect the "B" terminal wires to the load side of an ammeter. Connect the other side of the ammeter to the "B" terminal.

(f) Switch on the vehicle's headlamps to place a load on the battery.

(g) Start the engine and increase its speed to approximately 3,000 r.p.m.

(h) The ammeter should indicate the rated generator output ± 1 amp.

(i) Adjust the setting as required by turning the adjustment cam. Anti-clockwise to lower the setting; clockwise to raise it.

(j) Decrease and then increase engine speed to the figure specified before and recheck settings.

(k) Switch off engine, refit the terminals into the connecting block, and replace the block.

6. **Test the Cut-in Voltage**

(a) Remove the connecting block from the regulator.

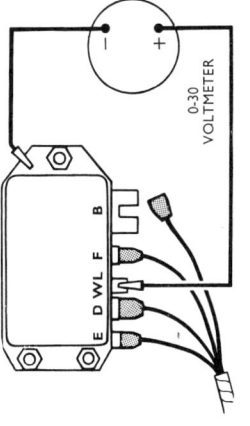

Current Regulator On-load Test

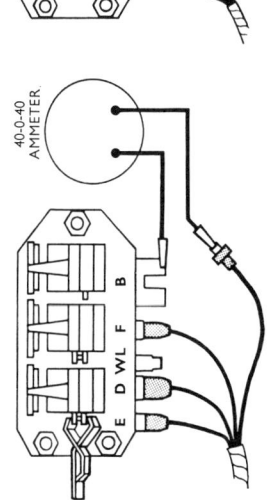

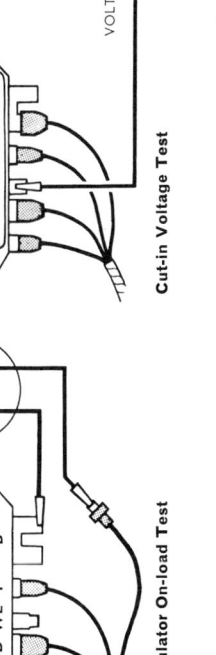

Open Circuit Voltage Test

Cut-in Voltage Test

(b) Insert the blade of a small screwdriver between the block and the terminal, depress the spring clip and remove each terminal in turn.

(c) Replace each wire onto the regulator.

(d) Connect a voltmeter between the "WL" terminal and the regulator base.

(e) Switch on the vehicle's headlamps.

(f) Start the engine and gradually increase its speed, observing the voltmeter needle which should climb steadily to 12.6–13.4 volts and then drop slightly as the points close.

(g) If the closure occurs outside the limits quoted, decrease the engine speed and rotate the cam slightly:—

Clockwise to raise the setting and anti-clockwise to lower it.

Increase the engine speed and recheck the setting.

7. **Adjust the Air Gap Settings** (if required)

(a) Remove the regulator from the car.

(b) Turn voltage regulator adjustment cam to the position giving minimum lift.

(c) Slacken the adjustable contact locknut and screw the contact out a few turns.

(d) Insert a 0.045 to 0.049 in. (1.14 to 1.24 mm.) feeler blade for Lucas regulators or a 0.24 to 0.28 in. (0.61 to 0.71 mm.) feeler blade for Autolite regulators between the armature and the copper coloured shim on top of the core face. Position the feeler blade as far back as the rivet heads will allow.

(e) Screw in the adjustable contact or bend the contact arm, as applicable, until it just traps the feeler blade. Retighten the locknut. (Lucas regulators only).

(f) Recheck the gap.

(g) Repeat the operation (items b, c, d, e and f) for the current regulator cam to obtain an air-gap of 0.045 to 0.049 in. (1.14 to 1.25 mm.) for Lucas regulators or 0.014 to 0.019 in. (0.36 to 0.48 mm.) for Autolite regulators.

(h) Insert a 0.015 in. (0.4 mm.) feeler blade for Lucas regulators or 0.020 in. (0.5 mm.) feeler blade for Autolite regulators between the top of the cut-out relay core and the armature. Press the armature onto the feeler blade to trap it. The contacts, which are between the cut-out and current relays, should just touch. If necessary, bend the fixed contact.

(i) The cut-out air-gap is 0.035 to 0.045 in. (0.9 to 1.1 mm.), for Lucas regulators or 0.025 to 0.37 in. (0.64 to 0.94 mm.) for Autolite regulators measured with a feeler gauge in the same way as for the current and voltage regulators. Adjustment is carried out by carefully bending the back stop as required.

(j) After setting the air-gaps replace the regulator in the car and carry out the electrical settings.

OP 10505-A   VOLTAGE REGULATOR UNIT – REMOVE AND INSTALL

To Remove

1. Disconnect the battery.
2. Remove the leads from the regulator, noting connections.
3. Unscrew two screws securing the unit.

To Install

4. Replace the regulator and secure it with two screws.
5. Reconnect the leads in the same positions as removed.
6. Reconnect the battery.

Voltage Regulator Air-gap Setting

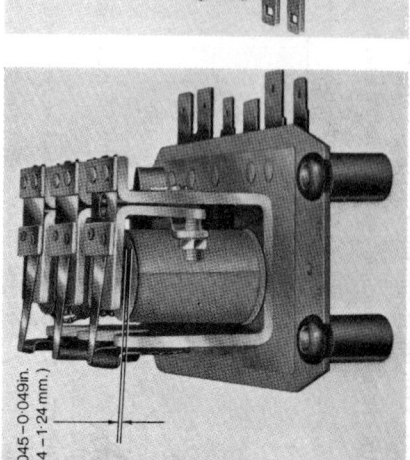

Cut-out Air-gap Setting

STARTING SYSTEM

GENERAL DESCRIPTION

The engine is started by the starter motor pinion engaging with the flywheel ring gear. Electricity for starting is supplied to the starter motor through a solenoid controlled by a "key-start" ignition switch mounted on the facia panel.

There are two types of starter motor:—

Type 1 Inertia type – This is a twenty-nine slot lap wound machine with a three hole fixing to the engine. It is fitted when "cold start" equipment is not requested.

Type 2 Pre-engaged type – This is a twenty-nine slot lap wound machine with a two hole fixing. It is fitted when "cold start" equipment is requested. Also, if a customer requires a pre-engaged starter but the car does not take "cold start" equipment as standard, this starter motor is available as an option.

The pre-engaged starter incorporates a "built-in" solenoid.

QUICK REFERENCE DATA

DATA

Type 1 – Inertia Starter Motor

Teeth on pinion	10
Teeth on ring gear	110
Gear ratio	11 : 1
Minimum brush length	0.4 in. (10.16 mm.)
Brush spring pressure	34 oz. (0.96 kg.)

Type 2 – Pre-engaged Starter Motor

Teeth on pinion	9
Teeth on ring gear	110
Gear ratio	12.2 : 1
Minimum brush length	0.4 in. (10.16 mm.)
Brush spring pressure	34 oz. (0.96 kg.)

SERVICE AND REPAIR OPERATIONS

OP 11001-A INERTIA STARTER MOTOR ASSEMBLY – REMOVE AND INSTALL

To Remove

1. Disconnect the battery.
2. Remove the starter motor lead from the starter.
3. Remove the upper securing bolt.
4. With the handbrake applied, jack up the front of the car and fit stands.
5. Slacken the starter motor lower mounting bolts.
6. Support the motor, remove the bolts and then remove the motor.

To Install

7. Replace the starter motor and refit the lower mounting bolts.
8. Fit the upper bolt and tighten the three bolts securely.
9. Lower the car to the ground.
10. Reconnect the lead to the starter.
11. Reconnect the battery.

OP 11001-A1 EXTRA: DRIVE COMPONENTS – REMOVE AND INSTALL (INERTIA STARTER MOTOR REMOVED)

To Remove

1. Compress the spring and remove the circlip.
2. Remove the main drive spring washer, pinion and barrel assembly.

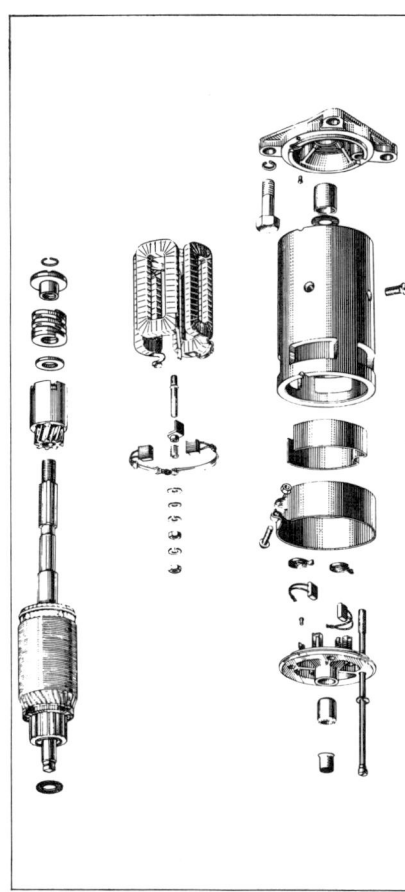

Inertia Starter Motor — Exploded View

CORTINA - LOTUS

3. Screw the sleeve out of the sleeve nut.
4. Remove the distance collar and cups, return spring and washer.

To Install

5. Refit the return spring, cups, and distance collar.
6. Reassemble the pinion and barrel assembly to the shaft.
7. Replace the main drive spring, spring cup; compress the spring and replace the circlip.

OP 11001-A2 EXTRA: BRUSHES – RENEW AND COMMUTATOR – CLEAN (INERTIA STARTER MOTOR REMOVED)

To Remove

1. Slacken the clamp and slide the cover along the body.
2. Raise the brush holding springs and withdraw the brushes.
3. Unscrew two through bolts and remove the starter motor drive end plate and armature. Clean off the commutator.
4. Remove the commutator end plate.
5. Unsolder the brush leads connected to the earthed holders.
6. Cut the brushes from the field coils, leaving approximately ¼ in. (7 mm.) copper wire attached to the field coils.

To Install

7. Solder new brushes to the ends of the copper wire still attached to field coils.

NOTE – Do not attempt to solder new brushes directly onto aluminium field coils. This requires special equipment.

8. Solder new brushes to the earthed holders.
9. Replace the starter motor drive end plate and armature.
10. Replace the commutator end plate and insert the brushes.
11. Relocate the cover band and tighten.

OP 11001-A3 EXTRA: INERTIA STARTER MOTOR – OVERHAUL (DRIVE COMPONENTS – BRUSHES AND COMMUTATOR REMOVED)

Tools Required

CPT.9504   Pole piece screwdriver
CPT.9509   Pole expander

1. Remove the four pole pieces in the yoke.
2. Install new field coils and replace the pole pieces in the yoke.
3. Test the coils for continuity.
4. Remove the two end plate bushes.
5. Replace the two end plate bushes.

OP 11001-B INERTIA STARTER MOTOR DRIVE COMPONENTS – REMOVE AND INSTALL
(Includes OPS 11001-A and A1)

OP 11001-C INERTIA STARTER MOTOR BRUSHES – RENEW AND COMMUTATOR – CLEAN
(Includes OPS 11001-A, A1 and A2)

OP 11001-D INERTIA STARTER MOTOR – OVERHAUL
(Includes OPS 11001-A, A1, A2 and A3)

OP 11001-A PRE-ENGAGED STARTER MOTOR ASSEMBLY – REMOVE AND INSTALL

To Remove

1. Disconnect the battery.
2. Remove the leads from the starter.
3. Remove the upper securing bolt.
4. With handbrake applied, jack up the front of the car and fit stands.
5. Slacken the starter motor lower mounting bolt.
6. Support the motor, remove the bolt and then remove the motor.

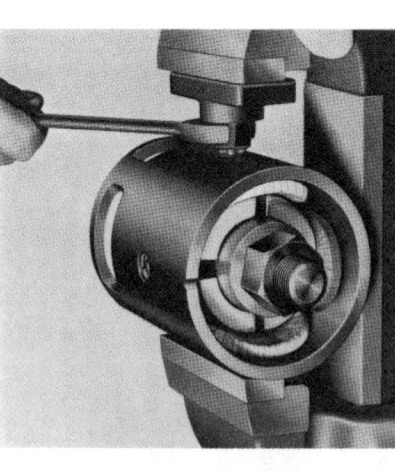

Field Coils in Position

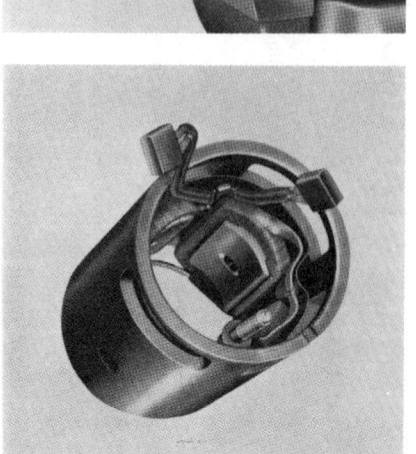

Pole Shoe Expander

To Install

7. Replace the starter motor and refit the lower mounting bolt.
8. Fit the upper bolt and tighten both bolts securely.
9. Lower the car to the ground.
10. Reconnect the leads to the starter.
11. Reconnect the battery.

OP 11001-A1 EXTRA: DRIVE COMPONENTS – REMOVE AND INSTALL

Tools Required

CP.4111   Differential bearing cone remover

To Remove

1. Remove the strap connecting solenoid to field post terminal.
2. Remove the two bolts securing solenoid to housing, remove solenoid and plunger spring.
3. Lift solenoid plunger clear of actuating lever and withdraw from housing.
4. Slacken the brush-cover band screw and remove the band.
5. Lift the brush springs with a piece of wire and withdraw the brushes.
6. Unscrew the two slot-headed thru-bolts from the com-end plate and remove the housing together with the armature and pinion assembly.
7. Release the locknut on the eccentric spindle retaining the actuating lever. Remove the spindle.
8. Remove the rubber grommet located in the housing.
9. Withdraw the armature and pinion assembly together with the actuating lever from the housing. Separate the lever from the pinion assembly.
10. Locate the armature, suitably protected, in a vice and tap down the circlip retaining cover. Remove the washer, circlip, cover and detach the pinion assembly.
11. Pull the actuating bush towards the pinion to expose the circlip, remove the circlip, bush, spring and large washer.

NOTE – Do not grip the one-way clutch, adjacent to the pinion, in a vice during the above operation as this will result in damage to the clutch.
The drive pinion and clutch are serviced as a complete unit therefore repairs to the unit are impractical.

To Install

12. Refit the large washer, spring and actuating bush to the drive pinion and clutch unit and secure with circlip.
13. Replace the assembly on the armature shaft with the pinion away from the armature.
14. Place the circlip cover over the armature shaft then fit the circlip. Using Tool No. CP.4111 pull the cover up over the circlip.

Pre-Engaged Starter Motor — Exploded View

CORTINA - LOTUS

15. Locate the plain washer on the armature shaft, at the drive end, the actuating lever within the bush, then position the assembly within the drive end housing.
16. Secure the actuating lever with the eccentric screw.
17. Locate the grommet in the drive end housing.
18. Position yoke and plate assembly over armature, ensure correctly located, then secure with slot-headed thru-bolts.
    NOTE – Care must be taken during the above operation to avoid damage to the commutator and brushes.
19. Raise the brush springs and refit the brushes.
20. Position the brush cover and tighten the clamping screw.
21. Locate the solenoid plunger in the actuating lever and the spring over the plunger.
22. Position the solenoid over the plunger and secure to the housing with two bolts and spring washers.

23. Fit strap to solenoid and com-end plate and secure with spring washers and nuts, brass nut at com-end plate and light alloy nut at solenoid.
24. Reset the pinion gear travel.
    (i) Remove the connection from the solenoid to the field post terminal.
    (ii) Energise the solenoid with an 8 volt supply. Slacken the eccentric pivot pin locknut and turn the pin until the correct setting of 0.005 to 0.150 in. (0.127 to 3.81 mm.) is obtained between the pinion and thrust washer. Note that the arc of adjustment is 180° as indicated by arrows on the casing.
    NOTE – This should not take more than approximately 2 minutes, otherwise the solenoid will overheat.
    (iii) Connect, through a switch, an 8 volt supply between the solenoid small unmarked terminal and the large terminal. Connect a test lamp between the switch side of the battery and the large blade terminal.
    (iv) Insert a stop of ¼ in. thickness between the pinion and drive end bracket to prevent the pinion moving fully outwards.
    (v) Close the switch, thus causing the 8 volt supply to energise the shunt winding. The test lamp should now light, indicating the solenoid contacts have closed satisfactorily.
    (vi) Switch off and remove the ¼ in. stop, restricting the pinion movement. Switch on again and hold the pinion in the fully engaged position by hand.
    (vii) Switch off and observe the test lamp, which should go out, indicating that the solenoid contacts have opened satisfactorily.

OP 11001-A2 EXTRA: BRUSHES – RENEW, AND COMMUTATOR – CLEAN

1. Remove nut, plain and fibre washer and split bush from the field terminal post to release the com-end plate.

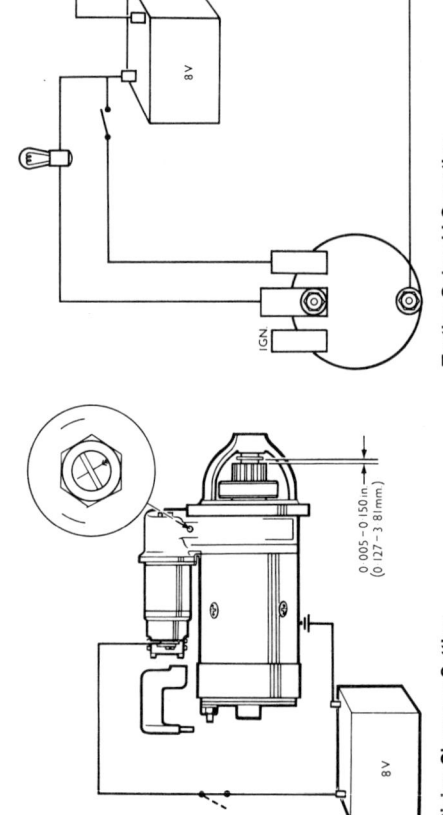

Pinion Clearance Setting

Testing Solenoid Operation

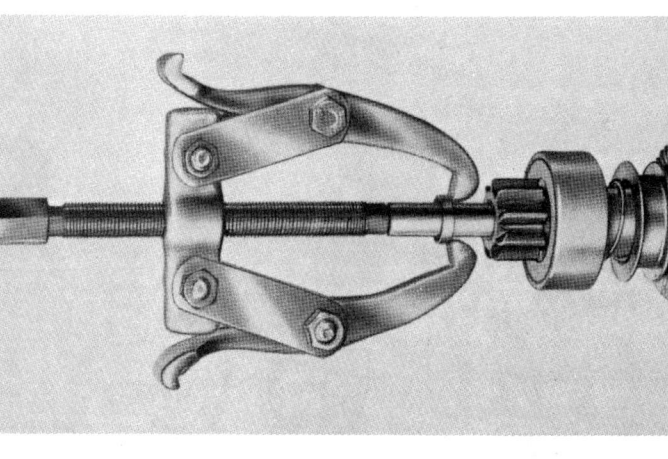

Removal

Drive Pinion Circlip Replacement

2. Carefully pull the commutator end plate from the starter motor, together with the earth brushes.

3. Cut the brush leads leaving ¼ in. (7 mm.) attached to the field coils and discard the old brushes. Do not attempt to unsolder the brushes because, as aluminium field coils are fitted, difficulty will be experienced in re-soldering unless special equipment is used.

4. Clean the commutator with a petrol-moistened cloth, or if necessary polish it with very fine glass paper.

   Do not use emery cloth and never undercut the mica insulation.

5. Check the brushes for freedom of movement in the brush holders and then solder the new brushes to the old leads. If necessary trim the new brush leads to the required length.
   The field coil or insulated brushes are longer than the earthed brushes and have a braided covering. Fit these brushes so that they both point towards the field coil terminal, when the starter motor yoke is viewed from the commutator end.

6. Before fitting the end plate, check the brush springs and renew if necessary. Take care to close the ends of the brush spring posts after fitting new springs.
   It is also advisable to check the insulated brush holders to ensure that they are not earthing. Use a battery and bulb for this test, or, if available, a multimeter.

7. If removed, locate nylon insulating sleeve on field post terminal.

8. Locate com-end plate on yoke and secure with split bush, fibre and plain washer and nut on terminal post.

OP 11001-A3 EXTRA: PRE-ENGAGED STARTER MOTOR – OVERHAUL

Tools Required

CPT.9504   Pole piece screwdriver
CPT.9509   Pole expander

1. Remove the four pole pieces in the yoke.
2. Install new field coils and replace the pole pieces in the yoke.
3. Test the coils for continuity.
4. Remove the end plate and housing bushes.
5. Replace the end plate and housing bushes.

GROUP OPERATIONS

OP 11001-B PRE-ENGAGED STARTER MOTOR DRIVE COMPONENTS – REMOVE AND INSTALL
(Includes OPS 11001-A and A1)

OP 11001-C PRE-ENGAGED STARTER MOTOR BRUSHES – RENEW, AND COMMUTATOR – CLEAN
(Includes OPS 11001-A, A1 and A2)

OP 11001-D PRE-ENGAGED STARTER MOTOR – OVERHAUL
(Includes OPS 11001-A, A1, A2 and A3)

OP 11450-A STARTER SOLENOID – REMOVE AND INSTALL

To Remove

1. Disconnect the battery.
2. Disconnect the leads from the starter solenoid.
3. Remove the starter solenoid.

To Install

4. Replace the starter solenoid.
5. Reconnect the leads to the starter solenoid.
6. Reconnect the battery and check the starter solenoid operation.

IGNITION SYSTEM

GENERAL DESCRIPTION

The ignition system consists of a ballast resistor coil, with a distributor and spark plugs.

The high tension leads are of the suppressor type.

Coil The wiring harness is designed to include a high resistance wire in the ignition coil feed circuit and thus it is essential that only a "ballast resistor" type coil is used. The resistance of the lead (which, at the coil end, has a pink insulation with a white tracer) between the ignition switch and the coil should be 1.4 to 1.6 ohms. Refer to the label on the coil to ensure that the correct type is fitted.

Distributor The distributor is mounted on the front right-hand side of the engine and is driven by a skew gear from the camshaft. The ignition advance is mechanically controlled, according to engine speed by governor weights within the distributor body.

QUICK REFERENCE DATA

PERIODIC SERVICE ATTENTION

Every 3,000 miles (5,000 km.)

    Clean sparking plugs and set gaps

Every 6,000 miles (10,000 km.)

    Examine and adjust distributor points

    Check and adjust the ignition timing

    Lubricate the distributor cam spindle with two drops of engine oil

    Lubricate the distributor cam with lithium base grease or grease obtainable in sachet form Part No. 68AB-19D533-AA.

    Clean the distributor cap, H.T. leads and coil

CAUTION – To avoid excess or insufficient cam lubrication coat the entire periphery of the cam profile with the specified grease such that when the shaft is rotated a fillet of $\frac{1}{16}$ in. (1.59 mm.) is built up on the back of the point set shoe. Points which have become dirty or contaminated with oil or grease should be cleaned with a stiff brush and a suitable grease solvent.

NOTE – The distributor contact breaker points should only be changed if they are worn, badly burnt, or if excessive metal transfer has occurred or have a "high resistance". Contacts showing a greyish colour and only slight signs of pitting need not be renewed. Metal transfer is considered excessive when it equals or exceeds the gap setting of 0.014 to 0.016 in. (0.36 to 0.41 mm.). The resistance is considered "high" when the voltage drop across the points exceeds 0.25 volts.

DATA

Distributor points gap	0.014 to 0.016 in. (0.36 to 0.41 mm.)
Dwell angle	57° to 63°
Firing order	1, 3, 4, 2
Rotation	Anti-clockwise
Initial advance (static)	12° B.T.D.C.
Dynamic advance	26° B.T.D.C. at 3,500 rev./min.
Spark plug, type and gap	Autolite AG22. 0.023 in. (0.59 mm.)

IGNITION TIMING

(a) Establish that the correct distributor is fitted. (Lucas Ref. 23D4).

(b) Check the octane rating of the fuel that is to be used with the engine, as this can affect the initial advance.

(c) The initial advance is "built-in" to the engine and when one of the marks on the crankshaft pulley aligns with the appropriate mark on the front cover timing pointer the initial advance setting is correct and no further adjustment is required at this stage.

Distributor Number 23D4

Compression Ratio	Octane Number	Star Rating	Initial Advance (Crankshaft Degrees)
9.5 : 1	100	5	12° B.T.D.C.

NOTE – If the car is normally operated at a high altitude the distributor setting may be **advanced** by 4° for each 2,000 ft. (600 m.) above sea level. Where the car is operated at varying altitudes the initial advance must, at all times, be set for the lowest altitude at which the car is operated.

Ignition Advance Settings

SERVICE AND REPAIR OPERATIONS

OP 12024-A IGNITION COIL – REMOVE AND INSTALL

To Remove

1. Disconnect the battery cables.
2. Remove the wires from ignition coil.
3. Undo the bolt securing the ignition coil mounting bracket mount.
4. Remove the coil from the bracket.

To Install

5. Reposition the ignition coil and bracket and secure in position with the bolt.
6. Reconnect the wires to the coil.
7. Reconnect battery cables.

OP 12100-A DISTRIBUTOR ASSEMBLY – REMOVE AND INSTALL

To Remove

1. Disconnect the high tension leads from the spark plugs.
2. Disconnect the high tension lead and the low tension lead from the coil.
3. Remove distributor cap.
4. Turn the engine crankshaft until the timing mark on the crankshaft pulley is in alignment with the 12° position on the front cover timing scale as No. 1 piston comes up on the compression stroke. The rotor should now point towards No. 1 spark plug.
5. Unscrew the bolt retaining the distributor clamp on the engine and carefully withdraw the distributor.

To Install

6. Fit the distributor with the low tension terminal adjacent to the cylinder block. Position the rotor, with the electrode towards the distributor cap rear clip, prior to inserting the distributor into the cylinder block. As the gears mesh the rotor will rotate clockwise into alignment with No. 1 H.T. electrode in the distributor cap.
7. Insert the distributor and, as the gears mesh, the rotor should rotate slightly. If necessary re-position the clamp, without turning the distributor, so that the hole is in line with the one in the cylinder block. Fit the retaining bolt and tighten.

A. To Adjust the Timing without the use of a Timing Light

(a) Slightly turn the distributor body as necessary until the contact breaker points are just opening when the rotor is adjacent to No. 1 H.T. electrode in the distributor cap.

NOTE – Excessive movement from the specified position would indicate that the gears are meshing one or more teeth out. Remove the distributor and refit if this occurs.

(b) Tighten the distributor body clamp bolt sufficiently to hold the distributor in position. DO NOT OVER-TIGHTEN.

8. Replace the distributor cap.
9. Reconnect the spark plug leads (firing order 1, 3, 4, 2 anti-clockwise rotation) and connect the grommet to the rocker cover bracket.
10. Reconnect the low tension lead to the coil.

B. To Adjust the Timing using a Timing Light

(a) Connect the leads of the timing light, using the clips provided, in accordance with the manufacturer's instructions.

(b) Check that the timing marks on the crankshaft pulley and front cover are visible and mark with chalk or paint if necessary.

(c) Start the engine and allow it to idle.

(d) Point the timing light at the timing pointer. Check that the mark on the crankshaft pulley is adjacent to the appropriate mark on the front cover timing pointer.

If the mark on the pulley is above and to the left of the correct timing mark, the engine is too far advanced. Slacken the distributor body clamp and turn body anti-clockwise slightly to retard the ignition.

Should the mark be below and to the right of the correct timing mark, the distributor body should be turned clockwise slightly to advance the ignition.

(e) After making an adjustment, tighten the clamp sufficiently to hold the distributor in position. DO NOT OVER-TIGHTEN.

The operation of the governor weights may be checked by opening and closing the throttle. As the throttle is gradually opened, the mark should move away from the indicator upwards; and as the throttle is closed the notch will move down in line with the indicator. Any tendency for erratic advance shown by the mark jumping suddenly away from the indicator shows that the governor weights are binding, or that the springs are weak.

11. A slight readjustment to the distributor may be necessary and should be carried out on the road in the following manner:—

(i) Warm up the engine to normal operating temperature.

(ii) Accelerate in top gear on wide throttle opening from 20 m.p.h. (32 k.p.h.) to 40 m.p.h. (64 k.p.h.).

CORTINA - LOTUS

(iii) If heavy pinking occurs, **retard** the ignition until a trace pink can just be heard under these conditions of acceleration.

NOTE – It is not necessary to advance the ignition beyond the initial setting (except under high altitude operating conditions previously detailed). Also, there is no need to use a fuel of a higher octane rating than that specified.

OP 12100-A1   EXTRA: DISTRIBUTOR CAP – REMOVE AND INSTALL

To Remove
1. Release the two spring clips and remove the cap and leads.
2. Remove the lead retaining screws, from within the cap, to release the leads.

To Install
3. Locate the leads in their correct aperture in the cap and secure with retaining screws.
4. Replace the cap and retain with the two spring clips.

OP 12100-A2   EXTRA: CONDENSER – RENEW (DISTRIBUTOR CAP REMOVED)

To Remove
1. Remove rotor.
2. Unscrew the condenser lead, and the condenser retaining screw.

To Install
3. Replace the condenser, retain in position and replace the wire.
4. Replace rotor.

OP 12100-A3   EXTRA: POINTS – RENEW (DISTRIBUTOR CAP REMOVED)

To Remove
1. Loosen the retaining screw, and remove the low tension, and condenser wires from the points.
2. Unscrew the retaining and adjusting screw, and remove the points.

To Install
3. Replace the points, and lightly retain in position.
4. Adjust the points to give a points gap of 0.014 to 0.016 in. (0.36 to 0.41 mm.).
5. Tighten the adjusting screw.
6. Replace the two wires and retain in position.

OP 12100-A4   EXTRA: DISTRIBUTOR BREAKER PLATE ASSEMBLY – REMOVE AND INSTALL (DISTRIBUTOR CAP REMOVED)

To Remove
1. Remove the two screws securing the breaker plate assembly to the distributor body.
2. Remove the breaker plate assembly.

To Install
3. Replace the breaker plate assembly, and secure it with the two screws.

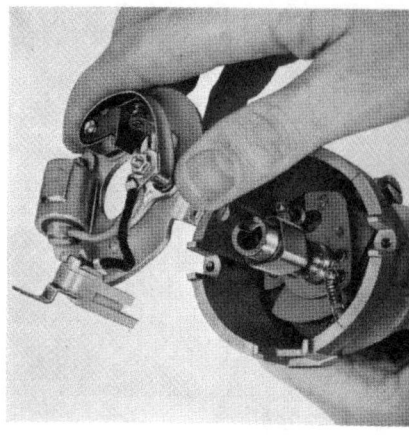

Replacing the Breaker Plate Assembly

Replacing the Low Tension and Condenser Wires

CORTINA - LOTUS

OP 12100-A6 EXTRA: GOVERNOR WEIGHTS AND SPRINGS – RENEW (BREAKER PLATE REMOVED)

To Remove

1. Note the location of the advance springs then disconnect and remove.
2. Unscrew the centre screw and withdraw the cam.
3. Remove the advance weights.

To Install

4. Locate the advance weights in the action plate, and fit springs.
5. Fit the cam, engaging the pivot pins in the weights. Secure the cam with the centre screw.
6. Fit the advance springs taking care not to stretch them while connecting them to the pegs on the action plate and the cam assembly.

OP 12100-A8 DISTRIBUTOR – OVERHAUL (GOVERNOR WEIGHTS REMOVED)

To Dismantle

1. Remove the skew gear and thrust washer after driving out the retaining pin.
2. Withdraw the distributor shaft from the distributor body.
3. Remove the spacer washer.
4. Using a suitable mandrel located at the body end, drive out the bush.

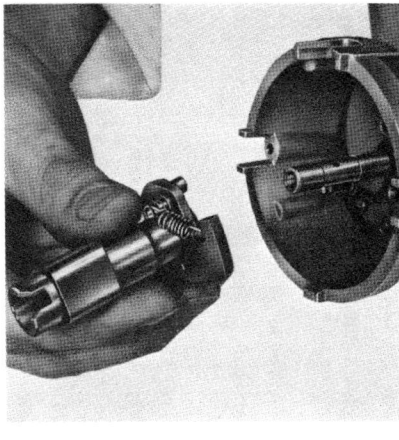

Replacing the Cam

Section 10 — 37

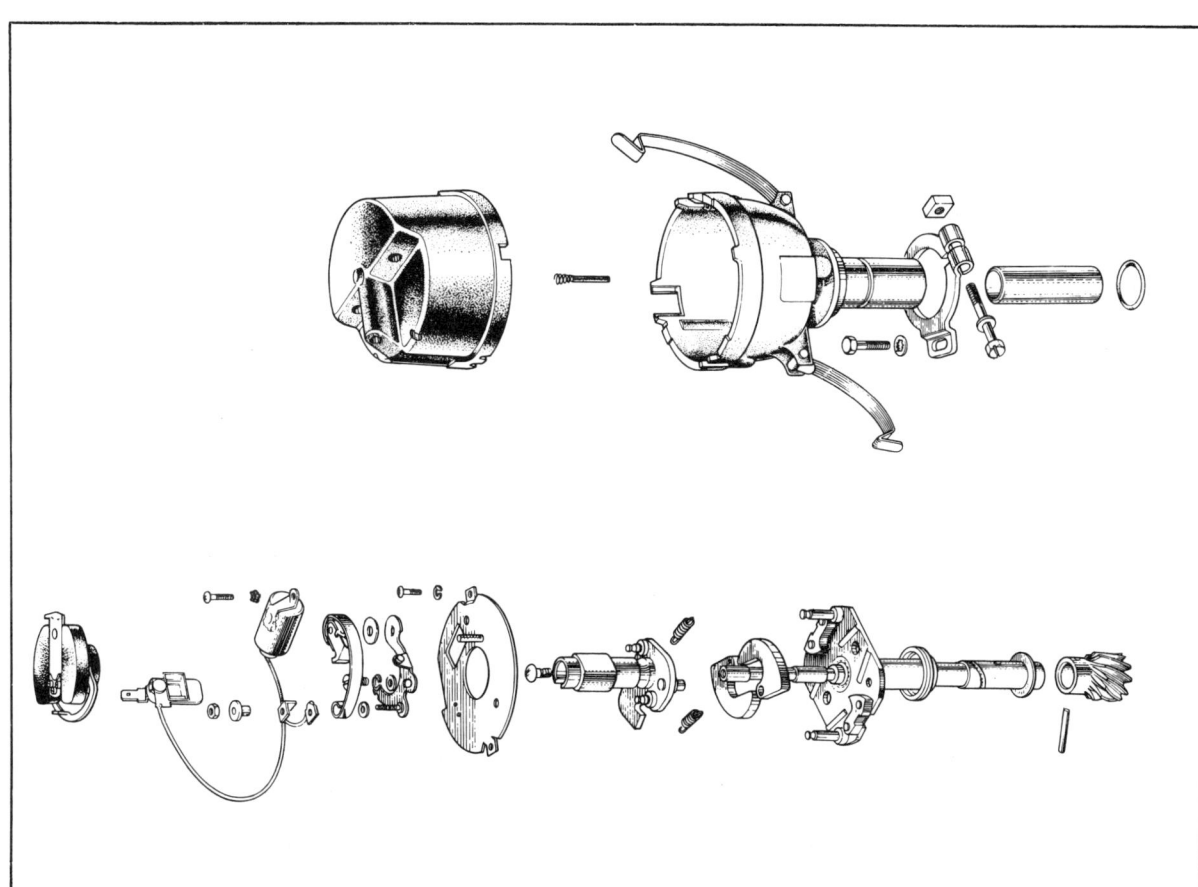

Distributor Assembly

Section 10 — 36

To Reassemble

5. Insert the small diameter end of the bush into the distributor body from the drive end and press fully home using a suitable mandrel. Drill the oil hole in bush, using the hole in the distributor body spigot as a guide.

   NOTE – The new bush should be soaked in engine oil for 24 hours prior to assembly.

6. Locate the spacer washer on the new distributor shaft.
7. Insert the distributor shaft into the body.
8. Locate a new thrust washer on the shaft with the pips towards the gear.
9. Press the new gear onto the distributor shaft until it nips the thrust washer.
10. With the assembly held tightly together drill the shaft with a No. 16 (0.177 in. (4.5 mm.)) drill using the hole in the skew gear as a guide. Fit the retaining pin and peen over the ends. Tap the end of the shaft with a hide mallet to flatten the thrust washer pips to establish the correct end-float.

OP 12100-B   GOVERNOR WEIGHTS AND SPRINGS – RENEW
(Includes OPS 12100-A, A1, A2, A3, A4 and A6)

OP 12100-D   DISTRIBUTOR – OVERHAUL
(Includes OPS 12100-A, A1, A2, A3, A4, A6 and A8)

OP 12100-E   DISTRIBUTOR ASSEMBLY – TEST AND ADJUST

Tools Required

Proprietary distributor tester.

The following instructions indicate the general principles to be followed for testing the distributor on a tester. The method of testing, however, may vary for machines of different manufacture: for specific instructions refer to the equipment manufacturer's handbook.

1. Mount the distributor on the tester, using an adaptor shaft, where necessary, to connect the drive from the machine to the distributor gear. Check that the distributor is free to rotate and that the adaptor shaft has the correct end-float, usually $\frac{1}{16}$ in. (1.59 mm.).
2. Make the necessary electrical connections and zero the instrument if required.
3. **Dwell Angle**
   (a) Turn the cylinder selector to the figure corresponding to the number of lobes on the cam of the distributor; in this case four.
   (b) Turn the test selector switch to the cam angle position and operate the distributor at approximately 1,000 rev./min. (crankshaft).
   (c) Adjust the distributor breaker point gap to a dwell angle of 57° to 63°.
   (d) Increase the speed up to a maximum of 5,000 rev./min. (crankshaft) and check the dwell reading, which must again be between 57° to 63°. If the reading changes more than 3° check for a worn distributor shaft or worn bushings.

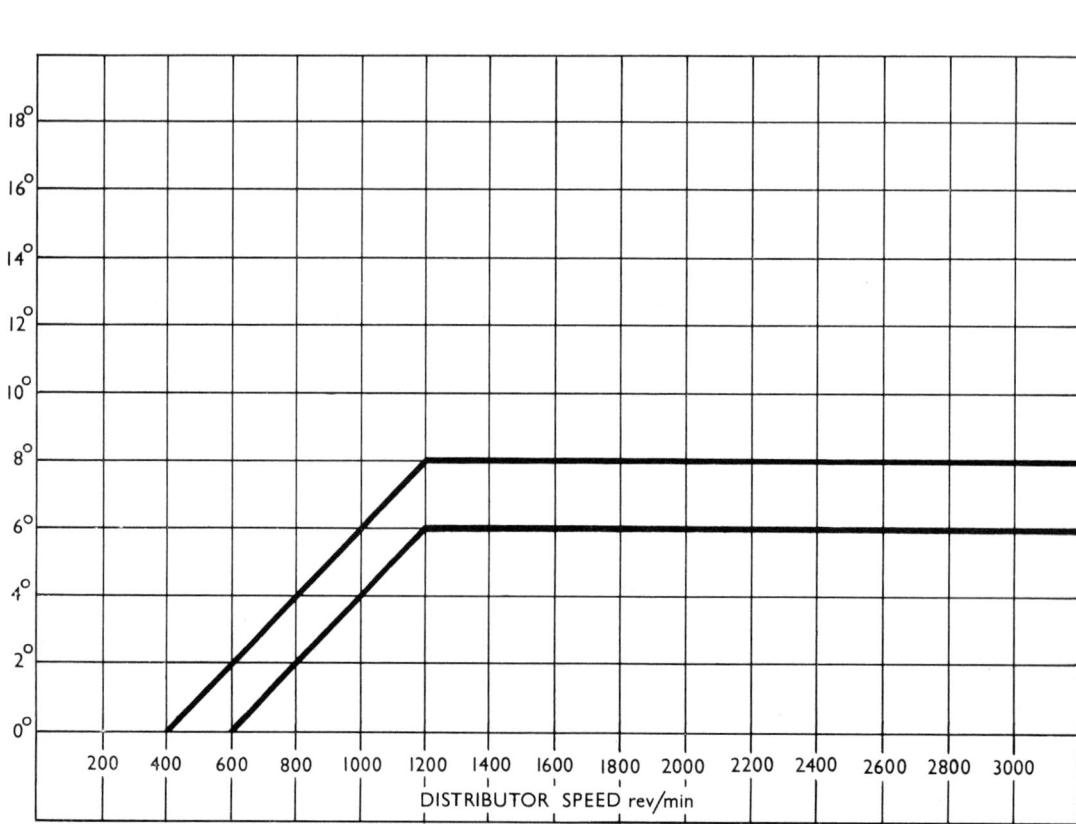

Distributor Advance

4. **Mechanical operation**

    (a) Make the necessary connections for the stroboscopic timing light or sparking protractor, refer to equipment manufacturer's handbook.

    (b) Adjust the speed control to vary the distributor speed between 400 and 5,000 rev./min. (crankshaft). Erratic or thin faint flashes of light preceding the regular flashes as the speed of rotation is increased can be due to weak breaker arm spring tension.

    (c) Operate the distributor at approximately 2,500 rev./min. (crankshaft).

    (d) Move the protractor scale with the adjustment control so that the zero degree mark on the scale is opposite one of the neon flashes. The balance of all the flashes should come within plus or minus 1°, evenly spaced around the protractor scale. A larger variation than 1° or erratic or wandering flashes may be caused by a worn cam or distributor shaft or a bent distributor shaft.

5. **Distributor spark advance**

    The spark advance is checked to determine if the ignition timing advances in proper relation to engine speed and load.

    Normally, this should not require adjustment as it is pre-set during manufacture. However, incorrect assembly, weakening of the advance springs or wear will change the advance curves and rectification will be necessary if engine performance is not to be affected.

    (a) Operate the distributor in the direction of rotation (anti-clockwise) and adjust the speed to 300 rev./min. (distributor). Move the protractor scale so that one of the flashes lines up with the zero degree mark.

    (b) Slowly increase the speed and check the advance at the other speeds quoted in the specification. Operate the distributor both up and down the speed range.

OP 12199-A   CONTACT BREAKER ASSEMBLY – REPLACE AND ADJUST

Tools Required

Dwell meter

To Remove

1. Remove the distributor cap.
2. Remove the rotor arm.
3. Remove the breaker arm.
4. Remove the adjustable contact.

To Replace

5. Locate a new adjustable contact on the pivot pin and loosely fit the retaining screw.
6. Locate fibre washers on the terminal post and the pivot pin and fit the breaker arm. Thread the insulating bush into the low tension and condenser lead eyelets before locating it on the terminal post with the end inside the spring eye.

7. Adjust the dwell angle to 57° to 63°, alternatively adjust the point gap to 0.014 to 0.016 in. (0.36 to 0.41 mm.). Use a screwdriver in the adjustable contact "V" notch to adjust the gap. Re-check the adjustment after tightening the retaining screw.

8. Fit the rotor and the distributor cap.

OP 12300-A   CONDENSER – RENEW (DISTRIBUTOR IN SITU)

1. Remove the distributor cap.
2. Unscrew the condenser lead, and the condenser retaining screw.
3. Reposition the condenser, retain in position and replace the wire.
4. Replace the distributor cap.

OP 12405-A   SPARK PLUGS – REPLACE

To Remove

1. Disconnect spark plug leads.
2. Remove spark plugs.

To Replace

3. Replace spark plugs.
4. Reconnect spark plug leads.

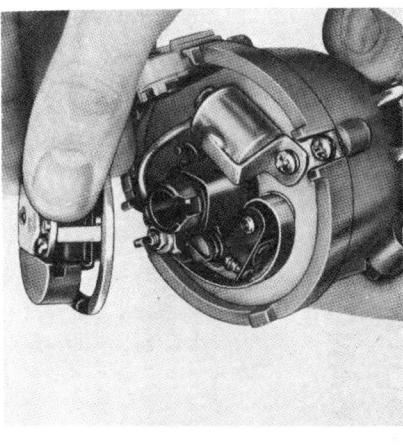

Replacing the Adjustable Contact

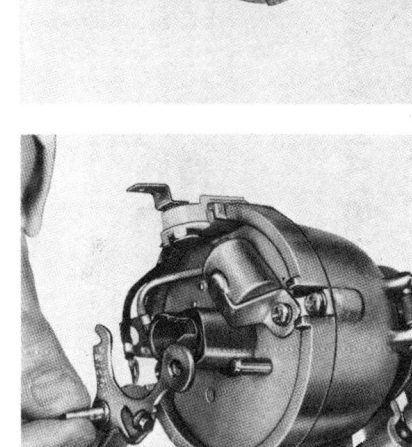

Fitting the Rotor

OP 12405-A. EXTRA TO CLEAN AND RESET SPARK PLUGS

Tools Required

Spark plug cleaner

1. Sandblast spark plugs.
2. File the centre electrode square.
3. Reset gaps to 0.023 in. (0.59 mm.) and pressure test.

OP 12405-B   SPARK PLUGS – CLEAN AND RESET
(Includes 12405-A and A1)

INSTRUMENTS, CONTROLS AND ANCILLARIES

GENERAL DESCRIPTION

A bank of four instruments, mounted on the facia, consists of fuel and temperature gauges, ammeter and a tube type oil pressure gauge. The fuel and temperature gauges operate at 5 volts which has the advantage of giving steady constant readings unaffected by current changes i.e. flashing the headlights or sounding the horn. The oil pressure gauge is operated via a small bore pipe from a sender unit tapped into the right-hand side of the engine. A speedometer and a tachometer, incorporating main beam and generator warning lights, complete the instrumentation.

The light switch, mounted on the facia, has three positions, OFF, sidelight and headlight operation.

Similarly, the heater blower switch, mounted on the heater controls bezel gives a two-speed control. Two levers mounted on the facia control the direction and temperature of the air delivered from the heater. The direction control lever allows cold, warm or hot air to be directed into the interior of the vehicle or on to the windscreen. The temperature control lever varies the amount of air passing through the heater element thus controlling the internal temperature. In addition adjustable "face level ventilators" mounted on the facia allow variable amounts of cold air to be circulated through the vehicle. The amount of air passing through the ventilators is controlled by a knob mounted in the centre of the movable "eyeball", rotating the knob opens or closes a butterfly type valve.

The windscreen wiper motor is also two-speed and is controlled by a two position rotary switch incorporating the windscreen washer pump. Direction indicators, horn, headlight dip, main beam and flasher switch are operated from a single lever mounted on the steering column.

Direction indicator warning lights and the ignition switch are on the facia. The ignition switch has three positions in addition to the OFF position. When the key is turned clockwise to the first position the ignition and auxiliary circuits become live. Turning the key further against the spring stop will operate the starter motor.

NOTE – If the engine does not start, the key must be returned to the OFF position before the starter motor can be operated again, thus preventing damage through operating the starter motor whilst the engine is running. The key turned anti-clockwise will operate certain accessory circuits.

An electric clock, mounted in the forward part of the centre console, can be adjusted for gain or loss by moving a small lever on the rear of the case. Moving the lever towards the + will speed up the clock and moving the lever towards the — will slow down the clock.

The driving lights consist of two headlamps, two side lights, two combined rear and stop lights and four direction indicator lights, two at the front and two at the back. The side lights and front direction indicator lights are mounted in the headlamp bezel. The headlamp is normally

CORTINA - LOTUS

a sealed beam unit but a semi-sealed unit with pre-focussed bulbs is also available. The bulbs for the rear/stop light and rear direction indicator lights can be removed from inside the luggage compartment.

All the lighting circuits are now fused. The fuses are grouped together in one fuse block mounted on the engine side apron panel and are rated at 8 amps. Where back-up lamps are fitted a separate fuse is provided, this fuse is located in the luggage compartment.

QUICK REFERENCE DATA

PERIODIC SERVICE ATTENTION

Weekly
Check operation of all lights

At first 3000 miles (5000 km.) or three months (whichever occurs first)
Check operation of all lights, instruments and controls

Every 3000 miles (5000 km.) or three months (whichever occurs first)
Check operation of all lights, instruments and controls

DATA

Horns

Type	4 in. (101.6 mm.) 'beep'
Current draw	$4\frac{1}{2}$ amp

Light Bulbs and Flasher Unit

Headlamp	60/45W* sealed beam
Side light	5W* wedge base
Front and Rear direction indicator	28W*
Flasher unit	56W*
Tail and Stop light	7/28W*
Licence plate	5W wedge base
Interior light	6W festoon
Warning lights	2.2W wedge base
Instrument panel light	2.2W wedge base
Clock	1.2W bayonet

*U.K.

Section 10 — 44

1 . 1969

CORTINA - LOTUS

SERVICE AND REPAIR OPERATIONS

OP 9280-A   FUEL GAUGE – REMOVE AND INSTALL

To Remove
1. Turn the instrument approximately $\frac{1}{3}$ turn anti-clockwise, using the tool detailed under OP. 17255-A, and pull the instrument out of the facia.
2. Make a note of the wiring positions, then disconnect the wires and illumination bulb from the gauge.

To Install
3. Reconnect the wires and illumination bulb.
4. Push the instrument into the facia and rotate approximately $\frac{1}{3}$ of a turn clockwise.

OP 10505-B   INSTRUMENT VOLTAGE REGULATOR – REMOVE AND INSTALL

To Remove
1. Remove the speedometer, see OP 17255-A.
2. Make a note of the wiring positions and remove the wires from the regulator.
3. Remove one bolt securing the regulator to the rear face of the speedometer and remove the regulator.

To Install
4. Position the regulator on the speedometer and secure it with one bolt.
5. Reconnect the wires to the regulator.
6. Replace the speedometer, see OP 17255-A.

OP 10670-A   AMMETER – REMOVE AND INSTALL
SEE **OP 9280-A**   FUEL GAUGE – REPLACE

OP 10880-A   OIL PRESSURE GAUGE – REMOVE AND INSTALL
See **OP 9280-A**   FUEL GAUGE – REPLACE

OP 10883-A   TEMPERATURE GAUGE – REMOVE AND INSTALL
SEE **OP 9280-A**   FUEL GAUGE – REPLACE

OP 11572-A   IGNITION SWITCH AND/OR LOCK BARREL – REMOVE AND INSTALL

To Remove
1. Disconnect the battery.
2. Unscrew the ignition switch bezel.
3. Push the ignition switch backwards and make a note of the various wiring positions.
4. Disconnect the wires and remove switch.

1 . 1969

Section 10 — 45

CORTINA - LOTUS

To Install

5. Reconnect the wires to the ignition switch and from underneath the facia position the switch.
6. Replace the bezel and tighten.
7. Reconnect the battery and check the operation of the ignition.

OP 11654-A LIGHTING SWITCH – REPLACE

To Remove

1. Disconnect battery cables.
2. Remove face-level vent, as in OP 18470-A.
3. Unscrew the switch bezel, and remove the switch from behind the facia panel.
4. Make a note of the wiring positions, and remove the wires.

To Replace

5. Reconnect the wires to the switch.
6. Replace the switch from behind the facia, and secure in position with the bezel.
7. Replace the facia vent and replace the battery cables.

OP 13000-A HEADLAMPS – ALL – ALIGN

The headlamps can be aligned with any suitable alignment equipment, but if this is not available, the following procedure should be carried out:—

1. Position the car on level ground 10 ft. (3 m.) in front of a suitable darkened board which is marked with a vertical and horizontal line. This board must be at right-angles to the car centre-line.

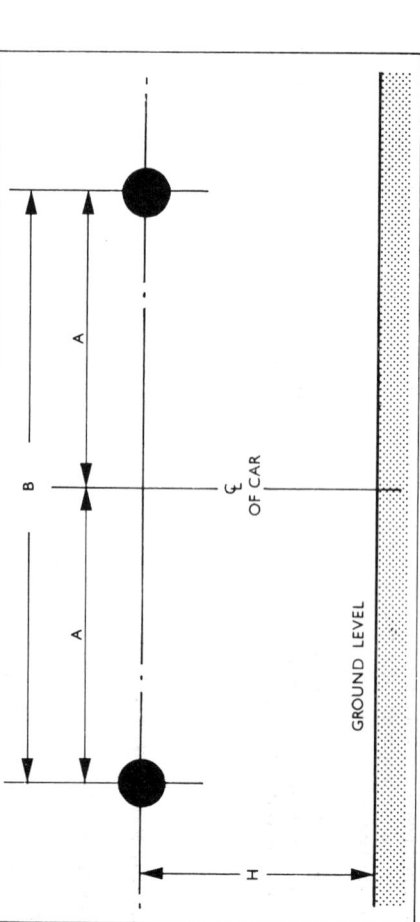

Headlamp Alignment Chart

2. Bounce the car to ensure correct settlement of the suspension and then measure the height 'H' from the ground to the centre of the headlamps.
3. Position the board so that the vertical line is exactly in line with the car centre-line. Position the board, also, so that the horizontal line is parallel to the ground and at a height 'H' from the ground.
4. Remove each headlamp outer bezel and switch on the headlamps.
5. By means of the horizontal and vertical adjusting screws, adjust each headlamp so that the centres of brightest illumination lie on the horizontal dividing line 43.2 in. (109.7 cm.) ('B') apart and equidistant from the vertical dividing line.
6. Switch off the headlamps and refit the outer bezels.

OP 13005-A HEADLAMP SEALED BEAM UNIT – REMOVE AND INSTALL

To Remove

1. Disconnect the battery.
2. Remove four screws securing the head and sidelamp bezel and pull the bezel forward.
3. Remove three screws securing the headlamp chrome bezel and remove the bezel.
4. Pull the sealed beam unit forward, disconnect the headlamp wiring plug and remove the sealed beam unit.

To Install

5. Plug the headlamp wiring plug into the sealed beam unit.
6. Fit the chrome bezel over the sealed beam unit and position them in the headlamp nacelle.
7. Secure the headlamp and bezel with three screws.
8. Replace the head and sidelamp bezel and secure with four screws.
9. Connect the battery and check the operation of the headlamps.

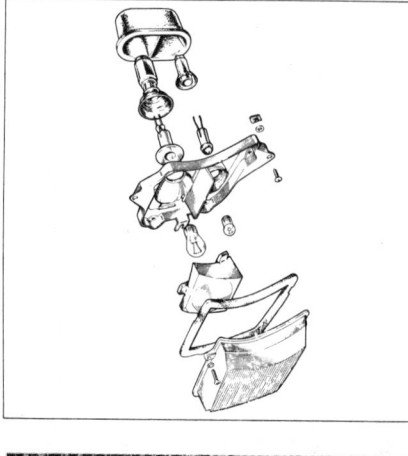

Side and Indicator Lamp

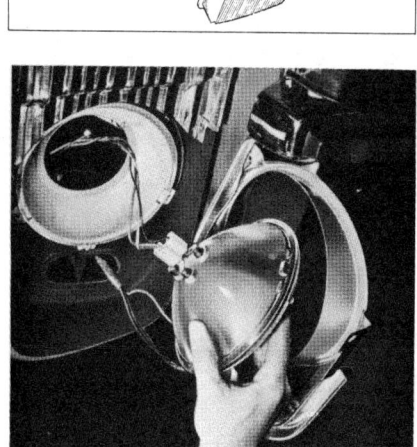

Removing Headlamp Sealed Beam Unit

OP 13200-A FRONT INDICATOR AND/OR SIDE LAMP ASSEMBLY – REMOVE AND INSTALL

To Remove

1. Disconnect the battery.
2. Remove four screws securing the head and sidelamp bezel and pull the bezel forward.
3. Disconnect the side and indicator spring-loaded bulb holders.
4. Slacken the screw securing the side and indicator lens assembly to the bezel and remove the lens assembly.

To Install

5. Position the side and indicator lens assembly in the bezel and secure with one screw.
6. Replace the side and indicator spring-loaded bulb holders.
7. Position the head and sidelamp bezel on the body and secure it with four screws.
8. Connect the battery and check the operation of the side and indicator lamps.

OP 13335-A DIRECTION INDICATOR, HEADLAMP FLASHER AND HORN SWITCH ASSEMBLY – REMOVE AND INSTALL

To Remove

1. Disconnect the battery.
2. Remove four screws securing the steering column shrouds and remove the shrouds.
3. Remove two screws securing the switch to the steering column.
4. Disconnect the multi-pin plug and remove the switch.

To Install

5. Connect the multi-pin plug to the harness.
6. Position the switch on the steering column and secure it with two screws.
7. Locate the steering column shrouds and secure them with four screws.
8. Connect the battery and check the operation of the switch.

OP 13350-A FLASHER UNIT – REMOVE AND INSTALL

To Remove

1. Disconnect the battery.
2. Remove the flasher unit from the clip on the steering column brace.
3. Remove the connectors to the flasher unit and remove the unit.

To Install

4. Replace the connectors to the flasher unit.
5. Position the flasher unit in clip on steering column brace.
6. Reconnect the battery and check the operation of the flasher unit.

OP 13404-A REAR DIRECTION INDICATOR AND/OR STOP/TAIL LAMP BULBS – REMOVE AND INSTALL

To Remove

1. Disconnect the battery.
2. Pull off rear lamp trim cover.
3. Pull out the spring-loaded rear and indicator bulb holders and remove the bulbs.

To Install

4. Replace the bulbs in the holders and clip the two bulb holders into the rear lamp assembly.
5. Replace the rear lamp trim cover.
6. Reconnect the battery.

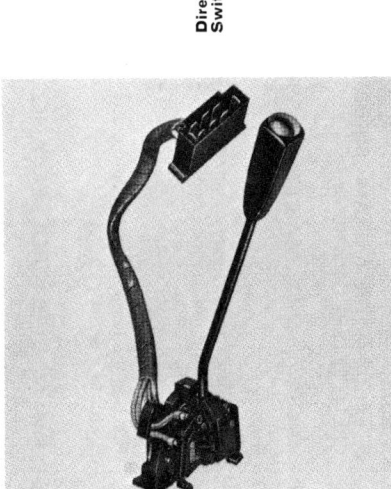

Rear and Indicator Lamp

CORTINA - LOTUS

OP 13404-A1 EXTRA: REMAINING REAR DIRECTION INDICATOR AND/OR STOP/TAIL LAMP BULBS – REMOVE AND INSTALL
(SEE **OP 13404-A**)

OP 13404-A2 EXTRA: REAR LAMP ASSEMBLY – ONE – REMOVE AND INSTALL

To Remove

1. From inside the luggage compartment remove four screws securing the lamp assembly to the body and remove the lamp.

To Install

2. Position the lamp assembly on the body and from inside the luggage compartment secure it with four screws.

OP 13404-A3 EXTRA: REAR LAMP BEZEL AND/OR LENS – REMOVE AND INSTALL

To Remove

1. Remove the rivets securing the lamp bezel.
2. Remove four screws securing the lens to the lamp backplate and remove the bezel, sealing gasket and lens.

To Install

3. Position the sealing gasket, lens and bezel on the backplate and secure them with four screws and rivets.

OP 13404-B REAR DIRECTION INDICATOR AND STOP/TAIL LAMP BULBS – ALL – REMOVE AND INSTALL
(Includes OPS 13404-A and A1)

OP 13404-C REAR LAMP ASSEMBLY – ONE – REMOVE AND INSTALL
(Includes OPS 13404-A and A2)

OP 13404-D REAR LAMP ASSEMBLIES – BOTH – REMOVE AND INSTALL
(Includes OPS 13404-A, A1 and A2 × 2)

OP 13404-E REAR LAMP BEZEL AND/OR LENS – ONE SIDE – REMOVE AND INSTALL
(Includes OPS 13404-A, A2 and A3)

OP 13404-F REAR LAMP BEZELS AND/OR LENS – BOTH SIDES – REMOVE AND INSTALL
(Includes OPS 13404-A, A1, A2 × 2 and A3 × 2)

OP 13480-A STOP LIGHT SWITCH – REMOVE AND INSTALL

To Remove

1. Disconnect the battery.
2. Make a note of the wires to the switch then disconnect them.
3. Unscrew the locknut and remove the stop light switch.

To Install

4. Locate the stop light switch in its mounting bracket and secure it with a locknut.
5. Reconnect the wires to the switch.
6. Connect the battery and check the operation of the stop light switch.

OP 13543-B REAR LICENCE PLATE LAMP – REMOVE AND INSTALL

To Remove

1. Disconnect the battery.
2. Disconnect the earth and feed wires inside the luggage compartment and pull the wires through the hole in the floor.
3. Press the lens retaining levers inwards from underneath the bumper and lift off the lens assembly.
4. Prise the lamp body out of the bumper bar.

To Install

5. Pass the earth and feed wires through the aperture in the bumper and press the lamp body into place.
6. Pass the wires through the hole in the luggage compartment floor and reconnect. Locate the sealing grommet.
7. Clip the lens assembly into place in the lamp body.
8. Reconnect the battery and check the operation of the lamp.

OP 13740-A INSTRUMENT PANEL LIGHT SWITCH – REMOVE AND INSTALL
(SEE **OP 11654-A**)

OP 13776-A INTERIOR LIGHT – REMOVE AND INSTALL

To Remove

1. Disconnect the battery.
2. Remove two screws securing the interior light and pull the light away from the headlining.
3. Make a note of the wiring positions then disconnect the wires.

To Install

4. Connect the wires to the interior light.
5. Position the light on the headlining and secure it with two screws.
6. Reconnect the battery and check the operation of the light.

OP 13801-A HORN – REMOVE AND INSTALL

To Remove

1. Disconnect the battery.
2. Remove ten screws, five each side, securing the headlamp bezels and remove the bezels.
3. Remove six rivets and two screws securing the radiator grille to the body and remove the grille.

CORTINA - LOTUS

4. Make a note of the wiring positions then disconnect the wires from the horn.
5. Remove two bolts securing the horn and remove the horn.

To Install

6. Position the horn in the car and secure it with two bolts.
7. Reconnect the wires to the horn.
8. Replace the radiator grille and secure it with two screws, and either pop-rivets or self-tapping screws.
9. Replace the headlamp bezels and secure them with ten screws, five each side.
10. Reconnect the battery and check the operation of the horn.

OP 13990-A   WARNING LIGHT BULB – REMOVE AND INSTALL

To Remove

1. From behind the instrument panel pull the bulb holder out of the lens socket.
2. Remove the bulb from the bulb holder.

To Install

3. Fit the new bulb (12 v, 2.2 w) into the bulb holder.
4. Push the bulb holder fully home in the lens socket.

OP 14067   FUSE BLOCK ASSEMBLY – REMOVE AND INSTALL

To Remove

1. Open the bonnet and pull off the two non-reversible wiring plugs on the fuse block.
2. Pull off the transparent fuse cover, remove two screws securing the fuse block to the body and remove the fuse block.

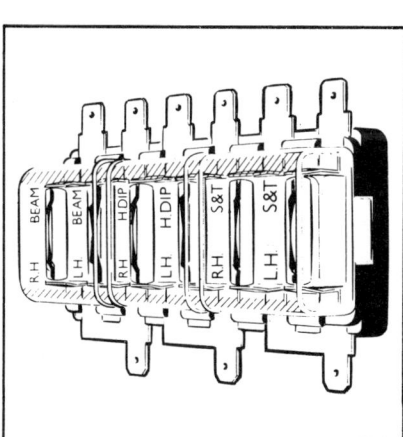

Fuse Block Assembly

3. Position the fuse block and secure it with two screws, then replace the transparent fuse cover.
4. Reconnect the two non-reversible wiring plugs to the fuse block, close the bonnet and check the operation of the lighting circuits.

OP 15000-A   CLOCK AND/OR BULB – REMOVE AND INSTALL

To Remove

1. Carefully prise the clock out of the console.
2. Remove the bulb.

To Install

3. Fit the bulb (12 v, 1.2 w) to the bulb holder.
4. Push the clock into the console and align as necessary.

OP 15055-A   CIGARETTE LIGHTER ASSEMBLY – REMOVE AND INSTALL

To Remove

1. Disconnect the battery.
2. Make a note of the wiring positions and remove the wires from the lighter.
3. Unclip the cigarette lighter illuminating bulb from the lighter body.
4. From behind the facia unscrew and remove the lighter body.
5. Remove the front section of the lighter through the facia.

To Install

6. Position the front section of the lighter in the facia.
7. From behind the facia replace the element body.
8. Clip the cigarette lighter illuminating bulb into the lighter body.
9. Replace the wires to the lighter.
10. Reconnect the battery and check the operation of the cigarette lighter and illuminating bulb.

OP 17255-A   SPEEDOMETER HEAD – REMOVE AND INSTALL

To Remove

1. Remove the tachometer.
2. Through the tachometer aperture, disconnect the speedometer cable and illumination bulb.
3. Turn the speedometer approximately $\frac{1}{3}$ of a turn anti-clockwise using the removal tool and remove the instrument and bezel from the facia.

To Install

4. Replace the bezel and speedometer in the facia and secure them by turning the speedometer approximately $\frac{1}{3}$ of a turn clockwise.

5. Reconnect the speedometer cable and illumination bulb.
6. Replace the tachometer.

OP 17262-A SPEEDOMETER INNER CABLE – REMOVE AND INSTALL

To Remove
1. Remove the tachometer.
2. Through the aperture in the facia disconnect the speedometer cable from the speedometer and withdraw the inner cable.
3. Remove the retaining plate or, alternatively, the circlip retaining the speedometer cable to the gearbox.

To Install
4. Through the aperture in the facia feed the speedometer inner cable into the outer cable.
5. Locate the inner cable square drive in the gearbox and secure the outer cable with a retaining plate and bolt or circlip as appropriate.
6. Replace the other end of the inner cable in the speedometer and secure the outer cable with the knurled retainer.
7. Replace the tachometer.

OP 17262-B SPEEDOMETER INNER AND OUTER CABLES – REMOVE AND INSTALL

To Remove
1. Jack up the front of the car and fit stands.
2. Remove the retaining plate or, alternatively, the circlip retaining the speedometer cable to the gearbox.
3. Open the bonnet and remove the speedometer cable clip.
4. Push the speedometer cable rubber grommet out of the bulkhead.
5. Remove the speedometer from the facia, disconnect the cable and remove the cable.

To Install
6. Replace the speedometer cable in the car, reconnect the speedometer and replace the speedometer in the facia.
7. Reposition the rubber grommet in the bulkhead.
8. Secure the speedometer cable to the bulkhead with a clamping clip.
9. Position the cable in the gearbox and secure it with a retaining plate and bolt or circlip as appropriate.
10. Remove the jack and stands.

OP 17360-A TACHOMETER ASSEMBLY – REMOVE AND INSTALL
(SEE **OP 17255-A**)

OP 17508-A WINDSCREEN WIPER ASSEMBLY – REMOVE AND INSTALL

To Remove
1. Disconnect the battery.
2. Remove the wiper arms and two nuts securing wiper spindles to body.
3. Remove the front parcel tray.
4. From under the facia remove the screw securing the motor bracket to the body.
5. Take a note of the wires and then remove the wires from the windscreen wiper motor.
6. Remove the wiper switch from the facia and disconnect the wires to the windscreen wiper motor.
7. Remove two screws securing the glove box catch and remove the catch.

Speedometer Head

Instrument Removal Tools

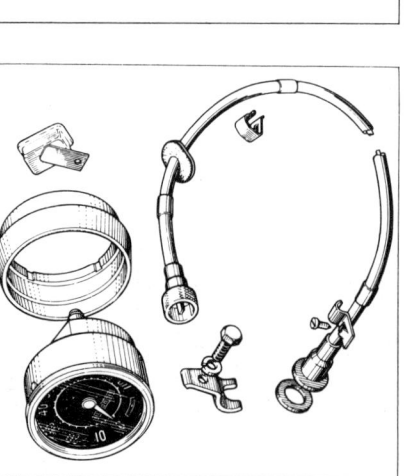

Removing the Windscreen Wiper Motor

CORTINA - LOTUS

8. Remove four screws securing the glove box lid and remove the lid.
9. Drill out nine "pop" rivets securing glove box to the facia and remove the glove box.
10. Pull the face level vent and demister flexible pipes off and remove the windscreen wiper motor.

To Install

11. Locate the windscreen wiper motor behind the facia.
12. Secure the motor to the body with the two wiper spindle nuts.
13. Replace the face level vent and demister flexible pipes.
14. Locate the glove box in the facia and secure it with nine "pop" rivets.
15. Replace the glove box lid and secure with four screws.
16. Replace the glove box catch and secure it with two screws.
17. Reconnect the wires from the windscreen wiper motor to the wiper switch and replace the switch.
18. Replace the feed wires to the wiper motor making sure that these are routed BELOW the speedometer cable.
19. Replace the screw securing the wiper bracket to the body.
20. Replace the wiper arms, front parcel tray and reconnect the battery.

OP 17535-A COMBINED WINDSCREEN WASHER AND WIPER SWITCH ASSEMBLY – REMOVE AND INSTALL

To Remove

1. Push a suitable rod into the hole on the underside of the washer/wiper knob until the retaining pin is released then pull off the knob.
2. Unscrew the washer/wiper assembly bezel and remove it.
3. From under the facia remove the wires to the wiper switch and the pipes to the washer pump, taking a note of their respective positions.
4. Remove the combined washer pump and wiper switch assembly from the vehicle.

NOTE – There are two types of windscreen wiper switch, one has an integral washer pump, the other has a detachable pump.

To Install

5. Connect the wires to the wiper switch and the pipes to the washer pump.
6. Position the washer/wiper assembly under the facia and secure it with its bezel.
7. Position the washer/wiper knob on its shaft, with the hole in the knob pointing downwards and push the knob in to engage the retaining pin.
8. Check the operation of the windscreen washer and wiper assembly.

Section 10 – 56

1 . 1969

OP 18309-A HEATER CONTROL ASSEMBLY – REMOVE AND INSTALL

To Remove

1. Release the two spring clips securing the two heater control cables to the heater operating arms and pull the cables out of the operating arms.
2. Remove the heater control bezel (see operation 18503-A).
3. Remove the heater controls mounting panel (see operation 18503-B).
4. Pull the heater control assembly forward out of the facia and remove two spring clips securing the heater control cable outers to the quadrant, pull the two inner cables off the quadrant and remove them.

To Install

5. Position the two heater control cables on the quadrant, loop the two cable inners on their respective levers and secure the outers with two spring clips.
6. Position the heater control assembly in the facia.
7. Replace the heater controls mounting panel (see operation 18503-B).
8. Replace the heater controls bezel (see operation 18503-A).
9. Position the two heater control cable inners in the operating arms and secure in plastic trunnions.
10. Secure the cable outers to the heater with two spring clips.

OP 18309-M HEATER CONTROL CABLES – ADJUST

The following operation should be carried out whenever the heater control cables are disconnected from the heater, or if the heater cannot be fully shut off.

1. Set the heater controls to the "OFF" and "HOT" positions.
2. Remove the spring clip (A) securing the distribution control cable to the heater.
3. Position the heater distribution valve firmly in the "OFF" position by raising the end of the lever (B) to the end of its travel.
4. Take up any slack in the distribution control outer cable and secure the cable to the heater with a spring clip.
5. Remove the spring clip (C) securing the temperature control cable to the heater.
6. Position the heater mixing valve (D) firmly in the "HOT" position by raising the end of the lever to the end of its travel.
7. Take up any slack in the temperature control outer cable and secure the cable to the heater with a spring clip.
8. Check the operation of the heater controls.

1 . 1969

Section 10 – 57

OP 18467-A HEATER ASSEMBLY – REMOVE AND INSTALL

To Remove

1. Disconnect the battery and drain the cooling system.
2. Remove the parcel tray.
3. From inside the engine compartment slacken two wire clips and disconnect the two heater pipes from the bulkhead.
4. Remove two screws and detach the heater pipe plate and sealing gasket from the bulkhead.
5. Remove the heater control cables, see OP 18309.
6. Make a note of the wiring positions and disconnect the wires from the resistor on motor case and motor lead.
7. Remove the face level vent and demister flexible pipes from the heater.
8. Remove four bolts and remove the heater.

To Install

9. Position the heater in the car and secure it to the bulkhead with four bolts.
10. Reconnect the face level vent and demister flexible pipes to the heater.
11. Replace the wires to the resistor on motor case and motor lead.
12. Replace the heater control cables, see OP 18309-A.
13. Replace the heater pipe plate and sealing gasket on the bulkhead and secure them with two screws.
14. Reconnect the two heater pipes and secure them with two wire clips.
15. Replace the parcel tray.
16. Reconnect the battery, fill the cooling system and check the operation of the heater.

OP 18467-A2 EXTRA: HEATER RADIATOR – REMOVE AND INSTALL

To Remove

1. Remove thirteen screws and remove the heater blower motor and mounting plate as an assembly.
2. Pull off the rear lower panel and remove the foam packing.
3. Carefully slide the radiator out of the heater.

To Install

4. Slide the radiator into the heater.
5. Position the foam packing and replace the rear lower panel.
6. Replace the heater blower motor and mounting plate and secure them with thirteen screws.

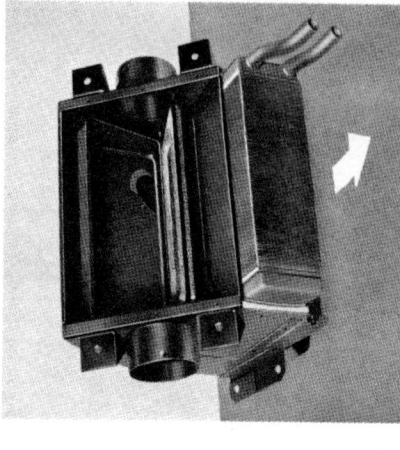

Removing the Heater Radiator

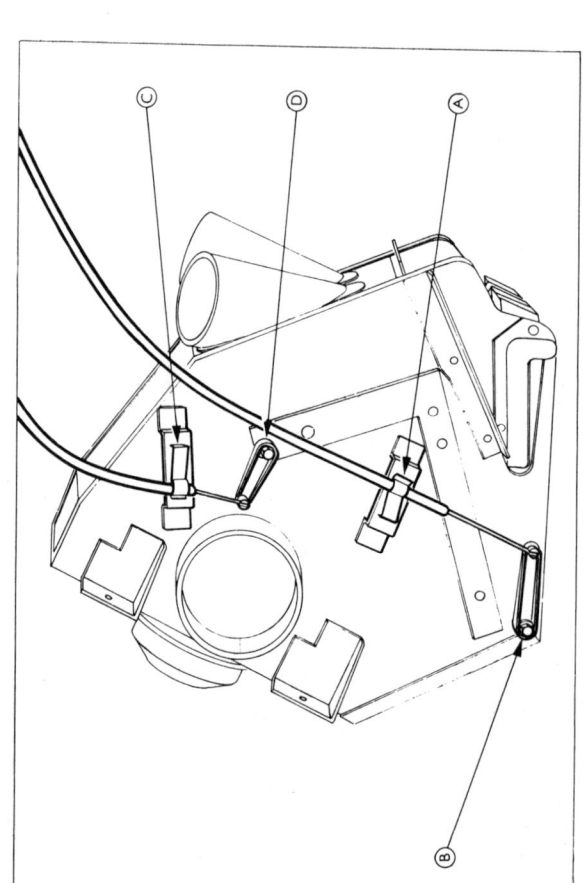

Heater Cable Adjustment

OP 18467-A3 EXTRA: HEATER MOTOR – REMOVE AND INSTALL

To Remove

1. Remove thirteen screws and remove the blower motor and mounting plate as an assembly.
2. Remove the sealing gasket from the mounting plate.
3. Remove the rubber gasket and pull the blower motor wires through the motor mounting plate.
4. Remove the spring clip and withdraw the blower motor fan.
5. Unscrew three screws securing the motor to the mounting plate and remove the motor.

To Install

6. Position the motor in the mounting plate and secure it with three screws.
7. Replace the blower motor fan and secure it with a spring clip.
8. Feed the blower motor wires through the motor mounting plate and replace the rubber gasket.
9. Replace the sealing gasket on the mounting plate.
10. Replace the blower motor and mounting plate on the heater and secure them with thirteen screws.

OP 18467-C HEATER RADIATOR – REMOVE AND INSTALL
(Includes OPS 18467-A and A2)

OP 18467-D HEATER MOTOR – REMOVE AND INSTALL
(Includes OPS 18467-A and A3)

OP 18470-A FACE LEVEL VENT – REMOVE AND INSTALL

To Remove

1. From under the facia pull the flexible air supply pipe from the vent.
2. Rotate the locking ring clockwise to disengage the three retaining pins and remove the vent from under the facia by rotating clockwise.

To Install

3. From under the facia position the vent and rotate it anti-clockwise to engage the three retaining flanges.
4. Clamp in position by rotating the locking ring anti-clockwise.
5. Connect the flexible air supply pipe to the back of the vent.

OP 18503-A HEATER CONTROLS BEZEL – REMOVE AND INSTALL

To Remove

1. Remove the cigar lighter (see operation 15055-A).

The Heater and Controls

CORTINA - LOTUS

2. Remove the combined windscreen washer and wiper knob and bezel (see operation 17535-A).

3. Push a suitable rod into the holes on the undersides of the heater control knobs until the retaining clips are released, then pull off the knobs.

4. Pull the heater controls bezel out of the facia and remove the bezel.

To Install

5. Position the bezel on the facia.

6. Push the heater control knobs on to their respective arms until the retaining clips engage.

7. Replace the combined windscreen washer and wiper knob and bezel (see operation 17535-A).

8. Replace the cigar lighter (see operation 15055-A).

OP 18503-B HEATER CONTROLS MOUNTING PANEL – REMOVE AND INSTALL

To Remove

1. Remove the cigar lighter (see operation 15055-A).

2. Remove the combined windscreen washer and wiper knob and bezel (see operation 17535-A).

3. Remove the heater controls bezel panel (see operation 18503-A).

4. Remove three screws securing the heater controls to the mounting panel and instrument panel and four screws securing the mounting plate to the facia.

5. Pull the mounting panel out of the facia and disconnect the wiring from the heater switch and remove the panel from the vehicle.

6. From behind the mounting panel, using a suitable flat-bladed screwdriver, depress the lug securing the top of the switch to the mounting panel and push the top of the switch out of the panel.

7. Repeat operation 6 for the bottom lug and remove the fan switch from the mounting panel.

To Install

8. Position the fan switch on the mounting panel and push it in to engage the retaining lugs.

9. Connect the wiring to the heater fan switch.

10. Position the mounting panel in the facia and secure it with four screws.

11. Replace the three screws securing the heater controls to the mounting panel and instrument panel.

12. Replace the heater bezel panel (see operation 18503-A).

13. Replace the combined windscreen washer and wiper knob and bezel (see operation 17535-A).

14. Replace the cigar lighter (see operation 15055-A).

Important

Following the completion of any service work, normal test procedures should be used to ensure that a satisfactory repair has been performed.

Heater Controls Mounting Panel

Heater Controls and Bezel

ELECTRICAL WIRING

ELECTRICAL WIRING

The loom connection illustrations have the wiring colour identified by a number.

The code for these numbers is as follows:—

Code	Colour		Code	Colour
1	Red		47	Blue/Light Green tracer
4	Red/Blue tracer		51	Black
5	Red/Black tracer		52	Black/Red tracer
6	Red/White tracer		57	Black/Light Green tracer
7	Red/Light Green tracer		61	White
8	Red/Brown tracer		62	White/Red tracer
11	Yellow		64	White/Green tracer
14	Yellow/Green tracer		65	White/Blue tracer
19	Yellow/Purple tracer		66	White/Black tracer
21	Green		68	White/Brown tracer
22	Green/Red tracer		71	Brown
23	Green/Yellow tracer		73	Brown/Yellow tracer
24	Green/Blue tracer		74	Brown/Green tracer
25	Green/Black tracer		77	Brown/White tracer
26	Green/White tracer		79	Brown/Purple tracer
31	Light Green		80	Brown/Orange tracer
34	Light Green/Blue tracer		81	Purple
35	Light Green/Black tracer		82	Purple/Red tracer
38	Light Green/Brown tracer		85	Purple/Blue tracer
41	Blue		86	Purple/Black tracer
42	Blue/Red tracer		87	Purple/White tracer
45	Blue/Black tracer		91	Pink/White tracer
46	Blue/White tracer			

NOTE – Where a connector is numbered, this indicates that the wires on each side of the connector are the same colour.

The wiring diagrams have the wiring colour identified by letters and the individual items in the diagram are identified by a number. The codes for these diagrams are overleaf.

Code	Wire Colour
R	Red
Bk	Black
Bl	Blue
W	White
Br	Brown
G	Green

Code	Wire Colour
Y	Yellow
LG	Light Green
P	Purple
O	Orange
Pk	Pink

Code	Item
1	R.H. Side Lamp (Front)
2	L.H. Side Lamp (Front)
3	R.H. Direction Indicator (Front)
4	L.H. Direction Indicator (Front)
5	R.H. Headlamp
6	L.H. Headlamp
7	R.H. Front Loom Connector
8	L.H. Front Loom Connector
10	Temperature Gauge Sender Unit
11	Coil
12	Distributor
13	Starter Motor
14	Generator
15	Body Earth
16	Battery
17	Horn
18	Fuses
19	Regulator
20	R.H. Bulkhead Multi-Way Connector
21	L.H. Bulkhead Multi-Way Connector
22	Starter Motor Solenoid
24	Blower Motor Ballast Resistor
25	Heater Motor
26	Back-up Lamp Switch (Manual Trans.)
28	Windscreen Wiper Motor

Code	Item
29	Stop Lamp Switch
31	R.H. Courtesy Light Switch
32	L.H. Courtesy Light Switch
33	R.H. Indicator Warning Light
34	L.H. Indicator Warning Light
35	Ignition Switch
36	Ammeter
37	Oil Pressure Gauge
38	Temperature Gauge
39	Fuel Gauge
40	Instrument Illumination Bulb
42	Generator Warning Light
44	Instrument Voltage Stabiliser
45	Speedometer
46	Tachometer
47	Clock
48	Turn Signal Unit
53	Heater Switch
54	Cigar Lighter
55	Windscreen Wiper Switch
56	Steering Column Connector
59	Panel Light Switch
61	Lighting Switch
62	Interior Light
63	Turn Signal Switch
64	Horn Switch
65	Column Dip Switch
66	Headlamp Flasher Switch
67	Fuel Tank Sender Unit

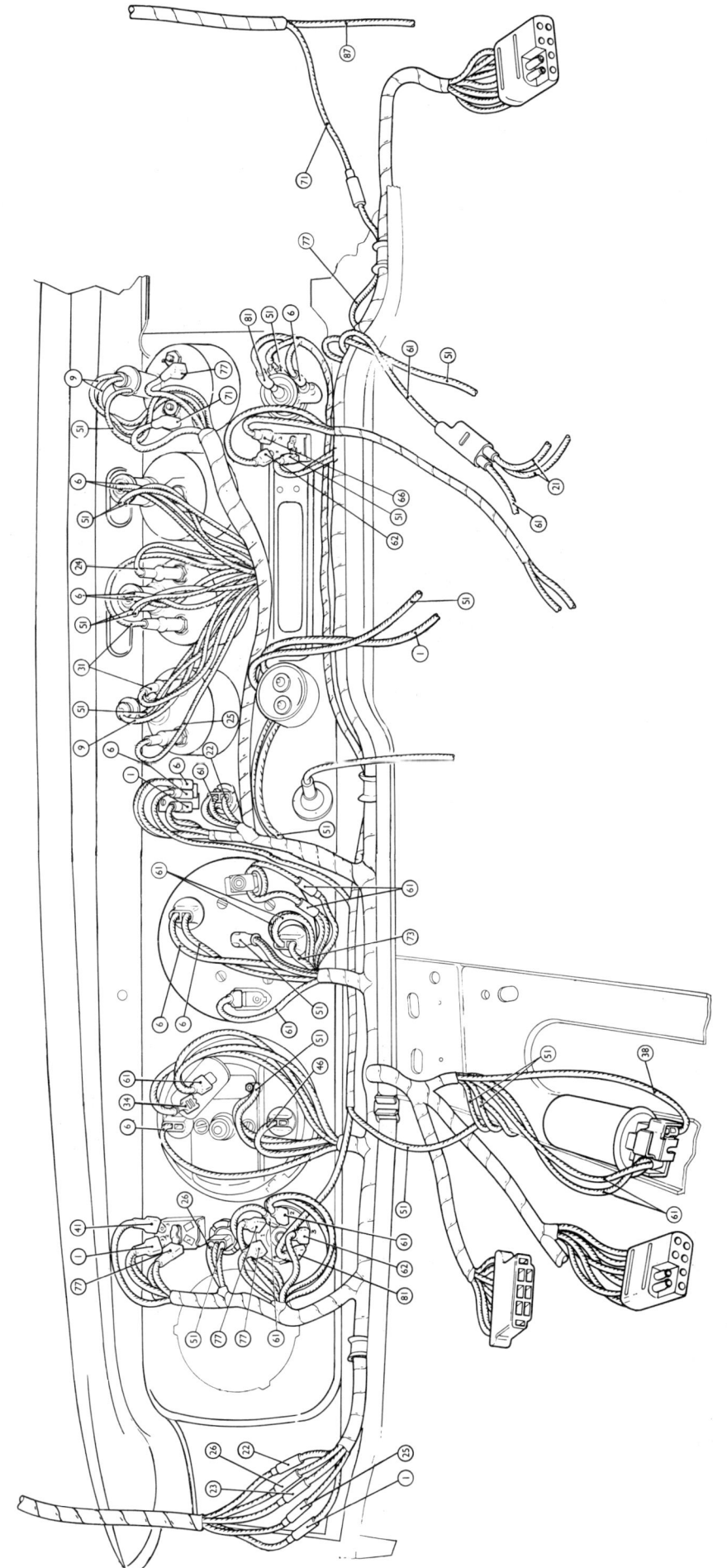
FACIA PANEL

CORTINA - LOTUS

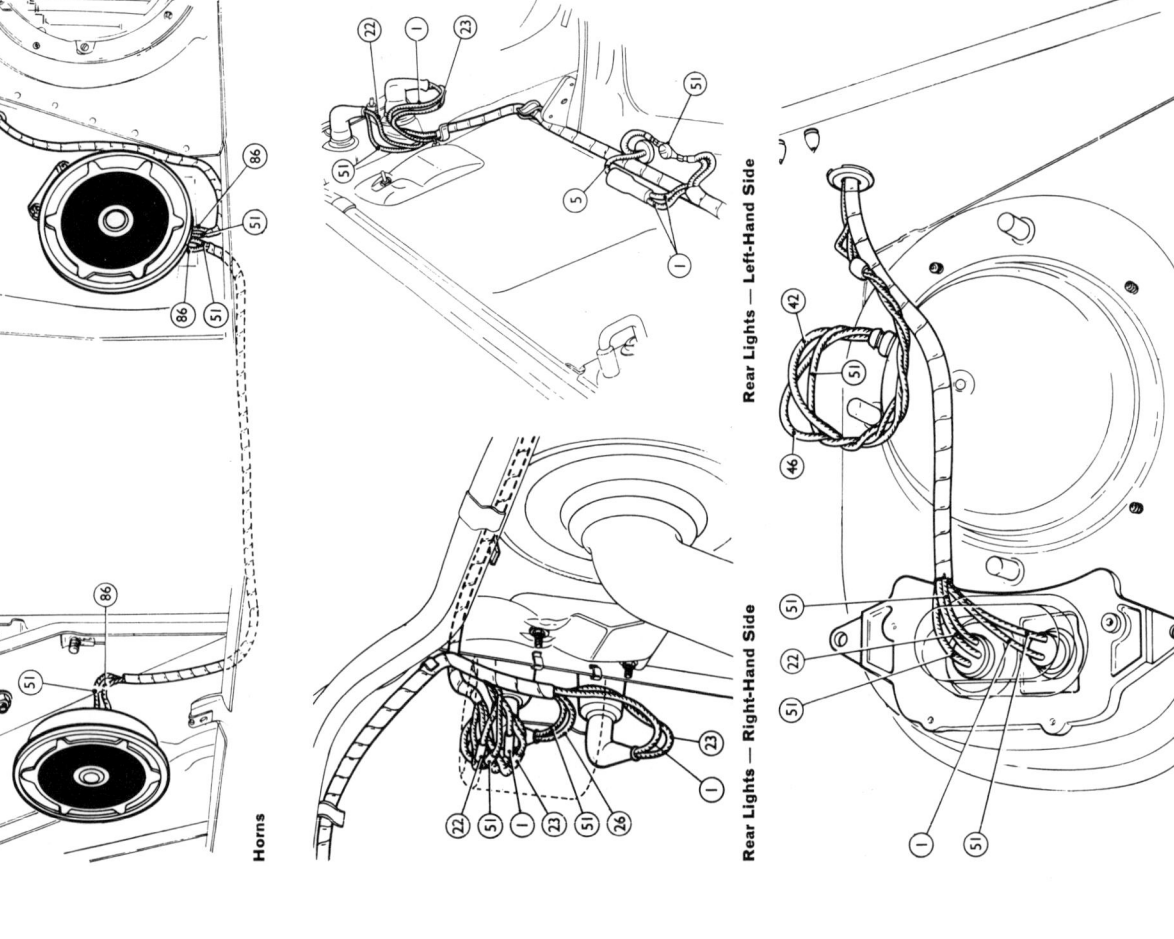

Horns

Rear Lights — Left-Hand Side

Rear Lights — Right-Hand Side

Front Lights

Code		Item	Code		Item
69	...	Rear Wiring Connector	77	...	L.H. Reversing Lamp (where fitted)
70	...	R.H. Stop Lamp	78	...	Licence Plate Lamp
71	...	L.H. Stop Lamp	79	...	Road Lamp Relay (where fitted)
72	...	R.H. Direction Indicator (Rear)	80	...	Tachometer Earth
73	...	L.H. Direction Indicator (Rear)	81	...	Speedometer Earth
74	...	R.H. Side Lamp (Rear)	82	...	Main Beam Warning Light
75	...	L.H. Side Lamp (Rear)	83	...	R.H. Road Lamp (where fitted)
76	...	R.H. Reversing Lamp (where fitted)	84	...	L.H. Road Lamp (where fitted)

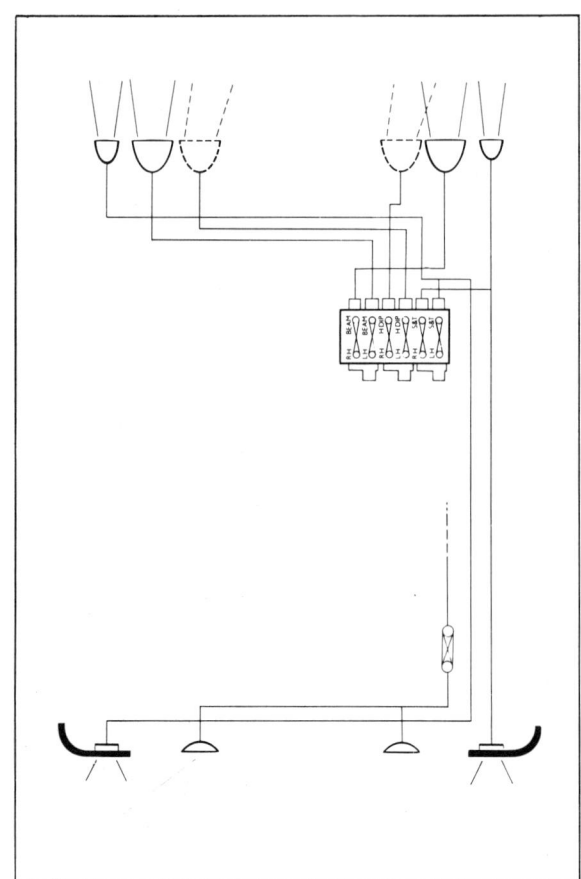

Fused Lighting Circuits

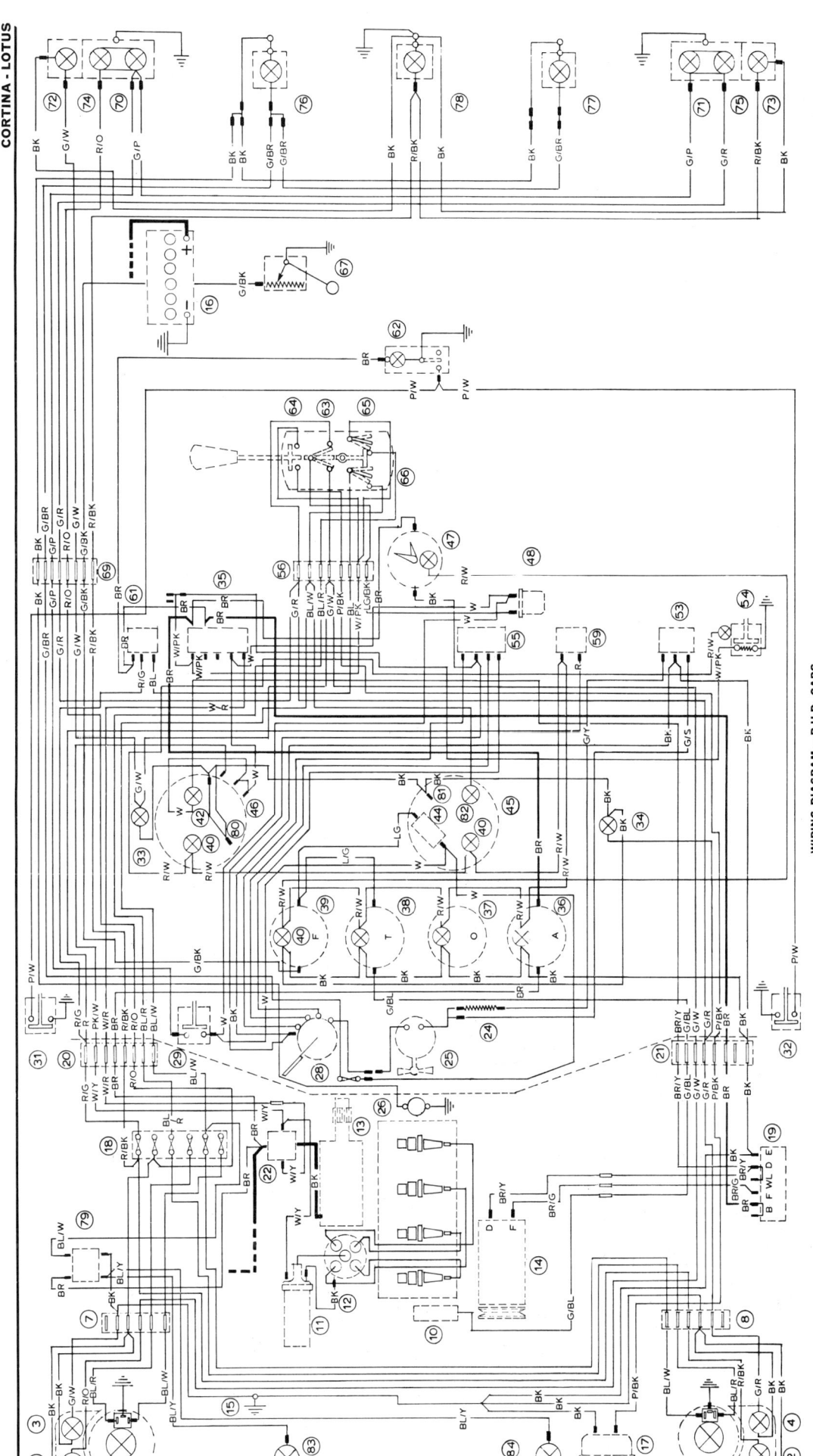

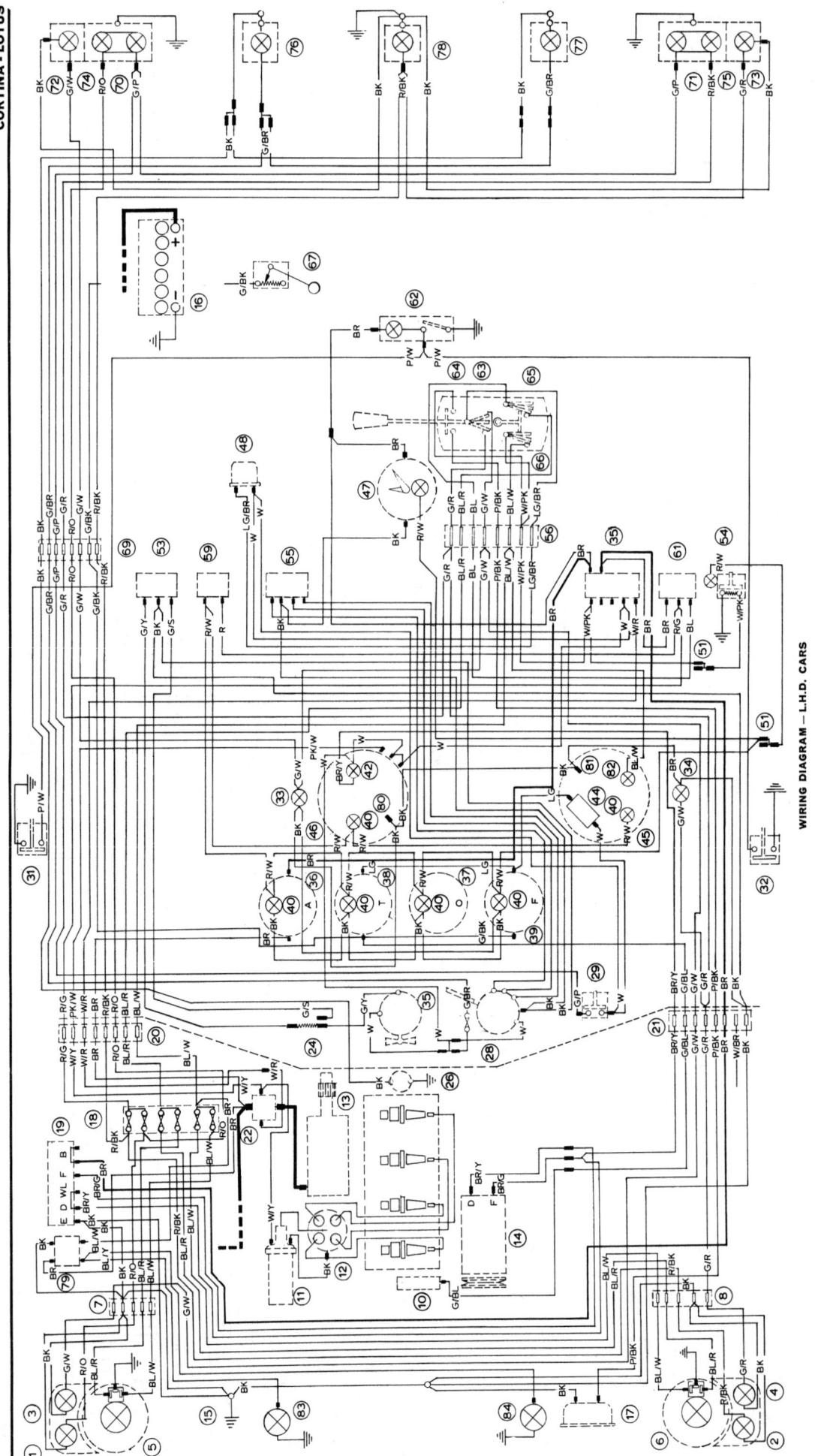

CORTINA - LOTUS

11/1
ACCESSORY FITTING INSTRUCTIONS

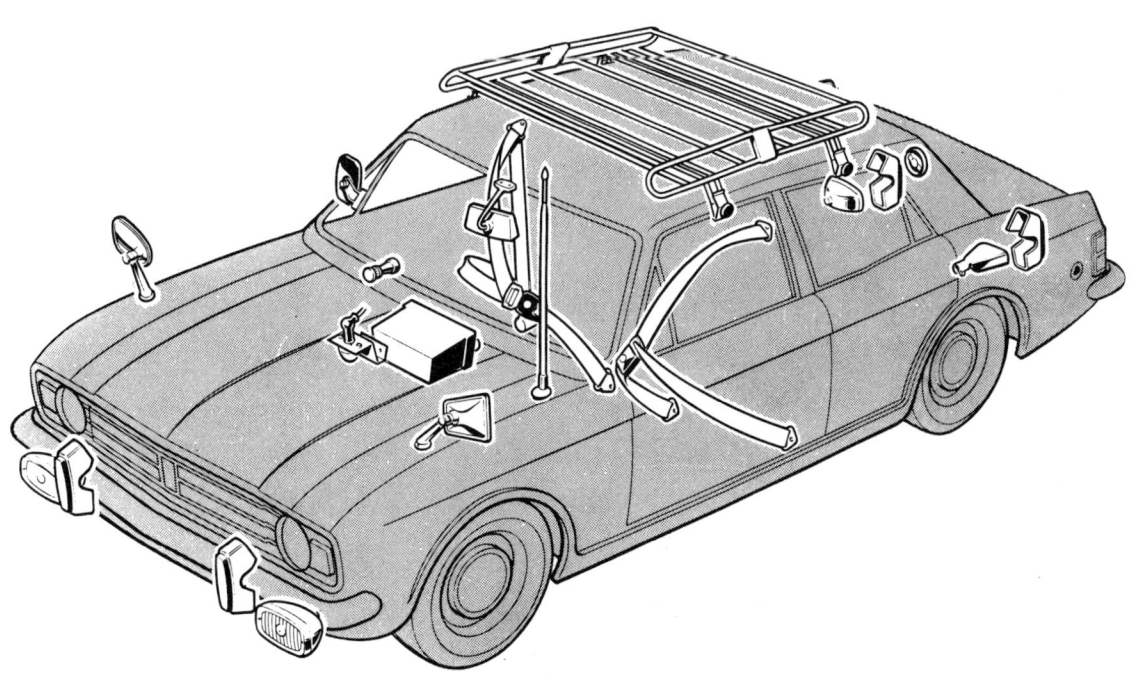

1 . 1969

Section **11/1** — 1

CORTINA - LOTUS

SECTION INDEX

FITTING INSTRUCTIONS

FOG AND ROAD LAMPS – Part Nos. 3014E-15200-A   Spot Lamp
                                 3014E-15K201-A   Fog Lamp
                                 3014E-15200-B   Spot Lamp-Iodine Quartz
                                 3014E-15K201-B   Fog Lamp-Iodine Quartz
                                 3014E-15204-A   Lamp Mounting Kit

RADIO PART I – AERIAL – Part Nos. 3024E-18828-A   Fully Retractable
                                    3006E-18828-A   Semi-Retractable

RADIO PART II – LOUDSPEAKER – Part No. 3034E-19A014-B – Speaker Kit

RADIO PART III – RECEIVER   – Part No. 3024E-18K810-A   } Manual Radio
              – Mounting Kit – Part No. 3034E-18814-A
              – RECEIVER   – Part No. 3006E-18K810-A   } Push Button Radio
              – Mounting Kit – Part No. 3014E-18814-E

RADIO PART IV – SUPPRESSOR

ENGINE COMPARTMENT LAMP – Part No. 3014E-15A700-A

LUGGAGE COMPARTMENT LAMP – Part No. 3014E-19K500-A

AUTOMATIC INERTIA REEL SEAT BELTS – Part No. 3008E-7361200-C

STANDARD SEAT BELTS – Part No. 3014E-7361200-A

OVER-RIDERS – Part No. 3014E-18419-A

DUAL HORN – Part No. 3016E-13800-A

BADGE BAR – Part No. 3014E-19M515-A

MUD FLAPS – Part No. 3014E-16412-A

WING MIRRORS – Part No. 3014E-17682-A

RADIATOR BLIND – Part No. 3016E-8B127-A

BONNET LOCK – Part No. 3014E-19242-A

REVERSING LAMP – Part Nos. 3014E-15499-A
                            3014E-15K494-A Twin Lamp Conversion

CIGAR LIGHTER – Part No. 3016E-15A044-A

CORTINA - LOTUS

FOG AND ROAD LAMPS – Part Nos. 3014E-15200-A   – Spot Lamp
                               3014E-15K201-A   – Fog Lamp
                               3014E-15200-B   – Spot Lamp-Iodine Quartz
                               3014E-15K201-B   – Fog Lamp-Iodine Quartz
                               3014E-15204-A   – Lamp Mounting Kit

The terms 'left', 'right', 'front' and 'rear', used in these fitting instructions are all relative to the normal driving position.

Where any holes are drilled in the body, it is important that the appropriate colour touch-up paint is applied to the bare metal edges to prevent rust formation.

Tools Required

Crosshead screwdriver
Rule or tape
Soft lead pencil
Centre punch
$\frac{9}{16}$ in. (14.29 mm.) dia. drill
$\frac{3}{16}$ in. (4.76 mm.) dia. drill
$\frac{5}{16}$ in. A/F spanner

Fitting Instructions

1. Open bonnet and luggage compartment and disconnect the battery.
2. Mark out the position of the lamp mounting hole as illustrated.
3. Centre punch and drill a $\frac{9}{16}$ in. (14.29 mm.) diameter hole.
4. Locate the mounting plate, with the single hole at the bottom, over the hole drilled in operation 3, and using it as a template, mark the position of the three holes.
5. Centre punch and drill three $\frac{3}{16}$ in. (4.76 mm.) diameter holes.
6. Clip a spring nut over each of the three holes in the mounting plate, position the plate behind the front panel and secure it with three crosshead screws.
7. Locate the lamp on the front panel and, from behind, secure with a nut and shakeproof washer. Feed the wire from the lamp through the grommet into the engine compartment.
8. Carefully prise the switch mounting panel from in front of the gear lever.
9. Mark and drill a $\frac{1}{2}$ in. (12.7 mm.) hole 1 in. (25.4 mm.) from the side of the panel and 1 in. (25.4 mm.) from the front of the panel.
10. Screw the thin nut on to the lamp switch and, from the underside, pass the switch through the switch mounting panel.
11. Secure the switch with the slotted chrome bezel.
12. Unscrew the bright metal bezel retaining the ignition switch, remove the black spacer, push the switch forward and pull down below the dash panel.
13. Connect the blade terminal on the short white wire to one of the No. 4 terminals on the back of the ignition switch.
14. Reposition the ignition switch and secure with the black spacer and bright metal bezel.
15. Connect the four-way connector to the bullet terminal on the short white wire.
NOTE – When a second lamp is being fitted, the feed should be taken from the four-way connector to the lamp switch.
16. From one end of the long blue/yellow wire, cut a length of wire sufficient to reach from the four-way connector to the lamp switch.

CORTINA - LOTUS

17. Fit a bullet terminal to the "cut" end of the shorter length of blue/yellow wire and connect to the four-way connector.
18. Connect the blade terminal end of the shorter length of blue/yellow wire to one terminal of the lamp switch, feeding the wire inside the centre console.
19. Connect the remaining blade terminal on the longer length of blue/yellow wire to the other terminal on the lamp switch. Replace the switch mounting panel.
20. Feed the blue/yellow wire through one of the grommets in the engine compartment rear bulkhead and through the loom clips in the right-hand side of the engine compartment. Where the lamp has been fitted to the L.H. side of the car, pass the wire across in front of the radiator.
21. Fit a bullet terminal to the end of the wire and connect it to the wire from the lamp with a snap connector.
22. Reconnect the battery, check the alignment of the lamp and fully tighten the securing nut.

RADIO PART 1 – AERIAL – Part Nos. 3024E-18828-A – Fully Retractable
3006E-18828-A – Semi-Retractable

The terms 'left', 'right', 'front', 'rear', etc. used in these fitting instructions are all relative to the driving position.

Where any holes are drilled in the body it is essential that the appropriate colour FoMoCo touch-up paint is applied to the bare metal edges to prevent rust formation.

Tools Required

7/8 in. (22.23 mm.) diameter drill
3/4 in. (19.05 mm.) diameter drill
Centre punch
Soft lead pencil
Ruler
9/16 in. A.F. spanner

Fitting Instructions

1. Mark, centre punch and drill a 7/8 in. (22.23 mm.) diameter hole in the front wing. (See illustration).
2. Remove the four press studs securing the parcel shelf and remove the parcel shelf.
3. Remove the forward self-tapping screw from the bright metal trim plate, situated on bottom edge of the front door aperture on the aerial side.
4. Remove the interior trim panel in front of the front door on the same side as the aerial by removing the two plastic plugs and straightening the securing tab.
5. Locate the pimple which is under the side trim panel, and in line with the top and slightly in front of the reinforcing pressing (see illustration).
6. Using a centre punch flatten the pimple and drill a 3/4 in. (19.1 mm.) diameter hole.
7. Clean the underside of the wing in the area of the hole drilled for the aerial to ensure a good earth contact for the aerial.
8. Remove the chrome nut, chrome capping, black plastic spacer and rubber from the aerial.
9. From underneath the wing, pass the aerial lead through the hole drilled in the side panel and position the grommet, which is on the aerial lead, in the hole.
10. Pass the aerial into position through the hole in the wing and replace the rubber washer, black domed spacer, bright metal plate and nut.
11. Ensure that the aerial lead is not kinked and follows a smooth path.
12. Extend the aerial, position it vertically and tighten the securing nut.
13. Replace the side trim panel, bend over the securing tab, replace the two plastic plugs and the forward self-tapping screw in the bright metal trim plate on the bottom edge of the front door aperture.

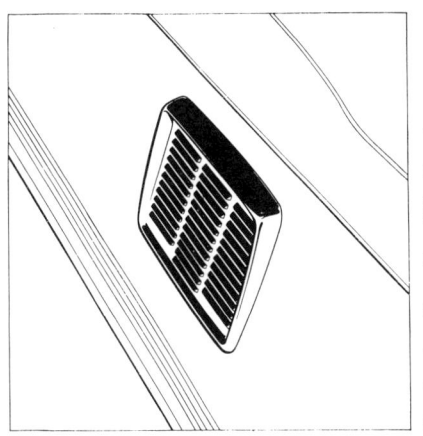

Speaker Location – Rear Trim Panel

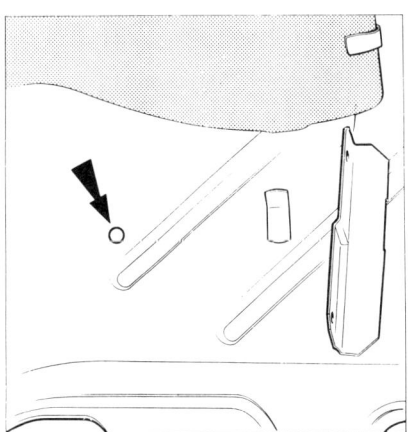

Aerial Lead Location

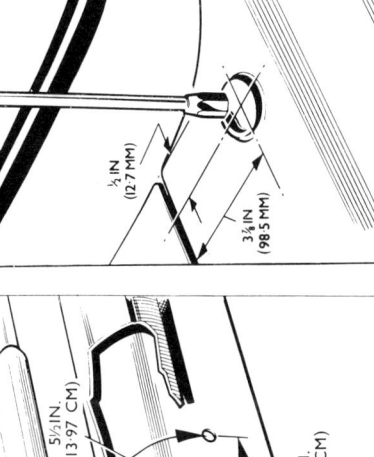

Location on Wing for Aerial

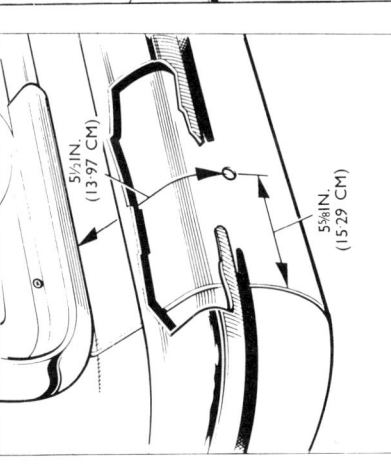

Fog/Road Lamp Location

14. Reposition the parcel shelf and secure with the four press studs.

RADIO PART II – LOUDSPEAKER – Part No. 3034E-19A014-B – Speaker Kit

1. Remove the rear seat cushion by lifting the front edge and pulling forward.
2. From inside the boot, remove the two nuts securing the top of the rear seat backrest.
3. From inside the car, remove the two set-screws securing the bottom of the rear seat backrest.
4. Prise up the metal tabs securing the front edge of the rear parcel shelf trim panel and remove the trim panel.
5. With a sharp knife, cut out the speaker aperture marked on the back of the panel.
6. Replace the trim panel and secure along the front edge by bending down the metal tabs.
7. Position the speaker grille, with the louvres facing rearwards, over the aperture and place the securing screws in the four holes in the grille. See illustration.
8. From inside the boot, locate the speaker over the securing screws, with the speaker leads on the L.H. side of the car. Secure the speaker and grille with four nuts and shakeproof washers.
9. Connect the speaker loom to the blade terminals on the speaker. The wires can be either way.
10. Remove the kick plate from the bottom of the L.H. side door, secured by crosshead screws.
11. Feed the speaker lead from the boot, down the L.H. side of the car, under the floor covering to the front of the parcel shelf and back to the centre of the facia.
12. Replace the kick plate and secure with crosshead screws.
13. Replace the rear seat backrest, secure the bottom with two set-screws and the top, from inside the boot, with two nuts.
14. Replace the seat cushion.

RADIO PART III – RECEIVER – MANUAL – Part Nos. 3024E-18K810-A 3034E-18814-A Mounting Kit

Tools Required

Crosshead Screwdriver
Sharp Knife

Fitting Instructions

1. Disconnect the battery.
2. Secure the receiver mounting bracket to the 2 rearward of the 4 holes on the underside of the facia with 2 self-tapping screws and spring nuts.
3. Fit the fuse holder on the end of the white/pink wire, and connect them to the fuse holder cap on the end of the white/pink wire from the receiver.

4. Remove the heater cover panel by removing 4 screws and washers. Then cut out and remove the area indicated by perforation on the back of the panel.
5. Replace the heater cover panel and secure with the 4 screws and washers.
6. Secure the black earth wire from the receiver to one of the forward holes on the underside of the facia with a self-tapping screw and spring nut.
7. Connect the single end of the red connecting loom to the bullet connector on the red pilot light wire from the receiver, connect the 2 speaker lead connections to the speaker loom on the receiver and plug in the aerial lead into the "front" of the receiver (see illustration).
8. Fit the black earth wire to the spade terminal to the "front" of the receiver.
9. Position the receiver in the mounting bracket, by first fitting 2 spacer washers on the receiver spindles, then pass the receiver spindles through the 2 holes in the mounting bracket and secure with the 2 internal tooth washers and locknuts.
10. Unscrew the bright metal bezel retaining the ignition switch, remove the black spacer, push the switch through the facia panel and pull down below the facia.
11. Connect the blade terminal on the end of the white/pink wire to the ACC or No. 4 terminal on the back of the ignition switch.
12. Replace the ignition switch with the white spacer behind the facia and the black spacer in front. Secure with the bright metal bezel.
13. Pull the bunch of snap connectors (tail lamp loom to main loom) down from the underside of the facia panel at the drivers side, and insert the red pilot light connecting loom between the red tail lamp loom and its snap connection.
14. Position the control facia on the radio spindles, then fit the black plastic shroud and secure it with the 2 flat washers and 2 locknuts, and fit the 2 control knobs to the spindles.
15. Reconnect the battery.

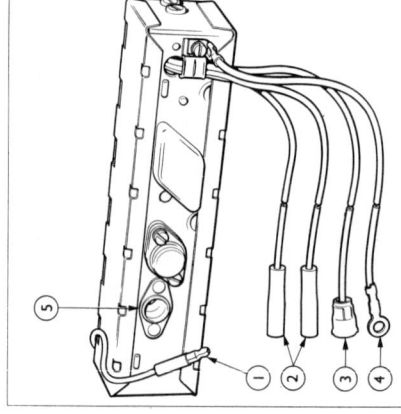

Receiver Connections

1. Pilot Light Wire
2. Speaker Wire Connections
3. Power Wire Connection
4. Earth Wire
5. Aerial Socket

CORTINA - LOTUS

16. Turn the ignition switch to the ACC position. Switch on the radio and select a weak station, about 200M medium wave band.

17. Pull off the wave band push button and adjust the fine tuning screw situated alongside the wave band control to obtain maximum volume. It should not be necessary to turn the screw more than 1/2 a turn in either direction.

18. Refit the wave band push button.

RADIO PART III – RECEIVER – Push Button – Part Nos. 3006E-18K810-A
3014E-18814-E – Mounting Kit

Tools Required
Crosshead Screwdriver
Knife

Fitting Instructions

1. Disconnect the battery.

2. Secure the receiver mounting bracket to the four holes on the underside of the facia with 4 self-tapping screws and spring nuts.

3. Fit the fuse to the fuse holder on the end of the white/pink wire, and then to the fuse holder cap on the end of the white/pink wire from the receiver.

4. Remove the heater cover panel by removing 4 screws and washers. Then cut out and remove the area indicated by perforations on the back of the panel.

5. Replace the heater cover panel and secure with 4 screws and washers.

6. Secure the black earth wire from the receiver to one of the forward holes in the underside of the facia with a self-tapping screw and spring nut.

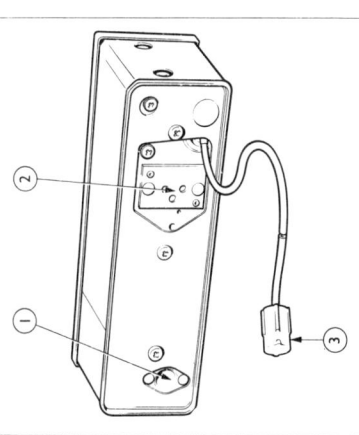

Pilot Light Loom Connection

7. Fit the aerial and speaker leads into the "front" of the receiver and fit the receiver to the mounting bracket with the 4 set screws and washers. Two each side.

8. Unscrew the bright metal bezel retaining the ignition switch, remove the black spacer, push the switch through the facia panel and pull down below facia.

9. Connect the blade terminal on the end of the white/pink wire to the ACC or No. 4 terminal on the back of the ignition switch.

10. Replace the ignition switch with the white spacer behind the facia and the black spacer in front. Secure with the bright metal bezel.

11. Reconnect the battery.

12. Fit the knobs to the radio spindles. Turn the ignition key to the ACC position, switch on the receiver, push in one of the push buttons and adjust the receiver to select a weak station around 200 metres medium wave band.

13. With a small screwdriver adjust the aerial tuning screw, situated on the right-hand side of the receiver (see illustration) to obtain maximum volume. It should not be necessary to turn the screw more than 1/2 a turn in either direction.

14. Remove the control knobs. Fit the two spacers and the black plastic cover over the spindles and secure with 2 washers and locknuts. Secure the "front" lower edge of the cover to the receiver with 2 set screws.

15. Fit the inner and outer knobs to the receiver and tune each push button in turn to the required station.

RADIO PART IV – SUPPRESSOR

1. Lift the bonnet and pull the blade connector from the large terminal on the generator. The terminal clip on the suppressor wire clips to the blade connector under the plastic cover. Replace the blade connector to the large terminal.

2. Secure the suppressor to the small hole in the generator end plate with a self-tapping screw.

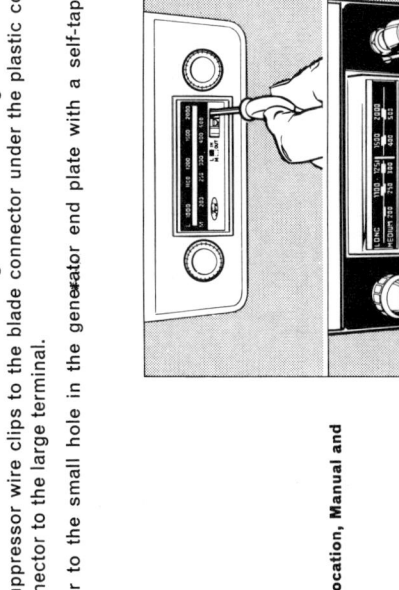

Fine Tuning Screw Location, Manual and Push Button

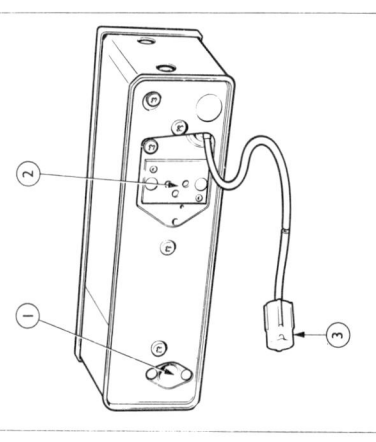

Receiver Connections
1. Aerial Socket
2. Speaker Socket
3. Power Wire Connection

CORTINA - LOTUS

ENGINE COMPARTMENT LAMP – Part No. 3014E-15A700-A

The terms 'left', 'right', 'front' and 'rear', used in these fitting instructions are all relative to the normal driving position.

Where any holes are drilled in the body, it is important that the appropriate colour touch-up paint is applied to the bare metal edges to prevent rust formation.

Tools Required

⅛ in. (3.17 mm.) dia. drill
Centre punch
Crosshead screwdriver
Small screwdriver
Knife

Fitting Instructions

1. Open the bonnet and boot, disconnect the battery and mark a line on the engine rear bulkhead as illustrated.

2. Position the lamp bracket, with the lamp pointing down, the bracket mounting holes on the line marked in operation one and mark the position of the holes. Centre punch these locations.

3. Using a ⅛ in. (3.17 mm.) drill, drill the holes centre punched in the previous paragraph and secure the lamp with the two self-tapping screws.

4. Position the switch bracket and, using it as a template, mark the position of the two mounting holes.

5. Centre punch and drill two ⅛ in. (3.17 mm.) diameter holes.

6. Secure the bracket to the bulkhead by the screws and washers provided so that the smooth face of the bracket is facing downwards.

7. Secure the switch to the bracket so that the plunger is uppermost.

8. Route the red wire from the lamp to the switch, using the clips provided to retain the wire to the edges of the sealing rubber channel.

9. Cut the wire to the length required, bare the end and secure to one of the screw terminals on the switch.

10. Bare the end of the remaining piece of wire and secure to the other switch terminal.

11. Pull the red wire from the four-way connector halfway along the L.H. engine side apron panel. Fit the nipple on the end of the wire from the switch and connect the wire removed from the four-way connector to the snap connector attached to the nipple end of the wire from the switch.

12. Reconnect the battery and test the lamp. It should go out when the bonnet is closed.

LUGGAGE COMPARTMENT LAMP – Part No. 3014E-19K500-A

The terms 'left', 'right', 'front' and 'rear', used in these fitting instructions are all relative to the normal driving position.

Where any holes are drilled in the body, it is important that the appropriate colour touch-up paint is applied to the bare metal edges to prevent rust formation.

Tools Required

Crosshead screwdriver
Small screwdriver
⅛ in. (3.17 mm.) dia. drill
Centre punch
Rule
Knife

Fitting Instructions

1. Open the boot and disconnect the battery.

2. The lamp bracket is located on the brace on the underside of the rear parcel tray support pressing. This is located immediately in front of the upper edge of the luggage compartment aperture and is in a central position.

3. Mark a line on the underside of the rear parcel tray as illustrated.

4. Using the bracket as a template, align the two mounting holes on the line drawn in operation 3 and mark the position of the holes.

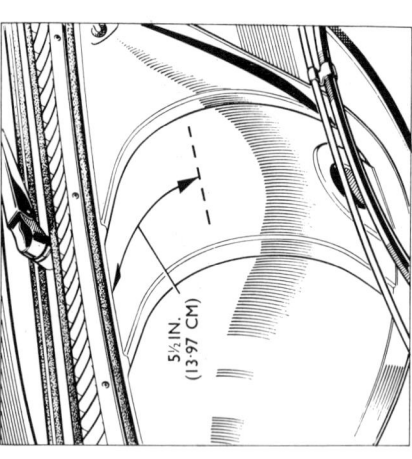

Engine Compartment Lamp Location

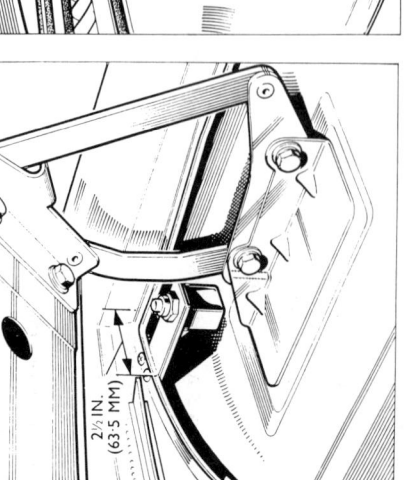

Engine Compartment Lamp Switch Location

5. Centre punch and drill two ⅛ in. (3.17 mm.) diameter holes.

6. Secure the bracket, with the lamp facing downward, with two self-tapping screws.

7. Lift the rubber sealing strip from the R.H. 'front' corner of the luggage compartment surround, locate the switch bracket and mark the position of the mounting holes.

8. Centre punch and drill two ⅛ in. (3.17 mm.) diameter holes.

9. Secure the bracket with two self-tapping screws and fit the switch to the bracket with the plunger upward.

10. Route the red wire from the lamp to the switch, using the clips provided to retain the wire to the edges of the luggage compartment reinforcement ribs.

11. Cut the wire to the length required, bare the end and secure to one of the screw terminals on the switch.

12. Bare the end of the remaining piece of wire and secure to the other switch terminal.

13. Pull out one of the red wires from the snap connector in the bunch of wires adjacent to the rear of the R.H. rear lamp assembly. Connect in its place the nipple on the end of the wire from the switch. Connect the wire previously disconnected to the snap connector attached to the nipple end of the wire from the switch.

14. Reconnect the battery and test the lamp. It should go out when the luggage compartment lid is closed.

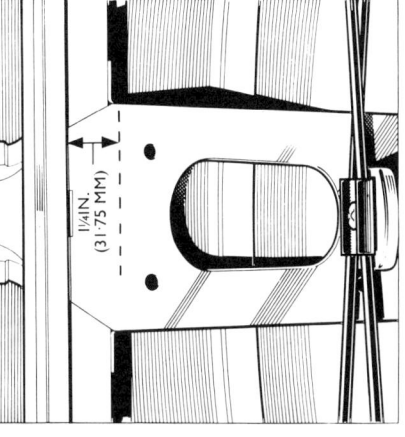

Luggage Compartment Lamp Switch Location

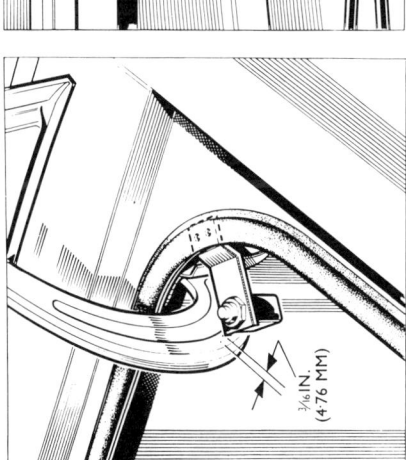

Luggage Compartment Lamp Location

AUTOMATIC INERTIA REEL SEAT BELTS – Part No. 3008E-7361200-C

The terms 'left', 'right', 'front' and 'rear' used in these fitting instructions are all relative to the normal driving position.

Where any holes are drilled in the body, it is important that the appropriate colour touch-up paint is applied to the bare metal edges to prevent rust formation.

Tools Required

Rule
Sharp knife
Small screwdriver
¾ in. A/F spanner
⅝ in. A/F spanner
11/16 in. open ended spanner
Torque wrench
½ in. (13 mm.) dia. drill (Cortina 2 door)

Fitting Instructions

NOTE – Seat belt mounting points are built into the car and are located as follows:—

(a) Short seat belt mounting—

On the side of the transmission tunnel, 7¼ in. (184 mm.) forward of the heel plate below the front of the rear cushion, and 3 in. (76 mm.) below the top of the drive shaft tunnel (dimensions D and E).

(b) Pillar loop mounting—

On the door lock pillar. A black plastic button is fitted to the pillar.

(c) Reel mounting—

At the base of the door lock pillar, 3¾ in. (95 mm.) in front of the heel plate below the front of the rear cushion, and 1 in. (25 mm.) below the bottom of the rear quarter trim panel (dimensions B and C).

To gain access to the mounting points described in these fitting instructions, it will be found easiest to cut a cross in the trim or carpet where indicated; the corners of the cut can then be turned back. When the seat belt has been fitted, the trim or carpet can be repositioned to give a neat appearance.

The seat belt is installed as follows:—

1. Carefully expose all mounting points as described.

2. Position the short seat belt and assemble to the floor in the following order: plain washer, stepped spacer, seat belt plate and spring washer. Secure with the ¾ in. A/F bright metal bolt and tighten to a torque of 15 to 20 lb. ft.

3. Position the reel mounting bracket in the following order, mounting bracket and toothed washer. Secure with the ⅝ in. A/F bolt and tighten to a torque of 40 to 45 lb. ft.
   NOTE – **Before fully tightening the reel mounting bracket securing bolt, ensure that the base plate of the bracket is horizontal.**

4. Position the reel on the mounting bracket and tighten the self-locking nut with an 11/16 in. A/F open ended spanner.

5. Position the pillar loop in the following order, stepped spacer, pillar loop plate and plain washer. Secure with the ¾ in. A/F bright metal bolt and tighten to a torque of 20 to 22 lb. ft.

CORTINA - LOTUS

MOST IMPORTANT – **This seat belt has been designed specifically as an item of safety equipment. Therefore, any attempt to modify either the way in which the belts are mounted or the mechanism itself will endanger the wearer. On no account should this belt be worn by children.**

NOTE – In case of any difficulty or queries arising in the wearing of the belts, your nearest Ford Dealer should be contacted.

STANDARD SEAT BELTS – Part No. 3014E-7361200-A

The terms 'left', 'right', 'front' and 'rear' used in these fitting instructions are all relative to the normal driving position.

Where any holes are drilled in the body, it is important that the appropriate colour touch-up paint is applied to the bare metal edges to prevent rust formation.

Tools Required

Rule
Sharp knife
Crosshead screwdriver
¾ in. A/F Ring spanner
⅝ in. A/F Ring spanner

Fitting Instructions

Seat belt mounting points are built into the car and are located as follows:—

(a) Short seat belt mounting point
    On the side of the transmission tunnel to the rear of the front seats.

(b) Lower pillar mounting point
    At the base of the door pillar, just to the rear of the front seats.

(c) Upper pillar mounting point
    At the top of the door pillar.

1. Lift the front edge of the rear seat cushion, pull forward to release and remove from the car.

2. Remove the crosshead screws securing the kick plates to the bottom of the door aperture(s) and remove the kick plate(s).

3. Lift the floor covering to expose the short belt and reel mounting points. Where carpets are fitted, they are secured at the edges with an adhesive and care is needed when easing the carpet away from the floor.

4. Prise the plastic buttons from the floor mounting. The short seat belt mounting plastic buttons are fitted from underneath the car, and are removed by pushing them out from inside the car. With a sharp knife carefully cut a small hole approximately ½ in. (1.27 cm.) diameter in the floor covering to line up with the short belt mounting hole.

5. Carefully cut a ½ in. (1.27 cm.) diameter hole in the floor covering to line up with the lower pillar mounting hole to allow the mounting bolt to pass through.

6. Reposition floor covering and replace the kick plates and rear seat cushion.

Automatic Inertia Reel Seat Belt

7. Lay the long seat belt over the seat and secure the lower end of the belt to the lower pillar mounting, with a spacer between the plate and the body.

8. Prise the plastic button from the upper part of the door pillar and secure the upper pillar mounting with a spacer under the mounting plate and the plastic hook under the bolt head.

MOST IMPORTANT – **This seat belt has been designed specifically as an item of safety equipment. Therefore, any attempt to modify either the way in which the belts are mounted or the mechanism itself, will endanger the wearer. On no account should this belt be worn by children.**

NOTE – In case of any difficulty or queries arising in the wearing of the belts, your nearest Ford Dealer should be contacted.

OVER-RIDERS – Part No. 3014E-18419-A

The terms 'left', 'right', 'front' and 'rear', used in these fitting instructions are all relative to the normal driving position.

Tools Required

½ in. A/F spanner

Fitting Instructions

1. Fit two lengths of rubber "anti-squeak" to each over-rider. This "anti-squeak" may be softened in warm water if it has become hard in storage.

2. Remove the two chrome headed bolts from the front bumper bar and secure the front over-riders, using the original nuts and washers and the new bolts.

   They should be fitted with the securing bolts sloping downwards.

3. Remove the two inner chrome headed bolts from the rear bumper and secure the rear over-riders, using the original nuts and washers and new bolts.

   The original bolts from the bumpers are not used again and may be discarded.

DUAL HORN – Part No. 3016E-13800-A

The terms 'left', 'right', 'front' and 'rear', used in these fitting instructions are all relevant to the normal driving position.

Tools Required

Crosshead screwdriver
¼ in. A/F spanner

Fitting Instructions

1. Open the bonnet and disconnect the battery.

2. Remove the ten crosshead screws, five each side, securing the headlamp bezels.

3. Remove the six rivets and two screws securing the radiator grille, and remove the grille.

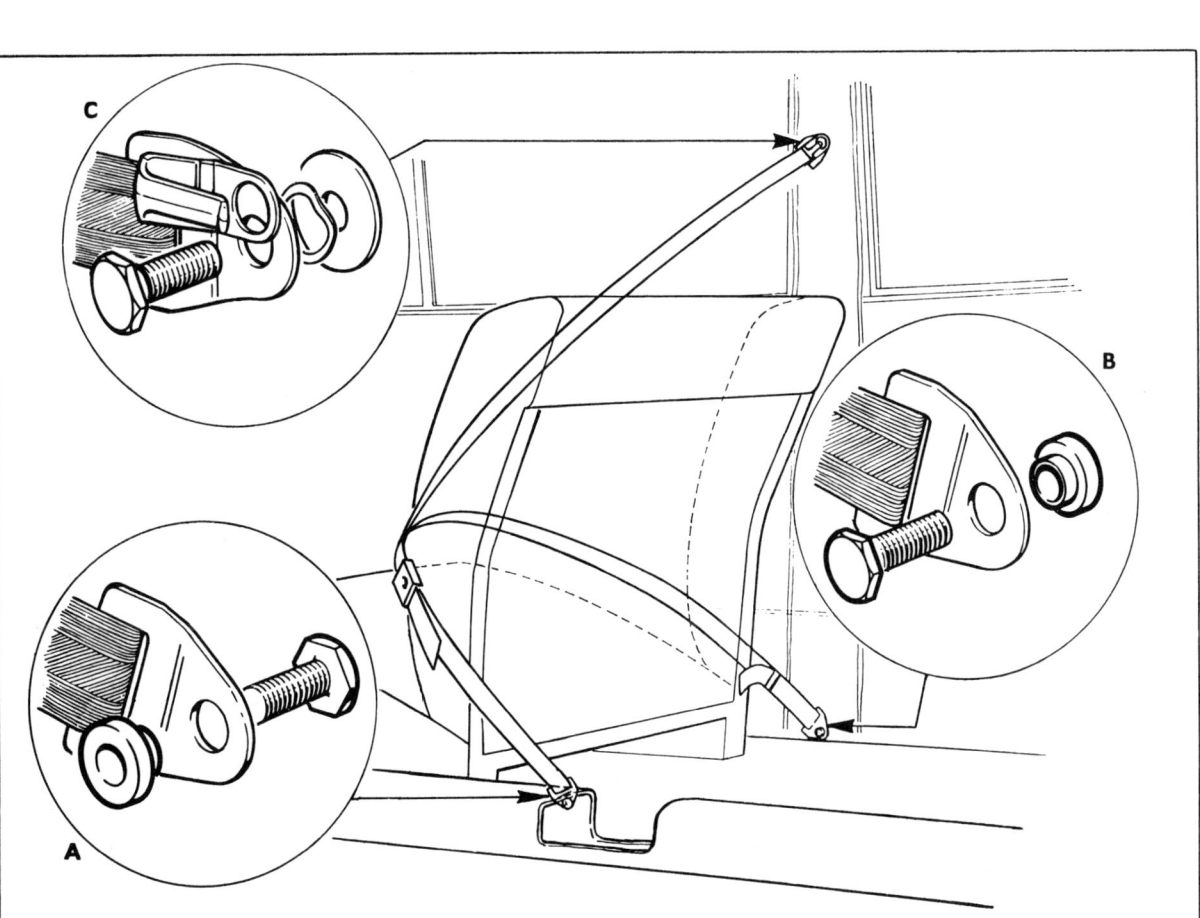

Standard Seat Belt

CORTINA - LOTUS

4. Disconnect two wires and remove the existing horn, secured with two bolts, and reposition in the same relative position on the opposite side of the grille opening. Secure to the body with the two bolts.

5. Connect the two wires in the loom supplied to the two terminals on this horn and route the loom across the grille opening, securing it with three loom clips.

6. Fit the new horn in place of the existing horn and secure it with two bolts.

7. Connect the new loom to the two outer terminals on the new horn, connect the original horn wires to the inner two terminals.

8. Reconnect the battery and test the operation of the horns.

9. Replace the radiator grille and secure with two crosshead screws, and either pop-rivets or self-tapping screws.

10. Replace the headlamp bezels and secure with ten crosshead screws, five each side.

BADGE BAR – Part No. 3014E-19M515-A

The terms 'left', 'right', 'front' and 'rear', used in these fitting instructions are all relative to the normal driving position.

Tools Required

½ in. A/F spanner

Fitting Instructions

1. Remove the two nuts and spring washers securing the front bumper bar.

2. Locate the badge bar with the two mounting lugs behind the bumper and over the bumper securing bolts.

3. Replace the nuts and washers and fully tighten.

MUD FLAPS – Part No. 3014E-16412-A

The terms 'left', 'right', 'front' and 'rear', used in these fitting instructions are all relative to the normal driving position.

Tools Required

$\frac{3}{16}$ in. (4.76 mm.) disc drill
Centre punch
Crosshead screwdriver
Rule

Fitting Instructions

1. Fit the metal brackets to the mud flaps and secure each bracket with a screw, flat washer, spring washer and nut.

2. Offer the mud flap to the car as illustrated, with the mud flap four to five inches from the near edge of the tyre and four to five inches from the ground.

3. Mark the centres of the mounting holes.

4. Jack up the car, fit stands and remove the rear wheels.

5. Centre punch the mounting hole positions and drill the four $\frac{3}{16}$ in. (4.76 mm.) diameter holes.

6. Secure the mud flaps with two screws, flat washers, spring washers and nuts each.

7. Replace the rear wheels and lower car to the ground.

WING MIRRORS – Part No. 3014E-17682-A

The terms 'left', 'right', 'front' and 'rear' used in these fitting instructions are all relative to the normal driving position.

Where any holes are drilled in the body, it is important that the appropriate colour touch-up paint is applied to the bare metal edges, to prevent rust formation.

Tools Required

Rule or tape measure
Centre punch
Adjustable spanner (with the jaws covered to prevent damage to chrome)
$\frac{11}{16}$ in. (17.46 mm.) dia. drill

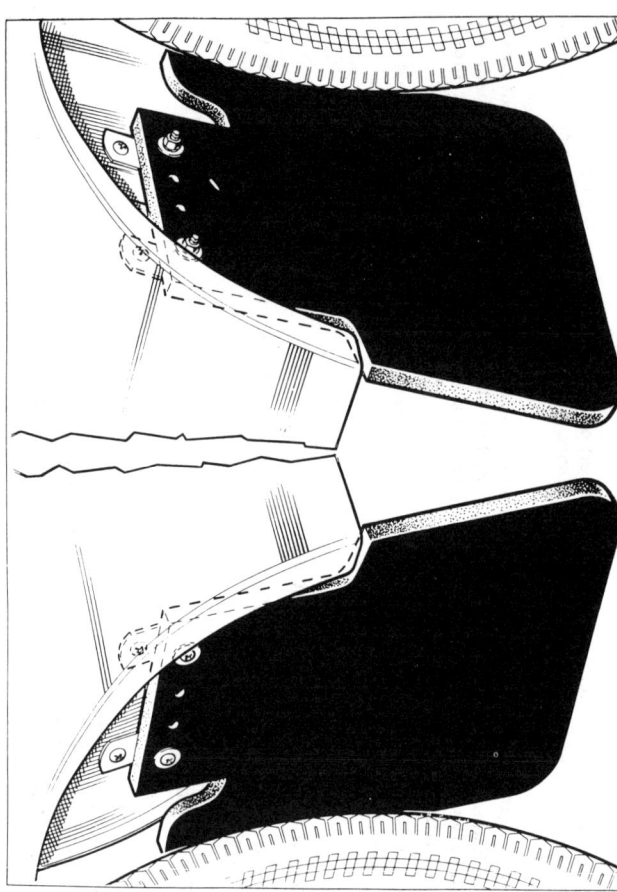

Mud Flap Location

Fitting Instructions

1. Mark out the hole location on the wing as illustrated.

   NOTE – The L.H. wing is shown, the R.H. wing is diametrically opposite.

2. Centre punch and drill an $\frac{11}{16}$ in. (17.46 mm.) dia. hole at each location.
3. Remove the nut and washer from the wing mirror stem.
4. Position the wing mirror in the hole, with the rubber washer on top of the wing.
5. From under the wing, replace the nut and washer. Hold the mirror in the required position and fully tighten the securing nut.
6. Slacken the nut securing the mirror to the stem, adjust the mirror to the required position and fully tighten the nut.

RADIATOR BLIND – Part No. 3016E-8B127-A

The terms 'left', 'right', 'front' and 'rear' used in these fitting instructions are all relative to the normal driving position.

Where any holes are drilled in the body, it is important that the appropriate colour touch-up paint is applied to the bare metal edges to prevent rust formation.

Tools Required

$\frac{1}{8}$ in. (3.18 mm.) dia. drill
$\frac{7}{32}$ in. (5.56 mm.) dia. drill
$\frac{3}{16}$ in. (4.76 mm.) dia. drill
Crosshead screwdriver
Rule
Centre punch

Wing Mirror Location

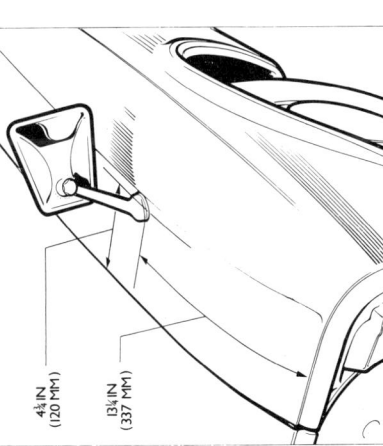

4¾ IN (120 MM)
13¼ IN (337 MM)

Fitting Instructions

1. Remove the four bolts securing the radiator to the engine compartment front bulkhead.
2. Thread the inner cable through the hole in the top of the blind frame. Slide the ferrule and outer cable over the inner cable.
3. Draw the radiator rearwards sufficiently to allow the blind assembly to be passed down in front of the radiator, with the operating cable pointing upwards. The lugs on the blind lower mounting holes should be facing forwards.
4. Line up the holes in brackets on the blind assembly with the holes in the bulkhead and the radiator mounting brackets. Secure the radiator and blind assembly with the four bolts supplied, using the original washers.
5. Prise open the two cable clips and fit them over the outer cable and squeeze the clips shut. Feed the cable through the front three wiring loom clips on the left-hand engine side apron panel.
6. Position the two clips on the radiator grille upper panel, allowing the cable to follow a natural curve. Using the clips as templates mark and drill two $\frac{1}{8}$ in. (3.18 mm.) dia. holes. Secure the clips with the self-tapping screws provided.
7. Using a pencil and rule, mark a horizontal line on the engine compartment rear bulkhead 2 in. (50 mm.) below the choke cable grommet.
8. Mark a vertical line on the bulkhead 2 in. (50 mm.) to the left of the choke cable grommet. Where the two lines meet, centre punch and drill a $\frac{7}{32}$ in. (5.56 mm.) dia. hole in the bulkhead.
9. Remove one nut from the adaptor unit and from the engine compartment fit the adaptor in the hole previously drilled. From inside the car replace the nut. Do not fully tighten at this stage.
10. Offer up the control bracket to the angled face of the facia panel immediately below the choke control knob, with the holes in the bracket 3 in. (76 mm.) "rearward" of the "forward" edge of this face.
11. Using the bracket as a template mark off and drill two $\frac{3}{16}$ in. (4.76 mm.) dia. holes and secure the bracket with the nuts and screws provided.
12. Pass the inner cable through the adaptor ensuring that the outer cable seats correctly in the adaptor.
13. Pass the ball chain through the large hole in the control bracket and connect it to the inner cable.
14. Adjust the adaptor until, with the blind in the fully down position, the operating ball is firmly positioned against the large hole in the bracket. If the adaptor does not provide enough adjustment remove two or three links of the ball chain.
15. Lubricate the cable and operate the blind several times. Revolve the adaptor until its cranked end lines up with the control bracket then finally tighten the securing nut.

BONNET LOCK – Part No. 3014E-19242-A

The terms 'left', 'right', 'front' and 'rear' used in these fitting instructions are all relative to the normal driving position.

CORTINA - LOTUS

Where any holes are drilled in the body, it is important that the appropriate colour touch-up paint is applied to the bare metal edges to prevent rust formation.

Tools Required

$\frac{7}{16}$ in. A/F spanner
$\frac{1}{2}$ in. A/F spanner
$\frac{9}{16}$ in. A/F spanner
$\frac{9}{64}$ in. (3.6 mm.) dia. drill
$\frac{3}{16}$ in. (4.76 mm.) dia. drill
$\frac{3}{8}$ in. (9.53 mm.) dia. drill
Centre punch
Rule

Fitting Instructions

1. Remove the two top bolts securing the radiator and slacken the two lower bolts. This will allow the radiator to be moved rearwards to gain access to the lock.

2. Remove the three bolts securing the lock to the front bulkhead and remove the lock.

3. Remove the two crosshead screws securing the chromium push button to the operating arm and remove the button.

4. Shorten the lock operating arm by cutting off a piece $2\frac{1}{4}$ in. (57 mm.) long.

5. The lock release handle bracket is located on the trim panel at the right-hand end of the parcel tray.

   Hold the wooden packing vertical, place the release handle bracket at the bottom of the packing, with the open part of the slot uppermost, mark on the packing the position of the two holes in the bracket and drill two $\frac{3}{16}$ in. (4.76 mm.) dia. holes.

6. Locate the packing on the trim panel with the top flush with the top of the trim panel and the 'rear' edge of the packing $2\frac{1}{4}$ in. (57 mm.) forward of the 'rear' edge of the trim panel. Mark the position of the two holes on the trim panel and drill two $\frac{9}{64}$ in. (3.6 mm.) dia. holes through the trim panel and the metal panel underneath.

7. Position the packing behind the trim panel and secure the bracket with self-tapping screws.

8. Mark off and drill a $\frac{3}{8}$ in. (9.53 mm.) hole on the right-hand side of the engine rear bulkhead, this hole should be $1\frac{1}{4}$ in. (32 mm.) in from the engine side apron panel and $\frac{3}{4}$ in. (19 mm.) above the top of the bracket to which the hydraulic brake pipes are clipped.

9. Mark off and drill a $\frac{3}{8}$ in. (9.53 mm.) hole in the right-hand radiator support panel directly above the lower of the two horn mounting nuts and 1 in. (25 mm.) below the top flange.

10. Pass the nipple end of the cable through the hole in the bulkhead, behind the loom clips and through the hole in the radiator support panel. Place the two grommets over the cable and locate one in each of the drilled holes. Slacken the release handle securing the nut, slide the cable into the bracket with the lockwasher to the "front" of the bracket and fully tighten the locknut.

11. Pass the inner cable through the clip in the end of the channel section fixed to the lock cover so that the inner cable lays in the channel.

12. Push the outer cable into the clip so that it is firmly held.

13. Attach the cable to the lock by passing the nipple end of the cable through the shackle. Unhook the return spring from the lock lever, insert the clevis pin through the shackle and lock lever, and hook the return spring through the lower hole of the clevis pin.

14. With a hacksaw, cut off the 'right' hand side of the chrome hood release button mounting flange. Hold the button in its original position and secure the remaining flange to the 'left' hand slot in the bonnet lock support panel, using the long screw, spacer, flat and spring washers.

15. Fit the lock into the lock cover and put the assembly into the original lock position and loosely replace the bolts.

16. Lower the bonnet to position the lock assembly. Open the bonnet and fully tighten the three bolts. (Because the lock assembly is slightly lower, it may be necessary to re-set the lock post in the bonnet. This can be done by slackening the large locknut and turning the post with a screwdriver until it is low enough to engage with the lock, and then retighten the locknut.)

17. Refit and tighten the radiator securing bolts, and tighten nut behind facia bracket.

18. Check the lock operation.

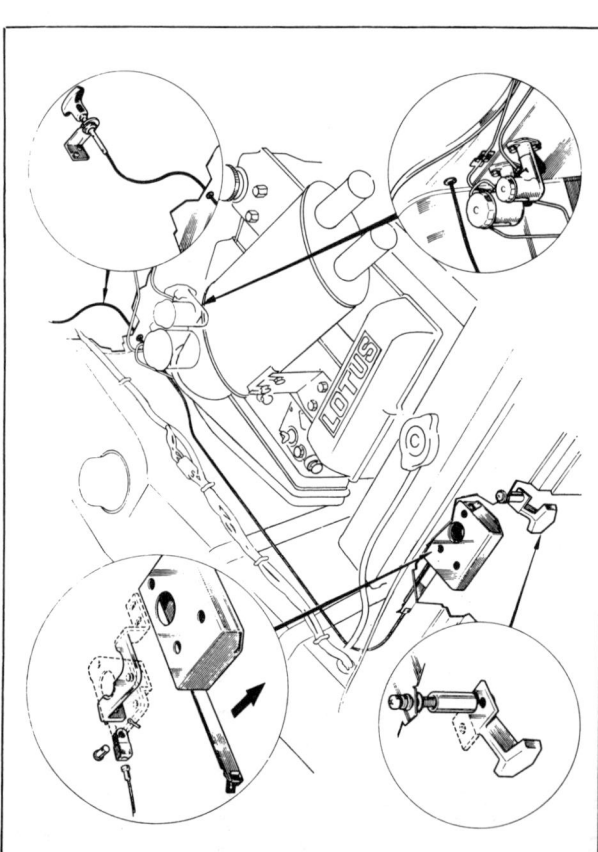

Bonnet Lock Cable

REVERSING LAMPS – Part Nos. 3014E-15499-A
3014E-15K494-A – Twin Lamp Conversion

The terms 'left', 'right', 'front' and 'rear' used in these fitting instructions are all relative to the normal driving position.

Where any holes are drilled in the body, it is important that the appropriate colour touch-up paint is applied to the bare metal edges to prevent rust formation.

Tools Required

Rule
Centre punch
Soft leaded pencil
$\frac{1}{4}$ in. (6.35 mm.) drill

Fitting Instructions

1. Open the boot and disconnect the battery.

2. Locate and remove the rubber grommet immediately below the bumper bar and approximately 3 in. from the R.H. support bracket.

   NOTE – If a second lamp is being fitted, the grommet is located in the same relative position on the L.H. side of the car.

3. Remove the nut, washer and one spacer from the lamp.

4. Feed the lamp wires and securing bolt through the mounting hole.

5. Replace the spacer, washer and nut. Position the spacer on the outside of the car with the thick portion at the bottom, and the spacer on the inside, with the thick portion at the top.

6. Clip the nipple on the end of one wire from the reversing lamp into the snap connector in the long reversing lamp lead.

   NOTE – If a second reversing lamp is fitted, the nipple end of one wire from this lamp fits into the other half of the same snap connector.

7. Secure the earth wire with the round connector to the earthing screw in the rear centre of the boot floor adjacent to the reinforcing strut.

8. Remove the rear seat cushion, by lifting and pulling forward the front edge, the R.H. door aperture kick plates, route the reversing lamp wire under the floor covering, to the 'front' of the facia.

9. Unscrew the bright metal bezel retaining the ignition switch, remove the black spacer, push the switch forward and pull down below the dash panel.

10. Locate the switch bracket underneath the R.H. ventilator, on the metal reinforcement along the rear edge, to the outside of the steering column. Mark and centre punch the position of the two mounting holes.

11. Drill a $\frac{1}{8}$ in. (3.17 mm.) diameter hole at each location.

12. Fit the two clip nuts over the bracket mounting holes, position the bracket and secure with two screws and washers.

13. Unscrew the bezel from the switch, position the switch in the R.H. hole in the bracket and secure with the bezel.

14. Push the warning lamp into the remaining hole in the bracket.

15. Connect the blade terminal with two wires from it on the reversing lamp lead to one terminal on the lamp switch.

16. Connect the blade terminal on the short wire connected to the reversing lamp lead to one terminal on the warning lamp.

17. Connect the blade terminal on the wire having a round terminal on the other end, to the remaining terminal on the warning lamp. Suitably earth the round terminal on this wire.

18. Connect one end of the remaining wire to the lamp switch, and the other end to the "Acc" or No. 4 terminal on the ignition switch.

19. Replace the ignition switch.

20. Reconnect the battery, test the operation of the lamps, align the lamps and fully tighten the lamp securing nuts.

CIGAR LIGHTER – Part No. 3016E-15A044-A

The terms 'left', 'right', 'front' and 'rear' used in these fitting instructions are all relative to the normal driving position.

1. Disconnect the battery.

2. Push a suitable rod into the hole on the underside of the windscreen washer/wiper control knob until the retaining pin is released then pull off the knob.

3. Similarly remove the heater control knobs.

4. Unscrew and withdraw the washer/wiper bezel.

5. Remove the heater facia.

6. Suitably support the L.H. side of the facia and tap out the weakened area of plastic.
   NOTE – This can be ascertained by locating the clear plastic bezel (supplied with kit) within the L.H. plastic area.

7. Unscrew the inner lighter body from the outer body.

8. Connect the black wire, having spade connection, and the lilac wire, having a snap connector to the outer lighter body and the indicating lamp respectively.

9. Locate the outer body within, and from behind, the facia mounting panel.

10. Replace the facia and secure with windscreen washer/wiper control knob bezel.

11. Fit the clear plastic bezel to the lighter inner body, then screw the body into the outer body.

12. Connect the red/white wire to the inner body.

13. Feed the wires across the body to the ignition switch.

14. Connect the red/white wire to No. 4 terminal of the ignition switch.

15. Secure the black wire to the indicator sender unit bracket.

16. Remove the spade terminal from the lilac wire and fit a bullet connection.

17. Connect the lilac wire to the red/white double snap connector adjacent to the rear of the R.H. vent.

18. Reconnect battery, push in lighter and check operation.

HAZARD WARNING LIGHT – Part No. 3034E-15B584-A

The terms 'left', 'right', 'front' and 'rear', used in these fitting instructions are all relative to the normal driving position.

Tools Required

$\frac{7}{16}$ in. A/F spanner
$\frac{7}{32}$ in. (5.56 mm.) diameter drill

Fitting Instructions

1. Disconnect battery.
2. Fit the metal clip type nut to the bracket on the flasher unit.
3. Connect the plastic plug in the wiring loom supplied to the flasher unit.
4. Drill a $\frac{7}{32}$ in. (5.56 mm.) diameter hole in the cowl side trim panel (located in front of the right hand front door). (See illustration.)
5. Secure the flasher unit with the screw and plain washer.
6. Pull the group of snap connectors connecting the main loom to rear loom from behind the top 'rear' of the trim panel.
7. Disconnect the green/white wires and connect the green/white wires from the loom supplied. (See illustration.)
8. Disconnect the green/red wires and connect the green/red wires in the loom supplied. (See illustration.)
9. Connect the green/white, green/red and light green/brown wires to the hazard warning light switch. (See illustration.)
10. Unscrew the bright metal bezel securing the ignition switch and feed the ignition switch out below the facia panel.
11. Connect the brown wire in the loom supplied to terminal No. 5 on the ignition switch.
12. Temporarily reconnect the battery, test the operation of the hazard warning flashers and disconnect the battery.
13. Replace the ignition switch and secure with the bright metal bezel.
14. Locate the hazard warning light switch through one of the holes in the lower edge of the facia panel to the right of the steering column and secure with the bright metal bezel.
15. Reconnect the battery.

NOTE – EMERGENCY FLASHERS MUST ONLY OPERATE WITH THE IGNITION SWITCH IN THE OFF POSITION.

VIEW IN DIRECTION OF ARROW OF COWLSIDE TRIM PANEL SHOWING FLASHER UNIT MOUNTING HOLE

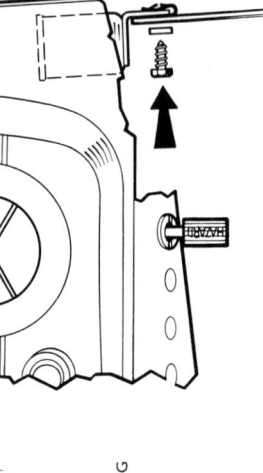

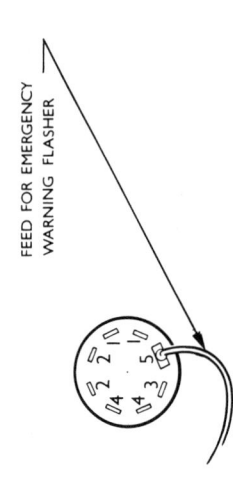

FEED FOR EMERGENCY WARNING FLASHER

REAR VIEW OF IGNITION SWITCH

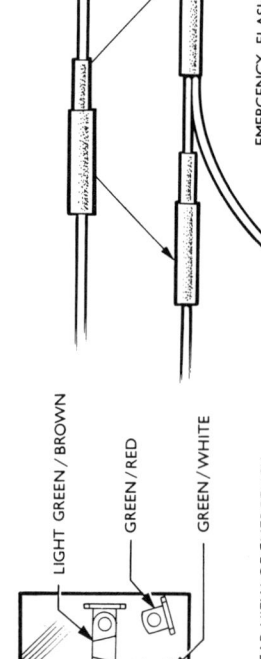

EMERGENCY FLASHER CONNECTIONS

LIGHT GREEN/BROWN
GREEN/RED
GREEN/WHITE

REAR VIEW OF EMERGENCY FLASHER SWITCH

CORTINA - LOTUS

OPERATING INSTRUCTIONS

The radio will only operate with the ignition switch in either the "auxiliary circuits" position or the "ignition and auxiliary circuits" position.

Switch on the radio by turning the left-hand inner control knob clockwise, a click will be heard as the switch operates. Select the required station by either manual or push-button selection.

Adjust the volume and tone controls to produce the best reception and to suit your individual requirements.

To switch the radio off turn the left-hand inner control knob anti-clockwise until a click is heard.

THE CONTROLS

1. Combined On/Off Switch and Volume Control

The left-hand inner control knob operates the on/off switch and varies the loudspeaker volume. The set is switched on by turning this knob clockwise, further movement in this direction progressively increases the volume; anti-clockwise rotation will reduce the volume and the set may be switched off by turning the control fully in this direction.

2. Tone Control

The left-hand outer control knob varies the tone. It should be turned fully clockwise for clearest reception of speech. Turning the control in an anti-clockwise direction will decrease the high note response, so producing a more mellow tone. Further rotation in the anti-clockwise direction will further reduce the high note response at the same time accentuating the bass.

3. Manual Tuning Control

Medium and long wave stations are tuned in by means of the right-hand inner control knob in conjunction with the illuminated tuning scale.

To change from the medium to the long wave band turn the right-hand outer control knob fully clockwise.

To change from the long to the medium wave band turn the right-hand outer control knob fully anti-clockwise.

Stations should always be tuned for maximum strength and the volume reduced, if necessary, by using the volume control.

4. Wave Change Control

The right-hand outer control knob operates the wave change switch. Anti-clockwise movement of the knob selects the medium wave band and clockwise movement the long wave band.

5. Push Button Controls

Any five medium wave or long wave stations may be pre-set for automatic selection by means of the push-button controls.

To set a push-button for a medium wave band station proceed as follows:—

1. Set the wave band switch to medium wave.
2. Pull out any one of the push-buttons to be set. This will unlock the selecting mechanism.
3. Using the manual tuning control knob accurately tune in the desired station.
4. Push the button in fully in order to set and lock the tuning.
5. Repeat the above procedure, selecting a different station for each push-button as required.

When a push-button has been set, as above, it is only necessary to depress this push-button to obtain automatically the required station.

To pre-set a station on the long wave band:—

Set the wave band to the long wave position and repeat the above operations.

It should be appreciated that only one station (either on the medium or long wave band) can be pre-selected by each button, that is a total of five stations.

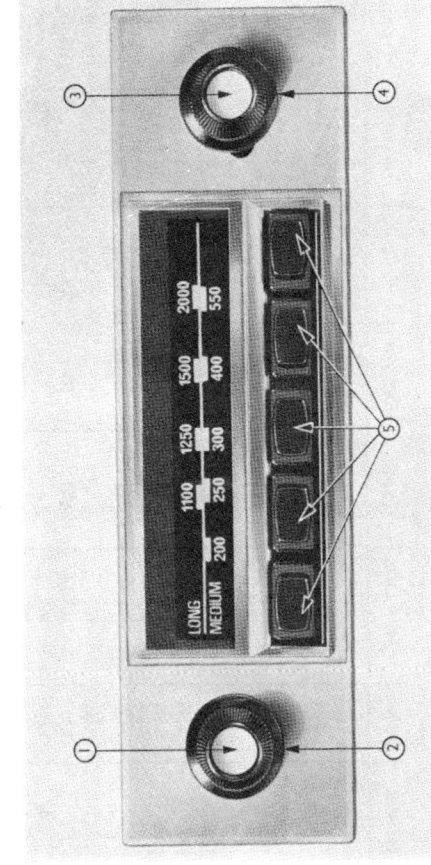

The Controls

CORTINA - LOTUS

DETAILED SPECIFICATION AND REFERENCE DATA

Type	8 Transistor, Superheterodyne
Power supply	12 volt, negative earth
Current consumption	0.25 amp at 13.5 volt input (No signal)
Fuse	1.5 amp cartridge type
Scale lamp	12v. 1.2 w. L.E.S.
Sensitivity	Under 15 microvolts for 0.5 w. output on M.W.
	Under 35 microvolts for 0.5 w. output on L.W.
A.G.C.	For 70 dB input change, the output changes by 10 dB
Power sound output	5 watts for 10% distortion at 400 Hz—13.5 volts input
Speaker	Moving coil 3 ohm impedance

Frequency Coverage
M.W.	520—1630 KHz 578— 184 metres
L.W.	150— 300 KHz 2000—1000 metres
Tuning	Push-button and Manual
Intermediate frequency	470 KHz

Transistors and Diodes
VT1 — BF 153 or BF 273	R.F. Amplifier
VT2 — BF 153	
VT3 — U 14711	Self-oscillating Mixer
VT4 — U 14711	... IF amplifier
VT5 — BC 113	
VT6 — BC 139	Audio amplifier
VT7 — AD 161	Audio driver
VT8 — AD 162	Complementary output amplifiers
D1 — BA 130	Detector Diode
D2 — BA 130	Damping Diode

Resistance of Components (Ohms)
Aerial filter choke	L1	0.25 ohms
Input filter choke	L16	0.6 ohms
Medium wave aerial coils	L2 and L3	L2=8 ohms L3=0.5 ohms
Medium wave R.F. coil	L7	6 ohms
Medium wave oscillator coil	L10	2.6 ohms
Long wave oscillator coil	L11	3.5 ohms
Aerial long wave loading coils	L4 and L5	L4=4.7 ohms L5=1.5 ohms
Long wave R.F. coil	L6	4 ohms

First I.F. Transformer (IFT 1)
Primary	L8	5.7 ohms
Secondary	L9	5.7 ohms

Second I.F. Transformer (IFT 2)
Primary	L12	3.4+2.6 ohms
Secondary	L13	5.6+0.2 ohms

Third I.F. Transformer (IFT 3)
Primary	L14	5.8 ohms
Secondary	L15	1.0 ohms

Resistors

Resistor	Value	Tolerance	Type
R1	82K	10%	9AP2
R2	3.3K	10%	9AP2
R3	6.8K	10%	9AP2
R4	560Ω	10%	9AP2
R5	560Ω	10%	9AP2
R6	18Ω	10%	9AP2
R7	15K	10%	9AP2
R8	6.8K	10%	9AP2
R9	1.5K	10%	9AP2
R10	2.7K	10%	9AP2
R11	6.8K	10%	9AP2
R12			
R13	1.5K	10%	9AP2
R14	1.8K	10%	9AP2
R15	82K	10%	9AP2
R16	5.6K	10%	9AP2
R17	680Ω	10%	9AP2
R18	12K	10%	9AP2
R19	560Ω	10%	9AP2
R20	1.8K	10%	9AP2
R21	5.6K	10%	9AP2
R22	68K	10%	9AP2
R23	120K	10%	9AP2
R24	6.8Ω	10%	9AP2
R25	1K	10%	9AP2
R26	1.2K	10%	9AP2
R27	47Ω	10%	9AP2
R28	15Ω±3Ω at 25°C	20%	VA1100 Thermistor
R29	180Ω	10%	9AP2

Capacitors

Capacitor	Capacity	Tolerance	Type	Voltage
C1	40—140 pf	—	Trimmer	—
C2	380 pf	±2½%	Styrene	350
C3	0.01 μf	±20%	Ceramic	50
C4	0.05 μf	+80%—20%	Ceramic	50
C5	0.1 μf	+80%—20%	Styrene	10
C6	190 pf	±5%	Styrene	350
C7	10—80 pf		Trimmer	—
C8	500 pf	±5%	Styrene	125
C9	3000 pf	±2½%	Styrene	125
C10	10 μf	+100%—0%	Electrolytic	6
C11	0.1 μf	±10%	Polyester	160
C12	0.1 μf	+80%—20%	Ceramic	10
C13	0.05 μf	+80%—20%	Ceramic	50
C14	1000 pf	±2%	Silver Mica	200
C15	6800 pf	±10%	Polyester	160
C16	220 pf	±5%	Ceramic	500
C17	2 pf	±½ pf	Part of IFT1	
C18	220 pf	±5%	Styrene	125
C19	450 pf	±5%	Trimmer	—
C20	40—140 pf		Styrene	350
C21	330 pf	±5%	Ceramic	50
C22	0.05 μf	+80%—20%	Electrolytic	16
C23	10 μf	+100%—0%		
C24	220 pf	±5%	Part of IFT2	—
C25	220 pf	±5%	Part of IFT2	—
C26	0.05 μf	+80%—20%	Ceramic	50
C27	0.05 μf	+80%—20%	Ceramic	50
C28	220 pf	±5%	Part of IFT3	—
C29	0.01 μf	±20%	Ceramic	50
C30	20 μf	+100%—20%	Electrolytic	16
C31	0.05 μf	+80%—20%	Ceramic	10
C32	0.2 μf	+80%—20%	Ceramic	50
C33	0.01 μf	±20%	Ceramic	10
C34	0.2 μf	+80%—20%	Ceramic	16
C35	0.2 μf	+100%—0%	Electrolytic	10
C36	10 μf	+100%—0%	Electrolyte	16
C37	125 μf	+100%—0%	Electrolyte	10
C38	2000 μf	±5%	Styrene	125
C39	0.2 μf	+80%—20%	Ceramic	20
C41	1000 μf	+100%—0%	Electrolytic	16
C42	1000 μf	+100%—0%	Electrolytic	16
C43	0.2 μf	+80%—20%	Ceramic	10
C44	2000 μf	+80%—20%	Feed Thru.	200
C45	2000 μf	+80%—20%	Feed Thru.	200
C46	20 pf	±5%	Ceramic	60

Table for Circuit Alignment

Waveband	Order of Alignment	Input Frequency KHz	Calibration Mark	Adjustment for Maximum Output			
				Oscillator Trimmer Core	R.F. Trimmer Core	Aerial Trimmer Core	
M	1	1600	1st on left	C20	—	C7	—	C1	—
	2	1300	2nd from left	—	L10	—	L7	—	L2
	3	Repeat operations 1 and 2 until no further improvement is obtained					
	4	1000	3rd from left	} Check Sensitivity			
	5	600	5th from left				
L	6	200	4th from left	—	L11	—	—	—	L4

FAULT DIAGNOSIS AND RECTIFICATION

Efficient radio servicing requires the quick and accurate diagnosis of any fault which might develop in some section of the installation. This section of the Bulletin details trouble diagnosis procedure which should be employed when investigating a complaint of poor or no reception.

The following aerial and main supply leads tests should be completed before removing the receiver or any other component from the car, but after having ascertained that all external connections are properly made, use the Trouble Diagnosis Charts included in this Bulletin to ensure that the checks and tests are performed in the correct sequence.

NOTE – Operation of a transistor receiver can be adversely affected by the supply voltage being too high or too low.

If too high, transistor failure may result.

If too low, the sensitivity of the receiver could be affected.

Aerial Efficiency Tests

Physically check the condition of the aerial and lead-in paying particular attention to the insulation of the lead-in from the car body and components.

NOTE – A duplicate aerial, consisting of a short length of suitably insulated wire connected to the receiver aerial socket and led out of the car, may be used for a quick test of aerial efficiency.

CIRCUIT DESCRIPTION

The car radio uses eight transistors in a superhet circuit designed to operate from a negative earth 12 volt D.C. supply.

The signal is obtained from a standard whip aerial and passed to the R.F. transistor via a permeability tuned aerial circuit on medium and long waves; the aerial lead capacity forms part of the tuned circuit. The signal developed across the R.F. collector load is inductively coupled to the mixer stage.

The oscillator circuit is a Colpitt's type for both medium and long wave bands.

The I.F. signal appearing at the mixer collector is transformer coupled to the base of the first I.F. amplifier, where it is amplified and transformer coupled to the base of the second I.F. amplifier where it is again amplified and coupled by a further transformer to the detector diode.

The audio signal appearing after the detector is passed via the volume control to the base of the audio amplifier; the collector of the audio amplifier is directly coupled to the base of the audio driver, and the collector of this transistor is directly coupled via a resistive network to the bases of the complementary audio output transistors. The emitters of these transistors are capacitively coupled to a 3 ohm speaker.

The A.G.C. is obtained from the diode load and applied via decoupling networks to the bases of the R.F. and I.F. amplifiers.

A fully variable tone control consisting of a 20K ohms log potentiometer in series with a 0.22 mfd capacitor is connected across the volume control.

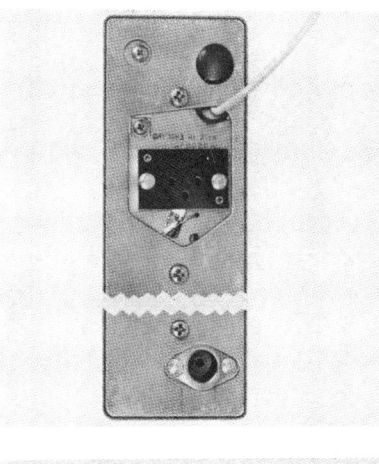

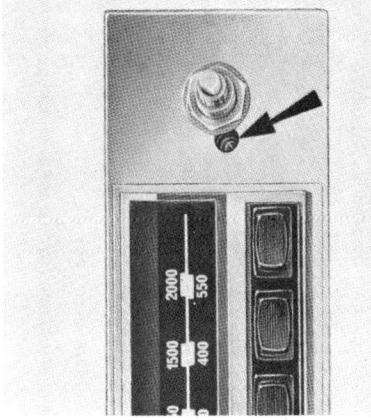

Aerial Trimmer

Rear View showing Speaker, Aerial Socket, Supply Lead, and Earth Tag

CORTINA - LOTUS

(a) Short Circuit Tests

1. Connect a suitable voltmeter and test battery in series with two test prods, as shown below. Check the battery voltage by momentarily connecting the two test prods.

2. Remove the aerial lead-in plug from the socket in the rear of the receiver and touch one test prod on the plug centre contact and the other prod on the outer diameter and rim of the plug. Any reading shown on the voltmeter would indicate a short circuit between the aerial lead-in and earth.

3. Further checks can be made by removing the test prod from the outer rim of the plug and touching various good earthing points on the car body, or by testing between the aerial mast and earth.

(b) Continuity Test

1. With the lead-in still disconnected from the receiver, touch one test prod on the centre contact of the plug.

2. Use the other test prod to contact the aerial mast as previously described. The voltmeter should record full battery voltage, as the voltage drop throughout the circuit should be negligible.

To check for intermittent breaks in the aerial circuit clip the prods in position and flex the lead-in throughout its length. The voltmeter reading should remain constant.

Radio Feed Lead Tests

With the receiver switched on, and supply and radio feed leads properly connected, the pilot lamp will be illuminated. If the lamp does not light, first check the 1.5 amp fuse and if this is in order, proceed as follows:—

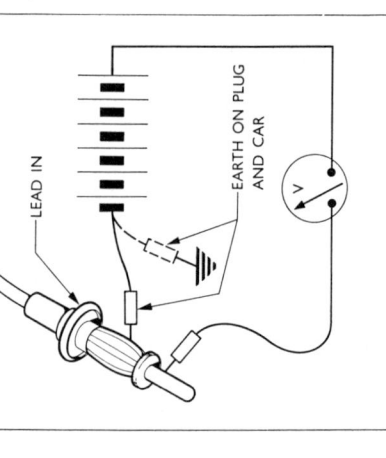

Aerial Short Circuit Test

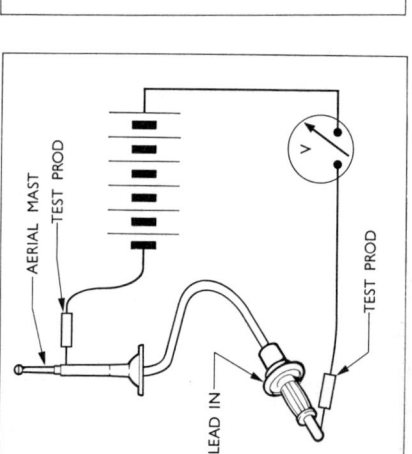

Aerial Continuity Test

Continuity Tests

1. Connect a suitable voltmeter in series with two test prods as shown below.

2. Disconnect the feed lead at the connector and remove the fuse.

3. Slip the connector casing down the lead when a test prod may be used to touch the contact formed on the supply lead end.

4. Earth the other prod, when the voltmeter should record full battery voltage, the volt drop through the leads being negligible.

5. Compare the reading obtained with the battery voltage. No reading on the voltmeter indicates a break in the supply circuit, while a drop in voltage indicates a poor connection resulting in a high resistance. Check back stage by stage, at the fuse, the ignition switch and the starter switch.

Loudspeaker Test

If after completing the aerial and supply lead tests it appears that a loudspeaker fault exists, disconnect the existing speaker, substitute a previously tested speaker and check for sound output.

Bench Checking the Receiver

WARNING – The radio must be connected to a negative earth supply system only. This is important.

The radio must not be operated without being connected to a 3 ohm speaker or an equivalent load of 3 ohms, otherwise the output transistors will fail.

It should be noted that the following test procedures apply only to a receiver no longer covered by Warranty. If it is still in Warranty, the receiver should not be dismantled, but returned unopened.

Before commencing to dismantle the radio receiver carry out the aerial and supply leads checks described previously. Make sure that all external connections to the set are correctly made.

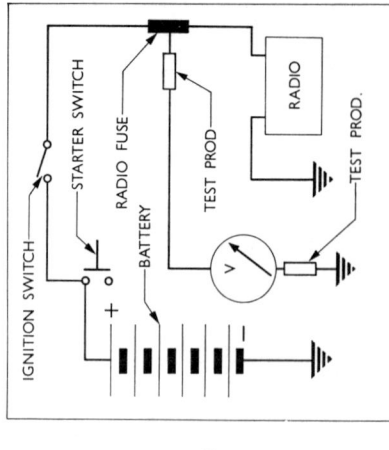

Radio Feed Lead Tests

CORTINA - LOTUS

Always use the "Trouble Diagnosis Charts" methodically when carrying out the checks detailed in this section.

1. Remove the radio from the car, as described in the appropriate Section of the workshop manual.
2. Unscrew the four screws retaining the cover of the receiver and remove the cover.
3. Connect a 3 ohm speaker into the loudspeaker socket at the rear of the receiver.
4. Plug a suitable test aerial into the receiver aerial socket.
5. Connect a suitable lead between the receiver chassis and the negative post of a test battery (12 volt) and, using a suitable lead incorporating a 1.5 amp fuse, connect the main supply lead to the negative post.

NOTE – The current consumption of the receiver is approximately 0.8 amps at full output, therefore it is important that a fuse of 1.5 amps should be used when checking the receiver.

Bench Test Fault Diagnosis Notes

In order to carry out effectively the tests and adjustments covered in this Section, the following instruments are required:—

A signal generator capable of covering the given test frequencies (amplitude modulated 30% at 400 c/s.).

An output meter, or low range meter such as a Model 8 Avometer used on 100 ma A.C. range.

An audio generator if it is not included in the signal generator.

Should the fault not be located by following the procedure laid down in the diagnosis charts "A" and "B", a more detailed examination must be undertaken. See the section on the Audio Amplifier stage and alignment instructions as these will assist in this examination.

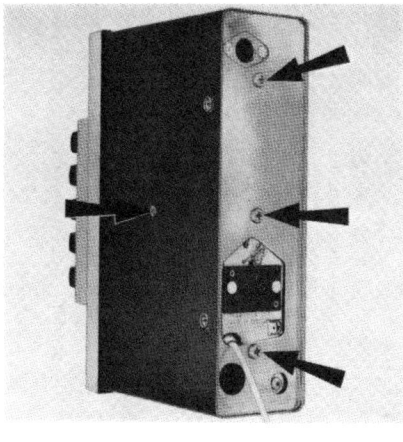

Location of Securing Screws

Printed Panel

T1 — to junction of L2 and L1
T2 — to L3
T3 — to L7
T4 — to L10
T5 — to chassis
T6 — to on/off switch
T7 — to T9 and T11
T8 — to on/off switch
T9 — to T7
T10 — to C45
T11 — T7

CORTINA - LOTUS

Printed Circuit Panel

In servicing the printed circuit panel, excessive heat can cause the printed circuit to lift from the panel and so result in the complete panel having to be replaced. A 25 to 50 watt pencil-bit soldering iron is recommended for this work. The tip of the iron should not be put directly on to the panel but applied to the tag or lead concerned and removed as soon as possible. The same care must be taken when removing or replacing components on the board and a small wire brush is very useful in clearing surplus solder during removal operations. Care must also be taken not to crack or break the panel as this would necessitate replacing the whole panel.

Reference should be made to the illustrations when removal or replacement of components is anticipated.

To obtain full access to all the components on the printed circuits carry out the following operations:—

1. Remove nuts and washers from the control rod spindles and remove escutcheon assembly.
2. Remove the nut and washer from the volume control spindle.
3. Set the wave band to long wave by turning it fully clockwise.
4. Unsolder the aerial filter choke from the aerial socket.
5. Remove the four self-tapping screws securing the front panel to the chassis.
6. Remove the tuning assembly by easing it forward thus exposing the entire printed circuit.

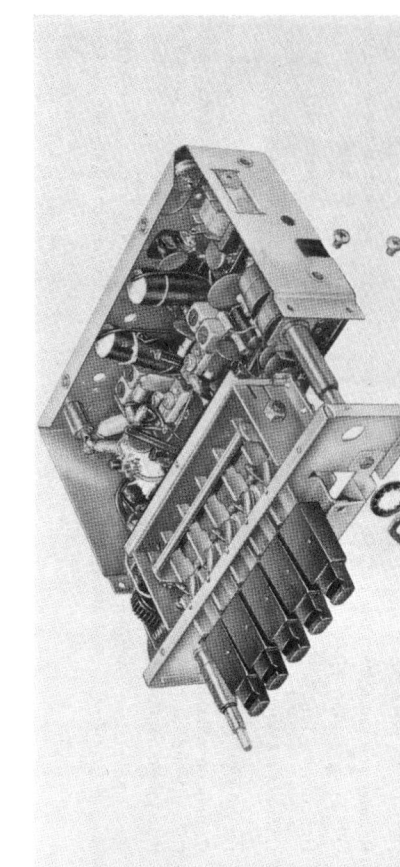

Removal of the Tuning Assembly exposing the entire Printed Circuit

General Layout of Components showing Resistors

CORTINA - LOTUS

Audio Generator Tests

If an audio generator is available the following tests can be applied to check the audio circuits and isolate any fault that may occur.

Connect the "earth" lead of the audio generator to the radio chassis and the "live" lead via a 0.1 µf capacitor to the following points, adjusting the generator output accordingly:—

1. Base of VT6 Transistor.
2. Base of VT5 Transistor.
3. Junction of C34 and C35.
4. Junction of R20 and R21.
5. Junction of D1 and R20.

Reference to the diagrams will identify the various test points.

I.F. Tests and Alignment

To check I.F. alignment proceed as follows:—

With the receiver on Medium wave, set the signal generator to 470KHz modulated 30% and connect between the chassis and the base of VT2 via a 50 pf capacitor. Turn the volume and the tone controls fully clockwise and adjust the generator output as required.

Adjust the cores of IFT3, IFT2 and IFT1 (in that order), with a suitable trimming tool for maximum output and repeat until no further improvement is obtained.

Keep the signal generator output to as low a level as will give a reading on the output meter.

Audio Amplifier Stages (VT5, VT6, VT7 and VT8)

The audio amplifier incorporated in the radio is the directly-coupled type without transformers. It should be noted that VT7 and VT8 (AD161/162) are a matched pair and both must be replaced if either fails; always use heat-sink compound (DP2623)—Midland Silicones Ltd.) on the mica washers when replacing these transistors.

When any transistor in the audio stage is replaced, a readjustment of the balance resistor RV3 may be necessary. This is carried out with the aid of an oscilloscope across the speaker connections.

Tune in a strong signal at 1 MHz. from the signal generator, modulated 30% by 400 Hz. and adjust the volume control until the sine wave shown on the oscilloscope just 'clips' at either or both top and bottom; adjust RV3 (and volume control as required) until even 'clipping' is achieved.

General Layout of Components showing Capacitors

CORTINA - LOTUS

R.F. Tests and Alignment

Set the signal generator to the mid-point of the wave band under test, i.e. 1,000 KHz for M.W. and 200 KHz for L.W. Apply the signal via a 0.1 μf capacitor to the following points in turn and tune the radio to signal, keeping the generator output as low as possible.

1. Base of VT2 transistor.
2. Base of VT1 transistor.
3. Junction of C3 and L3.
4. Each end, in turn, of the aerial filter choke (L1).

To carry out re-alignment, proceed as follows:—

Connect the generator output to the aerial socket via a dummy aerial of a 15 pf capacitor in series with the "live" lead and a 60 pf capacitor between the aerial socket and the chassis.

An attenuator pad, as supplied with most signal generators, can be used with the dummy aerial when carrying out R.F. alignment.

Remove the black scale backplate to expose the calibration marks and with the receiver fully connected, carry out the alignment procedure according to the table which appears on page 8 of this section.

To adjust the iron dust cores of L2, L7 and L10 hold a soldering iron against solder securing the wire of the iron dust core to the carriage. When the solder has melted, adjust the core with a pair of thin-nose pliers by means of the wire. When the optimum possible is reached, remove the soldering iron and allow the solder to cool. It is advisable to continue to grip the wire until the solder has set.

The I.F. trimming tool can be used to adjust the L.W. coils L11 and L4.

Transistor Tests

The audio and output transistors should be checked by carrying out the voltage measurements detailed under "Audio-Amplifier Stages".

R.F. and I.F. Transistors

The following tests can be carried out on the R.F. (VT1), Mixer (VT2) and I.F. (VT3 and VT4) transistors.

Connect the negative lead of the test voltmeter to the receiver chassis and check the voltage readings on the emitter base and collector of each transistor.

If any of these voltages vary by 25% from the readings given in the table below, the transistor concerned can be considered faulty provided the associated resistors are correct.

Transistor	VOLTAGE READINGS AT 13.5 VOLTS INPUT		
	Collector	Base	Emitter
VT1 (BF 153 or BF 273)	4.7	0.95	0.41
VT2 (BF 153)	7.9	2.7	2.15
VT3 (U 14711)	10.7	1.3	0.95
VT4 (U 14711)	10.7	1.65	1.4
VT5 (BC 113)	12.8	0.6*	7.0
VT6 (BC 139)	6.5	12.8	13.2
VT7 (AD 161)	13.2	6.5	6.4
VT8 (AD 162)	—	6.3	6.4

* Measured between emitter and base with negative lead to emitter.

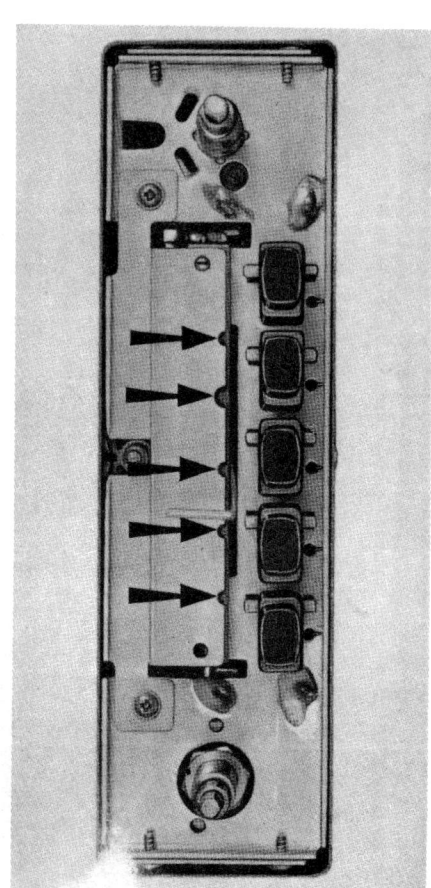

Calibration Marks

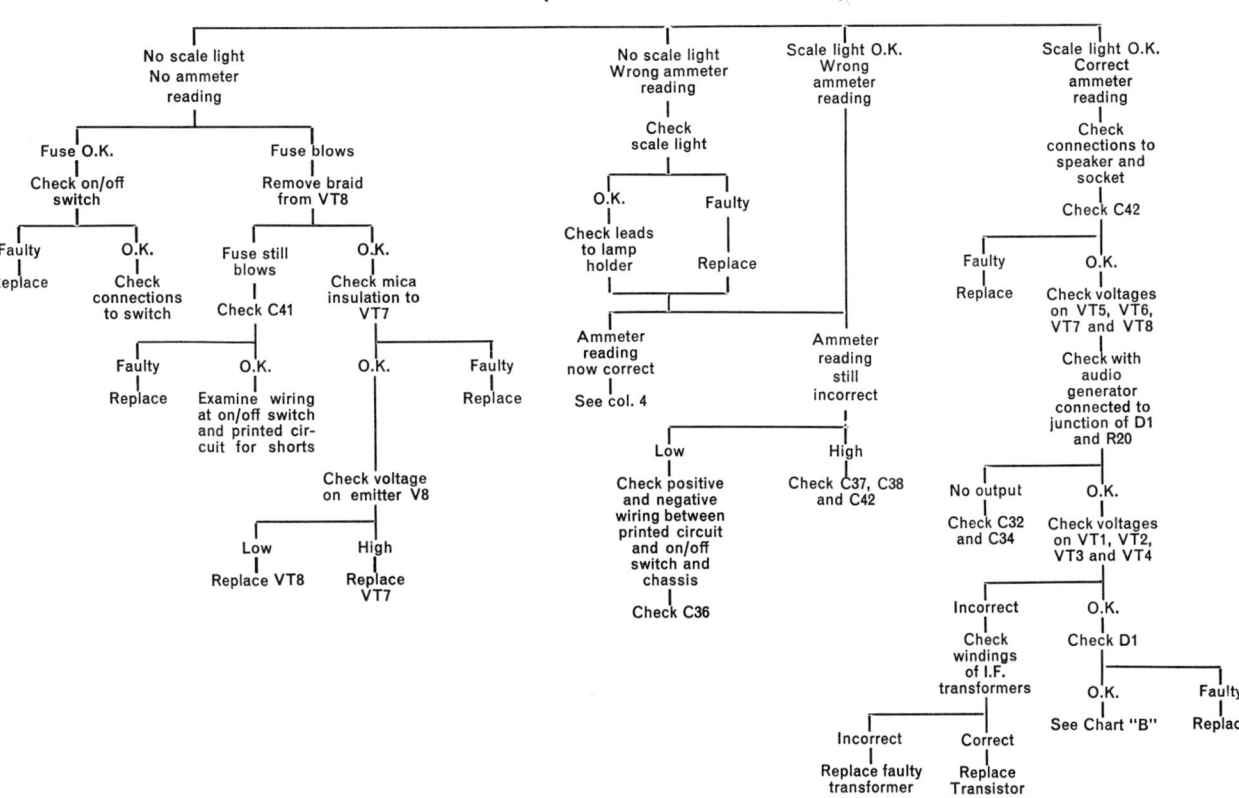

Diagnosis of Faults by the "Click Test" method

If no suitable test equipment is immediately available a simple and effective method of fault diagnosis can be carried out by carefully following the sequence of tests indicated on chart "A".

These tests are based on the assumption that the radio is "dead", but the scale light is operating and the total current consumption is correct.

1. Switch on the receiver.
2. Turn the volume and tone controls fully clockwise and select the medium wave band.
3. Connect one end of a resistor of the value indicated in the chart to the receiver chassis and then touch the lead from the other end of the resistor to the various points indicated in the order specified.

   If the circuits tested are satisfactory, a pronounced "click" will be heard in the loudspeaker corresponding to each time the lead touches a test point. If no "click" occurs, reference to the Diagnosis Chart will identify the faulty unit.

   NOTE – When there is reception on one wave band only, select the "dead" wave band and proceed with the click tests from the point marked "X".

4. Using the blade of a small screwdriver held in contact with a finger, continue the click tests in the sequence indicated in the second half of the chart.

Before replacing any unit, carefully examine the receiver for broken wires and short circuits.

Reference to the printed circuit panel will assist in identifying the required test points.

CHART "B" PUSH-BUTTON CAR RADIO "CLICK" TEST

Connect one end of a 100Ω resistor to chassis and the other end to:—

- Base VT7
  - Clicks O.K.
  - No clicks — Check speaker socket and speaker
- Base VT6
  - Clicks O.K.
  - No Clicks — Check C38
- Use 10,000Ω Resistor
- Base VT5
  - Clicks O.K.
  - No clicks — Check C37
- Use 1,000Ω Resistor
- Junction D1 and R20
  - Clicks O.K.
  - No clicks — Check C34, C32, R20
- Collector VT4
  - Clicks O.K.
  - No clicks — Check L14, L15 of IFT3; Check D1
- Collector VT3
  - Clicks O.K.
  - No clicks — Check L12 and L13 of IFT2
- Collector VT2
  - Clicks O.K.
  - No clicks — Check L8, L9 of IFT1
- Junction of C20 and L11

continued on the next page

continued from previous page

- Clicks O.K. on M.W. and L.W.
  - No clicks on both or either band — Check L10, C19 and C20 M.W.; Check L11, C21 L.W.; Check wave change switch
- Use finger on blade of small screwdriver and tune in station on M.W. or L.W.
- On Base of VT2
  - Signals O.K.
  - No signals — Check C11, C14, C15
- On Base of VT1
  - Signals O.K.
  - No signals — Check C6, C7, C8 and C9 and L8 M.W.; Check R6, L6 L.W.; Check wave change switch
- On aerial socket
  - Signals O.K.
  - No signals — Check C3, L3, L2, L1 and C1 on M.W.; Check L4, L5, C2 L.W.; Check wave change switch
- Check aerial

NOTE:
Before using this chart the voltages on all transistors must be checked in accordance with the instructions in the text. The "click" test is inconclusive, unless these voltages are correct.

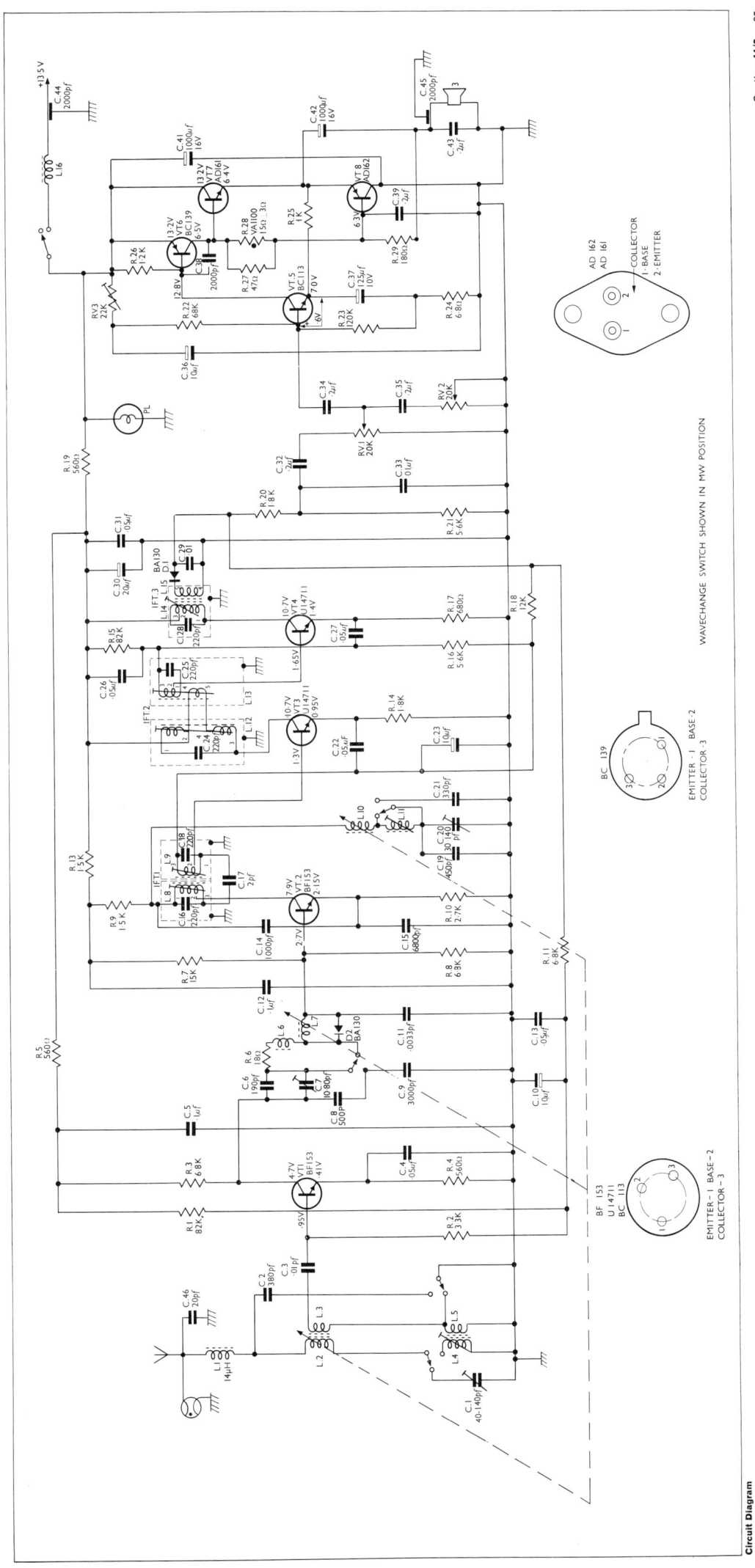

CORTINA - LOTUS

11/3

CAR RADIO

(FULLY TRANSISTORISED MANUAL TUNING)
PART NUMBER 3024E-18K810-A

CORTINA - LOTUS

SECTION INDEX

GENERAL DESCRIPTION
OPERATING INSTRUCTIONS
DETAILED SPECIFICATION AND REFERENCE DATA
TUNING INSTRUCTIONS
CIRCUIT DESCRIPTION
FAULT DIAGNOSIS AND RECTIFICATION
CIRCUIT DIAGRAM

CAR RADIO *(Fully Transistorised—Manual Tuning)*
OVERHAUL PROCEDURES

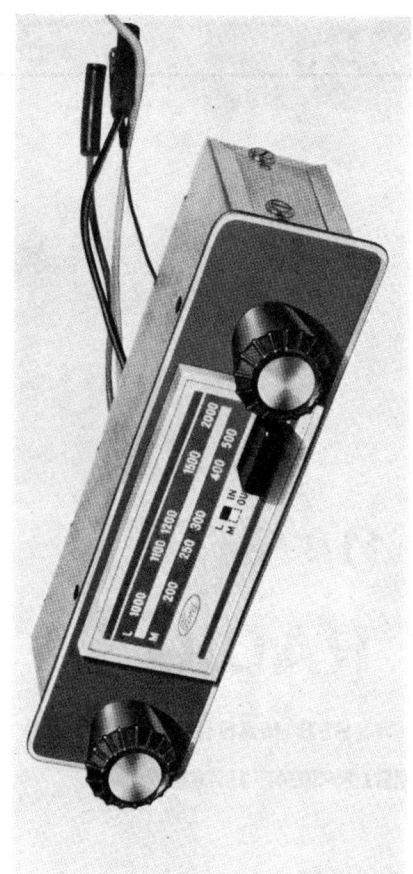

GENERAL DESCRIPTION

Car Radio Part No. 3024E-18K810-A is a 12 volt seven-transistor superheterodyne receiver. It covers both the medium and the long wave bands.

The receiver incorporates two controls in addition to the wave change button. The left-hand control operates the on/off switch and the volume, and the right-hand knob the manual tuning of the receiver.

This section of the Workshop Manual describes the radio receiver in detail, it also covers fault diagnosis and repair procedure. However, it should be remembered that if the receiver is still under Warranty, no attempt should be made to repair it. It should be returned unopened to the manufacturer.

NOTE – This receiver is suitable for negative earthed systems only.

CORTINA - LOTUS

DETAILED SPECIFICATION AND REFERENCE DATA

Type	...	7 Transistor, Superheterodyne
Power supply	...	12 volt negative earth
Current consumption	...	0.95 A at 13.5 volt input (No signal)
Fuse	...	1.5 amp cartridge type
Scale lamp	...	24v. 3w. Festoon
Sensitivity	...	Under 25 microvolts for 0.5 w. output on M.W.
		Under 80 microvolts for 0.5 w. output on L.W.
Speaker	...	Moving coil 3 ohm impedance

Frequency Coverage

M.W.	...	512—1622 KHz 585—185 Metres
L.W.	...	150— 380 KHz 2000—789 Metres
Tuning	...	Manual
Intermediate frequency	...	470 KHz

Transistors and Diodes

TS401	AF 126	...	R.F. amplifier
TS402	AF 126	...	Mixer/oscillator
TS403	AF 126	...	IF amplifier
TS404	AC 127	...	AF amplifier
TS405	AC 128	...	Driver
TS406a	AD 161	...	Power stage
TS406b	AD 162	...	
GR408	AA 119	...	A.V.C.
GR409	AA 119	...	Detector
GR410	AA 119	...	Damping

Resistors

Resistor	Wattage	Type	Tolerance	Value (ohms)
R481	⅛	Carbon	5%	1,000
R482	⅛	Carbon	5%	2,200
R483	⅛	Carbon	5%	3,300
R484	⅛	Carbon	5%	0.15 Ω
R485	⅛	Carbon	5%	1,500
R486	⅛	Carbon	5%	820
R487	⅛	Carbon	5%	3,300
R488	⅛	Carbon	5%	680
R489	⅛	Carbon	5%	27,000
R491	⅛	Carbon	5%	1,500
R493	⅛	Carbon	5%	5,100
R495	—	N.T.C.	—	0.15 Ω
R496	⅛	Carbon	5%	22,000
R497	⅛	Carbon	5%	82,000
R498	⅛	Carbon	5%	3,900
R499	⅛	Carbon	5%	820
R500	⅛	Carbon	5%	10,000
R501	⅛	Carbon	5%	1,000
R506	⅛	Carbon	5%	0.82 Ω
R507	⅛	Carbon	5%	17,000
R508	—	Volume	—	+5,000
R509	—	N.T.C.	—	0.15 Ω
R510	⅛	Carbon	5%	10,000
R511	⅛	Carbon	5%	47,000
R512	⅛	Carbon	5%	4.7
R513	⅛	Carbon	5%	1,000
R514	⅛	Carbon	5%	1,000
R515	⅛	Carbon	5%	100
R516	⅛	Carbon	5%	1,500
R517	⅛	Carbon	5%	68
R518	⅛	Carb. variable	—	220
R519	⅛	Carbon	5%	68
R520	—	N.T.C.	—	50
R521	⅛	Metal oxide	—	1
R522	⅛	Metal oxide	—	1
R523	⅛	Metal oxide	—	1
R524	⅛	Metal oxide	—	1
R525	⅛	Carbon	5%	220
R528	⅛	Carbon	5%	120
R529	⅛	Carbon	5%	680
R530	⅛	Carbon	5%	68
R531	⅛	Carbon	5%	18,000

Operating Instructions

The radio will operate with the ignition switch in either the "auxiliary circuits" position or the "ignition and auxiliary circuits" position. The scale illuminating bulb operates with panel lights.

Switch on the radio by turning the left-hand control knob clockwise, a click will be heard as the switch operates.

Select the wave length, by operating the wave change button and tune to the required station, using the right-hand control knob.

To switch the radio off, turn the left-hand control knob anti-clockwise until a click is heard.

THE CONTROLS

Combined On/Off Switch and Volume Control

The left-hand control knob operates the on/off switch and varies the loudspeaker volume. The set is switched on by turning this knob clockwise, further movement in this direction progressively increases the volume; anti-clockwise rotation will reduce the volume and the set may be switched off by turning the control fully in this direction.

Manual Tuning Control

Medium and long wave stations are tuned in by means of the right-hand control knob in conjunction with the tuning scale.

To change from the medium to the long wave band depress the wave change button. Further operation of the button will cause it to move outwards, thus returning to the medium wave length band.

Stations should always be tuned for maximum strength and the volume reduced, if necessary, by using the volume control.

Wave Change Button

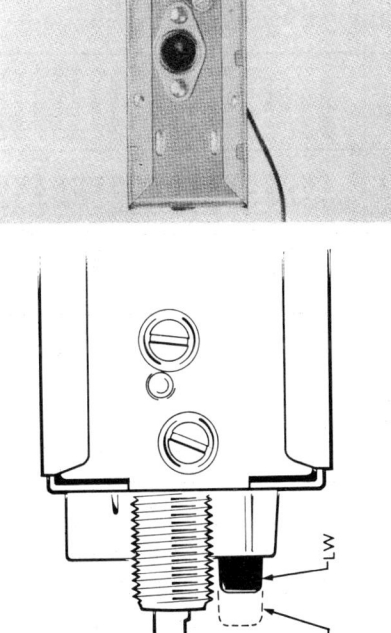

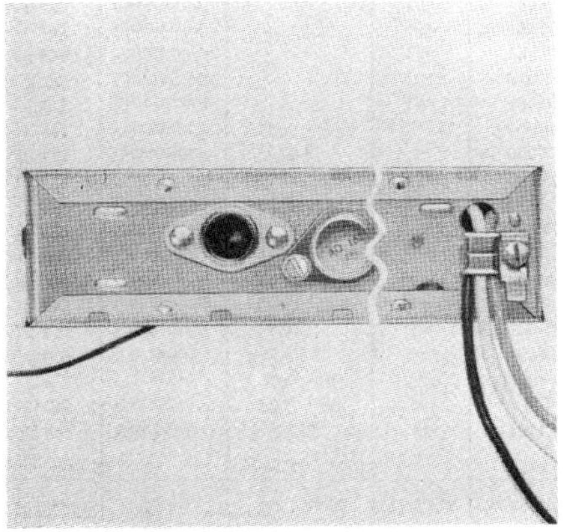

Rear View showing Aerial Socket and Supply and Speaker Leads

CORTINA - LOTUS

Tuning Instructions

(1) Tune in the cores of L425 a/b, L424 a/b and L424 c/d.
(2) Corresponds to minimum self-inductance of the tuning unit.
(3) Corresponds to maximum self-inductance of the tuning unit.
(4) C432 should be tuned to minimum capacitance and a 30 pf capacitor should be connected in parallel with it.
(5) Tune the set.
(6) **Apply the signal to the aerial socket via an artificial aerial, see below.**

NOTE – C432 serves to adapt the car aerial to be set. Slide out the aerial completely and tune the set to a weak station near 200 m. (MW) Tune C432 by ear to maximum output power.

Adjusting the collector current of the output transistors.

Connect an ammeter between the collector of TS406a and the "+". After three minutes warm up time the collector current should be 60MA. This can be adjusted with the aid of R518.

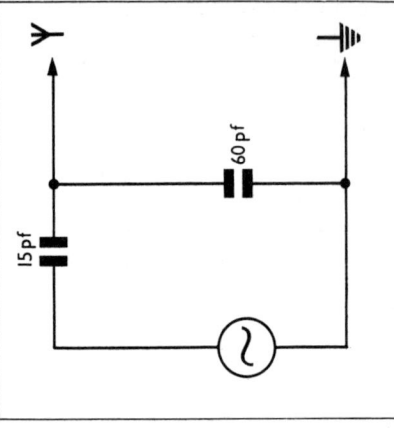

Aerial Trimmer

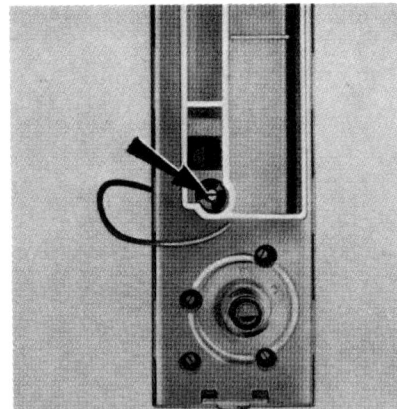

Artificial Aerial Layout

Capacitors

Capacitor	Type	Tolerance	Voltage	Capacity		Capacitor	Type	Tolerance	Voltage	Capacity
C430	Ceramic	5%	—	27pf		C452	Styroflex	2.5%	—	100pf
C431	Ceramic	5%	—	68pf		C456	Ceramic	−20%+80%	—	2,000pf
C432	Trimmer	—	—	60pf		C457	Ceramic	−20%+80%	—	5,000pf
C433	Ceramic	5%	—	18pf		C458	Ceramic	2%	—	82pf
C435	Ceramic	2%	—	33pf		C459	Ceramic	20%	—	82,000pf
C436	Ceramic	±0.5pf	—	10pf		C460	Ceramic	−20%+80%	—	5,000pf
C437	Styroflex	2.5%	—	820pf		C461	Ceramic	−20%+80%	—	5,000pf
C438	Ceramic	20%	—	47,000pf		C466	Polyester	10%	—	0.33μf
C439	Ceramic	20%	—	82,000pf		C467	Electrolytic	—	64v	0.64μf
C440	Electrolytic	—	16v	10μf		C468	Electrolytic	—	10v	200pf
C441	Polyester	10%	—	5,100pf		C469	Ceramic	20%	—	56,000pf
C442	Ceramic	±0.5pf	—	22pf		C470	Ceramic	−20%+80%	—	2,000pf
C443	Ceramic	20%	—	47,000pf		C471	Electrolytic	—	16v	125μf
C444	Polyester	2%	—	3,600pf		C472	Electrolytic	—	10v	640μf
C445	Styroflex	2.5%	160v	200pf		C473	Electrolytic	—	16v	125μf
C446	Polyester	5%	—	9,100pf		C476	Electrolytic	—	16v	125μf
C447	Styroflex	2.5%	—	390pf		C477	Electrolytic	—	16v	125μf
C448	Ceramic	±0.5pf	—	15pf		C478				

Table for Circuit Alignment

Wave Range	Tuning	Signal	Applied to	Trim	Indication
IF	MW	Min. (2)	470 KHz (1) via 33 kpf	b – TS403	L425 c, d, L425 a, b	
				c – TS402	L424 c, d	Max. output
				b – TS402	L424 a, b	
RF	MW	Max. (3)	508 KHz		L421 (4)	
		(5)	600 KHz	(6)	L420, L416 (4)	
			1450 KHz		C432	Max. output
	LW	Max.	145 KHz		L432	
		(5)	170 KHz	(6)	L417	
			260 KHz		L418	

CIRCUIT DESCRIPTION

The car radio uses seven transistors in a superhet circuit designed to operate from a negative earth 12 volt D.C. supply.

The signal is obtained from a standard whip aerial and passed to the R.F. transistor via a permeability tuned aerial circuit on medium and long waves; the aerial lead capacity forms part of the tuned circuit. The signal developed across the R.F. collector load is inductively coupled to the mixer stage.

The oscillator circuit is a modified Colpitt's type for both medium and long wave bands.

The I.F. signal appearing at the mixer collector is transformer coupled to the base of the I.F. amplifier, where it is amplified and transformer coupled to the detector diode.

The audio signal appearing after the detector is passed via the volume control to the base of the audio amplifier; the collector of the audio amplifier is directly coupled to the base of the audio driver, and the collector of this transistor is directly coupled via a resistive network to the bases of the complementary audio output transistors. The emitters of these transistors are capacitively coupled to a bifilar wound choke and finally to a 3 ohm speaker.

The A.G.C. is obtained from the diode load and applied via decoupling networks to the bases of the R.F. and I.F. amplifiers.

FAULT DIAGNOSIS AND RECTIFICATION

Efficient radio servicing requires the quick and accurate diagnosis of any fault which might develop in some section of the installation. This section of the Bulletin details trouble diagnosis procedure which should be employed when investigating a complaint of poor or no reception.

The following aerial and main supply leads tests should be completed before removing the receiver or any other component from the car, but after having ascertained that all external connections are properly made, use the Trouble Diagnosis Chart included in this section to ensure that the checks and tests are performed in the correct sequence.

NOTE – Operation of a transistor receiver can be adversely affected by the supply voltage being too high or too low.

If too high, transistor failure may result.

If too low, the sensitivity of the receiver could be affected.

Aerial Efficiency Tests

Physically check the condition of the aerial and lead-in paying particular attention to the insulation of the lead-in from the car body and components.

NOTE – A duplicate aerial, consisting of a short length of suitably insulated wire connected to the receiver aerial socket and led out of the car, may be used for a quick test of aerial efficiency.

(a) *Short Circuit Tests*

1. Connect a suitable voltmeter and test battery in series with two test prods, as shown below. Check the battery voltage by momentarily connecting the two test prods.

2. Remove the aerial lead-in plug from the socket in the rear of the receiver and touch one test prod on the plug centre contact and the other prod on the outer diameter and rim of the plug. Any reading shown on the voltmeter would indicate a short circuit between the aerial lead-in and earth.

3. Further checks can be made by removing the test prod from the outer rim of the plug and touching various good earthing points on the car body, or by testing between the aerial mast and earth.

(b) *Continuity Test*

1. With the lead-in still disconnected from the receiver, touch one test prod on the centre contact of the plug.

2. Use the other test prod to contact the aerial mast as previously described. The voltmeter should record full battery voltage, as the voltage drop throughout the circuit should be negligible.

To check for intermittent breaks in the aerial circuit clip the prods in position and flex the lead-in throughout its length. The voltmeter reading should remain constant.

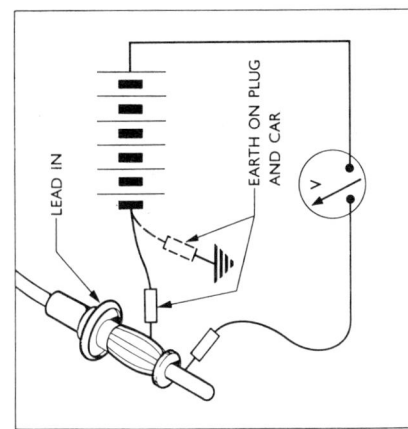

Aerial Short Circuit Test

Aerial Continuity Test

CORTINA - LOTUS

Radio Feed Lead Tests

With the receiver switched on, and supply and radio feeds properly connected, first check the 1.0 amp fuse and if this is in order, proceed as follows:—

Continuity Tests

1. Connect a suitable voltmeter in series with two test prods as shown below.

2. Disconnect the feed lead at the connector and remove the fuse.

3. Slip the connector casing down the lead when a test prod may be used to touch the contact formed on the supply lead end.

4. Earth the other prod, when the voltmeter should record full battery voltage, the volt drop through the leads being negligible.

5. Compare the reading obtained with the battery voltage. No reading on the voltmeter indicates a break in the supply circuit, while a drop in voltage indicates a poor connection resulting in a high resistance. Check back stage by stage, at the fuse, the ignition switch and the starter solenoid.

Loudspeaker Test

If after completing the aerial and supply lead tests it appears that a loudspeaker fault exists, disconnect the existing speaker, substitute a previously tested speaker and check for sound output.

Radio Feed Lead Tests

Bench Checking the Receiver

WARNING – The radio must be connected to a negative earth supply system only. This is important.

The radio must not be operated without being connected to a 3 ohm speaker or an equivalent load of 3 ohms, otherwise the output transistors will fail.

It should be noted that the following test procedures apply only to a receiver no longer covered by Warranty. If it is still in Warranty, the receiver should not be dismantled, but returned unopened.

Before commencing to dismantle the radio receiver carry out the aerial and supply leads checks described previously. Make sure that all external connections to the set are correctly made.

Always use the "Trouble Diagnosis Charts" methodically when carrying out the checks detailed in this section.

1. Remove the radio from the car, as described in the appropriate section of the Workshop Manual.

2. Remove the two receiver covers.

3. Connect a 3 ohm speaker to the loudspeaker leads.

4. Plug a suitable test aerial into the receiver aerial socket.

5. Connect a suitable lead between the receiver chassis and the negative post of a test battery (12 volt) and, using a suitable lead incorporating a 1.0 amp fuse, connect the main supply lead to the positive post.

NOTE – The current consumption of the receiver is approximately .95 amp at full output, therefore it is important that a fuse of 1.0 amp should be used when checking the receiver.

Bench Test Fault Diagnosis Notes

In order to carry out effectively the tests and adjustments covered in this Bulletin, the following instruments are required:—

A signal generator capable of covering the given test frequencies (amplitude modulated 30% at 400 c/s.).

An output meter, or low range meter such as a Model 8 Avometer used on 100 ma A.C. range.

An audio generator is not included in the signal generator. Should the fault not be located by following the procedure laid down in the diagnosis chart, a more detailed examination must be undertaken. See the section on the Audio Amplifier stage and alignment instructions as these will assist in this examination.

CORTINA - LOTUS

Printed Panel Circuit

In servicing the printed circuit panel, excessive heat can cause the printed circuit to lift from the panel and so result in the complete panel having to be replaced. A 15 watt pencil-bit soldering iron is recommended for this work. The tip of the iron should be put directly on to the panel and the component and soldering iron removed as soon as possible. The same care must be taken when removing or replacing components on the board and a small brush is very useful in clearing surplus solder during removal operations. Care must also be taken not to crack or break the panel as this would necessitate replacing the whole panel.

Audio Generator Tests

If an audio generator is available the following tests can be applied to check the audio circuits and isolate any fault that may occur.

Connect the "earth" lead of the audio generator to the radio chassis and the "live" lead via a 0.1 μf capacitor to the following points, adjusting the generator output accordingly:—

(1) Base of TS405 Transistor
(2) Base of TS404 Transistor
(3) Junction of R507 and R508

Reference to the diagrams will identify the various test points.

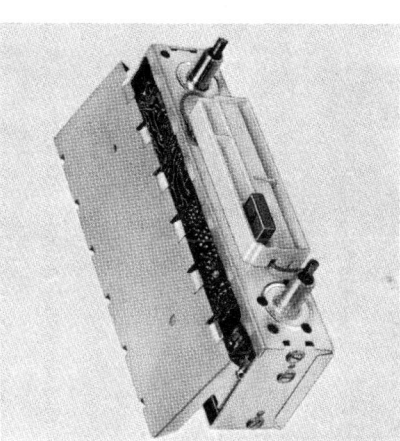

Location of Iron Dust Core Adjustment

Removal of Receiver Cover Sides

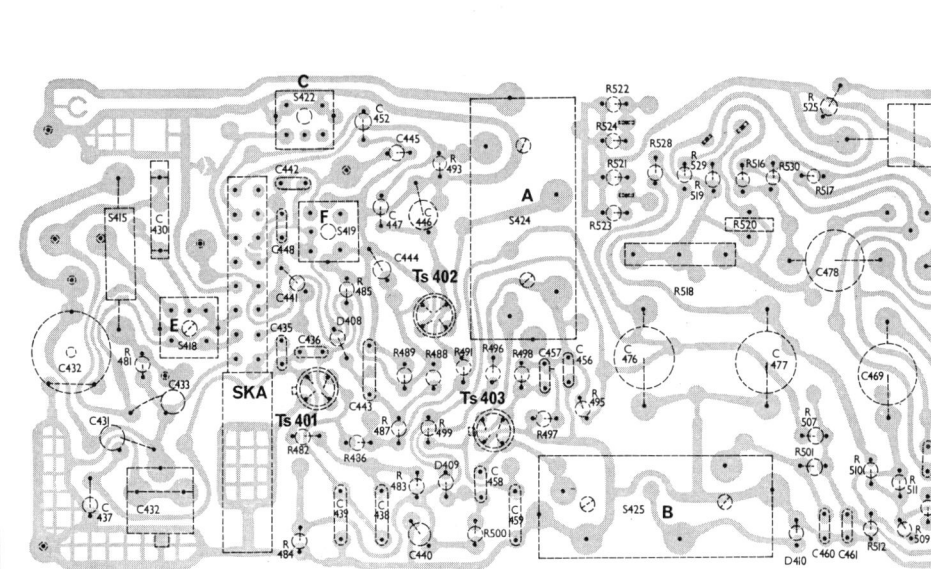

Printed Panel Circuit

CORTINA - LOTUS

I.F. Tests and Alignment

To check I.F. alignment proceed as follows:—

With the receiver on Long wave, set the signal generator to 470 Hz modulated 30% and connect between the chassis and the base of TS402 via a 50pf capacitor. Turn the volume control fully clockwise and adjust the generator output as required.

Adjust the cores of L425 L424 (in that order) with a suitable trimming tool for maximum output and repeat until no further improvement is obtained.

Keep the signal generator output to as low a level as will give a reading on the output meter.

Audio Amplifier Stages TS404, TS405, TS406a, TS406b

The audio amplifier incorporated in the radio is the directly-coupled type without transformers. It should be noted that VT7 and VT8 (AD161/162) are a matched pair and both must be replaced if either fails; always use heat-sink compound (DP2623—Midland Silicones Ltd.) on the mica washers when replacing these transistors.

When any transistor in the audio stage is replaced, a readjustment of the balance resistor R518 may be necessary. This is carried out with the aid of an oscilloscope across the speaker connections.

R.F. Tests and Alignment

Set the signal generator to the mid-point of the wave band under test, i.e. 1,000 KHz for M.W. and 200 KHz for L.W. Apply the signal via a 0.1 μf capacitor to the following points in turn and tune the radio to signal, keeping the generator output as low as possible.

(1) Base of TS402 transistor
(2) Base of TS401 transistor
(3) Each end, in turn, of the aerial filter choke L415

To carry out re-alignment, proceed as follows:—

Connect the generator output to the aerial socket via a dummy aerial or a 15 pf capacitor in series with the "live" lead and a 60 pf capacitor between the aerial socket and the chassis.

An attenuator pad, as supplied with most signal generators, can be used with the dummy aerial when carrying out R.F. alignment.

With the receiver fully connected, carry out the alignment procedure according to the table which appears in the Specification.

To adjust the iron dust cores of L416, L417, L420, L421, insert a suitable screwdriver in the holes around the tuning control which are exposed when the front escutcheon is removed.

The I.F. trimming tool can be used to adjust the L.W. coils.

View from Above

View from Below

General Views of the Components

CORTINA - LOTUS

Transistor Tests

The audio and output transistors should be checked by carrying out the voltage measurements detailed under "Audio-Amplifier Stages".

R.F. and I.F. Transistors

The following tests can be carried out on the R.F. (TS401), Mixer (TS402) and I.F. (TS403) transistors.

Connect the negative lead of the test voltmeter to the receiver chassis and check the voltage readings on the emitter base and collector of each transistor.

If any of these voltages vary by 25% from the readings given on the circuit diagram, the transistor concerned can be considered faulty provided the associated resistors are correct.

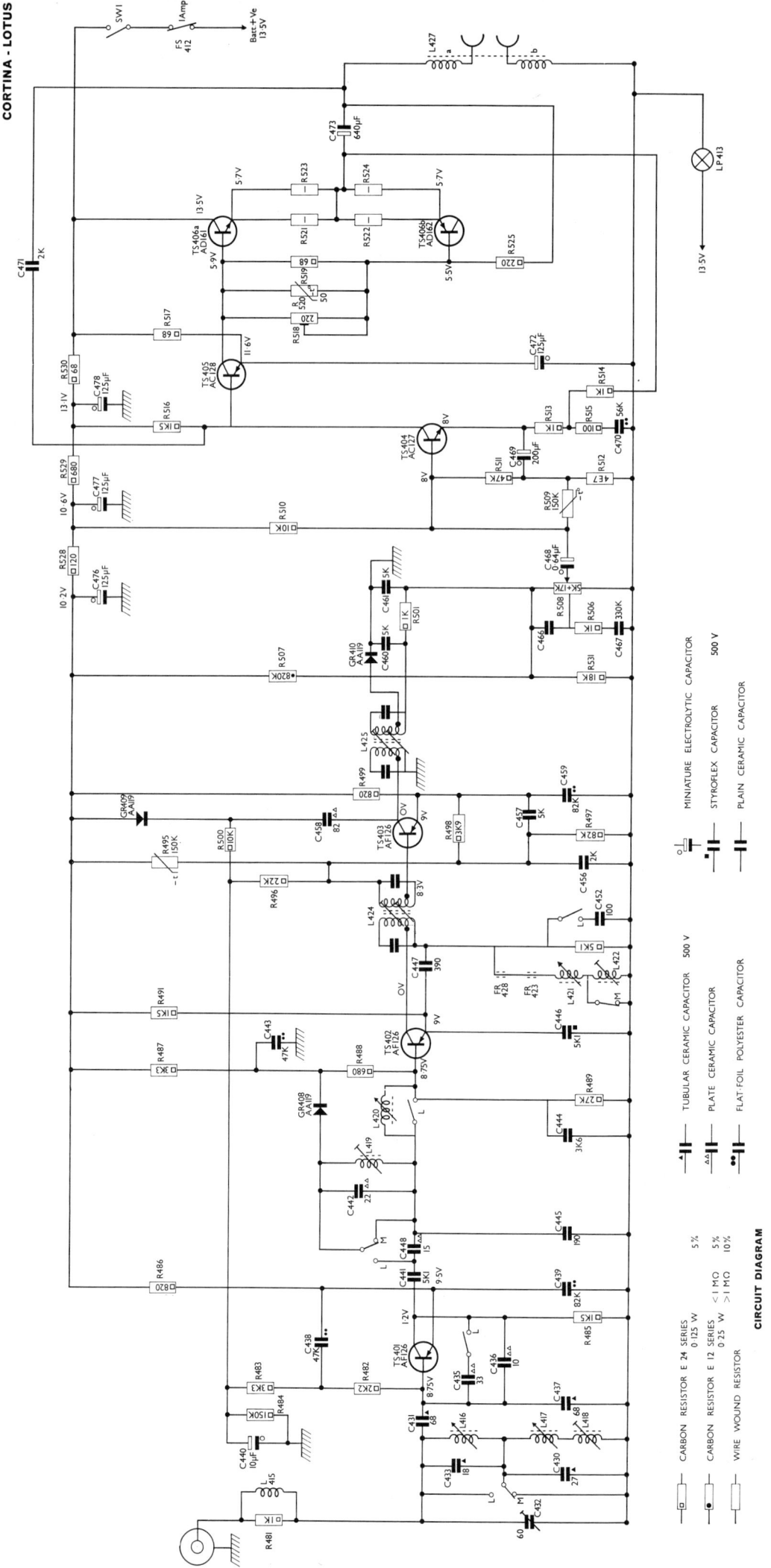

CORTINA - LOTUS

12
BODYWORK

CORTINA - LOTUS

SECTION INDEX

GENERAL DESCRIPTION

SERVICE AND REPAIR OPERATIONS

OPERATION	M 1005-A	BUMPER BAR – FRONT – REMOVE AND INSTALL
"	M 1005-A1	Extra: bumper bracket – each – remove and install
"	M 1010-A	BUMPER BAR – REAR – REMOVE AND INSTALL
"	M 1010-A1	Extra- bumper bracket – each – remove and install
"	M 1030-A	RADIATOR GRILLE – REMOVE AND INSTALL
"	M 1035-A	BONNET ASSEMBLY – REMOVE AND INSTALL
"	M 1035-A1	Extra: bonnet components – remove and install
"	M 1035-B	**BONNET – REMOVE AND INSTALL** (Includes OPS M 1035-A and A1)
"	M 1036-A	BONNET MOTIF – REMOVE AND INSTALL
"	M 1037-A	BONNET STRIKER POST – REMOVE AND INSTALL
"	M 1038-A	BONNET LOCK ASSEMBLY – REMOVE AND INSTALL
"	M 1039-A	BONNET SAFETY CATCH – REMOVE AND INSTALL
"	M 1040-A	BONNET SAFETY CATCH STRIKER PLATE – REMOVE AND INSTALL
"	M 1042-A	BONNET STAY – REMOVE AND INSTALL
"	M 1043-C	BONNET HINGE – ONE – REMOVE AND INSTALL
"	M 1060-A	DOOR ASSEMBLY – REMOVE AND INSTALL
"	M 1060-A2	Extra: door components – remove and install
"	M 1060-D	**DOOR SHELL – REMOVE AND INSTALL** (Includes OPS M 1060-A and A2)
"	M 1061-A	DOOR UPPER OR LOWER HINGE – REMOVE AND INSTALL
"	M 1062-A	DOOR APERTURE WEATHERSTRIP, OR WINDCORD/WEATHERSTRIP – ONE – RENEW
"	M 1062-A1	Extra: door aperture weatherstrip, or windcord/weatherstrip each additional – renew
"	M 1063-A	DOOR EXTERIOR HANDLE – REMOVE AND INSTALL
"	M 1063-A1	Extra: door exterior handle – overhaul
"	M 1063-B	**DOOR EXTERIOR HANDLE – OVERHAUL** (Includes OPS M 1063-A and A1)
"	M 1064-A	DOOR CHECK STRAP – REMOVE AND INSTALL
"	M 1066-A	DOOR STRIKER PLATE – REMOVE AND INSTALL AND/OR ADJUST
OPERATION	M 1067-A	DOOR INNER AND OUTER BELT WEATHERSTRIPS – REMOVE AND INSTALL
"	M 1069-A	DOOR REMOTE CONTROL HANDLE – REMOVE AND INSTALL
"	M 1070-A	WINDOW REGULATOR HANDLE – REMOVE AND INSTALL
"	M 1071-A	DOOR ARMREST – REMOVE AND INSTALL
"	M 1072-A	DOOR TRIM PANEL – REMOVE AND INSTALL
"	M 1072-B	WINDOW REGULATOR ASSEMBLY – REMOVE AND INSTALL
"	M 1072-C	DOOR LOCK ASSEMBLY – REMOVE AND INSTALL
"	M 1072-D	DOOR REMOTE CONTROL ASSEMBLY – REMOVE AND INSTALL
"	M 1072-E	FIXED VENT WINDOW GLASS AND/OR WEATHERSTRIP – REMOVE AND INSTALL
"	M 1072-H	DOOR WINDOW GLASS SILENT CHANNEL – REMOVE AND INSTALL
"	M 1072-I	DOOR WINDOW GLASS – REMOVE AND INSTALL
"	M 1105-B	LUGGAGE COMPARTMENT LID ASSEMBLY – REMOVE AND INSTALL
"	M 1107-A	LUGGAGE COMPARTMENT LID MOTIF – REMOVE AND INSTALL
"	M 1108-A	LUGGAGE COMPARTMENT STRIKER POST – REMOVE AND INSTALL
"	M 1109-A	LUGGAGE COMPARTMENT LOCK – REMOVE AND INSTALL
"	M 1109-A1	Extra: luggage compartment lock barrel – remove and install
"	M 1109-B	**LUGGAGE COMPARTMENT LOCK BARREL – REMOVE AND INSTALL** (Includes OPS 1109-A and A1)
"	M 1111-A	LUGGAGE COMPARTMENT TORSION BAR – ONE – REMOVE AND INSTALL
"	M 1111-A1	Extra: remaining torsion bar – remove and install
"	M 1111-A2	Extra: luggage compartment hinge – one – remove and install
"	M 1111-B	**LUGGAGE COMPARTMENT TORSION BARS – BOTH – REMOVE AND INSTALL** (Includes OPS M 1111-A and A1)
"	M 1111-C	**LUGGAGE COMPARTMENT HINGE – ONE – REMOVE AND INSTALL** (Includes OPS M 1111-A, A1 and A2)

CORTINA - LOTUS

OPERATION M 1111-D	**LUGGAGE COMPARTMENT HINGES – BOTH – REMOVE AND INSTALL** (Includes OPS M 1111-A, A1 and A2 × 2)
,, M 1205-A	WINDSCREEN GLASS AND/OR MOULDING – REMOVE AND INSTALL
,, M 1210-A	REAR WINDOW GLASS AND/OR MOULDING – REMOVE AND INSTALL
,, M 1215-A	REAR QUARTER WINDOW GLASS AND/OR MOULDING – REMOVE AND INSTALL
,, M 1220-A	"AEROFLOW" EXTERIOR EXTRACTION LOUVRE – REMOVE AND INSTALL
,, M 1220-A1	Extra: extraction box – remove and install
,, M 1305-A	FRONT PARCEL SHELF – REMOVE AND INSTALL
,, M 1305-A1	Extra: cowl side trim panel – remove and install
,, M 1305-B	**COWL SIDE TRIM PANEL – REMOVE AND INSTALL** (Includes OPS M 1305-A and A1)
,, M 1310-A	REAR PARCEL SHELF – REMOVE AND INSTALL
,, M 1315-A	REAR QUARTER TRIM PANEL – REMOVE AND INSTALL
,, M 1320-A	DRIP RAIL MOULDING — ONE SIDE REMOVE AND INSTALL
,, M 1325-A	REAR SEAT BACKREST – REMOVE AND INSTALL
,, M 1327-A	REAR SEAT CUSHION – REMOVE AND INSTALL
,, M 1330-A	BUCKET SEAT – FRONT – REMOVE AND INSTALL
,, M 1340-A	INTERIOR MIRROR – REMOVE AND INSTALL
,, M 1345-A	SUN VISOR – REMOVE AND INSTALL
,, M 1351-A	GLOVE LOCKER LID – REMOVE AND INSTALL
,, M 1355-A	HEATER TRIM PANEL – REMOVE AND INSTALL
,, M 1360-A	HEADLINING – RENEW
UNDERBODY REPAIR PROCEDURES	
UNDERBODY TOLERANCE CHARTS	

GENERAL DESCRIPTION

The body is of all steel welded construction with safety glass in all windows and single curvature windscreen and rear window.

The wide opening doors fitted with bar type check straps, incorporate zero torque door locks. An exterior key-operated lock is provided on the driver's door.

The well proved "Aeroflow" ventilation system is incorporated in the car, therefore fixed quarter lights are normally fitted to the front doors. Opening front door vent windows are available if required as an option.

Seating is within the wheelbase, bucket type front seats are provided with floor gear change with reclining seats available as an option.

The rear luggage compartment incorporates a self-locking counterbalanced lid, with the spare wheel mounted on the left-hand side of the compartment.

SEALING MATERIALS

There are four types of sealing materials used in the construction of the car, but only one is required for Service. This is "type 3" which is used for windscreen sealing and is water resistant, non-drying and remains permanently flexible. A material known to conform to these requirements is "SR-51-B" manufactured by Expandite Ltd., Chase Road, London, N.W.10.

WARNING

Before starting any operation which involves the use of an oxy-acetylene cutting or welding torch, ensure that the fuel lines or electrical wiring are removed from the area to prevent a fire hazard.

When carrying out repairs which involve the removal of the fuel tank, sender unit, carburettor or fuel line, ensure that any petrol fumes which may be present are dispersed by positioning the vehicle in a well ventilated area. In cases like these it is a wise precaution to disconnect the battery to prevent a possible electrical fault from starting a fire.

Paint Finish

The finish coats of paint are applied in two stages.

Stage 1 is the application of special primer paints, which are baked or stoved on the body for thirty minutes at 149°C (330°F), to produce a tough durable paint film with excellent adhesion properties. When hard this paint film is sanded to produce a really smooth surface, in preparation for stage 2.

Stage 2 is the application of Acrylic resin base enamel which is in turn baked or stoved for thirty minutes at 121°C to 129°C (250°F to 265°F). The end result is a hard high-gloss paint film with excellent colour retention and adhesion properties.

This type of finish is specially formulated so that any scratches which have not penetrated to the primer can be removed by polishing with fine compound followed by body polish.

The only attention the paintwork normally requires to retain its high gloss is regular washing. The use of FoMoCo shampoo and body polish is recommended for all models in the range.

Service Repairs

The acrylic enamel used is a "thermo-setting" type which, as previously stated, has excellent polishing characteristics so that minor surface defects such as surface scratches, dry spray, etc. can often be corrected by machine or hand polishing techniques.

But in addition, with this type of finish, the stoved paint film is a stable one so that it can be repaired in Service with any of the conventional repair materials, etc., nitro-cellulose lacquers (half-hour enamels), air drying acrylic lacquers or low-bake enamel. It is not essential to repair the factory acrylic finish with acrylic repair paint.

For Service repairs, by arrangement with the major paint manufacturers, specially tested repair paints which are known to possess durability and gloss characteristics similar to the original finish are available in all current body colours.

These approved repair paints carry a label on the lid "approved by Ford Motor Company Ltd." and should be used whenever possible to ensure customer satisfaction.

Beauty Care

The only attention the paintwork normally requires to retain its high gloss is regular washing. To obtain the best results, fill a bucket with warm water (Never Use Hot Water) and add a quantity of FoMoCo car shampoo. Using a soft sponge, apply the shampoo mixture generously over the car. Thoroughly wash off all traces of shampoo using copious amounts of clean warm water. When all traces of shampoo have been washed off, leather the car dry. If required, the paintwork lustre may be further enhanced by occasionally polishing the body after washing, using FoMoCo body polish.

Listed below is a selection from the extensive range of FoMoCo beauty aids, all of which are available from your local Ford Main Dealer.

	Part Number
Upholstery Cleaner	204E-19526
Paint Touch-in Pencil	E56-WY-1 to E96-WY-1
Paint Touch-in Aerosol	105E-19L531-A to 3008E-19L531-S
Paint Touch-in Tin and Brush	211E-19K500-B to 3004E-19K500-E49
Body Polish	M-230-A
Liquid Wax Polish	105E-19520
Polish Cream	211E-19530
Car Shampoo (Bottle)	204E-19524
Car Shampoo (Satchet)	211E-19524
Chrome Cleaner (Can)	105E-19522
Chrome Cleaner (Satchet)	EOA-19522
Vehicle Washing Brush	105E-19M510-A
Shampoo Tablets	105E-19M512-A
Cleaning Sponge	105E-19K508-A
Chamois Leather	211E-19517-G to 211E-19517-L
Engine Cleaner	3014E-19L531-A

SERVICE AND REPAIR OPERATIONS

OP M 1005-A BUMPER BAR – FRONT – REMOVE AND INSTALL

To Remove
1. Remove two bolts, one each side, from the inside of the front wings.
2. Remove two chrome headed bolts and nuts from the centre of the bumper.
3. Remove the bumper assembly.

To Install
4. Position the bumper and loosely retain it with the two centre bolts and nuts.
5. Replace the two bolts securing the ends of the bumper.
6. Finally tighten all four bolts.

OP M 1005-A1 EXTRA: BUMPER BRACKET – EACH – REPLACE

To Remove
1. Remove the two bolts securing the bumper bracket to engine side apron panel.
2. Remove the bracket.

To Install
3. Position the bracket on the engine side apron panel and secure it with two bolts.

OP M 1010-A BUMPER BAR – REAR – REMOVE AND INSTALL

To Remove
1. Remove the licence plate light, see OP 13543-A, section 10.
2. Remove four chrome headed bolts and nuts securing the rear bumper.
3. Remove the bumper.

To Install
4. Position the bumper on the bumper brackets.
5. Secure the bumper with four bolts and nuts.
6. Replace the licence plate lamp as in OP 13543-A, section 10.

OP M 1010-A1 EXTRA: BUMPER BRACKET – EACH – REPLACE

To Remove
1. Remove two bolts securing the bracket to the body.
2. Remove the bumper bracket.

To Install
3. Locate the bumper bracket on the body.
4. Secure the bracket with two bolts.

OP M 1030-A RADIATOR GRILLE – REMOVE AND INSTALL

To Remove

1. Remove four screws each side securing the two headlight bezels and move the bezels out of the way.
2. Remove two screws securing the bottom edge, then drill out the six rivets securing the top of the radiator grille, and remove the grille.

To Install

3. Position the radiator grille on the body, secure the top edge with six rivets and the bottom edge with two screws.
4. Replace the two headlight bezels and secure them with four screws each side.

OP M 1035-A BONNET ASSEMBLY – REMOVE AND INSTALL

To Remove

1. Open the bonnet, fit wing covers and place a cloth on the scuttle to prevent accidental damage.
2. If the bonnet lid alignment is satisfactory, scribe a line around the hinges on the underside of the bonnet lid.
3. Remove two bolts securing the bonnet lid to one of the hinges and rest the lid on the front scuttle panel.
4. Support the bonnet lid, remove two bolts from the second hinge and detach the bonnet lid.

To Install

5. Place the bonnet lid on the front scuttle panel and alternately lift each side and loosely replace two securing bolts to each hinge.
6. Align the hinges to the marks scribed on the underside of the bonnet lid and lightly tighten the bolts.
7. Close bonnet and check alignment. Adjust if necessary.
8. Fully tighten securing bolts.
9. Remove the protective cloth from the front scuttle and remove the wing covers.

OP M 1035-A1 EXTRA: BONNET COMPONENTS – REMOVE AND INSTALL
(Includes OPS M 1036-A, M 1037-A and M 1039-A)

OP M 1035-B BONNET – REMOVE AND INSTALL
(Includes OPS M 1035-A and A1)

OP M 1036-A BONNET MOTIFS – REMOVE AND INSTALL

To Remove

1. Carefully prise out the individual motifs from the bonnet.
2. Remove the spring clips from the lugs on the back of the motifs.

To Install

3. Fit the retaining spring clips on the lugs at the back of the motifs.
4. Press the motifs into the bonnet to engage the spring clips.

OP M 1037-A BONNET STRIKER POST – REMOVE AND INSTALL

To Remove

1. Open and prop up the bonnet.
2. Unscrew the adjusting bolt and remove the bolt, spring seats and spring.

To Install

3. Locate the spring seats, spring and bolt on the bonnet lid and tighten the bolt.
4. Close the bonnet and check the height of the bonnet to the surrounding body panels and adjust if necessary.

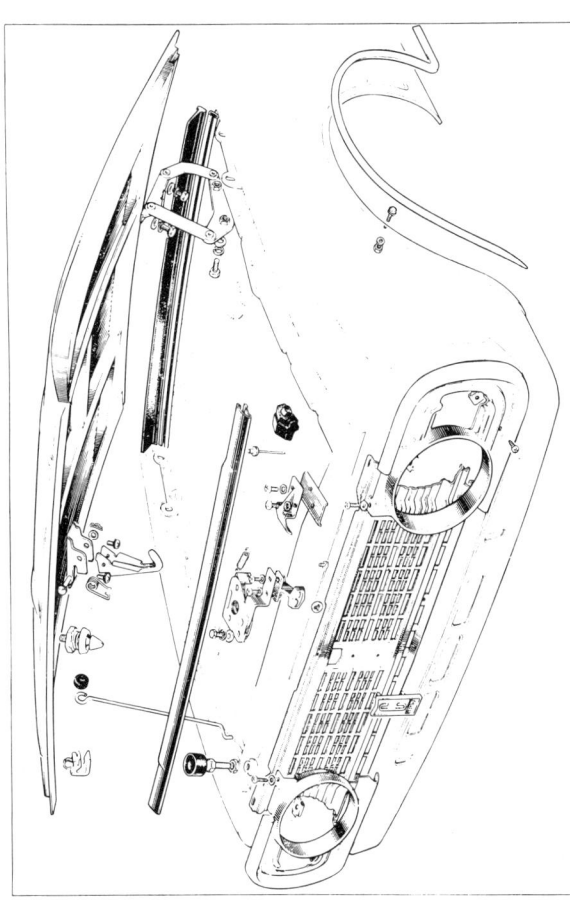

Radiator and Bonnet Lid

CORTINA - LOTUS

OP M 1038-A BONNET LOCK ASSEMBLY – REMOVE AND INSTALL

To Remove

1. Open and prop up the bonnet.
2. Remove ten screws, five each side, securing the two headlamp bezels and remove the bezels.
3. Drill out six rivets and remove the two screws securing the radiator grille and remove the grille.
4. Remove three bolts securing the lock mechanism, disconnect the operating cable and remove the lock assembly.

To Install

5. Locate the lock mechanism under the radiator upper panel, connect the operating cable and secure with three bolts.
6. If necessary slacken the two release handle securing screws and adjust the handle until it lines up with surrounding radiator mouldings, then tighten the securing screws.
7. Replace the radiator grille and secure it with two screws, and either pop-rivets or self-tapping screws.
8. Replace the two headlamp bezels and secure them with ten screws, five each side.

OP M 1039-A BONNET SAFETY CATCH – REMOVE AND INSTALL

To Remove

1. Open and prop up the bonnet.
2. Remove two bolts securing the safety catch to the bonnet and remove the catch.

To Install

3. Position the safety catch on the bonnet and secure it with two bolts.
4. Close the bonnet and check the operation of the safety catch.

OP M 1040-A BONNET SAFETY CATCH STRIKER PLATE – REMOVE AND INSTALL

To Remove

1. Remove the two screws securing the striker plate to the body and remove the striker plate.

To Install

2. Position the striker plate on the body and secure it with two screws.

OP M 1042-A BONNET STAY – REMOVE AND INSTALL

To Remove

1. Open and support the bonnet.
2. Release one end of the stay from its retaining clip and disconnect the other end from the rubber bushed mounting bracket on the bonnet.

To Install

3. Reconnect the stay to its mounting bracket on the bonnet and secure the other end in its retaining clip.
4. Close the bonnet.

OP M 1043-C BONNET HINGE – ONE – REMOVE AND INSTALL

To Remove

1. Open the bonnet, fit wing covers and place a cloth on the scuttle to prevent accidental damage.
2. Scribe a line around the hinge on the underside of the bonnet, and on the bracket welded to the front mudguard apron.
3. Remove the two bolts securing the bonnet lid, and rest the lid on the front scuttle panel.
4. Remove the two bolts securing the hinge to the apron bracket.

To Install

5. Loosely replace the four hinge bolts.
6. Align the hinge to the marks scribed on the underside of the bonnet lid, and the apron bracket, and lightly tighten the bolts.
7. Close the bonnet and check alignment. Adjust if necessary.
8. Fully tighten the securing bolts.
9. Remove the protective cloth from the front scuttle and remove the wing covers.

OP M 1060-A DOOR ASSEMBLY – REMOVE AND INSTALL

To Remove

1. Remove the circlip and withdraw the door check strap pivot pin.
2. Support the door and remove four bolts securing the door to the hinges.
3. Remove the door and anti-slip shims.

To Install

4. Replace the door and anti-slip shims, and secure lightly with four bolts.
5. Shut the door and check the alignment of the door to bodywork and adjust if necessary.
6. Carefully open the door and fully tighten the securing bolts.
7. Replace the check strap pivot pin and secure it with a circlip.
8. Carefully open the door and fully tighten the securing screws.
9. Replace the check strap pivot pin and secure it with a circlip.
10. Replace the door trim panels and handles.

CORTINA - LOTUS

OP M 1060-A2 EXTRA: DOOR COMPONENTS – REMOVE AND INSTALL

To Remove

1. Remove the trim panel and interior handles.
2. Remove the remote control assembly (see Op. M 1076-A).
3. Remove the door lock mechanism (see Op. M 1075-A).
4. Remove the door window glass (see Op. M 1050-A).
5. Remove the window regulator (see Op. M 1051-A).
6. Remove the vent window assembly (see Op. M 1052-A).
7. Remove the door exterior handle (see Op. M 1077-A).
8. Remove the door window glass lower stop, door bump rubbers and weatherstrip.

To Install

9. Replace the door window glass lower stop, door bump rubbers and weatherstrip.
10. Replace the door exterior handle (see Op. M 1077-A).
11. Replace the vent window assembly (see Op. M 1052-A).
12. Replace the window regulator (see Op. M 1051-A).
13. Replace the door window glass (see Op. M 1050-A).
14. Replace the door lock mechanism (see Op. M 1075-A).
15. Replace the remote control assembly (see Op. M 1076-A).
16. Replace the trim panel and interior handles.

OP M 1060-D DOOR SHELL – REMOVE AND INSTALL
(Includes OPS M 1060-A and A2)

OP M 1061-A FRONT DOOR UPPER OR LOWER HINGE – REMOVE AND INSTALL

To Remove

1. Remove the door trim panel and handles.
2. Remove the circlip and withdraw the pivot pin securing door check strap.
3. Support the door and remove four bolts securing the door hinges to the door.
4. Remove the door and anti-slip shims, and parcel shelf.
5. Pull the floor carpet back and remove the cowl side trim panel.
6. Remove the three nuts and washer plate securing each hinge to the body and remove the hinge and anti-slip shim.

To Install

7. Locate the hinge and anti-slip shim on the body, and secure them with washer plate and three nuts.
8. Lightly secure the door and anti-slip shims to the hinges with four bolts.
9. Shut the door and check the alignment of the door to the bodywork and adjust if necessary.
10. Carefully open the door and fully tighten the hinge securing bolts.
11. Replace the cowl side trim panel and carpet.
12. Replace the parcel shelf.
13. Locate the door check strap with the pivot pin and secure it with a circlip.
14. Replace the door trim panel and handles.

OP M 1063-A DOOR EXTERIOR HANDLE – REMOVE AND INSTALL

To Remove

1. Close the window and remove the door trim panel and handles.
2. Remove the screw securing the rear end of the handle to the door.
3. Through the door access hole remove the screw securing the front end of the handle.
4. Remove the handle and rubber sealing gaskets.

To Install

5. Position the rubber gaskets and handle on the door and secure the rear end with a screw.
6. Through the door access hole replace the screw securing the front of the handle.
7. Check the operation of the door handle plunger and replace the door trim panel and handles.

OP M 1063-A1 EXTRA: DOOR EXTERIOR HANDLE – OVERHAUL

To Dismantle

1. Remove the circlip from the end of the handle.
2. Withdraw the end plate, spring and lock barrel assembly.
3. Slacken the locknut and unscrew the adjusting bolt.
4. Remove the cast end of the lock barrel.
5. Insert the key in the barrel and withdraw the lock barrel from the chrome bezel.

To Reassemble

6. Insert the lock barrel in the chrome bezel and remove the key.
7. Replace the cast end of the lock barrel and secure it with the adjusting bolt and locknut.
8. Insert the lock barrel assembly, spring and end plate in door handle.
9. Replace the circlip in the end of the handle.

OP M 1063-B DOOR EXTERIOR HANDLE – OVERHAUL
(Includes OPS M 1063-A and A1)

CORTINA - LOTUS

OP M 1064-A DOOR CHECK STRAP – REMOVE AND INSTALL

To Remove

1. Remove the door trim panel and handles.
2. Remove the circlip and withdraw the check strap pivot pin.
3. Through door access hole remove two bolts securing the check strap to the door.
4. Remove the check strap through door access hole.

To Install

5. Locate the check strap in the door and secure it with two bolts.
6. Replace the check strap pivot pin and secure it with a circlip.
7. Replace the door trim panel and handles.

OP M 1066-A DOOR STRIKER PLATE – REMOVE AND INSTALL AND/OR ADJUST

To Remove

1. Remove three screws and remove the striker plate and anti-slip shim.

To Install

2. Position the anti-slip shim and striker plate on the door post and lightly secure them with three screws.
3. Shut the door ensuring the lock is fully engaged and not on safety and move it in or out until the door is flush with the bodywork.
4. Depress the door handle release button and carefully open the door.

Door Striker Plate Adjustment

Door Handle — Exploded

5. Check that the striker plate is vertical, then fully tighten the three striker plate securing screws.
6. Shut the door and check the alignment of the door to the bodywork.
7. Check push button effort and free play.

OP M 1067-A DOOR INNER AND OUTER BELT WEATHERSTRIPS – REMOVE AND INSTALL

To Remove

1. Wind the window down.
2. Using a thin-bladed screwdriver carefully prise out one end of the inner belt weatherstrip from its retaining clip.
3. Carefully pull the rest of the weatherstrip from its retaining clips.
4. Repeat operations 1 to 3 for the remaining weatherstrip.

To Install

5. Position the inner belt weatherstrip on the door and push downwards to engage the retaining clips.
6. Repeat operation 5 for the remaining weatherstrip.

OP M 1069-A DOOR REMOTE CONTROL HANDLE – REMOVE AND INSTALL

To Remove

1. Remove the central screw securing the door remote control handle and remove the handle.

To Install

2. Position the remote control handle on its control shaft and secure it with a screw.

OP M 1070-A WINDOW REGULATOR HANDLE – REMOVE AND INSTALL

To Remove

1. Remove the central screw securing the door regulator handle and remove the handle.

To Install

2. Position the door regulator handle on its control shaft and secure it with a screw.

OP M 1071-A DOOR ARMREST – REMOVE AND INSTALL

To Remove

1. Remove the two screws securing the armrest/door pull to the door.

To Install

2. Position the armrest/door pull and secure with two screws.

OP M 1072-A DOOR TRIM PANEL – REMOVE AND INSTALL

To Remove

1. Remove the window regulator handle, remote control handle and interior lock knob.
2. Remove two screws securing the combined armrest/door pull and remove the armrest.
3. Pull the trim panel away from the door inner panel to release the securing spring clips and remove the panel.

To Install

4. Position the trim panel on the door and press the spring clips into their appropriate holes in the door inner frame.
5. Replace the combined armrest/door pull and secure it with two screws.
6. Replace the window regulator handle, remote control and interior lock knob.

OP M 1072-B WINDOW REGULATOR ASSEMBLY – REMOVE AND INSTALL

To Remove

1. Lower the window and remove the door trim panel and handles.
2. Support the window glass and remove four screws securing the window regulator assembly to the door.
3. Disconnect the window regulator operating arm from the window glass lower run and remove the regulator assembly through the door access hole.

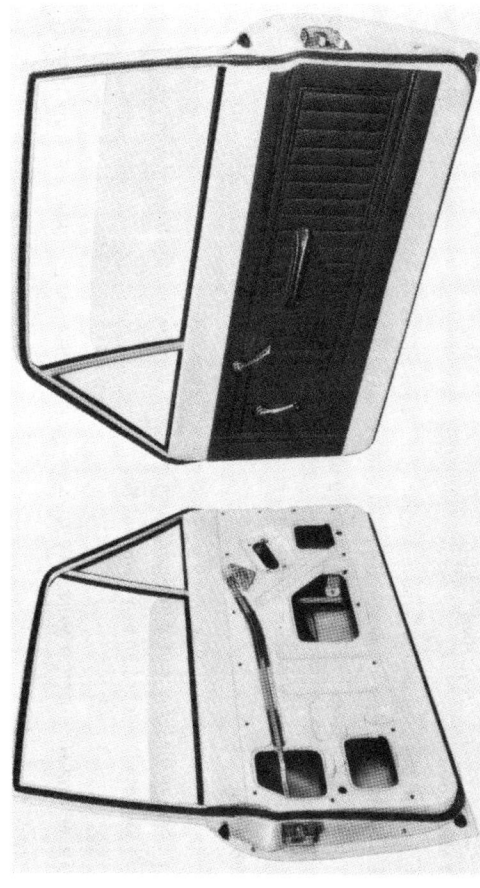

Front Door Components

CORTINA - LOTUS

To Install

4. Position the regulator assembly in the door and connect the operating arm to the window glass lower run.
5. Secure the regulator to the door with four screws.
6. Replace the door trim panel and handles.

OP M 1072-C DOOR LOCK ASSEMBLY – REMOVE AND INSTALL

To Remove

1. Close the window and remove the door trim panel and handles.
2. Remove the spring clip securing the door lock remote control arm to the lock mechanism and disconnect the arm.
3. Remove the screw securing the lower end of the window glass rear run.
4. Pull down the rear run to disengage the top end and remove the rear run through the door access hole.
5. Remove three screws securing the lock mechanism and dovetail plate to the door and remove the lock mechanism through the door access hole.

To Install

6. Position the lock mechanism in the door through the access hole.
7. Secure lock and dovetail plate to the door with three screws.
8. Replace window glass rear run, clip the top end in position and secure the lower end with a screw.
9. Connect the door lock remote control arm to the lock mechanism and secure it with a spring clip.
10. Check operation of lock, and push button free play. Adjust if necessary to $\frac{1}{16}$ in. (1.6 mm.).
11. Replace the door trim panel and handles.

OP M 1072-D DOOR REMOTE CONTROL ASSEMBLY – REMOVE AND INSTALL

To Remove

1. Remove the door trim panel and handles.
2. Remove the spring clip securing the remote control arm to the lock mechanism and disconnect the arm.
3. Remove three screws securing the remote control mechanism to the door and remove the remote control assembly.

To Install

4. Locate the remote control assembly on the door and secure it with three screws.
5. Connect the remote control operating arm to the lock mechanism and secure it with a spring clip.
6. Replace the door trim panel and handles.

OP M 1072-E FIXED VENT WINDOW GLASS AND/OR WEATHERSTRIP – REMOVE AND INSTALL

To Remove

1. Lower the window and remove the door trim panel and handles.
2. Remove the screw securing the lower end of the vent window dividing channel.
3. Drill out the two "pop" rivets securing the top end of the dividing channel and remove the channel from the door.
4. Pull out the vent window and weatherstrip.

To Install

5. Locate the vent window and dividing channel in the door.
6. Secure the lower end of the dividing channel with a screw and the top end with "pop" rivets.
7. Replace the door trim panel and handles.

OP M 1072-H DOOR WINDOW GLASS SILENT CHANNEL – REMOVE AND INSTALL

To Remove

1. Remove the screw securing the bottom of the window glass lower rear run and pull down the metal backing to disconnect the run from the door.
2. Pull the silent channel out of the door frame and lower rear run.

To Install

3. Replace the silent channel in the door frame starting at the top front edge.
4. Position the lower rear glass run metal backing over the silent channel and position the top edge in the door. Secure the lower edge with a screw.

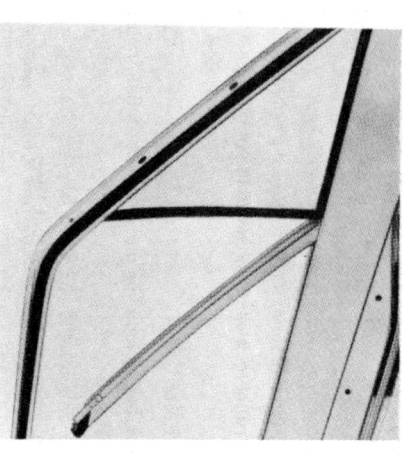

Removing Door Vent Window Dividing Channel

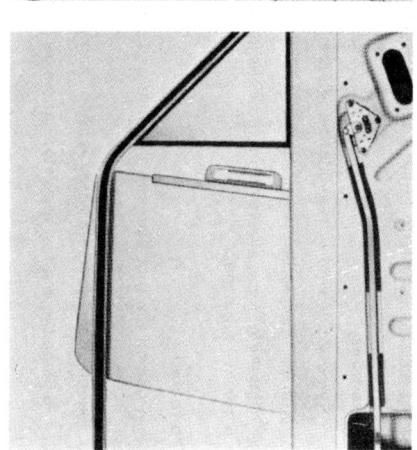

Removing Door Window Glass

OP M 1072-I FRONT DOOR WINDOW GLASS – REMOVE AND INSTALL

To Remove

1. Remove the door trim panel and handles.
2. Remove the door inner and outer belt weatherstrips.
3. Replace the window winder handle, close the window and remove the screw securing the lower end of the window glass rear run.
4. Pull down the rear run to disengage the top end and remove the run through the door access hole.
5. Open and support the window then remove four screws securing the window regulator to the door.
6. Disconnect the window regulator operating arm from the window glass lower run and remove the regulator through the door access hole.
7. Drill out the two "pop" rivets securing the top end of the vent window dividing channel.
8. Remove the screw securing the lower end of the vent window dividing channel and remove the channel from its door.
9. Rotate the window glass through 90° and remove it from the door.

To Install

10. Enter the window glass in the door.
11. Position the dividing channel in the door and locate the window glass in the dividing channel.
12. Secure the top end of the dividing channel with two "pop" rivets and the lower end with a screw.
13. Locate the window regulator in the door and connect the operating arm to the window glass lower run.
14. Secure the window regulator to the door with four screws.
15. Close the window and replace the window glass rear run, clip the top end in position and secure the lower end with a screw.
16. Replace the door inner and outer belt weatherstrips.
17. Replace the door trim panel and handles.

OP M 1105-B LUGGAGE COMPARTMENT LID ASSEMBLY – REMOVE AND INSTALL

To Remove

1. Cover the rear cowl panel with a cloth to prevent accidental damage.
2. If the luggage compartment lid alignment is satisfactory open the lid and scribe a line around the hinges on the underside of the lid.
3. Remove two bolts securing the luggage compartment lid to one of the hinges and rest the lid on the rear cowl panel.
4. Support the luggage compartment lid, remove two bolts from the second hinge and detach the lid.

To Install

5. Place the luggage compartment lid carefully on the rear cowl panel and alternately lift each side and loosely replace the two securing bolts to each hinge.
6. Align the hinges to the marks scribed on the underside of the lid and lightly tighten the bolts.
7. Close lid and check alignment. Adjust if necessary.
8. Fully tighten securing bolts.
9. Remove the protective cloth from the rear cowl.

OP M 1107-A LUGGAGE COMPARTMENT LID MOTIFS – REMOVE AND INSTALL

To Remove

1. Carefully prise out the individual motifs from the luggage compartment lid.
2. Remove the spring clips from the lugs on the back of the motifs.

To Install

3. Fit the retaining spring clips on the lugs at the back of the motifs.
4. Press the motifs into the luggage compartment lid to engage the spring clips.

OP M 1108-A LUGGAGE COMPARTMENT STRIKER POST – REMOVE AND INSTALL

To Remove

1. Open the luggage compartment lid and scribe a line around the catch.
2. Remove the two bolts securing the catch in position.

To Install

3. Replace the catch and lightly secure the two bolts.
4. Close the luggage compartment lid, check alignment and operation, adjust if necessary.
5. Fully tighten the bolts and close the luggage compartment lid.

OP M 1109-A LUGGAGE COMPARTMENT LOCK – REMOVE AND INSTALL

To Remove

1. Open the luggage compartment lid and remove the circlip from the end of the lock spindle.
2. Remove three bolts securing the lock assembly to the luggage compartment lid and remove the lock.

To Install

3. Position the lock assembly in the luggage compartment lid and secure it with three bolts.
4. Replace the circlip on the end of the lock spindle.

CORTINA - LOTUS

OP M 1109-A1 EXTRA: LUGGAGE COMPARTMENT LOCK BARREL – REMOVE AND INSTALL

To Remove

1. Open luggage compartment lid and remove the circlip from the end of the lock spindle.
2. Remove the lock assembly.
3. Carefully prise the lock barrel assembly from the luggage compartment lid.
4. Slacken the locknut and remove the lock spindle, cam, spacer and return spring.
5. Withdraw the lock barrel from the chrome bezel.

To Install

6. Insert the lock barrel in the chrome bezel.
7. Refit the return spring, spacer and cam.
8. Secure the assembly with the lock spindle and tighten the locknut.
9. Replace the lock barrel assembly in the luggage compartment lid.
10. Replace the lock assembly.
11. Refit the circlip on the end of the lock spindle.

OP M 1109-B LUGGAGE COMPARTMENT LOCK BARREL – REMOVE AND INSTALL
(Includes OPS M 1109-A and A1)

OP M 1111-A LUGGAGE COMPARTMENT TORSION BAR – ONE – REMOVE AND INSTALL

To Remove

1. Open and support the luggage compartment lid and remove one screw and clip securing the centre of the torsion bars to the body.
2. Using a suitable spanner release cranked end of the torsion bar from the mounting bracket on the rear cowl reinforcing panel.
3. Disengage the other end of the torsion bar from the hinge and remove the torsion bar.

To Install

4. Locate the torsion bar in the hinge.
5. Using a suitable spanner force the cranked end of the torsion bar into position on its mounting bracket on the rear cowl reinforcing panel.
6. Replace the clip and screw securing the centre of the torsion bars to the body, remove the support and close the luggage compartment lid.

OP M 1111-A1 EXTRA: REMAINING TORSION BAR – REMOVE AND INSTALL
(See OP M 1111-A)

OP M 1111-A2 EXTRA: LUGGAGE COMPARTMENT HINGE – ONE – REMOVE AND INSTALL

To Remove

1. Open the luggage compartment lid and remove one screw and clip securing the centres of the torsion bars to the body.
2. Note the position of the torsion bar in the mounting bracket on the rear cowl reinforcing panel, and release the cranked end with a suitable spanner.
3. Disengage the other end of the torsion bar from the hinge and remove.
4. Scribe a line around the hinge on the underside of the luggage compartment, and on the bracket welded to the rear wheel arch.
5. Remove the two bolts securing the hinge to the luggage compartment lid, and support the lid.
6. Remove the two nuts securing the hinge to the wheel arch bracket, and remove the hinge.

To Install

7. Loosely replace the four hinge bolts.
8. Align the hinge to the marks scribed, and partially tighten the bolts.
9. Close the luggage compartment lid and check alignment. Adjust if necessary.
10. Fully tighten the securing bolts.
11. Replace the end of the torsion bar in the hinge.
12. Fit the cranked end of the torsion bar in its correct position.
13. Replace the clip and screw the centres of the torsion bars to the body.

OP M 1111-B LUGGAGE COMPARTMENT TORSION BARS – BOTH – REMOVE AND INSTALL
(Includes OPS M 1111-A and A1)

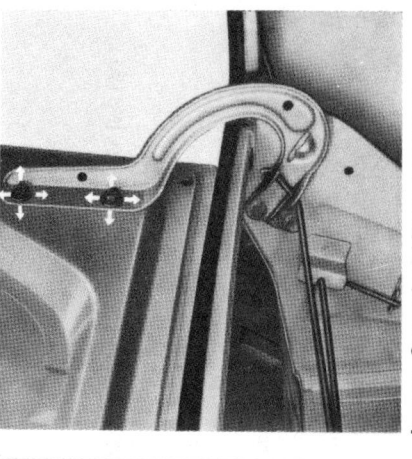

Luggage Compartment Hinge Adjustment

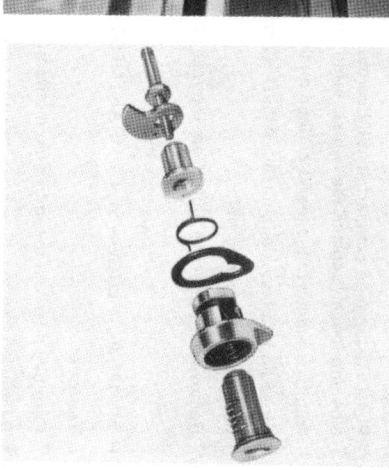

Luggage Compartment Lock – Exploded

CORTINA - LOTUS

OP M 1111-C LUGGAGE COMPARTMENT HINGE – ONE – REMOVE AND INSTALL
(Includes OPS M 1111-A, A1 and A2)

OP M 1111-D LUGGAGE COMPARTMENT HINGES – BOTH – REMOVE AND INSTALL
(Includes OPS M 1111-A, A1 and A2 × 2)

OP M 1205-A WINDSCREEN GLASS AND/OR MOULDING – REMOVE AND INSTALL

To Remove

1. Cover the bonnet and cowl top with a cloth to prevent accidental damage.
2. Remove the windscreen wiper arms.
3. Using a lipping tool, push the weatherstrip lip under the top and sides of the windscreen aperture flange, from inside the car.
4. From inside the car push out the windscreen and weatherstrip as an assembly.
5. Prise out the joint cover clips and pull the finish strip out of the weatherstrip and remove the weatherstrip from the glass.

NOTE – A second method of windscreen removal often employed by experienced operators is described below. However, if this method is used to remove a laminated windscreen, extreme care must be used as there is a certain element of risk involved.

Ensure that the operator is wearing light shoes, then:—

(a) Sit in the front seat.

(b) To avoid scratching the glass place a piece of soft cloth between the operator's shoes and the windscreen glass.

(c) Place both feet in the top corner of the windscreen and push firmly.

(d) When the weatherstrip is free of the body flange in that area, repeat the procedure at intervals along the top edge of the windscreen until, from outside the vehicle, the glass and weatherstrip can be removed together.

To Install

6. Refit the weatherstrip around the windscreen glass.
7. Fit a downcord in the weatherstrip rubber to body groove allowing a crossover of approximately 6 in. (15 cm.) at the bottom.
8. Using a sealer gun fitted with a $\frac{1}{8}$ in. (3 mm.) swan-necked nozzle, apply a suitable sealer (SR-51-B) to the weatherstrip rubber to body groove, with special attention to the corners.
9. Position the windscreen in the body aperture with the cord ends inside the car.
10. Pull out the cord whilst applying hand pressure to the outside of the glass.
11. Using a sealer gun fitted with a $\frac{1}{8}$ in. (3 mm.) swan-necked nozzle, apply a suitable sealer (SR-51-B) to the rubber to glass groove.
12. Replace the finish strip in its groove in the weatherstrip. The lipping tool will help in replacement. Replace the joint cover clips.
13. Replace the windscreen wipers.
14. Remove the cloth covering the bonnet and cowl top and remove any surplus sealer from around the windscreen.

OP M 1210-A REAR WINDOW GLASS AND/OR MOULDING – REMOVE AND INSTALL

To Remove

1. Cover the luggage compartment lid and rear quarter panels with a cloth to prevent accidental damage.
2. Using a lipping tool push the weatherstrip lip under the top and sides of the rear window aperture flange, from inside the car.
3. From inside the car push out the window and weatherstrip as an assembly.
4. Remove the joint clips and pull the finish strip out of the weatherstrip and remove the weatherstrip from the glass.

NOTE – A second method of screen removal is outlined in operation M 1205-A.

To Install

5. Refit the weatherstrip around the rear window glass.
6. Fit a drawcord in the weatherstrip rubber to body groove, allowing a crossover of approximately 6 in. (15 cm.) at the bottom.
7. Using a sealer gun fitted with a $\frac{1}{8}$ in. (3 mm.) swan-necked nozzle, apply a suitable sealer (SR-51-B) to the weatherstrip rubber to body groove, with special attention to the corners.
8. Position the rear window in the body aperture with the cord ends inside the car.
9. Pull out the cord whilst applying light hand pressure to the outside of the glass.

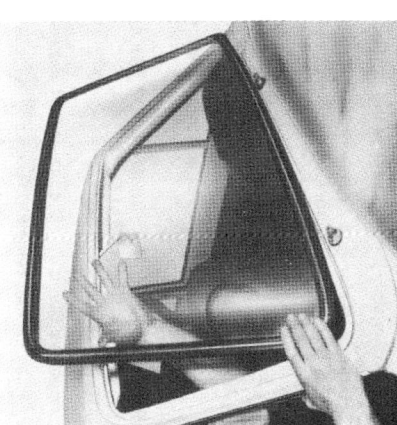

Replacing the Windscreen

Removing the Windscreen

CORTINA - LOTUS

10. Using a sealer gun fitted with a ⅛ in. (3 mm.) swan-necked nozzle, apply a suitable sealer (SR-51-B) to the rubber to glass groove.

11. Replace the finish strip in its groove in the weatherstrip. The lipping tool will help in replacement. Replace the joint clips.

12. Remove the cloth covering the bonnet and cowl top and remove any surplus sealer from replacement. Replace the joint clips.

OP M 1215-A REAR QUARTER WINDOW GLASS AND/OR MOULDING – REMOVE AND INSTALL

To Remove

1. Using a lipping tool push the weatherstrip lip under the top and sides of the body aperture flange.

2. From inside the car push out the window and weatherstrip as an assembly.

3. Carefully prise the metal moulding out of the weatherstrip.

   NOTE – If the glass is being replaced due to accident damage, remove all traces of hardened sealer and shattered glass from the weatherstrip and body aperture flange.

To Install

4. Replace the weatherstrip around the rear quarter glass.

5. Using a sealer gun fitted with a ⅛ in. (3 mm.) swan-necked nozzle, apply a suitable sealer (SR-51-B) to the rubber to glass groove.

6. Refit the metal moulding to the rear quarter window weatherstrip, starting at top front corner of the weatherstrip and working towards either end.

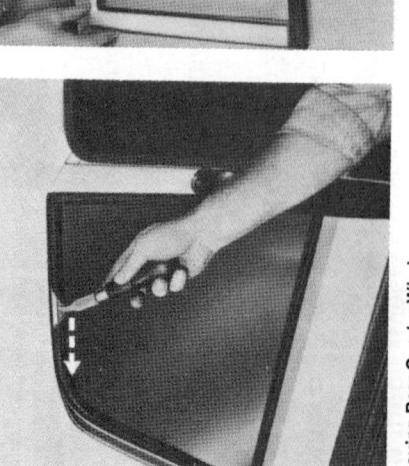

Removing Rear Quarter Window

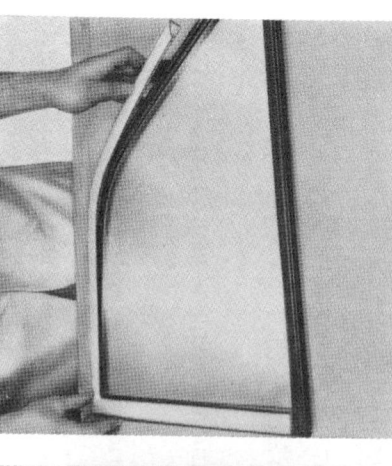

Rear Quarter Window and Moulding

7. Fit a drawcord in the weatherstrip rubber to body groove allowing a crossover of approximately 6 in. (15 cm.) at the bottom.

8. Using a sealer gun fitted with a ⅛ in. (3 mm.) swan-necked nozzle, apply a suitable sealer (SR-51-B) to the rubber to body groove of the weatherstrip, paying particular attention to the bottom corners.

9. Locate the rear quarter window in the body aperture with the cord ends within the car.

10. Pull out the cord whilst applying light hand pressure to the outside of the glass.

11. Remove any surplus sealer from around the rear quarter window.

OP M 1220-A "AEROFLOW" EXTERIOR EXTRACTION LOUVRE – REMOVE AND INSTALL

To Remove

1. Remove the rear seat cushion and backrest.

2. Prise up four retaining tangs and remove the rear parcel shelf trim pad.

3. Pull the rear door finish strip away in the area adjacent to the "Aeroflow" trim panel.

4. Pull down the top edge of the rear quarter trim panel and drill out the "pop" rivet securing the lower edge of the "Aeroflow" trim panel.

5. Pull the "Aeroflow" trim panel away from the roof to rear quarter panel and remove.

6. Remove five nuts securing the exterior louvre and bright metal garnish moulding and withdraw louvre and moulding from the car.

To Install

7. Clip the rear end of the garnish moulding over the stud on the rear edge of the roof to rear quarter panel.

8. Locate the exterior louvre and garnish mouldings in position and secure with five nuts.

Rear Extraction Louvre

Extraction Box

CORTINA - LOTUS

9. Replace the extraction trim panel and secure the lower edge with a "pop" rivet.
10. Replace the rear quarter trim panel and rear door finish strip.
11. Replace the rear parcel shelf and secure with four retaining tangs.
12. Replace the rear seat cushion and backrest.

OP M 1220-A1 EXTRA: EXTRACTION BOX – REMOVE AND INSTALL

To Remove

1. Remove one screw securing the top edge of the extraction box.
2. From outside the car remove five screws securing the extraction box and remove the box from the car.

To Install

3. Locate the extraction box in the roof to rear quarter panel and secure the top end with one screw.
4. Replace the five screws securing the extraction box.

OP M 1305-A FRONT PARCEL SHELF – REMOVE AND INSTALL

To Remove

1. Remove the heater trim panel (see OP M 1355-A).
2. Prise out the four trim clips, two each side, securing the parcel shelf to the heater and cowl side panel and remove the parcel shelf.

To Install

3. Position the parcel shelf on its mounting brackets and secure the shelf with four trim clips, two each side.
4. Replace the heater trim panel (see OP M 1355-A).

OP M 1305-A1 EXTRA: COWL SIDE TRIM PANEL – REMOVE AND INSTALL

To Remove

1. Prise the metal tag securing the front edge of the cowl side trim panel backwards.
2. Pull the front edge of the trim panel away from the cowl side, move the panel towards the rear of the vehicle and remove the panel.

To Install

3. Position the trim panel in the vehicle ensuring that the lip on the rear edge of the panel is correctly positioned over the door aperture flange.
4. Push the metal tag forwards to secure the front edge of the trim panel.

OP M 1305-B COWL SIDE TRIM PANEL – REMOVE AND INSTALL
(Includes OPS M 1305-A and A1)

OP M 1310-A REAR PARCEL SHELF – REMOVE AND INSTALL

To Remove

1. Remove the rear seat cushion and backrest (see OP M 1325-A).
2. (*Where fitted*) Remove four nuts securing the radio speaker and grille to the rear parcel shelf and remove the grille.
3. Prise up the four tags securing the front edge of the parcel tray and remove the parcel tray.

To Install

4. Position the rear parcel tray in the vehicle ensuring that the rear edge is tucked under the rear window weatherstrip flange.
5. Secure the front edge with four metal tags.
6. (*Where fitted*) Reposition the radio speaker and grille and secure them with four nuts.

OP M 1315-A REAR QUARTER TRIM PANEL – REMOVE AND INSTALL

To Remove

1. Remove the rear seat cushion.
2. Remove two screws from the lower edge of the backrest, two nuts securing the top edge of the backrest and remove the backrest.
3. Pull the trim pad away from the body to release the spring clips and remove the trim pad.

To Install

4. Position the trim pad and press the edges to engage the securing spring clips.
5. Replace the backrest and secure it with two screws and nuts.
6. Replace the rear seat cushion.

OP M 1320-A DRIP RAIL MOULDINGS – ONE SIDE – REMOVE AND INSTALL

To Remove

1. Starting from the front of the vehicle carefully prise off the front moulding from the drip rail.
2. Repeat operation 1 for the rear moulding.

To Install

3. Position the rear moulding on the drip rail and using a rubber mallet carefully tap the moulding over the drip rail.
4. Repeat operation 3 for the front moulding.

OP M 1325-A REAR SEAT BACKREST – REMOVE AND INSTALL

To Remove

1. Remove the rear seat cushion (see OP M 1327-A).

CORTINA - LOTUS

2. Remove two screws securing the bottom corners of the backrest.
3. From inside the luggage compartment remove two nuts securing the top corners of the backrest.
4. Remove the rear seat backrest.

To Install

5. Position the backrest in the vehicle.
6. From inside the luggage compartment secure the top corners of the backrest with two nuts.
7. Secure the bottom corners of the backrest with two screws.
8. Replace the rear seat cushion (see OP M 1327-A).

OP M 1327-A REAR SEAT CUSHION – REMOVE AND INSTALL

To Remove

1. Lift the front edge of the rear seat cushion upwards to disengage the cushion from the heel plate and remove the cushion from the vehicle.

To Install

2. Position the rear seat cushion in the vehicle and press the front edge downwards to engage the cushion frame in the heel plate.

OP M 1330-A BUCKET SEAT – FRONT – REMOVE AND INSTALL

To Remove

1. Pull back the floor carpet to expose the front seat mounting brackets.
2. Remove four bolts securing the mounting brackets and reinforcing plates to the floorpan.
3. Remove the seat, mounting brackets and reinforcing plates.

To Install

4. Position the seat, mounting brackets and reinforcing plates in the car and secure them with four bolts.
5. Replace the carpet.

NOTE – Additional leg room can be obtained by reversing the mounting brackets which are offset for this purpose.

OP M 1340-A INTERIOR MIRROR – REMOVE AND INSTALL

To Remove

1. Remove two screws securing the interior mirror to the windscreen header panel and remove the mirror.

To Install

2. Position the interior mirror on the windscreen header panel and secure it with two screws.
3. Screw the anti-vibration pad anti-clockwise until the pad is firmly pressing on the windscreen.

OP M 1345-A SUN VISOR – REMOVE AND INSTALL

To Remove

1. Remove three screws securing sun visor to the windscreen header panel.
2. Remove the sun visor.

To Install

3. Locate the sun visor in position on the windscreen header panel.
4. Secure the sun visor with three screws.

OP M 1351-A GLOVE LOCKER LID – REMOVE AND INSTALL

To Remove

1. Open the glove locker lid and remove four securing screws and detach the lid.

To Install

2. Locate the lid on its hinges and secure it with four screws.

OP M 1355-A HEATER TRIM PANEL (Where Fitted) – REMOVE AND INSTALL

To Remove

1. Remove four screws securing the top edge of the heater trim panel.
2. Carefully slide the trim panel from around the heater and remove the panel.

To Install

3. Position the trim panel around the heater.
4. Secure the top of the trim panel with four screws.

OP M 1360-A HEADLINING – RENEW

To Remove

1. Remove four screws and detach the sun visors.
2. Remove two screws and detach the rear view mirror.
3. Disconnect the battery, remove two screws and pull down the interior light and disconnect the wiring.
4. Pull the door finish strips away from the door aperture flanges adjacent to the headlining and extraction trim panel.
5. Remove the rear seat cushion and backrest.

6. Remove two screws and detach the two extraction trim panels.
7. Remove the wiper arms and blades.
8. Remove the windscreen and rear window.
9. Carefully free the edges of the headlining from the door aperture flanges, windscreen header panel and rear window header panel.
10. Disconnect the four listing wires from their nylon sockets and remove the headlining and listing wires as an assembly.

To Install

11. If a new headlining is being fitted, transfer the listing wires from the old headlining to the appropriate cotton sleeve in the new headlining.
    NOTE – The listing wires are colour coded with a daub of paint. Starting at the front the colours are: 1=Blue, 2=Green, 3=Brown, 4=Yellow.
12. Break the cotton sleeves to allow the listing wires to protrude approximately 4 to 6 in. (101 to 152 mm.), then locate the headlining complete with listing wires in the car and position the listing wires in their respective sockets.
13. Apply adhesive to the edges of the headlining and the windscreen header panel, door aperture flanges and rear window header panel.
14. Carefully position the front edge of the headlining on the windscreen header panel and temporarily secure with suitable spring clips.
15. Pull the rear edge of the headlining back and starting at the centre, position the headlining on the rear window header panel.
16. Carefully stick the sides of the headlining to the door aperture flanges and temporarily secure them with suitable spring clips.
17. Remove any creases in the headlining and then trim the headlining round the flanges allowing approximately ½ in. (12.7 mm.) of material to overlap the flanges.

Listing Wire Colour Code

18. Remove the spring clips and stick the edges of the headlining over the windscreen, rear window and body flanges.
19. Replace the windscreen and rear window.
20. Replace the wiper arms and blades.
21. Replace the extraction trim panels and secure with two screws.
22. Replace the rear seat cushion and backrest.
23. Refit the door aperture finish strips.
24. Cut a hole in the headlining and pull the interior light wires through and connect them to the interior light.
25. Position the interior light and secure it with two screws.
26. Replace the rear view mirror and secure it with two screws.
27. Replace the two sun visors and secure with six screws.
28. Reconnect the battery.

UNDERBODY REPAIR PROCEDURES

Dimensional drawings of the underbody and upper body framework are contained at the end of this section.

These drawings, used in conjunction with a level floor and suitable stands, will be of assistance when checking or undertaking repairs to the bodywork.

Alternatively, a body checking and repair fixture known as the Churchill "700" system, available from V. L. Churchill & Co. Ltd., may be used. This specialist piece of equipment is suitable for all Ford passenger cars produced from July 1954 onwards and is adaptable for all future models. To ensure accurate alignment of the front suspension and engine mounting points, a jig or checking fixture is available from V. L. Churchill & Co. Ltd. under Tool No. P 5517.

Outlined below are the recommended procedures to be followed when checking or replacing parts in the engine compartment and underbody.

CHECKING THE ENGINE COMPARTMENT

If the damaged car is still on its wheels and misalignment of the body within the engine compartment is suspected, carry out a full steering geometry check as outlined in Section 5.

WHEEL ALIGNMENT (unladen)

	STANDARD
Castor	−0° 45′ to +0° 45′
Camber	0° 15′ to 1° 45′
King Pin Inclination	7° 11′ to 8° 41′
Front lock wheel angle with back lock set at 20°	19° 00′ to 20° 30′

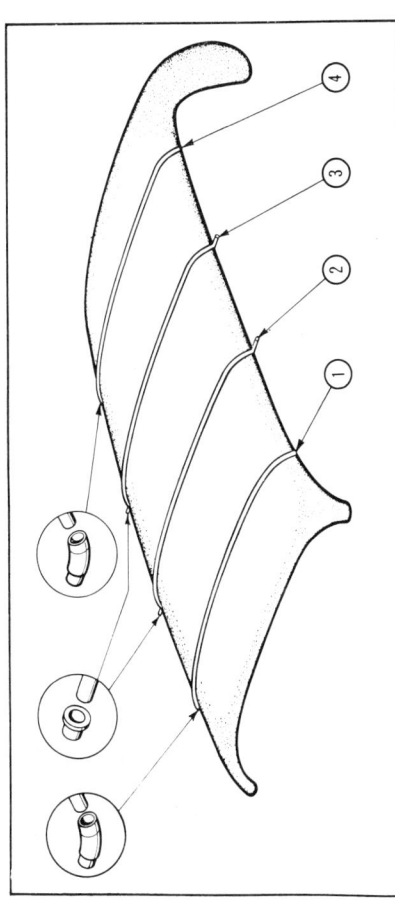

CORTINA - LOTUS

Toe-in (Service Setting)* ... ... 0.10 to 0.20 in. (2.8 to 5.2 mm.) (Nominal Fraction 3/32 in.)
Toe-in (Production Setting)* ... ... 0.06 to 0.25 in. (1.5 to 6.4 mm.) (Nominal Fraction 5/32 in.)

*The production toe-in setting should be used prior to the first 500 mile (1,000 Km.) service only. From and including that service, the Service Setting must be used.

If these checks indicate any discrepancy outside the allowed tolerances and no damage to the suspension or steering controls exists, then it will be necessary to investigate further in the following manner.

Remove the engine assembly, the front suspension crossmember, the front suspension, the stabiliser brackets and the steering box from the car.

Ensure that the undersides of the front body side members are clean at the stabiliser bracket mountings and attach the two jig adaptor plates, one either side. Remove the detachable locating pins and place the jig in position within the engine compartment, locating the lower portion between the front body side members.

From under the front mudguards, pass the longer threaded locating pins, one either side, through the foremost hole in each body side member for the steering box securing bolts. Engage the corresponding holes in the jig and tighten the locating pins.

Fit the locating pins through the holes in the jig adaptors to engage in the corresponding holes in the jig.

To check the alignment of the suspension unit upper mounting holes in the engine side apron panels, engage the set of three parallel pins. The pins should pass through the holes in the engine side apron panel and reinforcement and should not be forced into engagement.

Check the clearance between the upper face of the engine side apron panel and the underside of the jig top plate. A clearance of 1/8 in. (3.18 mm.) should be present.

If distortion to any major component affecting the front suspension (e.g. a distorted front side member or apron panel) is disclosed as a result of tests carried out, no attempt should be made to straighten members, etc. by using hydraulic body jacks with the jig in position, or damage to the jig will result. In view of the importance of the front end alignment, it is suggested that in general, serious distortion to these items should be corrected by replacing affected parts.

When rebuilding the engine compartment as part of an accident repair, the jig Tool No. P 5517 may be used to align and assemble any or all of the engine compartment components. Thus, if necessary, a complete engine compartment may be built and welded together as a sub-assembly around the jig. The two longest locating pins which are threaded may be used to hold the body side members and side apron panels to the jig during this operation. To establish a dimension between the stabiliser brackets, utilise the 1/16 in. (1.6 mm.) flats machined on the two tapered wedges supplied with the jig. The top of the engine side apron panels may be held in position with the smaller parallel pins located in the inner holes in the jig outriggers.

NOTE – If a complete engine compartment is built around the jig in this manner, offer the welded assembly in the body **with the jig in position.** When correctly aligned to the body, the sub-assembly may then be welded in position. Do not remove the jig until both front mud-guards and the radiator grille panel have been welded in position, otherwise distortion to the engine compartment may occur.

Once the assembly has been finally welded in position, remove the parallel pins locating the top edges of the engine side apron panels and using an 11/32 in. (8.7 mm.) drill, drill the six holes, three each side, for the front suspension units top mounting bolts.

Three drill bushes are provided in each outrigger to ensure accurate alignment of the holes.

CHECKING THE UNDERBODY

Two methods can be used to check the underbody for misalignment. The first method is to use the Churchill "700" system fixture as outlined below.

The fixture consists of two heavy steel girders which are let into an existing floor to provide a level working base. These are drilled and tapped to allow four steel stands to be bolted to them at pre-determined locations. Two adaptors to suit the model being checked are then secured to each stand, one either side. In this way, eight locations on the body within specified limits are established and any distortion to the underbody is at once indicated.

Since the car jacking positions are used as "pick up" points, only minor dismantling of mechanical parts is necessary to carry out a complete alignment check.

The "700" system body fixture ensures the correct alignment of critical underbody parts straightened or renewed during repair.

For example, if obvious extensive damage exists to the front end area of the car, it is usually possible to secure the car by the jacking points and rear longitudinal members. The undamaged section of the car is then held firmly within specified tolerances and the frontal area may be repaired.

When damaged sections have been cut away or straightened, the fixture front stand and adaptor is then fitted. By engaging locating pins, it is then possible to ensure correct alignment of new or straightened parts relative to the undamaged body section.

The second method, if the equipment described above is not available, is to use the dimensioned drawings contained at the end of this section. The checking operation should be carried out in the following manner:—

Support the body on suitable stands or adjustable supports on a level floor extending the full length of the car. Check the dimensions shown along the whole length of the body and note any discrepancy outside the allowed tolerances shown on the chart overleaf.

The diagonals marked on the plan view can be checked by using large callipers or a pair of trammels, or alternatively, they may be checked by using a plumb bob and line. The latter method enables a simple and accurate check to be made. Suspend the plumb bob from the appropriate reference points on the body and carefully mark the floor at each location. Connect these points by a chalk line and then draw a line through the intersecting points of the diagonals.

Finally, check the dimensions between the front and rear side members.

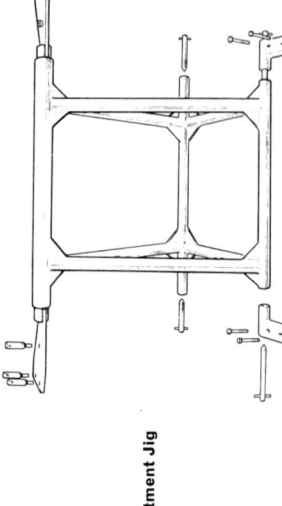

Engine Compartment Jig

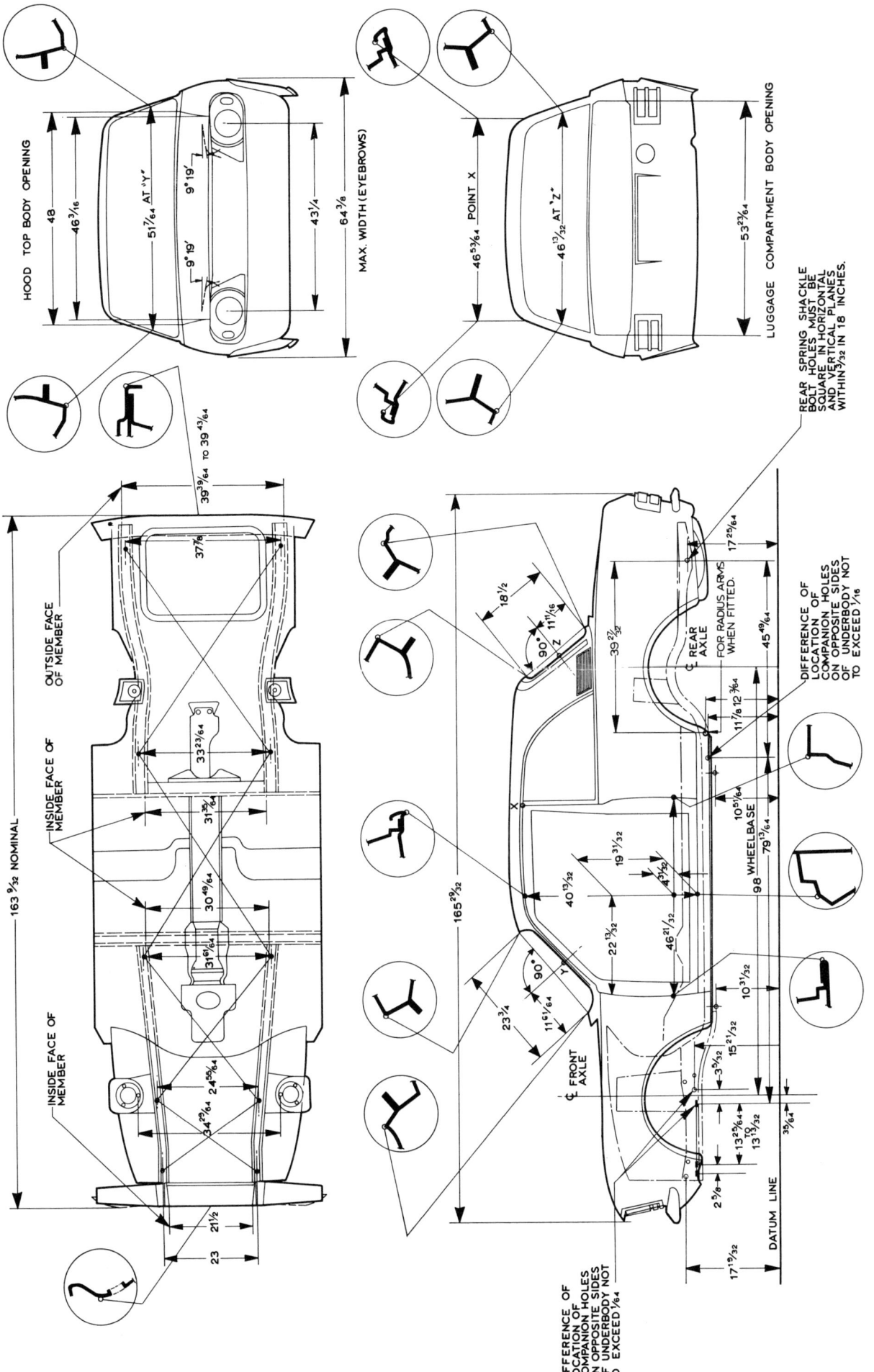

CORTINA - LOTUS

METRIC EQUIVALENTS (IN DIMENSIONAL SEQUENCE)

Inches	Centimetres	Inches	Centimetres	Inches	Centimetres
$\frac{1}{64}$	0.04	$17\frac{19}{32}$	44.69	$40\frac{13}{32}$	102.63
$\frac{1}{16}$	0.16	$18\frac{1}{2}$	46.99	$43\frac{1}{4}$	109.86
$\frac{3}{32}$	0.24	$19\frac{31}{32}$	50.72	$45\frac{49}{64}$	116.24
$\frac{35}{64}$	1.39	$21\frac{1}{2}$	54.61	$46\frac{3}{16}$	117.32
$2\frac{5}{8}$	6.67	$22\frac{13}{32}$	56.91	$46\frac{13}{32}$	117.87
$3\frac{5}{32}$	8.01	23	58.42	$46\frac{21}{32}$	118.51
$4\frac{31}{32}$	12.62	$23\frac{3}{4}$	60.32	$46\frac{53}{64}$	118.94
$10\frac{51}{64}$	27.42	$24\frac{55}{64}$	63.14	48	121.92
$10\frac{31}{32}$	27.86	$30\frac{49}{64}$	78.14	$51\frac{7}{64}$	129.82
$11\frac{11}{16}$	29.69	$31\frac{35}{64}$	80.13	$53\frac{23}{64}$	135.53
$11\frac{7}{8}$	30.16	$31\frac{61}{64}$	81.16	$64\frac{3}{8}$	163.51
$11\frac{61}{64}$	30.36	$33\frac{23}{64}$	84.73	$79\frac{13}{64}$	201.18
$13\frac{25}{64}$	34.01	$34\frac{29}{64}$	87.51	98	248.92
$13\frac{13}{32}$	34.05	$37\frac{7}{8}$	96.20	$163\frac{9}{32}$	414.73
$15\frac{21}{32}$	39.77	$39\frac{39}{64}$	100.61	$165\frac{35}{32}$	421.40
$17\frac{43}{64}$	44.17	$39\frac{49}{64}$	100.77		

CORTINA - LOTUS

13

LUBRICATION
AND
MAINTENANCE

SERVICING INTERVALS

    First Service: 600 Miles (1,000 Km.)   Second Service: 3,000 Miles (5,000 Km.)

        continuing at subsequent 3,000 Mile (5,000 Km.) intervals

CORTINA - LOTUS

LUBRICANTS AND FLUIDS

1. FRONT WHEEL BEARINGS ... ... Lithium base grease
2. DISTRIBUTOR ... ... Engine oil and grease Part No. 68AB-19D533-AA
3. ENGINE

SAE Grade	Use Between
5W-20, 5W-30	−40°F to +32°F
10W-30	−10°F to +70°F
10W-40	−10°F to +90°F
10W-50	−10°F to +120°F
20W-40	+25°F to +90°F
20W-50	+32°F to +120°F
20W-50 "All Season"	+25°F to +120°F
10W	−10°F to +32°F
20W-20	+25°F to +70°F
30	+32°F to +90°F
40	+50°F to +120°F

4. STEERING GEAR ... ... S.A.E. 90 E.P. Gear Oil
5. GEARBOX ... ... S.A.E. 80 E.P. Gear Oil
6. GENERATOR ... ... Engine Oil
7. REAR AXLE ... ... S.A.E. 90 Hypoid Oil
8. BRAKE FLUID RESERVOIR ME-3833-F
9. CLUTCH FLUID RESERVOIR ME-3833-F
10. ANTI-FREEZE ... ... Long-Life Part No. M97B18-C in 50% solution

APPROVED LUBRICANTS (applicable to U.K. only)

Manufacturer	Engine	Gearbox	Rear Axle
Amoco ('UK') Ltd.	(a) Amoco Permalube SAE 20-20W (b) Amoco Super Permalube SAE 10W-30 (c) Amoco Super Permalube SAE 20W-50	Amoco M.P. Gear Lubricant SAE 80	Amoco M.P. Gear Lubricant SAE 90
B.P. (Shell-Mex and B.P. Ltd).	(b) B.P. Super-Visco Static 10W-40 (c) B.P. Super-Visco Static 20W-50	B.P. Gear Oil SAE 80 EP	B.P. Gear Oil SAE 90 EP
Castrol Ltd.	(c) Castrol GTX*	Castrol Hypoy Light	Castrol Hypoy
Alexander Duckham Ltd.	(b) Duckham's Q5500 (c) Duckham's Q20-50*	Duckham's Hypoid 80	Duckham's Hypoid 90
Esso Petroleum Co. Ltd.	(a) Esso Motor Oil 20W (b) Esso Extra 10W/30 or Uniflo** (c) Esso Extra 20W/50	Esso Gear Oil GP 80	Esso Gear Oil GP 90/140
Mobil Oil Co. Ltd.	(b) Mobiloil Super** 10W/50 (c) Mobiloil Special 20W/50	Mobilube GX 80	Mobilube GX 90
Petrofina (G.B.) Ltd.	(b) Fina Supergrade SAE 10W/30 (c) Fina Supergrade SAE 20W/50*	Fina Pontonic MP SAE 80	Fina Pontonic MP SAE 90
Regent Oil Co. (Texaco Ltd.)	(a) Regent Havoline 20/20W (b) Regent Havoline 10W/30 (c) Regent Havoline 20W/50	Regent Multi-Gear EP 80	Regent Multi-Gear EP 90
Shell (Shell-Mex and B.P. Ltd.)	(b) Shell Super 101 (10W-30) (c) Shell Super 100 (20W-50)*	Shell Spirax 80 EP	Shell Spirax 90 EP

(a) Monograde 20W-20 oils
(b) Multigrade 10W-30, 10W-40 or 10W-50 oils
(c) Multigrade 20W-50 oils

*This oil is an "All Season" 20W-50 suitable for use in temperatures down to 25°F, i.e. 7° of frost.
**Oils marked thus conform to temperature requirements and are therefore completely suitable for year round use to a minimum temperature of −10°F, i.e. 42° of frost.

CORTINA - LOTUS

AT FIRST 600 MILES (1,000 KM.)

1 Top-up Brake and Clutch Fluid Reservoirs

Clean area around reservoir caps. Unscrew caps and check fluid level. If necessary top-up with correct FoMoCo fluid ME-3833-F and refit caps.

2 Check Door Operation, Adjust Striker where necessary

3 Tighten Cylinder Head Bolts to the Correct Torques

Tighten cylinder head bolts to 60 to 65 lb. ft. (8.29 to 8.98 kg.m.) torque with engine cold.

4 Check and Adjust Valve Clearances

Check clearances when engine cold.
Inlet   0.005 to 0.007 in. (0.13 to 0.18 mm.)
Exhaust
    Pre engine No. LP 9952
        0.006 to 0.008 in. (0.15 to 0.20 mm.)
    From engine No. LP 9952
        0.009 to 0.011 in. (0.228 to 0.279 mm.)

5 Check Timing Chain Tension

The chain tension should allow ½ in. (13 mm.) free movement at a point midway between the two camshaft sprockets. If necessary, adjust the tension by means of the adjuster located in the side of the front cover.

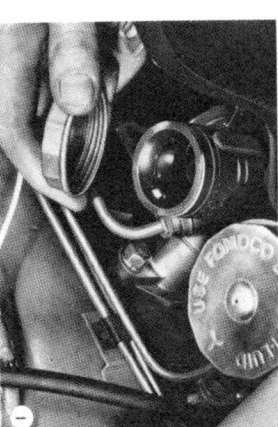

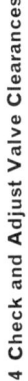

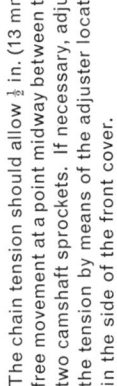

6 Tighten Sump Bolts (in alphabetical then numerical sequence shown) and Manifold Bolts to the Correct Torque

Manifold Nuts   15–18 lb. ft. (2.07–2.49 kg.m.)
Sump Bolts   7–9 lb. ft. (0.968–1.244 kg.m.)

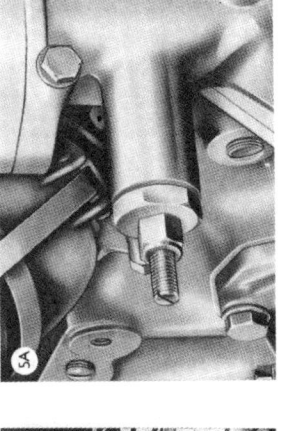

7 Inspect Brake Hoses and Lines for Chafing and Leaks

8 Check Fan Belt Tension and adjust if required

Free movement of ½ in. (13 mm.) measured midway between generator and water pump pulleys. If required, adjust by slackening front and rear lower mounting bolts and front adjusting bolt. Move generator to give correct belt tension. Tighten generator mounting bolts to a torque of 15 to 18 lb. ft. (2.08 to 2.49 kg.m.).

9 Road Test, Adjust Carburettor Idling and Ignition Timing

Remove the outer part of the air-box, secured by three bolts between the carburettors. Carefully screw each volume control screw in until it just contacts its seat, and then unscrew three-quarters of a turn.
Set the throttle stop screw so that the slow running is slightly higher than normal, and then adjust the synchronizing screw of the throttle spindle connecting linkage to match the throttle openings. This can be checked by placing one end of a piece of rubber tube against the ear and the other at the carburettor intake. Since the throttle plates in each carburettor are on a common spindle it is only necessary to place the tube in the intake of one barrel of each carburettor. If the hiss of one carburettor is louder than the other, adjust the throttle stop screw and synchronizing screw to equalise the intensity of the hiss and therefore the intake of air.

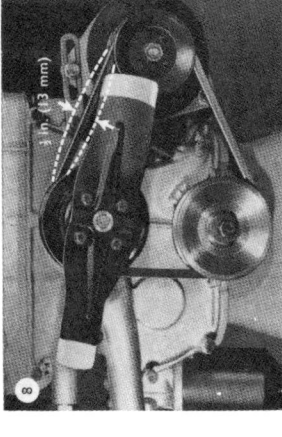

Each volume control screw should then, in turn, be screwed in or out an eighth of a turn at a time, allowing time for each adjustment to take effect, until the engine runs evenly. If a volume control screw is screwed in too far then that cylinder will cut-out and the screw must therefore be unscrewed until the cylinder functions again. When it is excessively unscrewed the engine will 'hunt' and therefore the needle must be turned back to bring it closer to its seat.

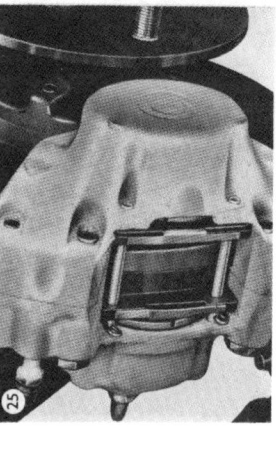

25 **Remove Road Wheels, check front brake pads for wear, examine rear brake shoes and self-adjusting mechanism and blow clean. Reposition road wheels if required.**

Check for pad wear. Jack up car, remove wheel and measure thickness of pad material. If between ⅛ to 1/16 in. (3.18 to 1.59 mm.) renew pads. Refer to operation 63 for packing front wheel bearings.

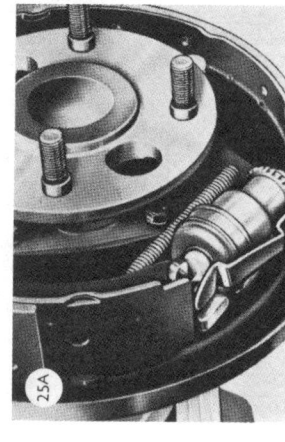

Remove the rear brake drums, after removing the road wheels and the brake drum securing screws, to check the condition of the brake linings and self-adjusting mechanism.

Reposition road wheels, if required. Spare to left front, left front to left rear, left rear to right rear, right rear to right front, right front to spare wheel compartment.

13 **Check and Adjust Valve Clearances**

14 **Check Timing Chain Tension**

15 **Road Test, Adjust Carburettor Idling and Ignition Timing**

16 **Top-up Clutch and Brake Fluid Reservoirs**

17 **Check operation of all Controls, Instruments and Lights (align if necessary)**

18 **Correct Tyre Pressures and Inspect Tyres**

19 **Check Fan Belt for Tension and Wear**

20 **Tighten Generator Mounting Bolts to Torque of 15 to 18 lb. ft. (2.08 to 2.49 kg.m.).**

21 **Inspect Radiator and Heater Hoses for Leaks and Deterioration**

22 **Check Engine for Water and/or Oil Leaks**

23 **Check Exhaust System for Leaks and Damage**

24 **Inspect all Brake Hoses and Lines for signs of Leaks and Chafing**

If, after setting the volume control screws, the engine idling speed is too fast (above 800/1000 r.p.m.) slacken the throttle stop screw until the idling is satisfactory and, if necessary, readjust the ignition setting. It may also be necessary to readjust the volume control screws and throttle stop screw when the outer part of the air box is refitted. Check that the spring washer gap is within the dimension indicated.

AT FIRST 3,000 MILES (5,000 KM.) OR THREE MONTHS (WHICHEVER OCCURS FIRST)

10 **Check Torque of Rear Spring 'U' Bolts**
Tighten bolts to a torque of 20 to 25 lb. ft. (2.76 to 3.46 kg.m.).

11 **Change Engine Oil and Renew Oil Filter Element**
Remove sump drain plug, allow oil to drain and replace plug. Unscrew oil filter centre bolt, remove casing, filter element and rubber sealing ring. Fit new sealing ring and element then refit cleaned casing and securing bolt. Tighten bolt. Remove oil filler cap and clean. Refill engine with approved oil to correct level. Replace oil filler cap. Start engine and check for oil leaks.

12 **Clean Sparking Plugs and Set Gaps**
The sparking plugs (14 mm. Autolite AG22) should be cleaned and the gaps set at 0.023 in. (0.584 mm.). For maximum efficiency it could be advantageous to renew the spark plugs every 10,000 miles (16,000 km.). Ensure that the sparking plugs insulators, distributor cap and leads are clean to prevent H.T. tracking.

CORTINA - LOTUS

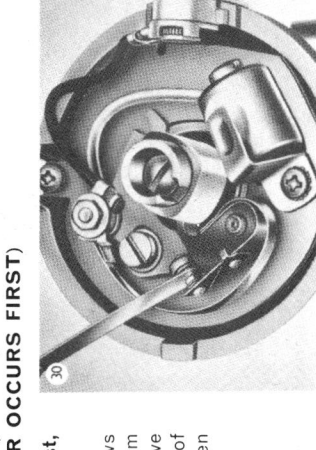
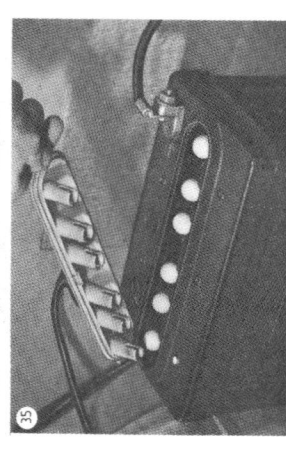

EVERY 6,000 MILES (10,000 KM.) OR SIX MONTHS
(WHICHEVER OCCURS FIRST)

30 Examine Distributor Points and Adjust, Clean Distributor Cap and Coil

To adjust point gap, slacken locking screws and ensure that moving contact breaker arm heel is on the highest point of cam. Move fixed contact point to give clearance of 0.014 to 0.016 in. (0.356 to 0.406 mm.), tighten locking screws and recheck gap.

31 Grease Distributor Cam

Apply a film of grease Part No. 681F-12107-AA to cam. Replace rotor and cap.

32 Lubricate Distributor and Generator Rear Bearing

Remove distributor cap and rotor. Apply two drops of engine oil to lubricating pad inside of cam body.
Inject a few drops of engine oil through aperture in centre of generator rear end plate.

33 Top-up Rear Axle

34 Top-up Gearbox

35 Top-up Battery and Check Connections

The electrolyte level should be ¼ to ⅜ in. (6 to 9 mm.) above tops of the plates. If below this level add distilled water to filling trough until the trough begins to fill with water. Check security of battery and coat terminals with petroleum jelly.

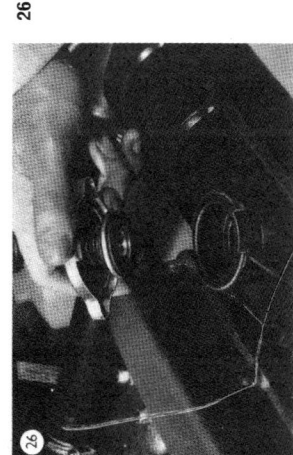

26 Top-up Radiator and Windshield Washer Reservoir

Remove radiator filler cap and top-up to the correct level with 50% water and 50% FoMoCo Longlife anti-freeze Part No. M97B18-C.
The windscreen washer fluid container should be topped-up with the solvent solution obtainable from your Authorised Dealer. Alternatively, use water.

27 Change Gearbox Oil

First 3,000 miles only, remove drain plug and allow oil to drain. Remove filler and level plug, refill gearbox with approved oil and replace plug. At 600 miles (1000 km.) and all other services (except first 3,000 miles (4000 km.)) check and top-up oil level only.

28 Top-up Steering Box

Remove rubber filler plug. Top-up with approved oil if necessary. Refit plug.

*29 Lubricate Drive Shaft Sliding Joint

Using a good quality molybdenum disulphide lithium base grease, lubricate the grease point on the drive shaft sliding joint just to the rear of the centre universal joint.

*Not applicable to 1969 model.

EVERY 3,000 MILES OR THREE MONTHS
(WHICHEVER OCCURS FIRST)

As for first 3,000 mile check except for item 11 which is substituted by:
Check and Top-up Engine Oil Level

And items 27 and 10 which are not applicable.

CORTINA - LOTUS

36 Check Tyre Pressures and Inspect Tyres

37 Check and Adjust Valve Clearances

38 Check Timing Chain Tension

39 Clean Sparking Plugs and Set Gaps

40 Lubricate Door Locks, Lock Cylinders, Bonnet Safety Catch Pivot, Around Door Striker Wedge and all oil can pivots

41 Check Operation of all Controls, Instruments and Lights

42 Lubricate Handbrake Linkage

43 Check Steering Linkages for Wear

44 Remove Road Wheels, Check Front Brake Pads for Wear, Examine Rear Brake Shoes and Self-Adjusting Mechanism and Blow Clean

45 Top-up Radiator and Windshield Washer Reservoir

46 Change Engine Oil and Renew Filter Element

47 Top-up Brake and Clutch Reservoir

48 Check Fan Belt for Tension and Wear

49 Tighten Generator Mounting Bolts to a Torque of 15 to 18 lb. ft. (2.08 to 2.49 kg.m.)

50 Inspect Radiator and Heater Hoses for Leaks and Deterioration

51 Tighten Manifold Bolts to Correct Torque

52 Check Engine for Water and/or Oil Leaks

53 Check Exhaust System for Leaks or Damage

54 Inspect Brake Hoses and Lines for Signs of Leaks or Chafing

55 Road or Roller Test, Adjust Carburettor Idling and Ignition Timing

56 Top-up Steering Box

EVERY 12,000 MILES (20,000 KM.) OR TWELVE MONTHS
(WHICHEVER OCCURS FIRST)

As for 6,000 mile service plus the following operations:

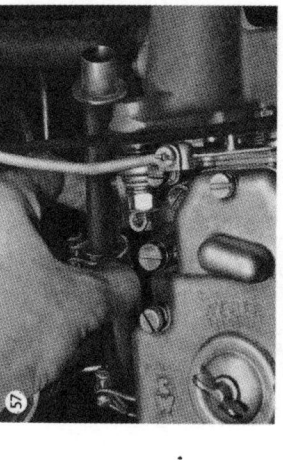

57 Clean Crankcase Emission Flame Trap
Slacken the clip retaining the flexible hose to the carburettor air box and remove hose from air box. The crankcase emission tube retained in the cylinder block by a rubber grommet should then be removed by pulling gently away from the block. The conical shaped wire gauze flame trap retained inside the emission tube should then be thoroughly washed in petrol. Replace emission tube, flexible hose and tighten retaining clip.

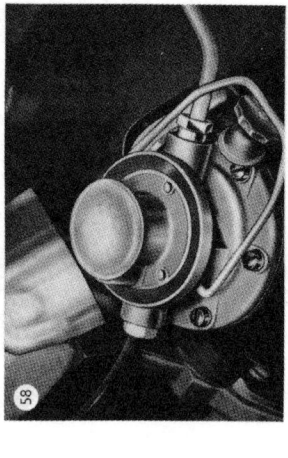

58 Clean Sediment from Fuel Filter
Unscrew clamp nut on top of pump, detach glass bowl and clean sediment from pump body and screen, using petrol. Check gasket. Replace filter screen glass bowl and tighten clamp nut.

59 Check Torque of Rear Spring 'U' Bolts
Tighten bolts to a torque of 20 to 25 lb. ft. (2.76 to 3.46 kg.m.).

60 Check Seat Belts for Security and Wear

EVERY 18,000 MILES (30,000 KM.) OR EIGHTEEN MONTHS
(WHICHEVER OCCURS FIRST)

As for 6,000 mile service plus the following operations:

61 Renew Air Cleaner Element
To renew filter element, slacken the four retaining nuts, release the two retaining clips and remove the air cleaner body. Unscrew the wing nut retaining the element to the air cleaner base and remove element.

62 Renew Fuel Line Filter (where fitted)

EVERY 24,000 MILES (40,000 KM.) OR TWO YEARS
(WHICHEVER OCCURS FIRST)

As for 6,000 and 12,000 mile service plus the following operations:

63 Repack and Adjust Front Wheel Bearings

Jack up the car, remove the wheel hub cap, road wheel, and then the dust cap by carefully levering it from the end of the hub.

Remove the retaining pin clips and the two retaining pins locating the brake pads in the calliper. The brake pads should be marked to identify their original positions, since the brake pads must be refitted in the same locations from which they were removed. The brake pads may now be removed with their shims.

From the suspension unit side of the calliper, release the calliper mounting bolt locking tabs.

The calliper can be moved aside, but must be supported to avoid tensioning the fluid pipe, after unscrewing the two mounting bolts.

The hub may be removed after withdrawing the split-pin, bearing adjusting nut retainer, bearing adjusting nut, thrust washer and outer bearing.

Wash these components together with the inner bearing and repack both bearings with a good quality lithium base grease, working the grease carefully into the rollers and cages.

DO NOT COMPLETELY PACK HUB — ALLOW FOR EXPANSION. Adjust bearings:— Tighten adjusting nut to 27 lb. ft. (3.7 kg.m.) torque, whilst rotating hub. Fit nut retainer. Slacken nut and retainer 90 or 2 castellation slots. Fit new split pin and replace dust cap. Refit calliper (use new tabwasher), reconnect fluid pipe and tighten connection. Tighten calliper bolts to 45 to 50 lb. ft. (6.22 to 6.91 kg.m.). Bend locking tabs. Refit brake pads, shims, retaining pins and clips. Replace wheel, tighten nuts, lower car to ground, check wheel nuts and refit hub cap. Bleed brakes.

64 Renew Brake Fluid

65 Renew Brake Servo Filter

Clean servo unit filter area, remove outer body retaining screw and carefully lift off body and filter. Clean filter-seat and body and fit new element. Replace body and tighten retaining screw.

EVERY 36,000 MILES (60,000 KM.) OR THREE YEARS
(WHICHEVER OCCURS FIRST)

66 Renew all Clutch, Brake Cylinder and Servo Seals, Brake Flexible Hoses and Brake Fluid

CORTINA - LOTUS
Supplement

13
LUBRICATION
AND
MAINTENANCE
(1967 - 1968)

CORTINA - LOTUS

LUBRICATION AND MAINTENANCE

As a result of engineering experience gained and the improved lubricants used, the details given in the foregoing section can be applied to the 1967-'68 range of vehicles.

CORTINA - LOTUS

14
SPECIFICATIONS, SERVICING AND REPAIR DATA

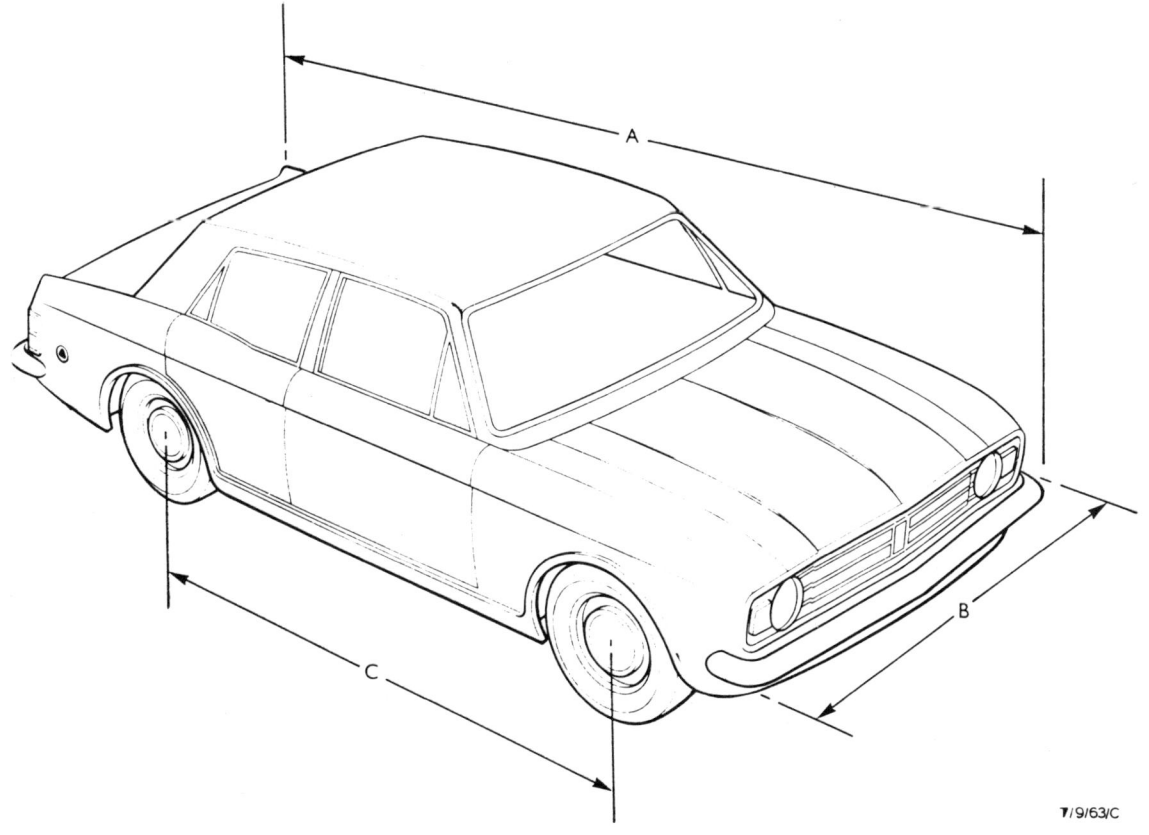

CORTINA - LOTUS

GENERAL DATA

1. WHEELS AND TYRES

Pressed steel with wide based rim 5½J×13. Tyres 165×13 radial ply tubeless.

2. BRAKES

Hydraulic operation. Front – disc type. Rear – internal expanding self-adjusting drum type, leading and trailing shoe. Suspended vacuum servo. Footbrake on all four wheels, hand brake on rear wheels only.

3. STEERING

High efficiency recirculatory ball type. Ratio 15.7:1. Two-spoke, dished steering wheel, diameter 15.5 in. (38.4 cm.).

4. REAR AXLE AND DRIVE SHAFT

Semi-floating, pressed steel banjo housing with hypoid final drive. Ratio 3.78:1. Two-piece drive shaft and centre bearing. The two-piece drive shaft fitted prior to 1969 incorporated a sliding spline.

5/1. FRONT SUSPENSION

Independent front wheel suspension; directly operated coil springs mounted on special double-acting hydraulic shock absorbers, integral with wheel spindle and located on body in rubber mounted upper bearing. Anti-roll torsion bar embodied.

5/2. REAR SUSPENSION

Longitudinal, asymmetrical semi-elliptic leaf springs with hydraulic double-acting shock absorbers. Two radius arms.

6/1. ENGINE

Four-cylinder, in line, twin overhead cam unit with a bore of 3.250 in. (82.550 mm.) and a stroke of 2.867 in. (72.821 mm.) giving a capacity of 95.20 cu. in. (1,560 c.c.). Compression ratio 9.5:1. B.H.P. (net) 109.5 at 6000 rev./min. Torque (net) 106 lb. ft. at 4,500 rev./min. Five bearing crankshaft with integral weights, dynamically balanced. Aluminium head with fully machined hemispherical combustion chambers. Three ring aluminium alloy pistons – two compression and one oil control. Chain driven camshafts and auxiliary shaft, mechanical chain tensioner. Three-point rubber suspension of engine and gearbox.

6/2. EXHAUST SYSTEM

Inlet and front muffler and rear muffler and tail pipe assembly. Suspension by two 'O' rings on brackets welded to floor pan and a support strap on the tail pipe. Four branch fabricated steel exhaust manifold.

7. CLUTCH AND GEARBOX

Self-adjusting 8.0 in. (20.4 cm.) diaphragm spring clutch, hydraulically actuated from a pendant clutch pedal.

Four-speed, fully synchronised gearbox controlled through a remote shift integral with gearbox extension housing.

8. COOLING SYSTEM

Pressurised, with thermostatic heat control. Belt driven fan and water pump. Tube and fin radiator. Long-life anti-freeze normally used in production.

9. FUEL SYSTEM

Mechanical fuel pump, two side-draught Weber type 40 DCOE-31 dual barrel carburettors. 10 gallon (45.4 litre) fuel tank. Paper element type air cleaner.

10. ELECTRICAL EQUIPMENT

Two brush, ventilated generator of 22 or 25 amp output, compensated voltage control, 'V' belt drive. One of two types of starter motor can be fitted, Inertia is standard equipment, Pre-engaged starter motor is fitted where cold start equipment is specified or as an option. Single tone horn operated by horn button on steering wheel antennae. Battery, negatively earthed 12 volt, 38 amp. hr. at 20 hr. rate (57 amp. hr. optional) located in boot. 12 volt oil-filled ballast resistor coil; distributor has automatic centrifugal advance and retard and an automatic cut-out. Autolite spark plugs.

Circular instruments located directly in front of driver are speedometer incorporating generator warning light, total distance recorder and "trip" distance recorder and tachometer incorporating main beam warning light. Four separate instruments in additional panel comprising ammeter, oil pressure, fuel and temperature gauges.

11. ACCESSORIES

A wide range of accessories are available. These are listed in Section 11/1 together with fitting instructions.

12. BODY, PAINT AND TRIM

All steel welded integral construction having two doors with safety glass in all windows. Seating within the wheelbase, bucket type seats and centre console. Rear luggage compartment with self-locking counterbalanced lid has capacity of 21 cu. ft. (0.59 cu. m.).

Full range of paint and matching trim colours including several colours in metallic finishes.

13. LUBRICATION and MAINTENANCE

Full details of lubricants and servicing procedures are contained in section 13. The first service is at 600 miles (1,000 km.), the second at 3,000 miles (5,000 km.) continuing at 6,000 miles (10,000 km.) intervals.

CORTINA - LOTUS

15. SPECIAL TOOLS

All the special tools needed are listed in this section, each with a brief description.

WEIGHTS AND DIMENSIONS

Wheelbase	98 in. (298.9 cm.)
Length	168.0 in. (426.7 cm.)
Width	63.8 in. (160.02 cm.)
Height (unladen)	56.0 in. (142.24 cm.)
Kerb weight	2,027 lb. (919.45 kg.)
Ground clearance (minimum)	4.94 in. (13.13 cm.)
Turning circle	31 ft. (914.4 cm.)
Track—Front	53.5 in. (133 cm.)
—Rear	52.0 in. (129 cm.)

General Recommended Tightening Torques, lb. ft. (kg.m.)

$\frac{1}{4}$ in.—20 UNC	5 to 7 (0.69 to 0.97)
$\frac{5}{16}$ in.—18 UNC	12 to 15 (1.66 to 2.07)
$\frac{5}{16}$ in.—24 UNF	12 to 15 (1.66 to 20.7)
$\frac{3}{8}$ in.—16 UNC	17 to 22 (2.35 to 3.04)
$\frac{3}{8}$ in.—24 UNF	22 to 27 (3.04 to 3.73)
$\frac{7}{16}$ in.—14 UNC	30 to 35 (4.15 to 4.84)
$\frac{7}{16}$ in.—20 UNF	40 to 45 (5.53 to 6.22)
$\frac{1}{2}$ in.—13 UNC	45 to 50 (6.22 to 6.91)
$\frac{1}{2}$ in.—20 UNF	50 to 60 (6.91 to 8.29)
$\frac{9}{16}$ in.—12 UNC	60 to 70 (8.29 to 9.67)
$\frac{9}{16}$ in.—18 UNF	65 to 75 (8.98 to 10.37)
$\frac{5}{8}$ in.—11 UNC	75 to 85 (10.37 to 11.75)
$\frac{5}{8}$ in.—18 UNF	100 to 110 (13.82 to 15.20)

1. WHEELS AND TYRES

Wheels

Size	$5\frac{1}{2}J \times 13$

Tyres

Size	165×13
Pressures — (normal speed conditions)	24 p.s.i. (1.69 kg./sq. cm.)
— (high speed conditions)	30 p.s.i. (2.09 kg./sq. cm.)

Tightening Torques, (lb. ft. kg.m.)

Wheel nuts	50 to 55 (6.91 to 7.60)
Brake calliper to front suspension unit	45 to 50 (6.22 to 6.94)

Section 14 — 4   1 . 1969

Front brake disc to hub	30 to 34 (4.15 to 4.70)
Front wheel bearing adjusting nut	See text
Axle shaft bearing retainer bolts	15 to 18 (2.08 to 2.48)

2. BRAKING SYSTEM

Type	Hydraulically operated, disc front drum, rear with servo assistance

Front Brakes

Disc diameter	9.620 in. (24.45 cm.)
Disc thickness	0.50 in. (12.70 mm.)
Disc run-out (max.)	0.0035 in. (0.089 mm.) T.I.R.
Disc effective radius	3.78 in. (9.59 cm.)
Pad thickness	0.40 in. (1.02 cm.)
Pad area (per pad)	5.16 sq. in. (33.31 sq. cm.)
Total pad area	20.64 sq. in. (133.32 sq. cm.)
Total swept area	189.50 sq. in. (1220 sq. cm.)
Pad material	Mintex M33
Colour code	Green
Cylinder diameter	2.125 in. (5.41 cm.)

Rear Brakes

Drum diameter	9.00 in. (22.9 cm.)
Drum width	1.75 in. (4.45 cm.)
Lining length	6.86 in. (17.42 cm.)
Lining thickness	0.187 in. (4.78 mm.)
Lining material	DON 242
Colour code	Red
Lining area (per shoe)	14.70 sq. in. (94.9 sq. cm.)
Total lining area	58.80 sq. in. (379.1 sq. cm.)
Total swept area	99.00 sq. in. (640.3 sq. cm.)
Cylinder diameter	0.70 in. (1.78 cm.)

Master Cylinder

Bore diameter — (Single line system)	0.75 in. (1.91 cm.)
— (Dual line system)	0.875 in. (2.22 cm.)
Stroke (max.)	1.375 in. (3.49 cm.)

Servo

Type	Suspended vacuum
Boost ratio	2.08 : 1

Section 14 — 5

CORTINA - LOTUS

General

Braking ratio	67.9% Front 32.1% Rear
Fluid type	ME-3833-F (amber)

3. STEERING

Type	Recirculatory
Lubricant	S.A.E. 90 E.P.
Lubricant capacity	420 c.c.
Ratio (steering box)	16.4 : 1
Number of turns of steering wheel, lock to lock	4½ approx.
Steering wheel diameter	15½ in. (40.4 cm.)
Steering shaft bearing adjustment	Shims
Steering shaft bearing pre-load	0.002 to 0.004 in. (0.051 to 0.102 mm.)
Steering shaft shim thickness and identification:	
105E-3592-B	Paper 0.010 in. (0.254 mm.)
105E-3595-A	Steel 0.004 in. (0.102 mm.)
105E-3595-B	Steel 0.010 in. (0.254 mm.)
3014E-3595-A	Steel 0.002 in. (0.051 mm.)
Rocker shaft end-float adjustment	Adjusting stud and locknut
Rocker shaft bush diameter (finish)	0.875 in. (21.225 mm.)
Number of balls in nut	13
Diameter	5/16 in. (7.94 mm.)
Number of balls in bearings (each)	12
Diameter	7/32 in. (5.56 mm.)

Tightening Torques, lb. ft. (kg.m.)

Drop arm nut	60 to 80 (8.3 to 11.1)
Steering wheel nut	25 to 30 (3.4 to 4.1)
Steering gear to sidemember	20 to 25 (2.7 to 3.4)
Steering gear top cover	18 to 20 (2.5 to 2.7)
Steering linkage to drop arm	25 to 30 (3.4 to 4.1)
Steering arm to suspension unit	30 to 35 (4.2 to 4.8)
Drop arm and idler arm to drop arm to idler arm rod	25 to 30 (3.5 to 4.1)
Idler arm joint	25 to 30 (3.5 to 4.1)
Track rod clamps	15 to 18 (2.1 to 2.4)
Track rod ball joints	18 to 22 (2.5 to 3.0)
Idler arm bracket to body sidemember	20 to 25 (2.8 to 3.5)

Wheel Alignment

Castor	$-0°\ 45'$ to $+0°\ 45'$
Camber	$0°\ 15'$ to $1°\ 45'$
King pin inclination	$7°\ 11'$ to $8°\ 41'$
Front lock wheel angle with back lock set at 20°	$19°\ 00'$ to $20°\ 30'$
Toe-in	0.06 to 0.25 in. (1.52 to 6.35 mm.)
Turning circle	31 ft. (9.44 m.)

4. REAR AXLE

Type	Semi-floating hypoid
Axle ratio	3.7 : 1
Number of teeth on crown wheel	34
Number of teeth on pinion	9
Crown wheel and pinion backlash	0.005 to 0.007 in. (0.1270 to 0.178 mm.)
Pinion bearing pre-load	9 to 11 lb. in. (0.104 to 0.127 kg.) excluding oil seal
Differential bearing pre-load (cap spread)	0.005 to 0.007 in. (0.127 to 0.178 mm.)
Differential pinion thrust washer thickness	0.030 to 0.032 in. (0.762 to 0.813 mm.)
Differential pinion inside diameter	0.628 to 0.630 in. (15.953 to 16.004 mm.)
Oil capacity	2 Imp. pints (2.4 U.S. pints, 1.1 litres)
Grade of oil (summer and winter)	S.A.E. 90 Hypoid
"Initial fill" lubricant	EM-2C-29
Grade of (topping-up) lubricant	S.A.E. 90 hypoid gear oil

Pinion Bearing Shims

105E-4672-A	0.1304 to 0.1308 in. (3.312 to 3.322 mm.)
105E-4672-B	0.1314 to 0.1318 in. (3.337 to 3.348 mm.)
105E-4672-C	0.1324 to 0.1328 in. (3.363 to 3.373 mm.)
105E-4672-D	0.1334 to 0.1338 in. (3.388 to 3.398 mm.)
105E-4672-E	0.1344 to 0.1348 in. (3.414 to 3.424 mm.)
105E-4672-F	0.1354 to 0.1358 in. (3.439 to 3.449 mm.)
105E-4672-G	0.1364 to 0.1368 in. (3.465 to 3.474 mm.)
105E-4672-H	0.1374 to 0.1378 in. (3.490 to 3.500 mm.)
105E-4672-J	0.1384 to 0.1388 in. (3.515 to 3.525 mm.)
105E-4672-K	0.1394 to 0.1398 in. (3.541 to 3.551 mm.)
105E-4672-L	0.1404 to 0.1408 in. (3.566 to 3.576 mm.)
105E-4672-M	0.1414 to 0.1418 in. (3.595 to 3.600 mm.)
105E-4672-N	0.1424 to 0.1428 in. (3.605 to 3.627 mm.)

Tightening Torques, lb. ft. (kg.m.)

Crown wheel to differential case bolts	50 to 55 (6.913 to 7.604)
Differential carrier to axle housing nuts	25 to 30 (3.5 to 4.0)
Differential bearing locking plate bolts	12 to 15 (1.659 to 2.074)
Differential bearing cap bolts	45 to 50 (6.221 to 6.931)
Axle shaft bearing retainer bolts	15 to 18 (2.074 to 2.489)

Universal joint flange to pinion flange	15 to 18 (2.074 to 2.489)
Rear axle filler plug	25 to 30 (3.460 to 4.148)
Axle shaft bearing assembly pressure (minimum)	1,200 lb. (544 kg.)
Axle shaft bearing retainer pressure (minimum)	1,500 lb. (680 kg.)

5/1. FRONT SUSPENSION

Type	Independent, coil springs with combined stabiliser bar
Load	600 lb.
Deflection rate	125.8 lb./in.
Wheel travel—Bounce	2.45 in. (6.217 cm.)
—Rebound	4.37 in. (11.10 cm.)
Mean diameter of coils	4.808 in. (11.19 cm.)
Wire diameter	0.492 in. (12.49 mm.)
Number of coils	6¾
Spring free length	10.33 in. (26.24 cm.)
Loaded length	5.93 in. (15.06 cm.)
Shock absorber type	Direct double-acting telescopic
Fluid capacity	350 c.c.
Fluid type	M-100502-E

Tightening Torques, lb. ft. (kg.m.)

Suspension unit upper mounting bolts	15 to 18 (2.07 to 2.49)
Spindle to top mount assembly	28 to 32 (3.9 to 4.4)
Track control arm ball stud nut	30 to 35 (4.15 to 4.84)
*Stabiliser bar attachment clamps	15 to 18 (2.07 to 2.49)
*Stabiliser bar to track control arm nut	15 to 45 (2.07 to 6.22)
*Track control arm inner bushing	22 to 27 (3.04 to 3.73)
Front suspension crossmember to body sidemember	25 to 30 (3.46 to 4.15)

*These to be tightened with the weight of the car on its wheels.

5/2. REAR SUSPENSION

Type	Semi-elliptic leaf spring with two trailing links (radius arms)
Length (between eye centres)	47.00 in. (119.38 cm.)
Number of leaves	5
Width of leaves	2.0 in. (5.08 cm.)
Rear shock absorber—Type	Double acting telescopic
—Capacity	Sealed
Spring rate	85 lb./in.

Tightening Torques, lb. ft. (kg.m.)

*Radius arm to axle	22 to 27 (3.04 to 3.73)
*Radius arm to body	45 to 50 (6.22 to 6.93)
*Shock absorber to body	15 to 20 (2.0 to 2.75)
Shock absorber to axle	40 to 45 (5.54 to 6.22)
*Rear spring "U" bolts	18 to 26 (2.50 to 3.60)
*Rear spring front securing nut	25 to 30 (3.46 to 4.15)
*Rear spring rear shackle nuts	8 to 10 (1.10 to 1.40)

*These items to be tightened with the component in the kerb weight position, i.e. the car must be resting on its wheels.

6/1. ENGINE

General

Type	4 cylinder in-line twin O.4.C.
Bore	3.2506 in. (82.565 mm.)
Stroke	2.867 in. (72.746 mm.)
Cubic capacity	95.2 cu in. (1,560 c.c.)
Compression ratio	9.5 : 1
Compression pressure (hot)	180 to 200 lb./sq. in. (12.66 to 14.06 kg./sq. cm.) at cranking speed 800 to 1,000 rev./min.
Maximum brake horsepower (nett)	109.5 at 6,000 rev./min.
Maximum torque (nett)	106.5 lb. ft. at 4,500 rev./min.
Firing order	1, 3, 4, 2
Location of No. 1 cylinder	Next to radiator
Idling speed	800 to 1,000 rev./min.
Engine mounting	3-point suspension on shear type bonded rubber mounting

Dimensions:

Weight (dry—less clutch and mountings)	264.4 lb. (120 kg.)
Length (fan to clutch face)	20.41 in. (51.84 cm.)
Width—overall	26.22 in. (66.80 cm.)
—without exhaust manifold	21.54 in. (54.71 cm.)
Height (crank ₵ to top of engine less air cleaner)	18.27 in. (46.41 cm.)

Camshafts

Material	Cast Alloy Iron
Drive	Chain
Thrust	Taken by integral disc in groove in cylinder head
End-Float	0.003 to 0.010 in. (0.076 to 0.254 mm.)
Number of bearings	5

Journal diameter	1.0000 to 1.0005 in. (2.540 to 2.541 cm.)
Total effective bearing length	3.181 in. (80.80 mm.)
Bearing clearance	0.0005 to 0.0020 in. (0.013 to 0.051 mm.)
Bearing—type	Steel-backed, white metal liners
Cam lift—exhaust and inlet	0.350 in. (8.890 mm.)
Cam heel to toe dimension	1.549 to 1.551 in. (3.934 to 3.939 cm.)

Connecting Rods and Big End Bearings

Type	'H' section steel forging
Material	EN15B Steel
Length (centre to centre)	4.799 to 4.801 in. (12.189 to 12.194 cm.)
Piston pin fit in rod	0.0005 to 0.002 in. (0.013 to 0.051 mm.)
End-float on piston pin	0.004 to 0.010 in. (0.102 to 0.254 mm.)
Big end bearings	Steel-backed copper/lead with lead indium overlay
Big end bore	2.0825 to 2.0830 in. (52.896 to 52.980 mm.)
Bearing liner wall thickness	0.0719 to 0.07225 in. (1.8262 to 1.8351 mm.)
Crankpin to bearing clearance	0.005 to 0.0022 in. (0.0127 to 0.0559 mm.)
Undersize bearings available	0.010 in. (0.25 mm.) 0.020 in. (0.51 mm.)
End-float on crankpin	0.004 to 0.010 in. (0.102 to 0.254 mm.)
Effective bearing length	0.83 to 0.87 in. (21.08 to 22.10 mm.)
Small end bush	Steel-backed lead/bronze
Small end bush inside diameter—Grade A (silver)	0.8125 to 0.8126 in. (20.637 to 20.640 mm.)
—Grade B (green)	0.8126 to 0.8127 in. (20.640 to 20.643 mm.)

Crankshaft and Main Bearings

Material	Nodular graphite cast iron
Number of main bearings	5
Main journal diameter	2.1253 to 2.1261 in. (00.00 to 00.00 mm.)
Regrind diameters	0.010 in. (0.25 mm.) u/s, 0.020 in. (0.51 mm.) u/s
Journal length—No. 1	1.219 to 1.239 in. (30.963 to 31.471 mm.)
—No. 2	1.273 to 1.283 in. (32.334 to 32.588 mm.)
—No. 3	1.247 to 1.249 in. (31.674 to 31.725 mm.)
—No. 4	1.273 to 1.283 in. (32.334 to 32.588 mm.)
—No. 5	1.300 to 1.330 in. (33.020 to 33.782 mm.)
Crankpin diameter	1.9370 to 1.9375 in. (49.200 to 49.213 mm.)
Centre journal fillet radius	0.70 in. (17.78 mm.)
Rear journal fillet radius	0.100 to 0.110 in. (2.54 to 2.794 mm.)
Intermediate and front journal fillet radius	0.096 to 0.110 in. (2.438 to 2.794 mm.)
Main bearings	Steel-backed copper/lead with lead/indium overlay
Main bearing clearance	0.0015 to 0.003 in. (0.038 to 0.076 mm.)
Bearing bore in cylinder block	2.2710 to 2.715 in. (57.683 to 57.696 cm.)

Crankpin journal diameter	1.9370 to 1.9375 in. (49.200 to 49.213 cm.)
Crankpin journal length	1.062 to 1.066 in. (26.975 to 27.076 cm.)
Crankshaft end-float	0.003 to 0.008 in. (0.076 to 0.203 mm.)
Crankshaft overall length	19.505 in. (49.54 cm.)
End-float thrust washer thickness	0.091 to 0.093 in. (2.311 to 2.362 mm.)

Cylinder Block

Type	Cylinder block cast integral with top half of crankcase
Material	Cast alloy iron
Water jackets	Full length
Cylinder bore diameter—Grade 1	3.2500 to 3.2503 in. (82.550 to 82.557 cm.)
—Grade 2	3.2503 to 3.2506 in. (82.557 to 82.563 cm.)
—Grade 3	3.2506 to 3.2509 in. (82.563 to 82.570 cm.)
—Grade 4	3.2509 to 3.2512 in. (82.570 to 82.577 cm.)

Cylinder Head and Valves

Type	Die cast aluminium alloy with inclined valves. Separate inlet and exhaust ports
Combustion chambers	Fully machined, hemispherical
Valve guides	Cast iron bushes

Size	Outside diameter	Bore diameter in cylinder head
Standard	0.5000 to 0.5005 in.	0.4990 to 0.4995 in.
	(12.700 to 12.713 mm.)	(12.675 to 12.687 mm.)
0.001 in. o/s	0.5010 to 0.5015 in.	0.5000 to 0.5005 in.
(0.025 mm.)	(12.725 to 12.738 mm.)	(12.700 to 12.713 mm.)
0.005 in. o/s	0.5050 to 0.5055 in.	0.5040 to 0.5045 in.
(0.127 mm.)	(12.827 to 12.840 mm.)	(12.802 to 12.814 mm.)
0.006 in. o/s	0.5060 to 0.5065 in.	0.5050 to 0.5055 in.
(0.152 mm.)	(12.852 to 12.865 mm.)	(12.827 to 12.840 mm.)

Valve guide inside diameter	0.3113 to 0.3125 in. (7.907 to 7.938 mm.)
Valve stem diameter	0.310 to 0.311 in. (7.874 to 7.899 mm.)

Valve stem to guide clearance			0.0025 to 0.003 in. (0.640 to 0.076 mm.)
Valve head diameter—Inlet			1.526 to 1.530 in. (38.760 to 38.862 mm.)
—Exhaust			1.321 to 1.325 in. (33.553 to 33.655 mm.)
Valve seat angle			45
Valve seat inserts			Cast iron

INLET

Size	Outside diameter	Cylinder head recess diameter
Standard	1.6235 to 1.6245 in. (41.237 to 41.262 mm.)	1.620 to 1.621 in. (41.148 to 41.173 mm.)
0.005 in. o/s (0.127 mm.)	1.6285 to 1.6295 in. (41.364 to 41.389 mm.)	1.625 to 1.626 in. (41.275 to 41.300 mm.)
0.010 in. o/s (0.254 mm.)	1.6335 to 1.6345 in. (41.491 to 41.516 mm.)	1.630 to 1.631 in. (41.402 to 41.427 mm.)
0.015 in o/s (0.381 mm.)	1.6385 to 1.6395 in. (41.618 to 41.643 mm.)	1.635 to 1.636 in. (41.529 to 41.554 mm.)

EXHAUST

Size	Outside diameter	Cylinder head recess diameter
Standard	1.4985 to 1.4995 in. (38.062 to 38.087 mm.)	1.495 to 1.496 in. (37.973 to 37.998 mm.)
0.005 in. o/s (0.127 mm.)	1.5035 to 1.5045 in. (38.214 to 38.240 mm.)	1.500 to 1.501 in. (38.100 to 38.125 mm.)
0.010 in. o/s (0.254 mm.)	1.5085 to 1.5095 in. (38.367 to 38.392 mm.)	1.505 to 1.506 in. (38.227 to 28.252 mm.)
0.015 in. o/s (0.381 mm.)	1.5135 to 1.5145 in. (38.519 to 38.545 mm.)	1.510 to 1.511 in. (38.354 to 38.379 mm.)

Bore for tappets			1.3750 to 1.3755 in. (34.925 to 34.946 mm.)

Section **14**—12

Flywheel and Ring Gear

Type			Cast iron with ring gear shrunk on
Number of teeth on ring gear			110
Maximum run-out			0.004 in. (0.1016 mm.)
Number of flywheel retaining bolts			6
Size			⅜ in.—24 U.N.F.
Clutch pilot spigot bearing—Pre-engine No. 18500			Sintered bronze bush
—From engine No. 18500			Needle roller bearing
Auxiliary jack shaft—material			Special Ford Cast Iron Alloy
—bearings			Steel-backed white metal
—journal diameter			1.5600 to 1.5605 in. (39.624 to 39.637 mm.)
Bearing—inside diameter			1.5615 to 1.5620 in. (39.662 to 39.675 mm.)
Bearing length—front			0.79 in. (20.066 mm.)
—centre			0.68 in. (17.272 mm.)
—rear			0.79 in. (20.066 mm.)
Bearing bore—cylinder block			1.6885 to 1.6895 in. (42.888 to 42.913 mm.)
End-float			0.002 to 0.007 in. (0.051 to 0.178 mm.)
Thrust plate thickness			0.176 to 0.178 in. (4.470 to 4.496 mm.)
Drive			Single roller chain with tensioner

Lubrication System

Type ... Wet sump, pressure feed system with full-flow filter. Main, camshaft auxiliary shaft and connecting rod big end bearings pressure fed. Piston pin and cylinder wall lubrication by splash and oil mist from squirt holes in connecting rods. Timing chain lubrication by metered jet of oil

Oil pressure			35 to 40 lb./sq. in.
Oil filter type			Full flow with replaceable element
Sump capacity (less oil filter)			6 Imp. pints
Oil filter capacity			½ Imp. pint

Grade of oil—*Temperature Range*   *S.A.E. Viscosity No.*

Under −10°F (−23°C)	5W/20
−10°F to +20°F (−23°C to −7°C)	10W/30
+20°F to +90°F (−7°C to 32°C)	20W/20 and 10W/30
Over +90°F (32°C)	20W/40 and 30

Section **14**—13

CORTINA - LOTUS

Piston Pins

Type	Fully floating, retained by end circlips
Material	Machined seamless steel tubing
Length	2.80 to 2.81 in. (71.120 to 71.145 mm.)
Outside diameter	0.8121 to 0.8123 in. (20.627 to 20.632 mm.)
Clearance in piston	0 to 0.0002 in. (0 to 0.005 mm.) selective
Clearance in small end bush	0.003 to 0.005 in. (0.076 to 0.127 mm.) selective

Piston Rings

Upper Compression Ring:

Material	Cast iron and chrome plated on periphery, cargraph plated (red) for identification and initial "bedding in"
Type	Tapered
Radial thickness	0.122 to 0.130 in. (3.099 to 3.302 mm.)
Width	0.077 to 0.078 in. (1.956 to 1.981 mm.)
Ring to groove clearance	0.0016 to 0.0031 in. (0.041 to 0.091 mm.)
Ring gap	0.009 to 0.014 in. (0.229 to 0.356 mm.)

Lower Compression Ring:

Material	Cast iron, copper plated for identification
Type	Externally stepped on lower face
Radial thickness	0.146 to 0.156 in. (3.708 to 3.962 mm.)
Width	0.077 to 0.078 in. (1.956 to 1.981 mm.)
Ring to groove clearance	0.0016 to 0.0036 in. (0.041 to 0.097 mm.)
Ring gap	0.009 to 0.014 in. (0.229 to 0.356 mm.)

Oil Control Ring:

Material	Cast iron, copper plated for identification
Type	"Micro land" scraper with slotted channel
Radial thickness	0.122 to 0.130 in. (3.099 to 3.302 mm.)
Width	0.186 to 0.1865 in. (4.724 to 4.737 mm.)
Ring to groove clearance	0.0015 to 0.0030 in. (0.038 to 0.076 mm.)
Ring gap	0.010 to 0.020 in. (0.254 to 0.508 mm.)
Oil pump	Eccentric bi-rotor driven by skew gear on auxiliary jack shaft
Oil pump capacity	2 Imp. galls./min. at 2,000 rev./min.
Pump body bore diameter	0.500 to 0.501 in. (12.700 to 12.725 mm.)
Drive shaft diameter	0.4980 to 0.4985 in. (12.649 to 12.662 mm.)
Drive shaft to body clearance	0.0015 to 0.0030 in. (0.038 to 0.076 mm.)
Inner and outer rotor clearance	0.006 in. (0.152 mm.) maximum
Outer rotor and housing clearance	0.010 in. (0.254 mm.) maximum
Inner and outer rotor end-float	0.005 in. (0.127 mm.) maximum

Pistons

Type	Solid skirt, valve recesses in crown
Material	Aluminium alloy
Piston diameter—Grade 1	3.2500 to 3.2503 in. (82.550 to 82.558 mm.)
—Grade 2	3.2503 to 3.2506 in. (82.558 to 82.565 mm.)
—Grade 3	3.2506 to 3.2509 in. (82.565 to 82.573 mm.)
—Grade 4	3.2509 to 3.2512 in. (82.573 to 82.580 mm.)
Number of rings	Two compression, one oil control
Width of ring grooves—compression rings	0.0796 to 0.0806 in. (2.022 to 2.047 mm.)
—oil control ring	0.1880 to 0.1890 in. (4.775 to 4.801 mm.)
Piston pin bore diameter	Graded
Grade—silver	0.8121 to 0.8122 in. (20.627 to 20.630 mm.)
—green	0.8122 to 0.8123 in. (20.630 to 20.632 mm.)
Piston pin bore offset	0.040 in. (1.016 mm.) towards thrust face

Tappets

Type	Piston
Outside diameter	1.3742 to 1.3745 in. (34.905 to 34.912 mm.)
Tappet to cylinder head bore clearance	0.0005 to 0.0014 in. (0.013 to 0.036 mm.)

Timing Chain

Type	Single roller
Pitch	0.375 in. (9.525 mm.)
Roller width	0.225 in. (5.715 mm.)
Roller diameter	0.25 in. (6.35 mm.)
Chain free movement	$\frac{1}{2}$ in. mid-way between camshaft sprockets

Valve Springs

Type	Coil, two per valve
Outer valve spring load at 1.17 in. (29.718 mm.)	45 lb. (20.412 kg.)
Outer valve spring load at 0.83 in. (21.082 mm.)	109 lb. (49.442 kg.)
Inner valve spring load at 0.92 in. (23.386 mm.)	12.4 lb. (5.625 kg.)
Inner valve spring load at 0.58 in. (14.732 mm.)	33.5 lb. (15.195 kg.)
Outer valve spring fitted length (valve closed)	1.130 to 1.185 in. (28.702 to 30.099 mm.)

Valve Timing and Clearances

Theoretical valve timing—Inlet opens ... 22° B.T.D.C.
—Inlet closes ... 62° A.B.D.C.
—Exhaust opens ... 62° B.B.D.C.
—Exhaust closes ... 22° A.T.D.C.

CORTINA - LOTUS

Valve lift—Inlet	...	0.35 in. (8.89 mm.)
—Exhaust	...	0.35 in. (8.89 mm.)
Valve clearance (cold)—Inlet	...	0.005 to 0.007 in. (0.127 to 0.178 mm.)
—Exhaust	...	0.006 to 0.008 in. (0.152 to 0.203 mm.)
—Exhaust from engine No. LP9952	...	0.009 to 0.011 in. (0.229 to 0.279 mm.)
Method of adjustment	...	Shims under tappets

Tightening Torques, lb. ft. (kg.m.)

Cylinder head bolts	...	60 to 65 (8.29 to 8.98)
Main bearing cap	...	55 to 70 (7.60 to 8.29)
Connecting rod big end	...	44 to 46 (6.08 to 6.36)
Flywheel	...	45 to 50 (6.22 to 6.91)
Oil filter centre bolt	...	12 to 15 (1.66 to 2.07)
Manifold nuts	...	12 to 15 (1.66 to 2.07)
Front cover—$\frac{1}{4}$ in. dia. bolts	...	5 to 7 (0.69 to 0.97)
—$\frac{5}{16}$ in. dia. bolts	...	10 to 15 (1.38 to 2.07)
Sump	...	6 to 8 (0.83 to 1.11)
Rear oil seal retainer	...	12 to 15 (1.66 to 2.07)
Crankshaft pulley	...	24 to 28 (3.32 to 3.87)
Oil pump	...	12 to 15 (1.66 to 2.07)
Auxiliary shaft thrust plate	...	5 to 7 (0.69 to 0.97)
Auxiliary shaft sprocket	...	12 to 15 (1.66 to 2.07)
Sump drain plug	...	20 to 25 (2.76 to 3.46)
Camshaft bearing cap nuts	...	9 (1.24)
Camshaft sprocket bolts	...	25 to 30 (3.46 to 4.15)
Chain tensioner sprocket pin	...	40 to 50 (5.53 to 6.91)
Chain tensioner retaining bolt	...	45 to 50 (6.22 to 6.91)
Chain tensioner pivot pin	...	40 to 45 (5.53 to 6.22)

6/2. EXHAUST SYSTEM

Tightening Torques, lb. ft. (kg.m.)

Clamp bolts (old type)—Manifold to inlet pipe	...	7 to 10 (0.97 to 1.38)
—Front muffler to rear inlet pipe	...	7 to 10 (0.97 to 1.38)
Clamp nuts (new type)—Manifold to inlet pipe	...	12 to 15 (1.66 to 2.07)
—Front muffler to rear inlet pipe	...	12 to 15 (1.66 to 2.07)
Tail pipe support strap—bracket to body—bolts	...	12 to 15 (1.66 to 2.07)
—strap to tail pipe—bolts	...	5 to 7 (0.70 to 0.97)

7. CLUTCH AND GEARBOX

Clutch

Type	...	Single dry plate, diaphragm spring self-adjusting
Actuation	...	Hydraulic

Master Cylinder

Bore diameter	...	0.75 in. (1.91 cm.)

Slave Cylinder

Bore diameter	...	0.875 in. (22.2 cm.)

Clutch Disc

Lining outside diameter	...	8.0 in. (20.4 cm.)
Lining inside diameter	...	5.75 in. (14.6 cm.)
Total friction area	...	47.36 sq. in. (304.1 sq. cm.)

Pressure Plate

Diameter	...	8.8 in. (22.35 cm.)

Gearbox

Ratios:		Gearbox
First	...	2.972
Second	...	2.010
Third	...	1.397
Top	...	1.000
Reverse	...	3.324

Main Drive Gear

Number of teeth	...	30
Inside diameter gear end	...	26
Mainshaft pilot end diameter	...	22
Countershaft:		Reverse 19
		Reverse 19
		17
Number of teeth	...	
End-float	...	0.008 to 0.020 in. (0.203 to 0.508 mm.)
Bore diameter	...	0.933 to 0.934 in. (2.370 to 2.372 cm.)
Thrust washer thickness	...	0.061 to 0.063 in. (0.155 to 0.160 cm.)
Number of rollers	...	40
Countershaft diameter	...	0.6818 to 0.6823 in. (1.732 to 1.733 cm.)

First Gear:

End-float	...	0.005 to 0.010 in. (0.127 to 0.254 mm.)
Internal diameter	...	1.376 to 1.377 in. (3.495 to 3.498 cm.)
Number of teeth	...	32

CORTINA - LOTUS

Second Gear:
End-float	...	0.005 to 0.010 in. (0.127 to 0.254 mm.)
Internal diameter	...	1.376 to 1.377 in. (3.495 to 3.3498 cm.)
Number of teeth	...	28

Third Gear:
End-float	...	0.005 to 0.016 in. (0.127 to 0.406 mm.)
Internal diameter	...	1.376 to 1.377 in. (3.495 to 3.498 cm.)
Number of teeth	...	23

Reverse Idler Gear:
Internal diameter	...	0.7500 to 0.7508 in. (1.905 to 1.907 cm.)
Shaft diameter	...	0.7465 to 0.7470 in. (1.896 to 1.898 cm.)
Number of teeth	...	22

Speedometer:
Number of teeth	...	23

Speedometer Driving Gear:
Number of teeth	...	7

Lubrication:
Oil capacity	...	1.75 Imp. pints (2.1 U.S. pints, 1 litre)
Grade of oil	...	S.A.E. 80 E.P.

Selective Circlips:

Mainshaft bearing to extension housing:

Part Number	Size—in. (mm.)	Colour Code
2824E-7030-A	0.0731 (1.860)	Yellow
2824E-7030-B	0.0720 (1.830)	Red
2824E-7030-C	0.0708 (1.800)	Blue
2824E-7030-D	0.0696 (1.770)	Violet
2824E-7030-E	0.0684 (1.737)	Green
2824E-7030-F	0.0682 (1.732)	Magenta
2824E-7030-G	0.0670 (1.702)	Plain

Mainshaft bearing to mainshaft:

Part Number	Size—in. (mm.)	Colour Code
2824E-7669-A	0.0707 (1.795)	Plain
2824E-7669-B	0.0719 (1.825)	Pink
2824E-7669-C	0.0731 (1.860)	Magenta
2824E-7669-D	0.0743 (1.890)	Violet
2824E-7669-E	0.0755 (1.920)	Green
2824E-7669-F	0.0767 (1.950)	Blue
2824E-7669-G	0.0779 (1.980)	Red
2824E-7669-H	0.0791 (2.010)	Yellow

Tightening Torques, lb. ft. (kg.m.)

Clutch pressure plate to flywheel	12 to 15 (1.66 to 2.07)
Clutch housing to transmission case	40 to 45 (5.53 to 6.22)
Transmission case drain and filler plugs	25 to 30 (3.46 to 4.15)
Transmission extension to transmission case	30 to 35 (4.15 to 4.84)

8. COOLING SYSTEM

Capacity:
Complete system—with heater	12.5 Imp. pints (7.10 litres, 15.2 U.S. pints)
—without heater	10.5 Imp. pints (5.96 litres, 12.6 U.S. pints)
Anti-freeze	FoMoCo Longlife anti-freeze, Part No. M97B18C in 50% solution of anti-freeze and water

Specific Gravity Readings at Constant Temperature 16°C (60°F)

Specific Gravity (providing no other additive is in the coolant)	Proportion of Anti-freeze (by volume)	Remains Fluid to °C	Remains Fluid to °F	Solidifies at °C	Solidifies at °F
1.080	50%	−37°	−34°	−58°	−72°
1.065	40%	−26°	−13°	−48°	−54°
1.050	30%	−16°	+3°	−39°	−38°
1.042	25%	−13°	+9°	−29°	−20°
1.034	20%	−9°	+15°	−19°	−3°
1.026	15%	−7°	+20°	−14°	+7°
1.016	10%	−4°	+25°	−8°	+17°

Radiator
Type	Modine high efficiency fin
Core width	17.25 in. (43.82 cm.)
Core height	14.12 in. (35.87 cm.)
Core depth	1.27 in. (3.23 cm.)
Frontal area	244 sq. in. (1574.3 sq. cm.)
Number of tubes	56
Cap release pressure	13 p.s.i. (0.914 kg./sq. cm.)

Fan
Number of blades—(metal)	2
—(plastic)	8
Diameter	11.0 in. (27.94 cm.)
Ratio—fan to engine	1 1:

Thermostat
Thermostat—Type	Wax
—Location	Cylinder head
—Starts to open	85° to 89°C (185° to 192°F)
—Fully open	99° to 102°C (210° to 216°F)

CORTINA - LOTUS

Fan Belt
Width	...	0.38 in. (9.7 mm.)
Outside length	...	29 in. (740 mm.)
Free play	...	½ in. (13 mm.)
Tension	...	40 to 50 lb. (18.14 to 22.18 kg.)

Tightening Torques, lb. ft. (kg.m.)
Thermostat housing	5/16 in. – 18 UNC	...	12 to 15 (1.66 to 2.07)
Fan blade	¼ in. – 20 UNC	...	5 to 7 (0.69 to 0.97)

9. FUEL SYSTEM

Fuel Tank
Capacity	...	10 Imp. gallons (12 U.S. gallons, 45.4 litres)

Carburettor
Type	...	Dual barrel, two venturis per barrel, horizontal Weber 40DCOE31
Jet sizes:		
Main venturi	...	30
Auxiliary venturi	...	4.5
Main jet	...	110
Idling jet	...	45/F9
Accelerator pump jet	...	35
Accelerator pump inlet valve bleed	...	40
Accelerator pump spring length	...	1.00 in. with 10.75 oz. lead
Progression holes	...	1×120, 2×100
Starting jet	...	100/F5
Emulsion tube	...	F11
Air corrector jet	...	155
Needle valve	...	1.75
Starting air jet	...	100
Float weight	...	26 gms.
Float level	...	8.5 mm. (including gasket)
Float stroke	...	6.5 mm.
Petrol level	...	29 mm.
Carburettor to inlet stub double coil spring washer gap (between coils)		0.040 in. (1.12 mm.)

Fuel Pump
Type	...	Mechanical
Inlet depression	...	8½ in. Hg. (21.59 cm.)
Delivery pressure	...	1¼ to 2½ lb./sq. in. (0.088 to 0.175 kg./sq. cm.)
Diaphragm spring test length	...	0.468 in. (11.883 mm.)
Diaphragm spring test pressure	...	3¼ to 3½ lb. (1.474 to 1.588 kg.)
Rocker arm spring test length	...	0.44 in. (11.18 mm.)
Rocker arm spring test pressure	...	5 to 5½ lb. (2.268 to 2.495 kg.)
Colour code	...	Red

Tightening Torques, lb. ft. (kg.m.)
Fuel pump retaining bolts	...	12 to 15 (1.66 to 2.07)
Exhaust manifold bolts	...	12 to 15 (1.66 to 2.07)

ELECTRICAL EQUIPMENT

Battery
Type	...	Lead acid
Voltage	...	12
Capacity (amp. hr.)—standard	...	38 at 29 hr. rate
—optional	...	57 at 20 hr. rate
Plates per cell—standard	...	9
—optional	...	13
Specific gravity (charged)	...	1.275 to 1.285
Low limit while discharging at 20 hr. rate	...	1.105
Electrolyte capacity	...	4.5 Imp. pints (5.4 U.S. pints, 2.5 litres)

Coil
Type	...	12 volt, oil filled type for use with ballast resistor
Resistance at 20°C (68°F)—Primary	...	3.1 to 3.5 ohms
—Secondary	...	4,750 to 5,750 ohms
Output	...	30 kV

Generator
	Standard	Optional
Type	C-40	C-40L
Speed (ratio to engine)		1.25 to 1
Brush length		0.718 in. (18.23 mm.)
Maximum charge	22 amps	25 amps
Maximum output	264 watts	300 watts
Fan belt tension (total free movement)		½ in. (13 mm.)

Regulator
	Lucas	Autolite
Cut-out—Cut-in voltage	12.6 to 13.4 volts	
—Drop-off voltage	9.25 to 11.25 volts	
—Armature to core air gap	0.035 to 0.045 in.	0.025 to 0.037 in.
	(0.9 to 1.1 mm.)	(0.64 to 0.94 mm.)
"Follow-through" of moving contact	0.010 to 0.020 in.	0.015 to 0.025 in.
	(0.3 to 0.5 mm.)	(0.38 to 0.64 mm.)

CORTINA - LOTUS

	Lucas	Autolite
Current regulator on-load setting	Maximum rated generator output ± 1½ amps.	
Armature to core air gap	0.045 to 0.049 in. (1.14 to 1.24 mm.)	0.014 to 0.019 in. (0.36 to 0.48 mm.)
Voltage regulator open circuit setting	14.4 to 15.6 volts at 20°C (68°F)	
Armature to core air gap	0.045 to 0.049 in. (1.14 to 1.24 mm.)	0.024 to 0.028 in. (0.61 to 0.71 mm.)

Atmospheric Temperature	Setting Voltage	Checking Voltage
10°C (50°F)	14.9 to 15.5	14.5 to 15.8
20°C (68°F)	**14.7 to 15.3**	**14.4 to 15.6**
30°C (86°F)	14.5 to 15.1	14.3 to 15.3
40°C (104°F)	14.3 to 14.9	14.2 to 15.1

Resistance of shunt windings	...	Cut-out 8.8 to 9.6 ohms
		Voltage Regulator 10.8 to 12.0 ohms
"Swamp" resistor ...	resistance measured between centre tag and base	13.25 to 14.25 ohms
	resistance measured between tag ends before fitting	53 to 57 ohms
Field resistor ...	either 55 to 65 ohms (identification colour—Red)	
	or 37 to 43 ohms (identification colour—Yellow)	

Table Showing Relationship Between Regulator, Generator and Battery

Equip-ment	Regulator			Generator			Battery	
	Part No.	Make	Identification No.	Part No.	Identification No.	Rated Output	Part No.	Capacity
Standard	3004E-10505-A	Autolite		2701E-10002-A	C.40	22 amp	113E-10658-A	38 A/H
	3004E-10505-C	Lucas	†GR5000 *37344					
Optional	3004E-10505-B	Autolite		2730E-10002-A	C.40L	25 amp	113E-10658-D	57 A/H
	3004E-10505-D	Lucas	†GR5001 *37342					

† On Regulator Cover
* Stamped on Regulator Base

Distributor

Type	...	Single pair contact breaker point
Automatic advance	...	Mechanical
Drive	...	Skew gear from jack shaft
Rotation	...	Anti-clockwise
Initial advance	...	12° B.T.D.C.
Condenser capacity	...	0.21 to 0.25 microfarad
Contact breaker points gap	...	0.014 to 0.016 in. (0.36 to 0.41 mm.)

Section 14—22   1.1969

CORTINA - LOTUS

Dwell angle	...	57° to 63°
Firing order	...	1, 3, 4, 2
Dynamic advance	...	26° B.T.D.C. at 3,500 rev./min.
Breaker arm spring tension	...	17 to 21 ozs. (481.9 to 567.0 gms.)
High tension lead resistance	...	5000 to 9000 ohms/ft. (164 to 295 ohms/cm.)

Sparking Plugs

Size	...	14 mm.
Type	...	Autolite AG22
Gap	...	0.023 in. (0.59 mm.)

Inertia Starter Motor

Ampere draw (zero r.p.m.)	...	340 at 7.4 volts
Ampere draw (1,000 starter r.p.m.)	...	245 at 7.8 volts
Gear ratio	...	11 : 1
Teeth on pinion	...	10
Teeth on ring gear	...	110
Lock torque	...	6.4 lb. ft. (0.884 kg.m.)

Pre-Engaged Starter Motor

Gear ratio	...	12 : 1
Teeth on pinion	...	9
Teeth on ring gear	...	110

Horns

| Type | ... | 4 in. (101.6 mm.) 'beep' |
| Current draw | ... | 4½ amp |

Light Bulbs and Flasher Unit

Headlamp	...	60/45W* sealed beam
Side light	...	5W* wedge base
Front and Rear direction indicator	...	28W*
Flasher unit	...	56W*
Tail and Stop light	...	7/28W*
Licence plate	...	5W wedge base
Interior light	...	6W festoon
Warning lights	...	2.2W wedge base
Instrument panel light	...	2.2W wedge base
Clock	... *U.K.	1.2W bayonet

Tightening Torques, lb. ft. (kg.m.)

Generator—mounting bolts	...	15 to 18 (2.08 to 2.49)
—adjusting strap	...	12 to 15 (1.66 to 2.08)
Spark plug	...	24 to 28 (3.32 to 3.87)

Section 14—23   1.1969

CORTINA - LOTUS

15
SPECIAL TOOLS

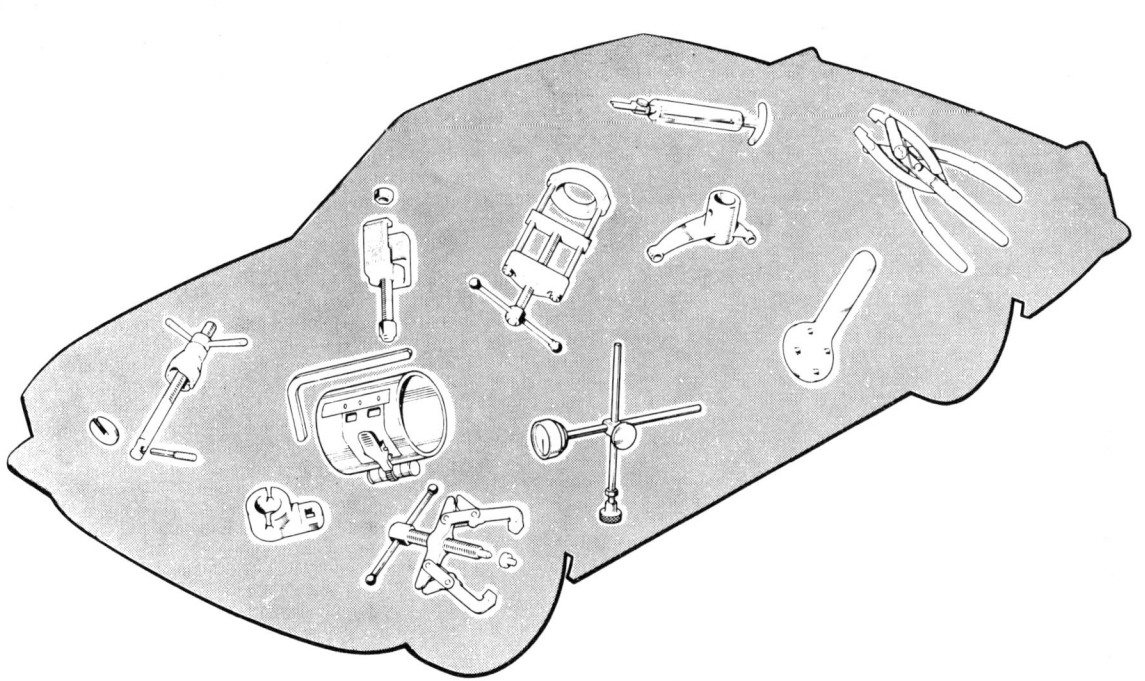

CORTINA - LOTUS

GENERAL DESCRIPTION

This Repair Manual issued for the "Cortina-Lotus" contains repair and overhaul procedures which, in some cases, require the use of special service tools to ensure the completion of a satisfactory and economical repair.

The purpose of this section is to list all tools of this kind numerically within sections.

Note is also made as to whether or not the tool is "existing", or a new tool introduced specifically for the "Cortina-Lotus".

Enquiries regarding the supply of these tools should be directed to:—

U.K. V. L. Churchill and Co. Ltd.,
London Road,
Daventry,
Northants.

Overseas Locations Room 1/330
Ford Motor Company Limited,
Warley,
Brentwood,
Essex,
England.

or through local Churchill Tool Distributors.

HUBS AND DRUMS

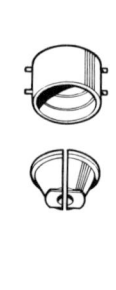 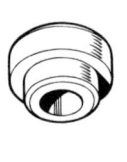   

PT 1024 Front Hub Bearing Cups Remover/Replacer (Main Tool)

Used in conjunction with P 1024-3 Adaptor Set.
(Existing Tool)

P 1024-3 Front Hub Bearing Cups Remover/Replacer (Adaptor)

Consists of three split ring adaptors and one solid ring adaptor. These are used in conjunction with PT 1024 to remove and replace the front hub bearing cups.
(Existing Tool)

P 1029 Front Hub Grease Retainer Replacer

Used in conjunction with the 550 Driver Handle and ensures square replacement of the front hub grease retainer.
(Existing Tool)

BRAKES

P 2006 Brake Bleeder Tubes

A set of four transparent flexible plastic tubes used when bleeding hydraulic brakes on current models.
(Existing Tool)

P 2012 Brake Line Plugs

A set of six brass plugs used to seal the open ends of hydraulic lines when dismantling hydraulic brakes.
(Existing Tool)

P 2031 Brake Line Plugs (Large)

A set of four plugs used as above in dual line systems.
(Existing Tool)

STEERING GEAR

 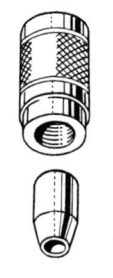 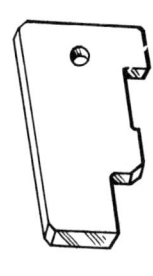 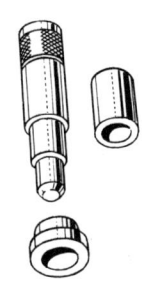

P 3041C Drop Arm Remover

Specifically designed to remove the drop arm correctly and without fear of damaging component parts. The thrust pad provided should be used at all times.
(Existing Tool)

P 3073-9 Steering Ball Joint Separator

Reduces possibility of damage when breaking taper joints.
(Existing Tool)

P 3085 Steering Rocker Shaft Oil Seal Replacer

Enables the rocker shaft oil seal to be replaced "in situ". It consists of one driver and one pilot sleeve.
(Existing Tool)

P 3086 Spindle Body Gauge

An accurately machined gauge of $\frac{1}{4}$ in. thick steel plate, used to check spindle bodies suspected of distortion.
(Existing Tool)

P 3087 Steering Rocker Shaft Bush Oil Seal Replacer

Only suitable for replacing the oil seal and bushes. For details of removing, see the appropriate Service Bulletin.
(Existing Tool)

SPRINGS AND FRONT SUSPENSION

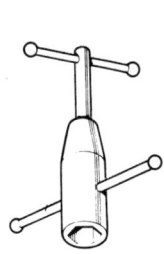  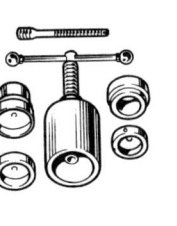 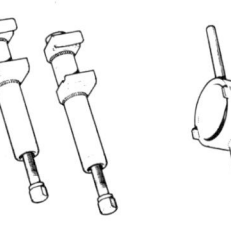  

P 5025 Front Suspension Unit Thrust Bearing Locknut Wrench (1967-68)

The screwdriver unit is used to hold the front suspension unit piston whilst the outer wrench is used to tighten the thrust bearing locknut.
(Existing Tool)

P 5026 Front Suspension Unit Torque Wrench (Adaptor) (1967-68)

Used in conjunction with Tool No. P 5025 and a suitable torque wrench to enable the locknut to be tightened to its specified torque loading.
(Existing Tool)

CP 5029A Rear Spring Front Shackle Bush Remover and Replacer

Used to ensure simple removal and square positive replacement of the shackle bush.
(Existing Tool)

P 5045 Coil Spring Adjustable Restrainers

These tools have adjustable jaws which grip the coils of the spring to keep it compressed.
(Existing Tool)

P 5041A Front Suspension Unit Bump Stop Platform Wrench

Consists of a single-sided peg wrench. (Existing Tool)

P 5042 Front Suspension Unit Gland and Bush Guide

This tool is essential to ensure that the gland and bush assembly is replaced on the piston rod without damage.
(Existing Tool)

CORTINA - LOTUS

P 3089A Steering Rocker Shaft Broaching Kit

The dummy cover plate details "e" and "d" are not required for the Cortina-Lotus Steering Box. The support detail "b" has a $\frac{7}{8}$ in. wide flat on the side to clear the lug on the steering box.   (Existing Tool)

REAR AXLE

P 3072 Slide Hammer (Main Tool)

Used in conjunction with adaptor sets for the removal of rear hubs, axle shafts and axle shaft oil seals.
(Existing Tool)

P 3072-4A Rear Hub and Axle Shaft Remover (Adaptor)

The slotted holes in the body are so arranged that the tool may be fitted to all current passenger cars, whether of the 4 or 5 stud type.   (Existing Tool)

CP 4000 Hand Press (Main Tool)

Used in conjunction with different adaptor sets to assist in the completion of various rear axle operations.   (Existing Tool)

P 4000-27A Differential Bearing Cone Remover (Adaptor)

Consists of:—One split ring adaptor and one thrust pad. Used in conjunction with Main Tool No. CP 4000 to remove the differential bearing cone.   (Existing Tool)

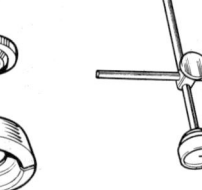
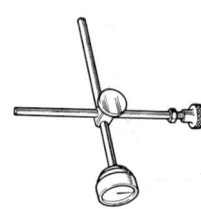

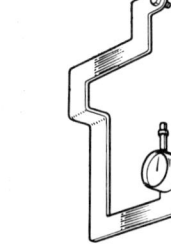

P 4000-28 Pinion Bearing Cone Remover/Replacer

Consists of:—One split ring adaptor, one replacer adaptor and one guide. Used in conjunction with either the Main Tool No. CP 4000 or the 370 Universal Taper Base.   (Existing Tool)

P 4008 Crown Wheel and Pinion Backlash Gauge (Main Tool)

Used to measure the backlash between pinion and crown wheel.   (Existing Tool)

P 4008-1 Crown Wheel and Pinion Backlash Gauge (Adaptor)

Consists of:—One mounting post and one locking nut. Used in conjunction with Main Tool P 4008 to measure the backlash between pinion and crown wheel.   (Existing Tool)

P 4009 Differential Bearing Preload Gauge

Secured to the differential bearing cup on one side, the dial gauge on the opposite side indicates the preload.
(Existing Tool)

P 4013A Drive Pinion Bearing Cups and Oil Seal Replacer (Main Tool)

Used together with the necessary adaptor to replace the drive pinion bearing cups. By reversing the top pad the oil seal may also be replaced.   (Existing Tool)

P 4013-3 Drive Pinion Cups and Oil Seal Replacer (Adaptor)

Used in conjunction with Main Tool No. P 4013A to effect easy replacement of the drive pinion bearing cups and oil seal.   (Existing Tool)

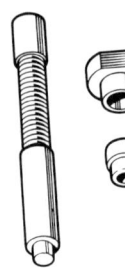

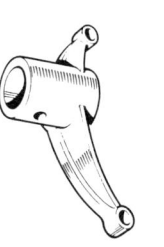

P 4015A Drive Pinion Bearing Cups Remover

The spring loaded legs in the tool are placed behind the cup and the handle is struck with a mallet to remove the cup.   (Existing Tool)

P 4028 Flange Holding Wrench

Fits in the holes of the drive pinion flange.
(Existing Tool)

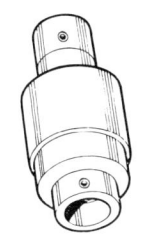

CP 4030 Drive Pinion Bearing Preload Gauge

Used in conjunction with a suitable adaptor, the sliding weight is moved along the beam until it rests at the correct figure.   (Existing Tool)

P 4030-1 Preload Gauge Adaptor

Used in conjunction with the Main Tool P 4030, this tool enables the final pinion preload setting to be obtained.
(Existing Tool)

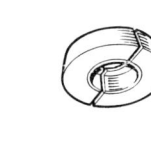

P 4075 Drive Pinion Depth Gauge (Main Tool)

Used in conjunction with P 4075-4 adaptor set, indicates precisely the thickness of the spacer shim required when reassembling.   (Existing Tool)

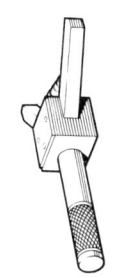

P 4075-4 Drive Pinion Depth Gauge Adaptor

This consists of a dummy pinion and two spacers, used with P 4075.   (Existing Tool)

P 4077A Mounting Adaptor

Used in conjunction with the 200A or B Engine Stand, provides a convenient means of holding the differential carrier when stripping or rebuilding.   (Existing Tool)

P 4079 Differential Bearing Adjusting Nut Wrench

A specially designed peg-style wrench which is used to turn the differential bearing adjusting nut when carrying out final adjustments.   (Existing Tool)

P 4080 Differential Bearing Cone Replacer

Used in conjunction with the 550 Universal Handle, to ensure square replacement of the differential bearing cone without fear of damage.   (Existing Tool)

P 4084 Spring Indicator for Press

Consists of:—A spring-loaded barrel assembly accurately graduated in 200 lb. steps. It enables the correct fitting pressure to be gauged when replacing the axle shaft bearing and inner retainer.
(Existing Tool)

P 4090-2 Axle Shaft Bearing and Inner Retainer Replacer

Used in conjunction with the 370 Taper Base and a workshop press, to replace the axle shaft bearing and inner retainer.   (Existing Tool)

P 4090-6 Rear Axle Bearing and Retainer Remover

Used in conjunction with the 370 Taper Base, this tool is essential for efficient removal of the bearing and bearing retainer.   (Existing Tool)

ENGINE

PT 4063A Cylinder Head Locating Studs

Used to assist in the location of the cylinder head without risk of damaging the cylinder head gasket. PT 4063 can be reworked to PT 4063A by filing the flat to extend as far as the tapered end.
(Modified Tool)

P 6031 Auxiliary Shaft Bush Remover and Replacer (Main Tool)

Used with P 6031-3 adaptors to remove and replace the auxiliary shaft bushes.
(Existing Tool)

P 6031-3 Auxiliary Shaft Bush Remover and Replacer (Adaptor)

Used with P 6031 main tool to remove and replace the auxiliary shaft bushes.
(Existing Tool)

CP 6032B Crankshaft Sprocket Replacer

Used to replace the crankshaft gears and sprockets as fitted to the range of passenger cars.
(Existing Tool)

CP 6041 Crankshaft Pulley Remover

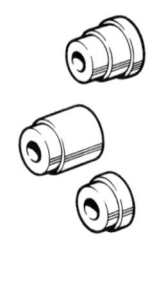

A two-legged puller with specially shaped feet to fit closely to the belt groove of the pulley.
(Existing Tool)

CP 6054 Valve Guide Remover/Replacer

Used to remove and replace valve guides.
(Existing Tool)

P 6064 Drain Plug Wrench

Used to remove the gearbox and rear axle drain and/or level plugs. Size 0.52 in. A/F double square socket. 0.583 in. A/F double square socket and $\frac{3}{8}$ in. and $\frac{7}{16}$ in. square plugs.
(Existing Tool)

P 6107 Engine Bracket

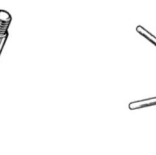

Enables the engine to be mounted on the 200A or B Engine stand.
(Existing Tool)

P 6110 Main Bearing Liner Remover and Replacer

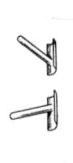

A set of two pins coloured as follows: Rear journal pin – Black. Front and Centre journals pin – Blue.
(Existing Tool)

P 6116 Crankshaft Sprocket Remover

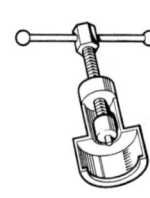

A robust tool, to enable the crankshaft sprocket to be removed with ease.
(Existing Tool)

P 6150 Crankshaft Front Oil Seal Aligner

Ensures the correct alignment of the crankshaft front oil seal with the crankshaft when replacing the front cover.
(Existing Tool)

P 6161 Crankshaft Front Cover Oil Seal Remover and Replacer

This is a double-ended tool used with the 550 handle.
(Existing Tool)

CORTINA - LOTUS

P 6165 Crankshaft Rear Oil Seal Remover and Replacer

This is a double-ended tool used with the 550 handle.
(Existing Tool)

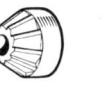

CP 6173 Crankshaft Rear Oil Seal Alignment

Ensures the correct alignment of the crankshaft rear oil seal when assembling the rear oil seal retainer.
(Existing Tool)

38-U-3 Piston Ring Compressor

This consists of an adjustable spring steel cylinder and an adjusting key.
(Existing Tool)

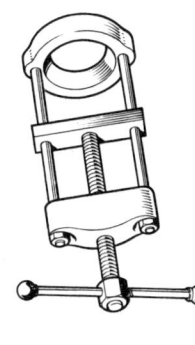

316 X Valve Seat Cutter (Main Tool)

Used with 316-10, 317-25, 317-G-25, 317-P-25 and 317-T-25 to cut the valve seats.
(Existing Tool)

316-10 Valve Seat Cutter Pilot

Used with 316X, 317-25, 317-G-25 and 317-P-25.
(Existing Tool)

317-25 Valve Seat Cutter

This cutter has an angle of 45°.
(Existing Tool)

317-G-25 45° Valve Seat Glaze Breaker

This cutter has an angle of 45°.
(Existing Tool)

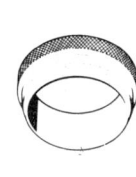

317-P-25 75° Valve Seat Narrowing Cutter

This cutter has an angle of 75°.
(Existing Tool)

317-T-28 15° Valve Seat Narrowing Cutter

This cutter has an angle of 15°.
(Existing Tool)

GEARBOX

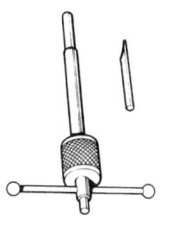

CP 4000 Hand Press (Main Tool)

Used in conjunction with different adaptor sets to assist in the completion of various rear axle operations.
(Existing Tool)

P 4000-3 Main Drive Gear Bearing Replacer

Consists of a replacer adaptor and a floating guide, used in conjunction with P 4090-6, the 370 taper base or CP 4009 and a workshop press. (Existing Tool)

CORTINA - LOTUS

 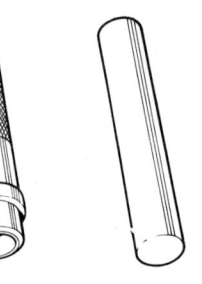 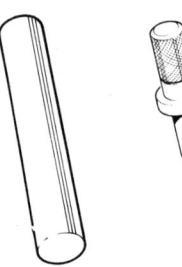 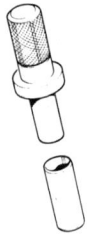 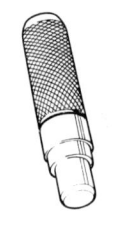   

P 7093 Interlock Plunger Assembly Tool

A small guide rod which facilitates the assembly of the interlock plunger to the gearbox housing.
(Existing Tool)

P 7095 Transmission Mainshaft Oil Seal Replacer

This driver-type tool ensures correct location of the transmission mainshaft oil seal without danger of damage to components.
(Existing Tool)

P 7113 Dummy Countershaft

An accurately machined shaft which provides a means of removing the cluster gear assembly complete from the gearbox.
(Existing Tool)

P 7136 Main Drive Gear Oil Seal Replacer

This tool ensures the accurate replacement of the oil seal.
(Existing Tool)

P 7137 Spigot Bearing Replacer

This tool eliminates the danger of damaging the bearing during replacement.
(Existing Tool)

P 7146 Mainshaft Nut Wrench

A specially shaped wrench with a ½ in. square drive used in conjunction with a suitable tension wrench to ensure correct torque loading of the mainshaft nut.
(New Tool)

P 7149 Transmission Extension Housing Bush Remover

This tool enables the extension housing bush to be removed with the gearbox in situ.
(Existing Tool)

P 7150 Transmission Extension Housing Bush Replacer

Ensures accurate replacement of the extension housing bush with the gearbox in situ.
(Existing Tool)

CORTINA - LOTUS

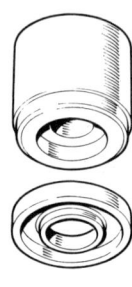 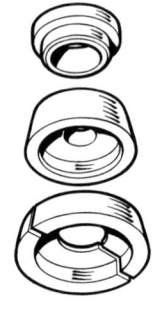   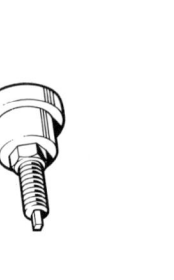 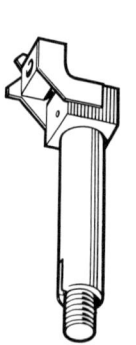

P 4000-31A Mainshaft Bearing and Hub Replacer

Consists of a ring and a collar, used with the 370 taper base and a workshop press.
(Existing Tool)

P 4090-7A Mainshaft Bearing and First Gear Remover

Use detail "a" only to remove first gear and mainshaft bearing.
(Existing Tool)

P 4090-9 Synchroniser Hub Remover

Used in conjunction with the 370 taper base and consists of a pair of split rings.
(Existing Tool)

P 7038 Transmission Extension Bearing Remover/Replacer

Used to replace the transmission extension bearing and is essential particularly when thin shell bearings are fitted.
(Existing Tool)

P 7043 Mainshaft Ball Bearing and Idler Shaft Remover

Used to remove the mainshaft ball bearing and idler shaft.
(Existing Tool)

P 7089 Gearbox Bracket

Used in conjunction with the 200A or B Engine Stand, this bracket enables the gearbox to be held in a rigid position for stripping or rebuilding.
(Existing Tool)

CORTINA - LOTUS

P 8008-A Slave Ring

Used with 370 Universal Taper base and P 8000-4B adaptors. (Modified Existing Tool)

CP 4111-A Water Pump Pulley Hub Remover

Used to remove the pulley hub. The thrust button is not required. (Existing Tool)

FUEL SYSTEM

P 9082 Fuel Tank Sender Unit Lock Ring Wrench

This tool simplifies removal and replacement of the fuel tank sender unit. (Existing Tool)

ELECTRICAL SYSTEM

CP 9504 Pole Piece Screwdriver

A generator is mounted in a vice complete with the body of the tool and a suitable bit. It is then easy to turn the bit with a spanner and slacken off the screw. (Existing Tool)

CP 9507-T End Plate Bushing Remover/Replacer

Used to remove and replace the bearing from the commutator end bracket. (Existing Tool)

CP 9509 Pole Piece Expander

Expands the pole pieces and firmly presses them into position in the generator without fear of damage or distortion.

CP 9521 Wiring Loom Retaining Clip Replacer

Necessary to replace the "fir tree" retaining clip. This tool has two prongs which locate in the blind holes provided in the clip enabling it to be squarely pressed home. (Existing Tool)

1 . 1969

Section 15 — 17

CORTINA - LOTUS

P 7152 Selector Rail Oil Seal Replacer

Used in conjunction with 575 light handle. (Existing Tool)

P 7154 Circlip Assessment Gauge

Used to assess mainshaft bearing circlip thickness. (Existing Tool)

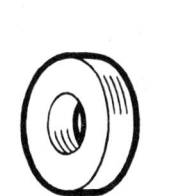

7600-A or B Flywheel Bearing Remover (Main Tool)

Used in conjunction with a suitable adaptor, facilitates removal of the flywheel bearing. (Existing Tool)

CP 7600-7 Flywheel Bearing Remover (Adaptor)

A split collet used in conjunction with 7600-A or B Main Tool (Existing Tool)

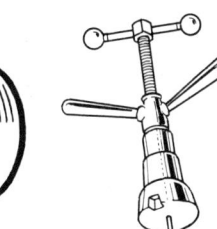

7657 Mainshaft Oil Seal Remover (Main Tool)

Used in conjunction with a suitable adaptor to remove the mainshaft oil seal without the need of removing the gearbox from the car. (Existing Tool)

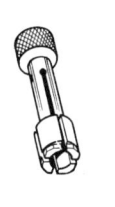

P 7657-4 Mainshaft Oil Seal Remover (Adaptor)

Used in conjunction with 7657 Main Tool the hollow centre allows the tool to be applied over the shaft, thus allowing the seal to be removed. (Existing Tool)

COOLING SYSTEM

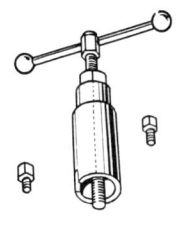 

P 8000-4B Water Pump Overhaul Tool (Adaptors)

Set of adaptors to enable the water pump to be dismantled and reassembled. (Existing Tool)

Section 15 — 16

1 . 1969

CORTINA - LOTUS

BODY

P 5517 Front Suspension Unit Mounting Alignment Jig

Of fibre-glass construction. This jig is an essential item for efficient accident repair.   (Existing Tool)

GENERAL TOOLS

200A Engine Stand

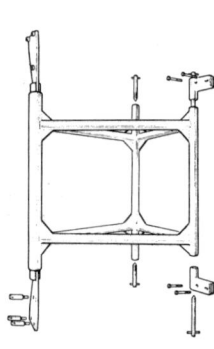

Using the appropriate mounting brackets, the various engines, gearboxes and differential assemblies can be mounted on this stand to facilitate dismantling and reassembly.   (Existing Tool)

370 Universal Taper Base

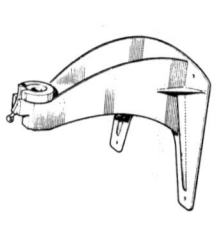

This tool fits most workshop presses to take the various split adaptors, etc., in place of the hand tools.   (Existing Tool)

550 Universal Handle

This tool has a spigot at one end onto which various oil seal replacers, etc., will fit.   (Existing Tool)

575 Light Universal Handle

This tool is a lighter version of the 550 universal handle.   (Existing Tool)

7066 Circlip Pliers

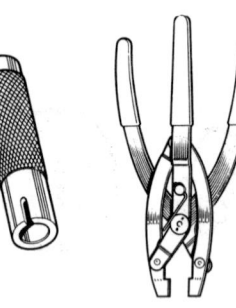

Medium sized circlip pliers for removal and replacement of various circlips, etc.

Section 15 — 18

1 . 1969